# Graph Theory

*Ashay Dharwadker*

*Shariefuddin Pirzada*

# Preface

Graph theory, as a branch of mathematics, has a glorious history: from Euler's seven bridges of Königsberg in 1756, to the elusive proof of the four colour theorem in 2000, and beyond. Graph theory is of practical importance to computer scientists for designing efficient algorithms, because many of the hardest (NP-complete) problems are essentially graph theoretic in nature. In recent times, graph theory has found diverse and often unexpected applications in science, engineering and technology. Seemingly difficult problems become easy to solve when expressed in the proper graph theoretical context. The powerful combinatorial methods of graph theory are being used to discover and prove significant new results in a variety of areas of pure mathematics itself. Indeed, the past few decades have witnessed a rising level of interest and growing activity among mathematicians, scientists and engineers in graph theory. As a result, one finds graph theory as a vital component of the mathematics curriculum in colleges and universities all over the world. In India, the model syllabus for graduate level mathematics proposed by the University Grants Commission includes graph theory as a recommended course.

This book has grown from our experience over the past several years in teaching various topics in graph theory, at both the graduate and undergraduate levels. As the number of students opting for graph theory is rapidly increasing, an attempt has been made to provide the latest and best available information on the subject. Our aim is to present the basics of graph theory in such a way that an average student can acquire as much depth and comprehension as possible in a first course.

The book is primarily intended for use as a textbook at the graduate level (for students pursuing masters in mathematics and computer science), with the first eleven chapters forming a one year course. However, the first eight chapters may be used as a one semester course at the undergraduate level for students of computer science and engineering. The final sections of many chapters introduce advanced topics and unsolved problems that are the object of current research in graph theory. Thus, the book can also be used by students pursuing research work in M. Phil and Ph. D. programmes.

There are many new topics in this book that have not appeared before in print: new proofs of various classical theorems, signed degree sequences, criteria for graphical sequences, eccentric sequences, matching and decomposition of planar graphs into trees. Scores in digraphs appear for the first time in print and the climax of the book is a new proof of the famous four colour theorem.

Many earlier books, monographs and articles have been used in the preparation of this book and we have included a comprehensive bibliography at the end of the book.

We would like to thank the Canadian Mathematical Society and the Math Forum at Drexel University for announcing the new proof of the four colour theorem in 2000. We are extremely grateful to the University of Kashmir and the Institute of Mathematics (Gurgaon) for their support during the writing of this book. Our sincere thanks go to Merajuddin (AMU, Aligarh), M. A. Sofi (University of Kashmir), Petrovic Vojislav (Novi Sad University), Ivanyi Antal (Eotvos Lorand University, Hungary), Zhou Guofei (Nanjing University, China), M. R. Sridharan (IIT, Kanpur), V. Krishnamurthy (BITS, Pilani), Niels Karlsson (Akureyri University), John-Tagore Tevet and Jüri Martin (Eurouniversity, Tallinn), Anita Pasotti (Universita degli Studi di Brescia) and Vladimir Khachatryan (SUNY, Stony Brook) for their valuable suggestions and encouragement. We thank all our friends and colleagues, especially T. A. Chishti (University of Kashmir) and all the members of the Institute of Mathematics (Gurgaon) for supporting our work. We thank our research students, especially T.A. Naikoo for help in drawing the figures and carefully proof reading the manuscript. We are grateful to our families for their love and support during the time this book was being written. Finally, it is a pleasure to thank the management and staff of Orient Longman and Universities Press (India) for their interest, cooperation, and fine workmanship.

*Ashay Dharwadker*          *Shariefuddin Pirzada*

# Contents

1. Introduction    1

2. Degree Sequences    37

3. Eulerian and Hamiltonian Graphs    64

4. Trees    84

5. Connectivity    111

6. Planarity    140

7. Colourings    169

8. Matchings and Factors    205

9. Edge Graphs and Eccentricity Sequences    242

10. Graph Matrices    273

11. Digraphs    307

12. The Four Colour Theorem    347

13. Graph Algorithms    378

14. Score Structure in Digraphs    415

Bibliography    459

# 1. Introduction

Graph theory owes its evolution to the study of some physical problems involving sets of objects and binary relations among them. It is difficult to pinpoint its formulation to a single source; in fact, graph theory can be said to have been discovered many times, each discovery being independent of the other. The earliest known studies appear in the works of Euler, Kirchoff, Cayley and Hamilton. The twentieth century witnessed considerable activity in this area, with new discoveries and proofs being proposed as solutions to classical problems including the celebrated four colour problem.

## 1.1  Basic Concepts

Let $V$ be a nonempty set. The cartesian product of $V$ with itself, denoted by $V \times V$, is the set of all unordered pairs of elements of $V$. That is, $V \times V = \{(u, v) : u, v \in V\}$. We denote by $V_{(2)}$ the set of unordered pairs of distinct elements of $V$, by $V_{(3)}$ the set of unordered pairs of elements of $V$, not necessarily distinct, and by $V_{(4)}$ the set of ordered pairs of distinct elements of $V$.

**Definition:**  A simple graph (or briefly, a graph) $G$ is a finite nonempty set $V$ together with a symmetric, irreflexive relation $R$ on $V$. The elements of the set $V$ are called the *vertices* of the graph and the relation $R$ is called the *adjacency relation*. If $u$ is related to $v$ by $R$, then $u$ is said to be *adjacent* to $v$ and we write $uRv$. An example of a simple graph is given in Figure 1.1(a).

Since $R$ is a symmetric relation, it defines a subset $E$ of $V_{(2)}$. The elements of the set $E$ are called the *edges* of the graph.

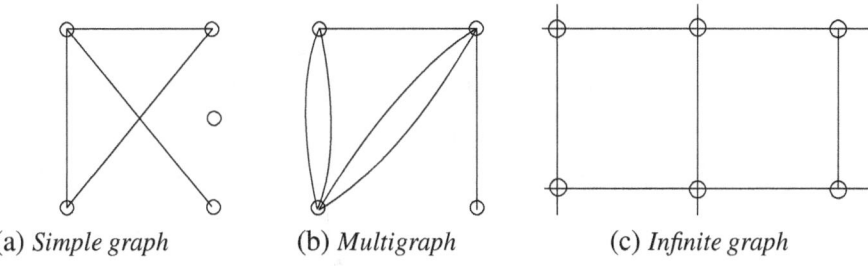

(a) *Simple graph*        (b) *Multigraph*        (c) *Infinite graph*

**Fig. 1.1**

From the above definition, we observe that a graph $G$ is a pair $(V, E)$, where $V$ is a nonempty set whose elements are called the vertices of $G$ and $E$ is a subset of $V_{(2)}$ whose elements are called the edges of $G$.

We observe that there is an incidence relation $I$ between the vertex set $V$ and the edge set $E$ of a graph. If the element $e \in E$, then there is a pair of distinct vertices $u$ and $v$ such that $e = \{u, v\}$. The vertices $u$ and $v$ are called *end vertices* of $e$, and $u$ and $v$ are said to be *incident* with $e$ ($uIe$ and $vIe$). Also, $e$ is said to be incident with $u$ and $v$, and in that case we write $eI'u$ and $eI'v$, where $I'$ is the relation converse to $I$.

A graph with a finite number of vertices and finite number of edges is called a *finite graph*, otherwise it is an *infinite graph* (Figure 1(c)).

We represent a graph $G$ with vertex set $V$ and edge set $E$ by $(V(G), E(G))$. Since we are only going to deal with finite graphs, we write $V(G) = \{v_1, v_2, \ldots, v_n\}$, $E(G) = \{e_1, e_2, \ldots, e_m\}$.

We define $|V| = n$ to be the *order* of $G$ and $|E| = m$ to be the *size* of $G$. Such a graph is called an $(n, m)$ graph. If there is an edge $e$ between the vertices $u$ and $v$, we briefly write $e = uv$ and say edge $e$ joins the vertices $u$ and $v$. A vertex is said to be *isolated* if it is not adjacent to any other vertex.

**Multigraph:** A multigraph is a pair $(V, E)$, where $V$ is a nonempty set of vertices and $E$ is a multiset of edges, being a multi-subset of $V_{(2)}$. The number of times an edge $e = uv$ occurs in $E$ is called the *multiplicity* of $e$ and edges with multiplicity greater than one are called *multiple edges*. An example of a multigraph is given in Figure 1(b).

**General graph:** A general graph is a pair $(V, E)$, where $V$ is a nonempty set of Moment and $E$ is a multiset of edges, being a multi-subset of $V_{(3)}$. An edge of the form $e = uu$, $u \in V$ is called a *loop*. An edge which is not a loop is called a *proper edge* or *link*. The number of times edge $e$ occurs is called its multiplicity, and proper edges with multiplicity greater than one are called multiple edges. Loops with multiplicity greater than one are called *multiple loops* (Fig. 1.2).

If $u, v \in V$ in a general graph or multigraph $G$, then the multiplicity of the edge $uv$ is denoted by $q_G[u, v]$. If $uv$ is not an edge, then $q_G[u, v] = 0$. Similarly, if $u$ is a vertex of a general graph $G$, then the number of loops at $u$ in $G$ is denoted by $q_G[u]$.

The graph obtained by replacing all multiple edges by single edges in a multigraph $G$ is called the *underlying graph* of $G$. Similarly, if $G$ is a general graph, the graph $H$ obtained by removing all its loops and replacing all multiple edges by single edges is called the underlying graph of $G$.

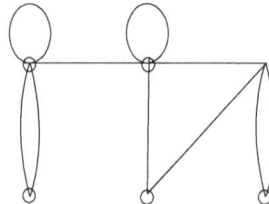

**Fig. 1.2** *General graph*

## 1.2 Degrees

The degree of a vertex $v$ in a graph $G$ is the number of edges of $G$ which are incident to $v$ and is denoted by $d(v)$ or $d(v|G)$. We have $d(v) = |\{e \in E : e = uv, \text{ for } u \in V\}|$. The *minimum degree* and the *maximum degree* of a graph $G$ are denoted by $\delta(G)$ and $\Delta(G)$ respectively. In the graph of Figure 1.3(a), $d(v_1) = d(v_3) = d(v_6) = 3$, $d(v_2) = 1$, $d(v_4) = 0$ and $d(v_5) = 2$.

A graph is said to be *regular* if all its vertices are of same degree and *k-regular* if all its vertices are of degree $k$. A 3-regular graph is also called a *cubic graph*. A vertex with degree zero is an *isolated vertex*, a vertex with degree one is a *pendant vertex* and the unique edge incident to a pendant vertex is a *pendant edge*. A vertex of odd degree is an *odd vertex* and a vertex of even degree is an *even vertex*. In Figure 1.3(a), $v_4$ is an isolated vertex and $v_2$ is a pendant vertex. The graph shown in Figure 1.3(b) is 2-regular and that given in Figure 1.3(c) is 3-regular, i.e., cubic.

In a general graph $G$, a loop incident to a vertex $v$ is counted as two edges incident to $v$. Therefore, $d(v)$ is the number of non-loop edges incident to $v$ plus twice the number of loops at $v$.

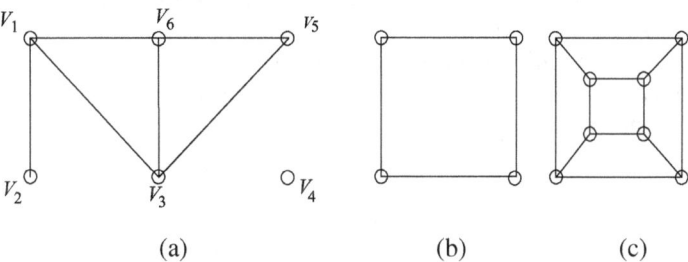

(a)        (b)        (c)

**Fig. 1.3** *Regular graphs*

The following result is due to Euler [76].

**Theorem 1.1**  The sum of the degrees of a graph is even, being twice the number of edges.

**Proof**  Let $m$ be the number of edges in a graph $G = (V, E)$. Since each edge contributes two to the degrees, one at the beginning vertex and one at the end vertex of the edge, the sum of the degrees is even and equal to twice the number of edges. Hence,

$$\sum_{v \in V} d(v) = 2m. \qquad \qquad \square$$

**Theorem 1.2**  In any graph there is an even number of vertices of odd degree.

**Proof**  Let $G = (V, E)$ be a graph and $d(v)$ be the degree of the vertex $v \in V$. Let $|E| = m$.

Then, $\sum_{v \in V} d(v) = 2m$, and therefore,

$$\underset{\substack{odd\ degree \\ vertices}}{\sum d(v_j)} \quad + \quad \underset{\substack{even\ degree \\ vertices}}{\sum d(v_k)} = 2m.$$

$$\underset{\substack{\text{I} \qquad\qquad\qquad \text{II}}}{}$$

(1.2.1)

Since the right hand side of (1.2.1) is even, and (II) in (1.2.1) is also even, therefore (I) in (1.2.1) is even. Hence,

$$\underset{\substack{odd\ degree \\ vertices}}{\sum d(v_j)} = \text{even}.$$

This is only possible when the number of vertices with odd degree is even. ❑

## 1.3 Isomorphism

Let $G$ and $H$ be general graphs. Let $f$ be a one–one mapping of $V(G)$ onto $V(H)$, and $g$ be a one–one mapping of $E(G)$ onto $E(H)$. Let $\theta$ denote the ordered pair $(f, g)$. $\theta$ is an isomorphism of $G$ onto $H$, when the vertex $x$ is incident with the edge $e$ in $G$ if and only if the vertex $fx$ is incident with the edge $ge$ in $H$ (Fig. 1.4). If such an isomorphism $\theta$ exists, then the graphs $G$ and $H$ are said to be *isomorphic*, and is denoted by $G \cong H$. We have, $|V(G)| = |V(H)|$ and $|E(G)| = |E(H)|$.

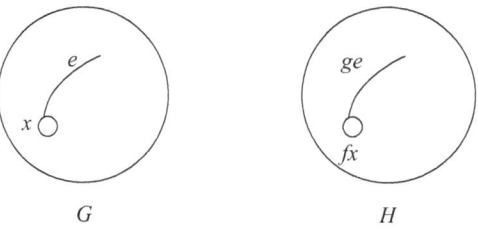

Fig. 1.4

We may think of $\theta$ as an operation transforming $G$ into $H$ and write $\theta G = H$. Also, we write $\theta v = fv$ and $\theta e = ge$ for each vertex $v$ and each edge $e$ of $G$. Clearly, $G$ and $H$ can be represented by the same diagram. The representative of an edge or vertex $x$ of $G$ can be reinterpreted as the representative of $\theta x$ in $H$.

An isomorphism of a graph $G$ onto itself is called an *automorphism* of $G$. Any graph $G$ has the *identical* or *trivial automorphism I* such that $Ix = x$ for each edge or each vertex $x$ of $G$.

Clearly, two graphs $G$ and $G'$ are *isomorphic* to each other if there is a one–one correspondence between their vertices, and between their edges such that the incidence relationship is preserved. For example, the graphs shown in Figure 1.5(a), Figure 1.5(b) and Figure 1.5(c) are isomorphic.

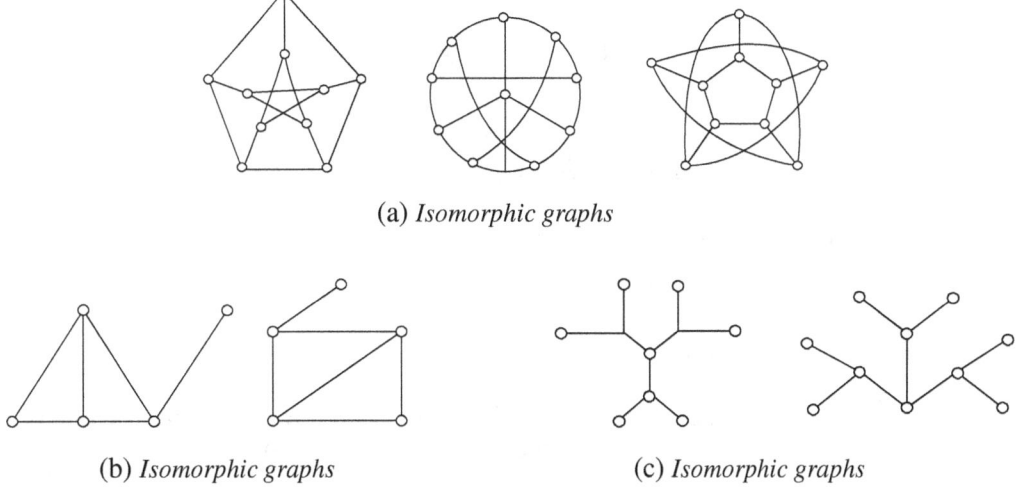

(a) *Isomorphic graphs*

(b) *Isomorphic graphs*          (c) *Isomorphic graphs*

**Fig. 1.5**

It follows from the definition of isomorphism that two isomorphic graphs have

    i. the same number of vertices,

    ii. the same number of edges, and

    iii. an equal number of vertices with a given degree.

However, these conditions are not sufficient. To see this, consider the two graphs given in Figure 1.6.

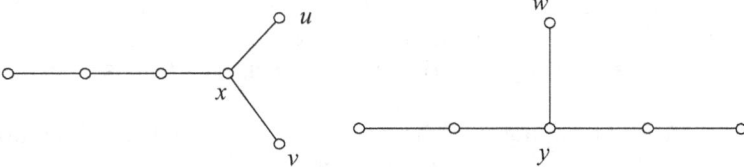

**Fig. 1.6** *Non-isomorphic graphs*

   These graphs satisfy all the three conditions, but they are not isomorphic because the vertex $x$ in (a) corresponds to vertex $y$ in (b) as there are no other vertices of degree three, and in (b) there is only one pendant vertex $w$ adjacent to $y$, while in (a) there are two pendant vertices $u$ and $v$ adjacent to $x$.

   The graphs in Figure. 1.7 also satisfy conditions (i), (ii) and (iii) but are not isomorphic.

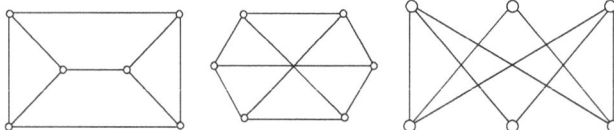

**Fig. 1.7** *Non-isomorphic graphs*

We have the following result on isomorphism of graphs.

**Theorem 1.3**    The relation isomorphism in graphs is an equivalence relation.

**Proof**    The relation of isomorphism between graphs is reflexive because of the trivial automorphisms.

Let $\theta = (f, g)$ be an isomorphism of a graph $G$ onto a graph $H$, so $G \cong H$. Then there is an inverse isomorphism $\theta^{-1} = (f^{-1}, g^{-1})$ of $H$ onto $G$. So $H \cong G$. Therefore $\cong$ is symmetric.

Now, let $\theta = (f, g)$ be an isomorphism of $G$ onto $H$, and $\phi\theta = (f_1 f, g_1 g)$ of $G$ onto $K$. Here $f_1 f$ is a mapping obtained by applying first $f$ and then $f_1$. Similarly, $\phi\theta$ is the isomorphism obtained applying first $\theta$ and then $\phi$. Thus $\cong$ is transitive.

Hence the relation isomorphism is an equivalence relation.                                ❏

**Remark**    The multiplication of the isomorphism defined above is associative.

Since the relation isomorphism is an equivalence relation, it partitions the class of all graphs into disjoint nonempty subclasses called *isomorphism classes*, such that two graphs belong to the same isomorphism class if and only if they are isomorphic.

**Theorem 1.4**    Let $G$ and $H$ be graphs and let $f$ be a one–one mapping $V(G)$ onto $V(H)$ such that two distinct vertices $x$ and $y$ of $G$ are adjacent if and only if the corresponding vertices $fx$ and $fy$ of $H$ are adjacent in $H$. Then there is a uniquely determined one–one mapping $g$ of $E(G)$ onto $E(H)$ such that $(f, g)$ is an isomorphism of $G$ onto $H$.

**Proof**    Let $e$ be any edge of $G$, having distinct ends $x$ and $y$. By hypothesis, there is a uniquely determined edge $e'$ of $H$ whose ends are $fx$ and $fy$. We define a one–one mapping $g$ by the rule $ge = e'$, for each edge $e$ of $G$. It is then clear that $(f, g)$ is an isomorphism of $G$ onto $H$.

Conversely, let $g$ be a mapping such that $(f, g)$ is an isomorphism of $G$ onto $H$. Then for each edge $e$ of $G$, there is an edge $e'$ of $H$ such that $ge = e'$.                                ❏

An isomorphism of a graph $G$ onto a graph $H$ is defined as a one–one mapping of $V(G)$ onto $V(H)$ that preserves adjacency. This specialisation can be regarded as an application of Theorem 1.4.

**Definition:** Two graphs $G = (V, E)$ and $H = (U, F)$ are label-isomorphic if and only if $V = U$, and for any pair $u, v$ in $V$, $uv \in E$ if and only if $uv \in F$. The graphs of Figure 1.8 are isomorphic, but not label-isomorphic.

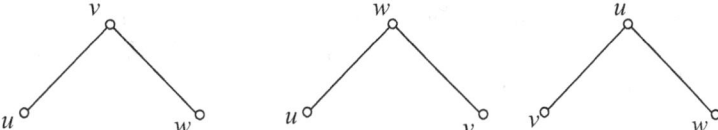

**Fig. 1.8** *Isomorphic but not label-isomorphic graphs*

**Definition:** A graph invariant is a function $f$ from the set of all graphs to any range of values (numerical, vectorial or any other) such that $f$ takes the same value on isomorphic graphs. When the range of values is numerical (real, rational or integral) the invariant is called a *parameter*. The order and size of a graph are *graph parameters*.

## 1.4 Types of graphs

**Simple directed graph or simple digraph:** A simple digraph (or simply digraph) $D$ is a pair $(V, A)$, where $V$ is a nonempty set of vertices and $A$ is a subset of $V_{(2)}$ whose elements are called *arcs* of $D$.

**Multidigraph:** A multidigraph $D$ is a pair $(V, A)$, where $V$ is a nonempty set of vertices, and $A$ is a multiset of arcs of $V_{(2)}$. The number of times an arc occurs in $D$ is called its multiplicity and arcs with multiplicity greater than one are called multiple arcs of $D$.

**General digraph:** A general digraph $D$ is a pair $(V, A)$, where $V$ is a nonempty set of vertices, and $A$ is a multiset of arcs, being a multisubset $V \times V$. An arc of the form $uu$ is called a *loop* of $D$ and arcs which are not loops are called *proper arcs* of $D$. The number of times an arc occurs is called its multiplicity. A loop with multiplicity greater than one is called a *multiple loop*.

An arc $(u, v) \in A$ of a digraph is denoted by $uv$, implying that it is directed from $u$ to $v$, $u$ being the *initial vertex* and $v$ the *terminal vertex*. Clearly, a digraph is an irreflexive binary relation on $V$.

Various types of digraphs are shown in Figure 1.9.

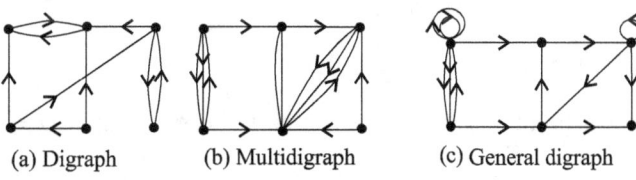

(a) Digraph     (b) Multidigraph     (c) General digraph

**Fig. 1.9**

If $D = (V, A)$ is a digraph, the graph $G = (V, E)$, where $uv \in E$ whenever $uv$ or $vu$ or both are in $A$, is called the underlying graph of $D$ (also called the covering graph $C(D)$ of $D$).

If $D = (V, A)$ is a general digraph, the digraph $D_1 = (V, A_1)$ obtained from $D$ by removing all loops, and by replacing all multiple arcs by single arcs is the *digraph underlying D*. The underlying graph of $D_1$ is the underlying graph of $D$.

**Mixed graph:**  A mixed graph $G = \{V, A \cup E\}$ consists of a nonempty set $V$ of vertices, a set $A$ of arcs ($A \subseteq V_{(2)}$), and a set $E$ of edges ($E \subseteq V_{(2)}$), such that if $uv \in E$ then neither $uv$, nor $vu$ is in $A$.

We represent a general, multi and simple graph by $g$-graph, $m$-graph and $s$-graph respectively.

**Subgraphs:**  A subgraph of a graph $G = (V, E)$ is a graph $H = (U, F)$ with $U \subseteq V$ and $F \subseteq E$. We denote it by $H < G$ ($G$ is also called the *super graph of H*.) If $U = V$, then $H$ is called the *spanning subgraph* of $G$, and is denoted by $H \leq G$. Here $G$ is called the *spanning super graph* of $H$, and is denoted by $G \geq H$.

If $F$ consists of all those edges of $G$ joining pairs of vertices of $U$, then $H$ is called the *vertex induced subgraph* of $G$, and is denoted by $H = <U>$. If $F \subseteq E$, and $U$ is the set of end vertices of the edges of $F$, then $H = (U, F)$ is called an *edge induced subgraph* of $G$, and is denoted by $H = <F>$. These definitions are illustrated in Figure 1.10.

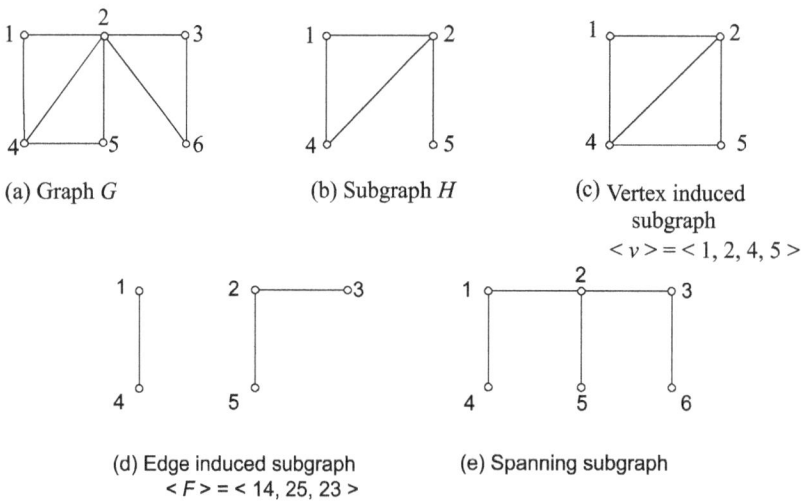

(a) Graph $G$                    (b) Subgraph $H$                    (c) Vertex induced
                                                                         subgraph
                                                                    $<v> = <1, 2, 4, 5>$

(d) Edge induced subgraph                    (e) Spanning subgraph
$<F> = <14, 25, 23>$

**Fig. 1.10**

If $S$ and $T$ are two disjoint subsets of the vertex set $V$ of a graph $G = (V, E)$, we define $[S, T] = \{uv \in E : u \in S, v \in T\}$.

Also, $q[S, T] = |[S, T]|$.

If $D = (V, A)$ is a digraph and $S$ and $T$ are disjoint subsets of $V$, we denote $(S, T) = \{uv \in A : u \in S \text{ and } v \in T\}$ and $q(S, T) = |(S, T)|$.

**Complete graph:** A graph of order $n$ with all possible edges $\left( m = \dfrac{n(n-1)}{2} \right)$ is called a complete graph of order $n$ and is denoted by $K_n$. A graph of order $n$ with no edges is called an *empty graph* and is denoted by $\overline{K_n}$. Each graph of order $n$ is clearly a spanning subgraph of $K_n$. Some examples of complete graphs are shown in Figure 1.11.

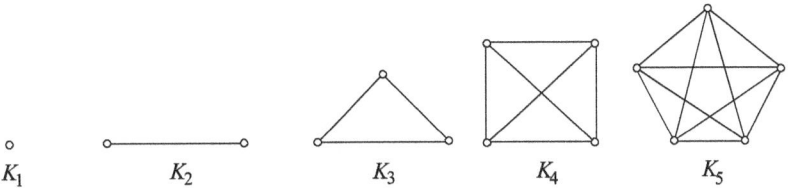

**Fig. 1.11** *Complete graphs*

**$r$-partite graph:** A graph $G = (V, E)$ is said to be $r$-partite (where $r$ is positive integer) if its vertex set can be partitioned into disjoint sets $V_1, V_2, \ldots, V_r$ with $V = V_1 \cup V_2 \cup \ldots \cup V_r$, such that $uv$ is an edge of $G$ if $u$ is in some $V_i$ and $v$ in some $V_j$, $i \neq j$. That is, every one of the induced subgraphs $< V_i >$ is an empty graph. We denote $r$-partite graph by $G = (V_1, V_2, \ldots, V_r, E)$.

If an $r$-partite graph has all possible edges, that is, $uv \in E$, for every $u \in V_i$ and every $v \in V_j$, for all $i$, $j$, then it is called a *complete $r$-partite graph*. If $|V_i| = n_i$, we denote it by $K_{n_1, n_2, \ldots, n_r}$.

**Bipartite graph:** A graph $G = (V, E)$ is said to be bipartite, or 2-partite, if its vertex set can be partitioned into two different sets $V_1$ and $V_2$ with $V = V_1 \cup V_2$, such that $uv \in E$ if $u \in V_1$ and $v \in V_2$. The bipartite graph is said to be *complete* if $uv \in E$, for every $u \in V_1$ and every $v \in V_2$. When $|V_1| = n_1$, $|V_2| = n_2$, we denote the complete bipartite graph by $K_{n_1, n_2}$. For example, $K_{2,2}$ and place $K_{2,3}$ are shown in Figure 1.12(a) and (b).

The complete bipartite graph $K_{1,n}$ is called an *$n$-star* or *$n$-claw*. For example, 3-star can be seen in Figure 1.12(e).

**Complement of a graph:** The complement $\overline{G} = (V, \overline{E})$ of a graph $G = (V, E)$ is the graph having the same vertex set as $G$, and its edge set $\overline{E}$ is the complement of $E$ in $V_{(2)}$, that is, $uv$ is an edge of $G$ if and only if $uv$ is not an edge of $G$. In Figure 1.12(c) and (d) the complements of $K_4$ and place $K_{2,3}$ respectively are shown.

A graph $G$ is said to be *self-complementary* if $\overline{G} \cong G$. The complement $\overline{K_n}$ of the complete graph of order $n$ is clearly the empty graph of order $n$.

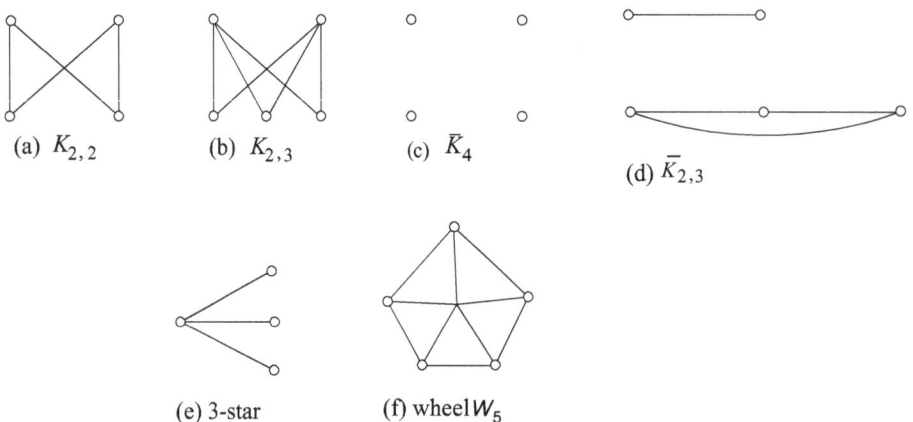

(a) $K_{2,2}$          (b) $K_{2,3}$          (c) $\bar{K}_4$

(d) $\bar{K}_{2,3}$

(e) 3-star          (f) wheel $W_5$

**Fig. 1.12** *Complete bipartite graphs*

**Removal of edges:** Let $G = (V, E)$ be a graph and let $F \subseteq E$. The graph $H = (V, E - F)$ with vertex set $V$ and edge set $E - F$ is said to be obtained from $G$ by removing the edges in $F$. It is denoted by $G - F$. If $F$ consists of a single edge $e$ of $G$, the graph obtained by removing $e$ is denoted by $G - e$.

Now, $G - F$ may contain isolated vertices which are not isolated vertices of $G$. The graph obtained by removing these newly created isolated vertices from $G - F$ is denoted by $G \mid F$. Similarly, the graph obtained by removing isolated vertices from $G - e$ is denoted by $G \mid e$. Figure 1.13 illustrates this operation.

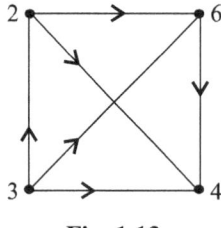

**Fig. 1.13**

**Removal of vertices:** Let $G = (V, E)$ be a graph and let $v \in V$. Let $Ev$ be the set of all edges of $G$ incident with $v$. The graph $H = (V - \{v\}, E - Ev)$ is said to be obtained from $G$ by the removal of the vertex $v$ and is denoted by $G - v$.

If $U$ is a subset of $V$, the graph obtained by removing the vertices of $G$ which are in $U$ is denoted by $G - U$. If $H$ is a subgraph of $G$, we denote $G - V(H)$ by $G - H$, and $G - E(H)$ by $\bar{H}(G)$. $\bar{H}(G)$ is called the *relative complement* of $H$ in $G$ (Fig. 1.14).

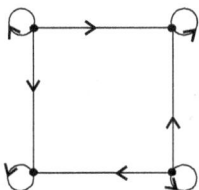

**Fig. 1.14**

**Addition of edges:**   Let $G = (V, E)$ be a graph and let $f$ be an edge of $\overline{G}$. The graph $H = (V, E \cup \{f\})$ is said to be obtained from $G$ by the addition of the edge $f$ and is denoted by $G + f$. If $F$ is a subset of edges of $\overline{G}$, the graph obtained from $G$ by adding the edges of $F$ is denoted by $G + F$ (Fig. 1.15(a) and (b)).

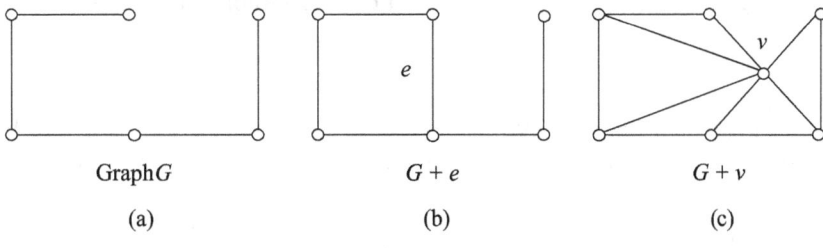

|     Graph $G$     |     $G + e$     |     $G + v$     |
|     (a)     |     (b)     |     (c)     |

**Fig. 1.15**

**Addition of vertices:**   Let $G = (V, E)$ be a graph and let $v \notin V$. The graph $H$ with vertex set $V \cup \{v\}$ and edge set $E \cup \{uv,$ for all $u \in V\}$ obtained from $G$ by adding a vertex $v$, is denoted by $G + v$. Thus $G + v$ is obtained from $G$ by adding a new vertex and joining it to all vertices of $G$. For illustration, refer to Figure 1.15(c).

**Join of graphs:**   Let $G = (V, E)$ and $H = (U, F)$ be two graphs with disjoint vertex sets $(V \cap U = \Phi)$. The join of $G$ and $H$ denoted by $GVH$ is the graph with vertex set $V \cup U$ and edge set $E \cup F \cup \{V, U\}$. So the join is obtained from $G$ and $H$ by joining every vertex of $G$ to each vertex of $H$ by an edge. Clearly, $G + v = GVK_1 = GV\overline{K_1}$. This operation is illustrated in Figure 1.16.

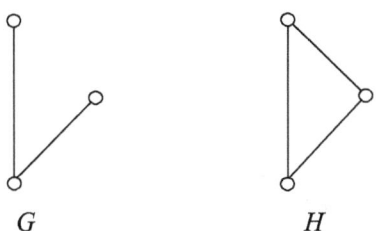

$G$ $\qquad\qquad$ $H$ $\qquad\qquad\qquad\qquad$ $GVH$

**Fig. 1.16**

## 1.5  Graph properties

**Parametric property:**  A property $P$ is called a parametric property of a graph if for a graph $G$ having property $P$, every graph isomorphic to $G$ also has property $P$. A graph with property $P$ is denoted as $P$-graph. A subgraph $H$ of a graph $G$ is said to be *maximal* with respect to a property $P$ (or $P$-maximal) if $H$ is a $P$-graph, and there is no subgraph $K$ of $G$ having property $P$ which properly contains $H$. So $H$ is $P$-maximal if $H$ is a $P$-graph, and for every $e \in E(G) - E(H)$, $H + e$ is not a $P$-graph, that is, the addition of any edge to $H$ destroys the property $P$ of $H$. A subgraph $H$ of a graph $G$ is said be *minimal* with respect to a property $P$ (or $P$-minimal) if $H$ is a $P$-graph and $H$ has no proper subgraph $K$ which is also a $P$-graph. So $H$ is minimal if and only if $H$ is a $P$-graph, and for every $e \in E(H)$, $H - e$ is not a $P$-graph, that is, if and only if the removal of any edge destroys the property $P$.

**$P$-critical:**  A graph $G$ is said to be $P$-critical if $G$ is a $P$-graph and for every $v \in V$, $G - v$ is not a $P$-graph.

**Hereditary property:**  A property $P$ is said to be a hereditary property of a graph $G$ if a graph $G$ has the property $P$, then every subgraph of $G$ also has the property $P$. It is called an *induced-hereditary property* if every induced subgraph of $G$ also has the property.

**Monotone property:**  A property $P$ is said to be a monotone property of a graph $G$, when $G$ has the property $P$, then for every $e \in E(\overline{G})$, $G + e$ also has the property $P$.

A vertex subset $U$ of a graph $G$ is said to be an *independent set* of $G$ if the induced subgraph $< U >$ is an empty graph. An independent set of $G$ with maximum number of vertices is called a *maximum independent set* (MIS) of $G$. The number of vertices in a *MIS* of $G$ is called the *independent number* of $G$ and is denoted by $\alpha_0(G)$.

A maximal complete subgraph of $G$ is a *clique* of $G$. The order of a maximum clique of $G$ is the *clique number* of $G$ and is denoted by $w(G)$. Consider the graph given in Figure 1.17. Here $\{1, 4\}$ is independent set, $\{1, 3, 5\}$ is an *MIS*, $\alpha_0(G) = 3$, $\{1, 2, 6\}$ is a clique, $\{2, 3, 4, 6\}$ is a maximum clique and $w(G) = 4$.

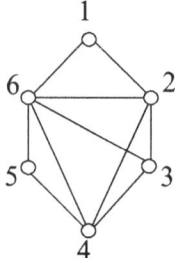

**Fig. 1.17**

The property of being a complete subgraph is an induced hereditary property. The property of being a planar graph is hereditary and the property of being a nonplanar graph is a monotone property.

## 1.6  Paths, Cycles and Components

The incidence relation $I$ between the elements of $V$ and the elements of $E$ induced by the adjacency relation $R$ further induces an adjacency relation among the edges, namely, two edges $e$ and $f$ are adjacent if and only if they have an end vertex in common. This relation is denoted by $L$.

Two edges $e$ and $f$ of a graph $G = (V, E)$ are adjacent if and only if $e = uv$ and $f = vw$, for some three vertices $u$, $v$, $w$. If $e$ and $f$ are adjacent, it is denoted by $eLf$.

**Walks:**  An alternating sequence of vertices and edges, beginning and ending with vertices such that no edge is traversed or covered more than once is called a walk. A vertex may appear more than once in a walk. $v_1 e_1 v_2 e_2 v_3 \dots v_k e_k v_{k+1}$ is called a $v_1 - v_{k+1}$ walk $W$, $v_1$ and $v_{k+1}$ are called the *initial* and *terminal vertices* of the walk $W$, and $k$, the number of edges in $W$ is called *length* of $W$. The walk is said to be *open*, if $v_1$ and $v_{k+1}$ are distinct, and *closed* if $v_1 = v_{k+1}$. If $v_r$ and $v_s$ are two vertices in $W$, $s > r$, then the walk $v_r e_r v_{r+1} e_{r+1} \dots e_{s-1} v_s$ is called a *subwalk* of $W$, or $v_r - v_s$ *section* of $W$, and is denoted by $W v_r - v_s$. The walk $v_{k+1} e_k v_k \dots e_1 v_1$ is called the *reverse walk* of $W$ and is denoted by $W^{-1}$.

**Paths:**  A path in a graph is an open walk in which no vertex (and therefore no edge) is repeated. A closed walk in which no vertex (and edge) is repeated is called a *cycle*. A path of length $n$ is called an *n-path* and is denoted by $P_n$. A cycle of length $n$ is called an *n-cycle* and is denoted by $C_n$. A loop is 1-cycle and a pair of edges joining two vertices form a 2-cycle. An *n*-cycle is proper only if $n \geq 3$.

As the edges and vertices in a path or cycle are not repeated, these are denoted by the sequence of vertices only. For example $u_1 u_2 \dots u_k$ where $u_i \in V$ is a $k - 1$ path.

Two distinct vertices $u$ and $v$ of a graph $G$ are said to be *connected* or *joined* if there is a $u - v$ walk in $G$. By convention, a vertex is connected to itself. A graph is said to be connected if every two of its vertices are connected, otherwise it is *disconnected*. The relation of connectedness is an equivalence relation on the vertex set $V$ of a graph. The graphs induced on the equivalence classes of this relation are called the *components* of the graph.

**Component of a graph:**  A maximal connected subgraph of a graph $G$ is called a component of $G$. A component which is $K_1$ is called a *trivial component*. The number of components of a graph $G$ is denoted by $k(G)$. A component of $G$ with an odd (even) number of vertices is called an *odd (even) component* of $G$. The number of odd components of $G$ is denoted by $k_0(G)$. Consider the graph shown in Figure 1.18. Here, $v_1 e_1 v_2 e_2 v_3 e_3 v_4 e_4 v_2 e_5 v_5 e_6 v_6$ is a walk, $v_1 e_1 v_2 e_2 v_3 e_3 v_4$ is a path and $v_2 e_2 v_3 e_3 v_4 e_4 v_2$ is a 3-cycle.

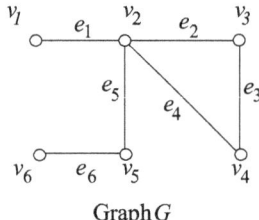

Graph G

**Fig. 1.18**

**Union of graphs:**  Let $G = (V, E)$ and $H = (U, F)$ be two graphs with $V \cap U = \Phi$. The union of $G$ and $H$, denoted by $G \cup H$, is the graph with vertex set $V \cup U$ and edge set $E \cup F$. Clearly, when $G$ and $H$ are connected graphs, $G \cup H$ is a disconnected graph whose components are $G$ and $H$ (Fig. 1.19).

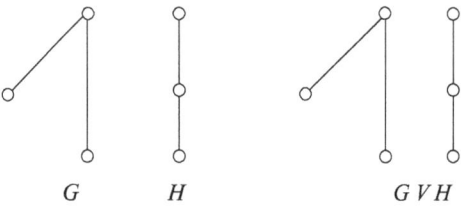

G          H                        G V H

**Fig. 1.19**

A graph $G$ which has $k$ components, all isomorphic to $H$, is $H \cup H \cup \ldots \cup H$, or $G = kH$. We write $G = k_1 G_1 \cup k_2 G_2 \cup \ldots \cup k_r G_r$, if $G$ has $k_i$ components isomorphic to $G_1$, $1 \le i \le r$.

**Theorem 1.5**    A graph $G$ is disconnected if and only if its vertex set $V$ can be partitioned into two nonempty, disjoint subsets $V_1$ and $V_2$, such that there exists no edge in $G$ whose one end vertex is in $V_1$ and the other in $V_2$.

**Proof**    Let $G$ be a graph whose vertex set can be partitioned into two nonempty disjoint subsets $V_1$ and $V_2$, so that no edge of $G$ has one end in $V_1$ and the other in $V_2$. Let $v_1$ and $v_2$ be any two vertices of $G$ such that $v_1 \in V_1$ and $v_2 \in V_2$. Then there is no path between vertices $v_1$ and $v_2$, since there is no edge joining them. This shows that $G$ is disconnected.

Conversely, let $G$ be a disconnected graph. Consider a vertex $v$ in $G$. Let $V_1$ be the set of all vertices that are joined by paths to $v$. Since $G$ is disconnected, $V_1$ does not contain all vertices of $G$. Let $V_2$ be the set of the remaining vertices. Clearly, no vertex in $V_1$ is joined to any vertex in $V_2$ by an edge, proving the converse.                                ❑

**Theorem 1.6**    If a graph has exactly two vertices of odd degree, they must be connected by a path.

**Proof**    Let $G$ be a graph with all its vertices of even degree, except for $v_1$ and $v_2$ which are of odd degree. Consider the component $C$ to which $v_1$ belongs. Then $C$ has an even number of vertices of odd degree. Therefore $C$ must contain $v_2$, the only other vertex of odd degree.

Thus $v_1$ and $v_2$ are in the same component, and since a component is connected, there is a path between $v_1$ and $v_2$. ❏

**Lemma 1.1**   For any set of positive integers $n_1, n_2, \ldots, n_k$

$$\sum_{i=1}^{k} n_i^2 \leq \left(\sum_{i=1}^{k} n_i\right)^2 - (k-1)\left(2\sum_{i=1}^{k} n_i - k\right).$$

**Proof**   We have, $\sum_{i=1}^{k}(n_i - 1) = \sum_{i=1}^{k} n_i - k.$

Squaring both sides, we get

$$\left[\sum_{i=1}^{k}(n_i-1)\right]^2 = \left[\sum_{i=1}^{k} n_i - k\right]^2,$$

or $[(n_1-1)+(n_2-1)+\ldots+(n_k-1)]^2 = \left(\sum_{i=1}^{k} n_i\right)^2 - 2k\sum_{i=1}^{k} n_i + k^2,$

or $(n_1-1)^2+(n_2-1)^2+\ldots+(n_k-1)^2+\sum_{i=1}^{k}\sum_{\substack{i=1\\i\neq j}}^{k}(n_i-1)(n_j-1)=\left(\sum_{i=1}^{k} n_i\right)^2 - 2k\sum_{i=1}^{k} n_i + k^2.$

Therefore, $\sum_{i=1}^{k} n_i^2 - 2\sum_{i=1}^{k} n_i + k + \sum_{i=1}^{k}\sum_{\substack{i=1\\i\neq j}}^{k}(n_i-1)(n_j-1) = \left(\sum_{i=1}^{k} n_i\right)^2 - 2k\sum_{i=1}^{k} n_i + k^2,$

or $\sum_{i=1}^{k} n_i^2 \leq 2\sum_{i=1}^{k} n_i - k + \left(\sum_{i=1}^{k} n_i\right)^2 - 2k\sum_{i=1}^{k} n_i + k^2,$

or $\sum_{i=1}^{k} n_i^2 \leq \left(\sum_{i=1}^{k} n_i\right)^2 - (k-1)\left(2\sum_{i=1}^{k} n_i - k\right).$ ❏

**Theorem 1.7**   A graph with $n$ vertices and $k$ components cannot have more than $\frac{1}{2}(n-k)(n-k+1)$ edges.

**Proof**   The number of components is $k$ and let the number of vertices in $i$th component be $n_i$, $1 \leq i \leq k$.

So,   $n_1+n_2+\ldots+n_k = \sum_{i=1}^{k} n_i = n.$

Now, a component with $n_i$ vertices will have the maximum possible number of edges when it is complete, and in that case it has $\frac{1}{2}n_i(n_i-1)$ edges.

Thus the maximum number of edges in $G = \frac{1}{2}\sum_{i=1}^{k} n_i(n_i-1)$

$$= \frac{1}{2}\sum_{i=1}^{k} n_i^2 - \frac{1}{2}\sum_{i=1}^{k} n_i \le \frac{1}{2}\left\{ \left(\sum_{i=1}^{k} n_i\right)^2 - (k-1)\left(2\sum_{i=1}^{k} n_i - k\right) \right\} - \frac{1}{2}\sum_{i=1}^{k} n_i$$

$$= \frac{1}{2}\left\{ n^2 - (k-1)(2n-k) \right\} - \frac{1}{2}n = \frac{1}{2}\left[ n^2 - 2nk + k^2 + n - k \right] = \frac{1}{2}(n-k)(n-k+1). \qquad \square$$

The following result is due to Turan [247].

**Theorem 1.8**  The maximum number of edges among all $n$ vertex graphs with no triangles is $\left[\frac{n^2}{4}\right]$, where $\left[\frac{n^2}{4}\right]$ is the greatest integer not exceeding the number $\frac{n^2}{4}$.

**Proof**  The result is proved by induction, taking the cases of $n$ odd and $n$ even separately.

Let $n$ be even. The result is obvious for small values of $n$. Assume the result to be true for all even $n \le 2p$. We then prove it for $n = 2p + 2$. So assume $G$ is a graph with $n = 2p + 2$ vertices and no triangles. Since $G$ is not totally disconnected, there are adjacent vertices $u$ and $v$. The subgraph $G' = G - \{u, v\}$ has $2p$ vertices and no triangles, so that by the induction hypothesis, $G'$ has at most $\left[\frac{4p^2}{4}\right] = p^2$ edges.

Now, there can be no vertex $w$ such that $u$ and $v$ are both adjacent to $w$, for then $u$, $v$ and $w$ form a triangle of $G$. Therefore, if $u$ is adjacent to $k$ vertices of $G'$, $v$ is adjacent to at most $2n - k$ vertices. Then $G$ has at most

$$p^2 + k + (2p - k) + 1 = p^2 + 2p + 1 = (p+1)^2 = \frac{n^2}{4} = \left[\frac{n^2}{4}\right] \text{ edges.}$$

It can be shown that for all even $p$, there exists a $(p, \frac{p^2}{4})$ graph with no triangles. Such a graph is formed as follows. Take two sets $V_1$ and $V_2$ of $\frac{p}{2}$ vertices each, and join each vertex of $V_1$ with each vertex of $V_2$.                                                                    $\square$

**Example**  When $n = 6$, the graph formed is shown in Figure 1.20.

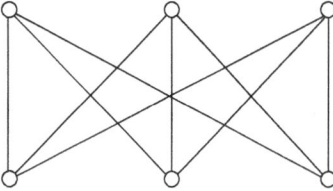

**Fig. 1.20**

The following result is due to Konig [136].

**Theorem 1.9**    A graph is bipartite if and only if it contains no odd cycles.

**Proof**    Let $G$ be a bipartite graph. Then its vertex set $V$ can be partitioned into two sets $V_1$ and $V_2$, so that every edge of $G$ joins a vertex of $V_1$ with a vertex of $V_2$. Thus every cycle $v_1 v_2 \ldots v_n v_1$ in $G$ has its oddly subscripted vertices in $V_1$, say, and the others in $V_2$, so that its length is even.

Conversely, we assume, without loss of generality, that $G$ is connected, for otherwise we can consider the components of $G$ separately. Take any vertex $v_1 \in V$, and let $V_1$ consist of $v_1$ and all vertices of $G$ whose shortest paths from $v_1$ (in terms of the number of edges) contain an even number of edges, while $V_2 = V_- V_1$. Since $G$ has no odd cycles, every edge of $G$ joins a vertex of $V_1$ with a vertex of $V_2$ (Fig. 1.21).

For suppose there is an edge $uv$ joining two vertices of $V_1$. Now, the length of $v_1$ to $u$ is even, the length of $v_1$ to $v$ is even and the union of edges from $v_1$ to $u$ and from $v_1$ to $v$ together with the edge $uv$ contains an odd cycle, which is a contradiction. Hence $G$ is bipartite.    ❑

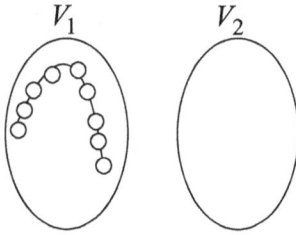

**Fig. 1.21**

**Theorem 1.10**    A graph and its complement are not both disconnected.

**Proof**    Let $G$ be a disconnected graph and $u$, $v$ be any two vertices of $\overline{G}$, the complement of $G$. If $u$ and $v$ belong to different components of $G$, then $u$ and $v$ are adjacent in $\overline{G}$, by definition. If $u$ and $v$ belong to the same component, say $G_i$ of $G$, then let $w$ be a vertex of some other component, say $G_j$, of $G$. Now, by definition both $u$ and $w$, and $v$ and $w$ are adjacent in $\overline{G}$. In either case $u$ is connected to $v$ by a path in $\overline{G}$. Thus, $\overline{G}$ is connected.    ❑

**Theorem 1.11**    For any graph $G$ with six vertices, $G$ or $\overline{G}$ contains a triangle.

**Proof**    Let $v$ be a vertex of a graph $G$ with six vertices. Since $v$ is adjacent either in $G$ or $\overline{G}$ to the other five vertices of $G$, we assume without loss of generality, that there are three vertices $u_1$, $u_2$ and $u_3$ adjacent to $v$ in $G$. If any two of three vertices are adjacent, then they are two vertices of a triangle whose third vertex is $v$. If no two of them are adjacent in $G$, then $u_1$, $u_2$ and $u_3$ are the vertices of a triangle in $\overline{G}$ (Fig. 1.22).    ❑

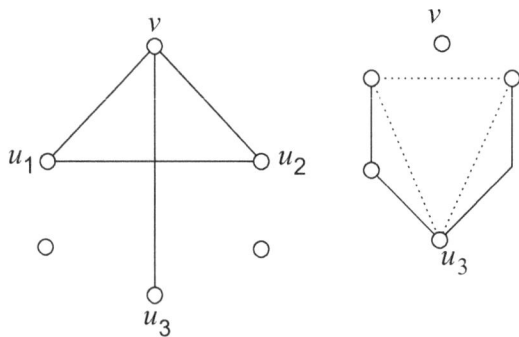

**Fig. 1.22**

**Theorem 1.12**   For any graph $G$ of order $n$, $\Delta(G) \leq n - 1$.

**Proof**   Since a vertex in a graph can be joined to at most $n - 1$ other vertices, the maximum degree a vertex can have is $n - 1$. So, $\Delta(G) \leq n - 1$.                              ❑

**Theorem 1.13**   Every graph $G$ contains a bipartite spanning subgraph whose size is at least half the size of $G$.

**Proof**   Let $B$ be a bipartite spanning subgraph of $G$ with $B$ being of maximum size and the bipartition of its vertex set be $V = V_1 \cup V_2$. Clearly, all the edges of $G$ with one end in $V_1$ and the other in $V_2$ must be in $B$.

Now, if any vertex $u \in V_1$ is joined to $p$ vertices of $V_1$, then $u$ is joined to at least $p$ vertices of $V_2$. That is, the number of vertices of $V_2$ joined to $u$ is greater or equal to the number of vertices of $V_1$ joined to $u$ (Fig. 1.23).

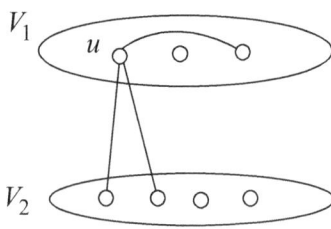

**Fig. 1.23**

If $u \in V_1$ is joined to more vertices of $V_1$ than of $V_2$, then $V_1 - \{u\}$, $V_2 \cup \{u\}$ is a bipartition of $V$ giving a bipartite subgraph of $G$ with larger size, which is a contradiction to our assumption (Fig. 1.24).

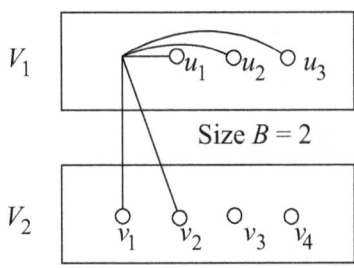 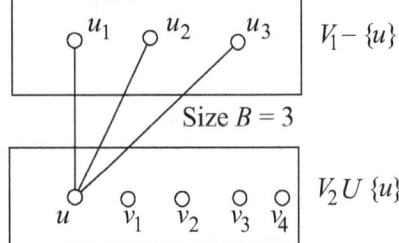

**Fig. 1.24**

Therefore, $d(u|B) \geq \frac{1}{2}d(u|B)$, for each $u \in V$.

Summing over all $u \in V$, we get $m(B) \geq m(G)$. ❑

**Theorem 1.14** If $G$ is a graph with at least $n_0$ vertices and at least $n_0.n(G) - \binom{n_0+1}{2} + 1$ edges, then $G$ contains a subgraph $H$ with $\delta(H) \geq n_0 + 1$, where $n_(G)$ is the number of vertices in $G$ and $n(G) \geq n_0$.

**Proof** Let $G_{n_0}$ be the set of all such graphs $G$ with

$$G_{n_0} = \left\{ G : n(G) > n_0, \ m(G) \geq n_0.n(G) - \binom{n_0+1}{2} + 1 \right\}.$$

If $G \in G_{n_0}$, then $n(G) > no$, because if $n(G) = no$, then

$$m(G) \geq n_0.n(G) - \binom{n_0+1}{2} + 1 \text{ gives } m(G) \geq n_0.n_0 - \binom{n_0+1}{2} + 1.$$

Therefore, $m(G) \geq n_0^2 - \frac{(n_0+1)n_0}{2} + 1 = n_0^2 - \left\{ n_0^2 - \frac{n_0(n_0-1)}{2} \right\} + 1$

$$= \frac{(n_0-1)n_0}{2} + 1 = \binom{n_0}{2} + 1.$$

Thus, $m(G) \geq \binom{n_0}{2} + 1$, which is a contradiction, as the number of edges can be at most $\binom{n_0}{2}$.

Now, if $\delta(G) \leq= n_0$, let $u$ be a vertex with $d(u) \leq n_0$. Then it is easily verified that $G - u$ also belongs to $G_{n_0}$. By repeating this process of removing vertices with degree smaller than $n_0 + 1$ (i.e. $= n_0$), we arrive at a graph $H \in G_{n_0}$ with $\delta(H) = n_0 + 1$. ❑

**Example**   Consider the graph $G$ of Figure 1.25. Here $n(G) = 6$. Let $n_0 = 3$. Clearly, $m(G) \geq$
$3 \times 6 - \begin{pmatrix} 4 \\ 2 \end{pmatrix} + 1 = 13$. In $G - u$, we have $n(G - u) = 5$ and again let $n_0 = 3$. Then $m(G - u) \geq$
$3 \times 5 - \begin{pmatrix} 4 \\ 2 \end{pmatrix} + 1 = 10$, and thus satisfies given conditions. Hence, $G - u \in G_{n_0}$.

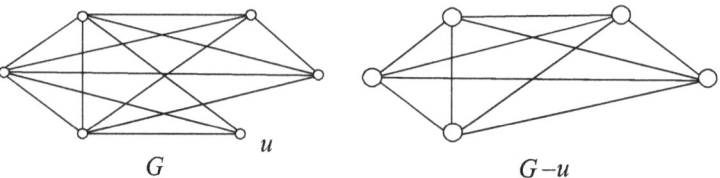

$$G \qquad\qquad G-u$$

**Fig. 1.25**

**Theorem 1.15**   Any graph $G$ has a regular supergraph $H$ of degree $\Delta(G)$ such that $G$ is an induced subgraph of $H$.

**Proof**   We have to prove that for a graph $G$, there exists a graph $H$ which is regular, and the degree of every vertex of $H$ is $\Delta(G)$, and $H$ is the supergraph of $G$, $\Delta(G)$ being the maximum degree in $G$. If $G$ is regular, there is nothing to prove.

Now let $G$ be not regular, and $\Delta = \Delta(G)$ be the maximum degree in $G$. Let $d_i$ be the degree of the vertex $v_i$ in $G$.

Let   $d = \sum_{i=1}^{n} (\Delta - d_i)$.

**Case 1**   If d is even, take $\frac{d}{2}$ disjoint copies of the graph $K_{\Delta+1} - e$ (where $e$ is an edge). For $1 \leq i \leq n$, join $v_i$ to $\Delta - d_i$ different vertices from these copies having degree $\Delta - 1$. We get a graph $H$ which is $\Delta$-regular. Clearly, $G$ is an induced subgraph of $H$.

**Example**   Consider the graph of Figure 1.26. Here $\Delta = 2$, $d = 2 - 1 + 2 - 1 + 2 - 2 = 2$ (even). We take $\frac{d}{2} = \frac{2}{2} = 1$ copy of $K_{2+1} - e = K_3 - e$. Join $v_i$ to $2 - d_i$ different vertices from this copy having degree $2 - 1 = 1$. Join $v_1$ to $2 - 1 = 1$ vertex of this copy having degree 1. Join $v_2$ to $2 - 1 = 1$ vertex.

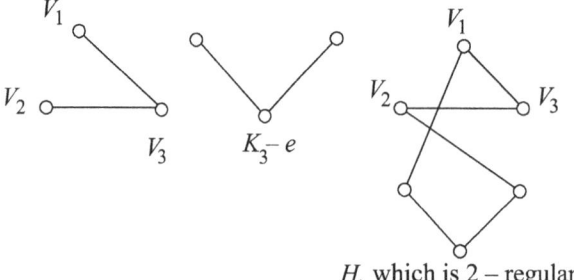

$$V_3 \qquad K_3 - e$$

$$H, \text{ which is } 2 - \text{regular}$$

**Fig. 1.26**

**Case (ii)**  If $d$ is odd take $\frac{(d-1)}{2}$ disjoint copies of the graph $K_{\Delta+1} - e$. For $1 \leq i \leq n$, join $v_i$ to $\Delta - d_i$ different vertices of degree $\Delta - 1$ from these copies. This can be done for all but one vertex of $G$, say $v_j$. Then, $d(v_j) = \Delta - 1$ in the graph $H_1$ so constructed. Now, take another copy of $H_1$ say $H_1'$ with $d(v_j') = \Delta - 1$. Then $H$ is obtained by joining $v_j$ and $v_j'$ by an edge. Hence $H$ is $\Delta$ − regular and $G$ is an induced subgraph of $H$.

**Example**  Consider the graph $G$ of Figure 1.27. Here $\Delta = 3$, $d = 3 - 1 + 3 - 1 + 3 - 1 + 3 - 2 = 2 + 2 + 2 + 1 = 7$, $\frac{(d-1)}{2} = \frac{(7-1)}{2} = 3$. We take 3 copies of $K_{3+1} - e = K_4 - e$.  ❏

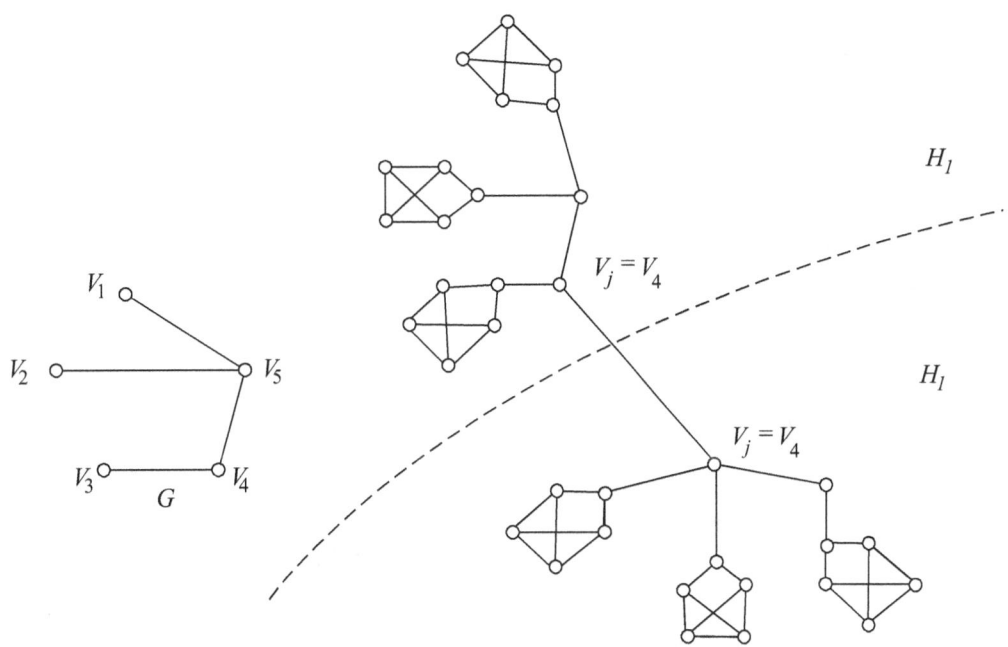

Graph $H$, which is 3 − regular

**Fig. 1.27**

**Theorem 1.16**  If $G = (X, Y, E)$ is a bipartite graph with maximum degree $\Delta$, then $G$ has a $\Delta$-regular bipartite supergraph $H = (U, V, F)$ with $X \subseteq U$, $Y \subseteq V$, $E \subseteq F$.

**Proof**  Let $\Delta$ be the maximum degree in $G = (X, Y, E)$. Take $\Delta$ copies of $G$ (each isomorphic to $G$) say $G_i = (X_i, Y_i, E_i)$, $1 \leq i \leq \Delta$. For each $x \in X_1$ with $d(x) < \Delta$, take $r = \Delta - d(x)$ vertices say $x_1, x_2, \ldots x_r$. Let $Y_o$ be the set of all such vertices as $x$ ranges over $X_1$ and let $V = Y_o \cup Y_1 \cup \ldots \cup Y\Delta$. Join each vertex $x_i$ to one vertex $x(d(x) < \Delta)$ from each of the copies $X_1, X_2, \ldots, X_\Delta$.

Similarly, for each $y \in Y_1$ with $d(y) \leq \Delta$, take $s = \Delta - d(y)$ vertices say $y_1, y_2, \ldots y_s$. Let $X_o$ be that set of all such vertices as $y$ ranges over $Y_1$ and let $U = X_0 \cup X_1 \cup \ldots \cup X_\Delta$. Join each vertex $y_j$ to one vertex $y(d(y) < \Delta)$ from each of the copies $Y_1, Y_2, \ldots, Y_\Delta$. The resulting bipartite graph $H$ has clearly $\Delta$ induced bipartite subgraphs $G_i$ each isomorphic to $G$ (Fig. 1.28).                                                                                          ❏

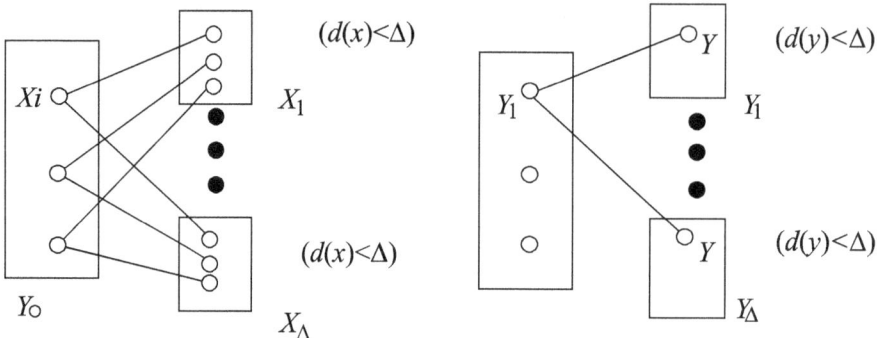

**Fig. 1.28**

**Example 1**    Consider the graph $G$ of Figure 1.29. Here $\Delta = 3$, $d(u) = 2 < 3$, $d(x) = 2 < 3$, $d(y) = 2 < 3$, $d(z) = 1 < 3$. Now, $r = \Delta - d(u) = 3 - 2 = 1$ and $s = \Delta - d(y) + \Delta - d(y) + \Delta - d(z) = 3 - 2 + 3 - 2 + 3 - 1 = 4$.

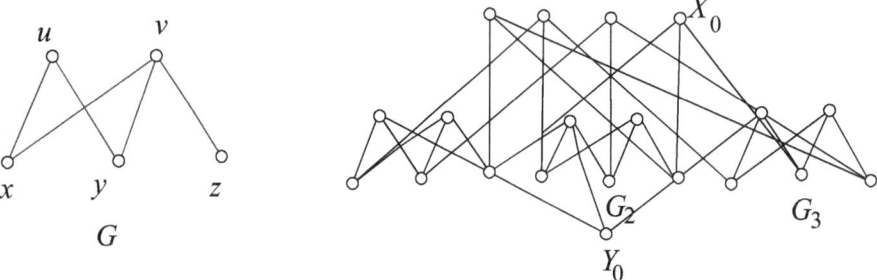

**Fig. 1.29**

**Example 2**    In the graph $G$ of Figure 1.30, $\Delta = 2$ and $d(v) = 0 < 2$, $d(x) = 1 < 2$, $d(y) = 1 < 2$ so that $r = \Delta - d(v) = 2 - 0 = 2$ and $s = 2 - 1 + 2 - 1 = 2$.

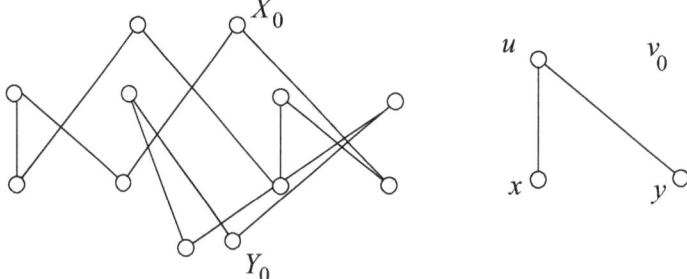

**Fig. 1.30**

## 1.7  Operations on Graphs

Let $G_1(V_1, E_1)$ and $G_2(V_2, E_2)$ be two graphs. The union of $G_1$ and $G_2$, denoted by $G_1 \cup G_2$, is the graph whose vertex set is $V_1 \cup V_2$ and edge set is $E_1 \cup E_2$. The *intersection* of $G_1$ and $G_2$, denoted by $G_1 \cap G_2$, is the graph consisting only of those vertices and edges that are both in $G_1$ and $G_2$. Clearly, $G_1 \cap G_2 = (V_1 \cap V_2, E_1 \cap E_2)$.

The *ring sum* of $G_1$ and $G_2$, denoted by $G_1 \oplus G_2$, is a graph whose vertex set is $V_1 \cup V_2$ and whose edges are that of either $G_1$ or $G_2$, but not of both. Examples of union, intersection and ring sum are given in Figure 1.31.

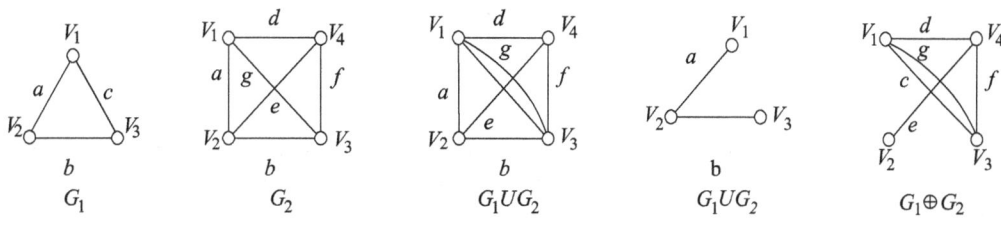

**Fig. 1.31**

We observe that these three operations are commutative, that is,

$$G_1 \cup G_2 = G_2 \cup G_1, \; G_1 \cap G_2 = G_2 \cap G_1, \; G_1 \oplus G_2 = G_2 \oplus G_1.$$

If $G_1$ and $G_2$ are edge disjoint, then $G_1 \cap G_2$ is a null graph and $G_1 \oplus G_2 = G_1 \cup G_2$. If $G_1$ and $G_2$ are vertex disjoint, then $G_1 \cap G_2$ is empty. For any graph $G$, $G \cup G = G \cap G = G$ and $G \oplus G =$ null graph.

If $H$ is a subgraph of $G$, then $G \oplus H$ is by definition, that subgraph of $G$ which remains after all the edges in $H$ have been removed from $G$. We write, $G \oplus H = G - H$, whenever $H \subseteq G$. $G \oplus H = G - H$ is also called complement of $H$ in $G$. Figure 1.32 illustrates this operation.

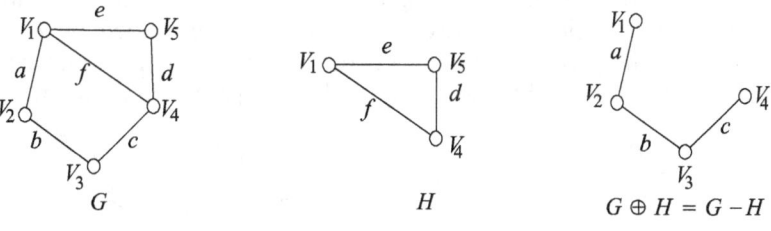

**Fig. 1.32**

**Decomposition:**  A graph $G$ is said to be decomposed into two subgraphs $G_1$ and $G_2$ if $G_1 \cup G_2 = G$ and $G_1 \cap G_2 =$ a null graph. In other words, every edge of $G$ occurs either in $G_1$ or in $G_2$, but not in both, while as some of the vertices can occur in both $G_1$ and

$G_2$. In decomposition, isolated vertices are disregarded. In Figure 1.33, graph $G$ has been decomposed into subgraph $G_1$ and $G_2$.

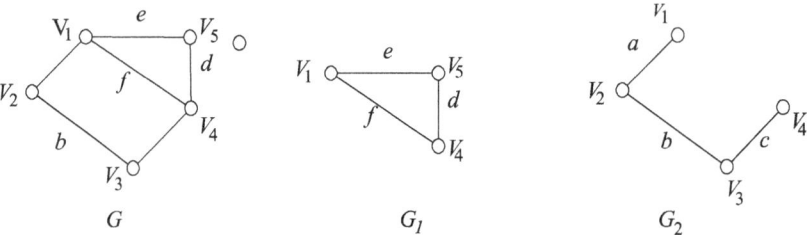

**Fig. 1.33**

**Deletion:** Let $G$ be a graph and $v$ be any vertex in $G$. Then $G - v$ denotes the subgraph of $G$ by deleting vertex $v$, and all the edges of $G$ which are incident with $v$. An example of deletion of a vertex is given in Figure 1.34.

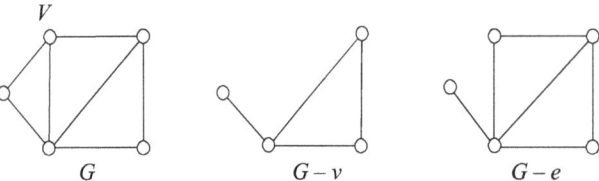

**Fig. 1.34**

If $e$ is any edge of $G$, $G - e$ is a subgraph of $G$ obtained by deleting e from $G$ (Fig. 1.34). Deletion of an edge does not imply deletion of its end vertices. Therefore $G - e = G \oplus e$ .

**Fusion:** A pair of vertices $u$ and $v$ in a graph are said to be fused (merged or identified) if $u$ and $v$ are replaced by a single new vertex such that every edge incident on $u$ or $v$ is incident on this new vertex. Therefore, fusion of vertices does not alter the number of edges, but reduces the number of vertices by one. In Figure 1.35, vertices $u$ and $v$ are fused to a single vertex $w$.

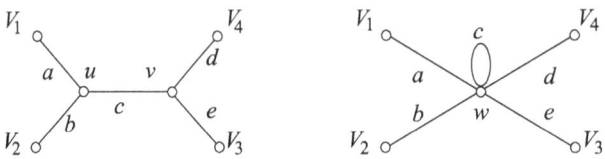

**Fig. 1.35**

Let $G_1 = (V_1, E_1)$ and $G_2 = (V_2, E_2)$ be two graphs with $V_1 \cap V_2 = \Phi$ and $E_1 \cap E_2 = \Phi$. Let $G$ be the compound graph whose vertex set is $V_1 \cup V_2$ and edge set is $E_1 \cup E_2 \cup E_3$, where $E_3$ is a subset of the set of edges $\{v_i v_j : v_i \in V_1, v_j \in V_2\}$. We represent $E_3$ by the symmetric binary relation $\pi \subseteq V_1 \times V_2$, and the compound graph $G$ by $G_1 \pi G_2$. If $\pi = \Phi$, then we get the union of $G_1$ and $G_2$, denoted by $G_1 V G_2$. If $\pi = V_1 \times V_2$ (that is all edges between $V_1$ and $V_2$), we get the join of $G_1$ and $G_2$, denoted by $G_1 \cup G_2$. If $\pi$ is a function from $V_1$ to $V_2$, we have the *function graph* $G_1$ f $G_2$. If $\pi$ defines a homomorphism $\Phi$ from $G_1$ to $G_2$, we have the *homomorphism graph* $G_1 \Phi G_2$. If $G \cong G_2 = G$ (say) and $\pi$ is a bijection $\alpha$ of $V_1$ to $V_2$, we have a *permutation* graph $G \alpha G$. This is also denoted by $P_\alpha(G)$. If $G \cong G_2 = G$ (say) and $\pi$ is an automorphism $\alpha$ of $G$, then $G \alpha G$ is an automorphism graph. These operations are illustrated in Figure 1.36.

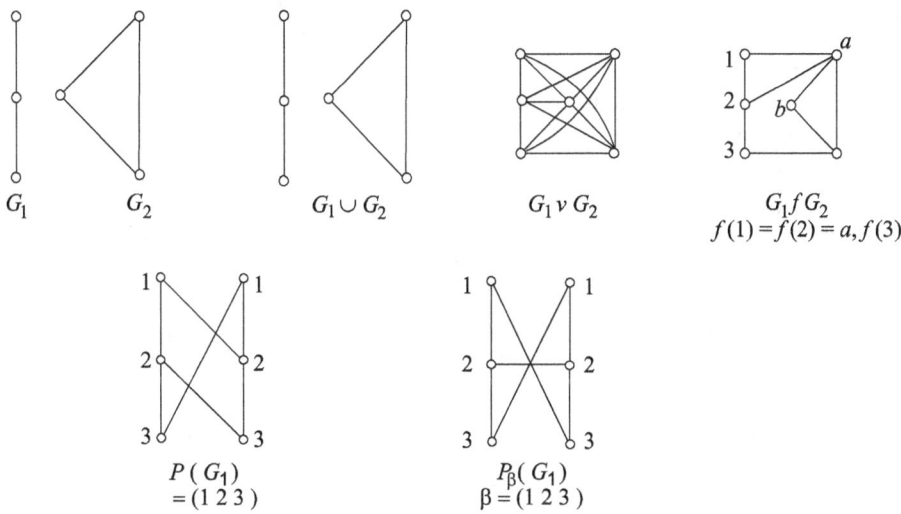

**Fig. 1.36**

Let $G_1 = (V_1, E_1)$ and $G_2 = (V_2, E_2)$ be two graphs, and let the vertex sets $V_1$ and $V_2$ be labelled by the same set of labels. We note here that each of $V_1$ and $V_2$ need not have all the labels.

The *sum* of $G_1$ and $G_2$, denoted by $G_1 + G_2$, is the graph $G = (V, E)$, where $V = V_1 \cup V_2$ and an edge $v_i v_j \in E$ if and only if $v_i v_j$ is an edge of $G_1$ or $G_2$ or both. The *direct sum* denoted by $G_1 (+) G_2$ is the graph $G = (V, E)$ with $V = V_1 \cup V_2$ and an edge $v_i v_j \in E$ if and only if $v_i v_j$ is an edge of $G_1$ or $G_2$, but not of both. The *superposition* graph $G = G_1$ s $G_2$ has vertex set $V = V_1 \cup V_2$ and the edge set $E$ contains all the edges of $G_1$ and $G_2$ with the identity of edges of $G_1$ and $G_2$ in $G$ being preserved by assigning two different labels to these edges. So, if $v_i v_j$ is an edge in both $G_1$ and $G_2$, then there are two edges $v_i v_j$ in $G$ with different labels. For illustration of these operations, see Figure 1.37. We can extend these operations to a finite number of graphs.

Let $G_1 = (V_1, E_1)$ and $G_2 = (V_2, E_2)$ be two graphs. We form the compound graphs $G$ from $G_1$ and $G_2$ with vertex sets $V_1 \times V_2$.

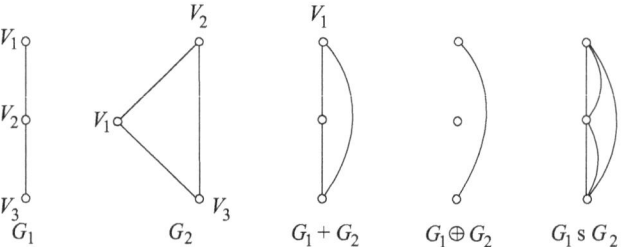

**Fig. 1.37**

The cartesian product of $G_1$ and $G_2$, denoted by $G = G_1 \times G_2$, is the graph whose vertex set is $V = V_1 \times V_2$ and for any two vertices $w_1 = (u_1, v_1)$ and $w_2 = (u_2, v_2)$ in $V$, $u_1, u_2 \in V_1$ and $v_1, v_2 \in V_2$, there is an edge $w_1 w_2 \in E(G)$ if and only if either (a) $u_1 = u_2$ and $v_1 v_2 \in E_2$, or (b) $v_1 = v_2$ and $u_1 u_2 \in E_1$.

The *tensor product* (conjunction), denoted by $G = G_1 \wedge G_2$, is the graph with vertex set $V = V_1 \times V_2$ and for any two vertices $w_1 = (u_1, v_1)$ and $w_2 = (u_2, v_2)$ in $V$; $u_1, u_2 \in V_1$ and $v_1, v_2 \in V_2$, there is an edge $w_1 w_2 \in E(G)$ if and only if $u_1 u_2 \in E_1$ and $v_1 v_2 \in E_2$.

The *normal product* (strong product), denoted by $G = G_1 \circ G_2$, is the graph with vertex set $V = V_1 \times V_2$ and for any two vertices $w_1 = (u_1, v_1)$ and $w_2 = (u_2, v_2)$ in $V$; $u_1, u_2 \in V_1$ and $v_1, v_2 \in V_2$, there is an edge $w_1 w_2 \in E(G)$ if and only if one of the following holds:

(a) $u_1 = u_2$ and $v_1 v_2 \in E_2$ (b) $v_1 = v_2$ and $u_1 u_2 \in E_1$ (c) $u_1 u_2 \in E_1$ and $v_1 v_2 \in E_2$.

It is easy to see that $G_1 \times G_2$ and $G_1 \wedge G_2$ are spanning subgraphs of $G_1 \circ G_2$ and also $G_1 \circ G_2 = (G_1 \times G_2) \oplus (G_1 \wedge G_2)$. Figure 1.38 illustrates these operations.

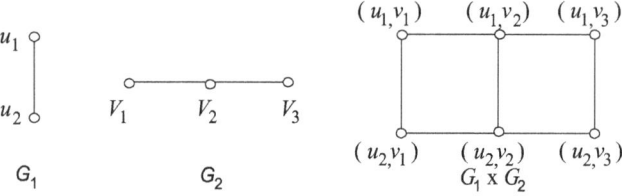

**Fig. 1.38**

The *composition* of $G_1$ and $G_2$, denoted by $G = G_1[G_2]$, is the graph with vertex set $V = V_1 \times V_2$, and for any two vertices $w_1 = (u_1, v_1)$ and $w_2 = (u_2, v_2)$ in $V$; $u_1, u_2 \in V_1$ and $v_1, v_2 \in V_2$, there is an edge $w_1 w_2 \in E(G)$ if and only if either (a) $u_1 u_2 \in E_1$ or (b) $u_1 = u_2$ and $v_1 v_2 \in E_2$ (Fig. 1.39).

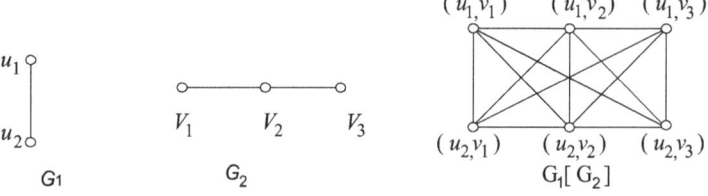

**Fig. 1.39**

## 1.8 Topological Operations

**Definition:** A *subdivision* of the edge $e = uv$ of a graph $G$ is the replacement of the edge $e$ by a new vertex $w$ and two new edges $uw$ and $wv$. This operation is also called an *elementary subdivision* of G (Fig. 1.40).

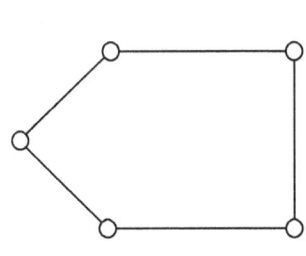 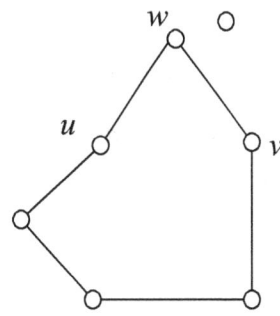

**Fig. 1.40**

A graph $H$ obtained by a sequence of elementary subdivisions from a graph $G$ is said to be a *subdivision graph* of $G$, or to be homeomorphic (or a homeomorph) of $G$. Two graphs $H_1$ and $H_2$ which are homeomorphs of the same graph $G$ are said to be *homeomorphic* to each other (or homeomorphically reducible to $G$). A graph $G$ is *homeomorphically irreducible* if whenever a graph $H$ is homeomorphic to $G$, then $H$ is homeomorphic from $G$. In Figure 1.41, $H_1$ and $H_2$ are homeomorphs of $K_4$ and $K_4$ itself is homeomorphically irreducible.

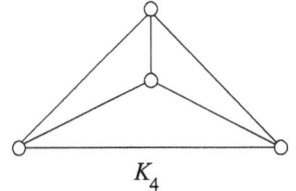 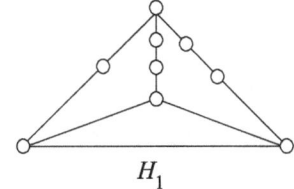 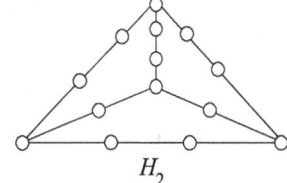

**Fig. 1.41**

The relation being homeomorphic to each other is an equivalence relation and each equivalence class contains a unique homeomorphically irreducible graph which is taken as the representative of the class.

**Identification of the vertices:** Let $G = (V, E)$ be a general graph and let $U$ be a subset of the vertex set $V$. Let $H$ be the general graph obtained from $G$ by the following operations.

i. Replace the set of vertices $U$ by a single new vertex $u$.

ii. Replace the edge $e = ab$ with $a \in U$ and $b \in V - U$ by a corresponding edge $e' = ub$.

iii. Replace each edge $e = ab$ with $a, b \in U$ ($b$ possibly being the same as $a$) by a loop at $u$.

Let $K$ be the multigraph obtained from $H$ by dropping all loops and $L$ be the graph obtained from $K$ by replacing each multiple edge by a single edge. Then $H$, $K$, $L$ are respectively said to be obtained from $G$ by a *general, multiple, simple identification* of the vertices of $U$. We write $H = G * U$, $K = G : U$, $L = G.U$. This operation is illustrated in Figure 1.42.

**Remarks**

1. The operation in (ii) may result in the generation of multiple edges and that in (iii), in the generation of loops even when $G$ is a simple graph.

2. For $b \neq u$ in $H$, $q_H[a, b] = \sum \{q_G[a, b] : a \in U\}$.

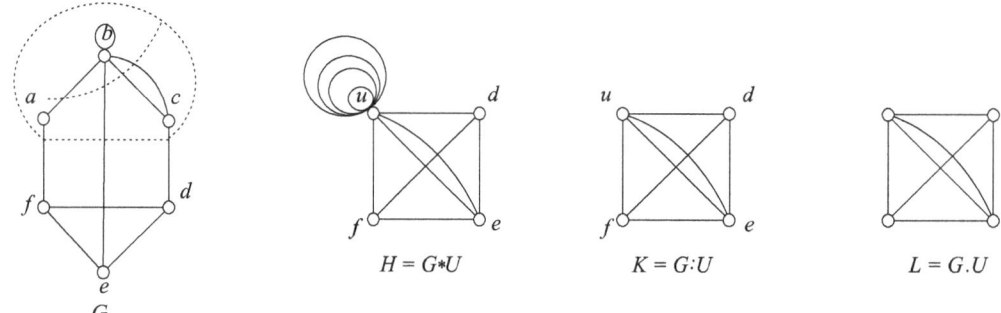

$$H = G * U \qquad K = G:U \qquad L = G.U$$

**Fig. 1.42**

**Coalescence of G:**   Let $G = (V, E)$ be a general graph and $V = U_1 \cup U_2 \cup \ldots \cup U_r$ be a partition $\pi$ of the vertex set $V$ with $U_i \cap U_j \neq \Phi$, $i \neq j$. Let $H = (U, E')$ be the general graph defined by replacing

i. each $U_i$ by a single new vertex $u_i$,

ii. each edge $e = v_i v_j$ with $v_i \in u_i$ and $v_j \in u_j$ $(i \neq j)$ by a corresponding edge $e' = u_i u_j$ and

iii. each edge $e = ab$ with $a$, $b \in U_i$ by a loop at $u_i$.

Let $K$ be the multigraph obtained from $H$ by dropping the loops and $L$ be the graph obtained from $K$ by replacing the multiple edges by single edges. Then $H$, $K$, $L$ are respectively called the *general/multiple/simple coalescence* of $G$ and we write $H = G * U$, $K = G : U$ and $L = G.U$. Coalescence of a graph is illustrated in Figure 1.43.

**Remarks**

1. Clearly $H$ is obtained from $G$ by sequentially identifying the vertices in $U_i$, $1 \leq i \leq r$ in any order.

2. To coalescence a partition $\pi_1$ of a proper subset $V_1$ of $V$, we may adopt the above definition by augmenting $\pi_1$ to a partition of $V$ by adjoining the singleton sets corresponding to the vertices in $V - V_1$.

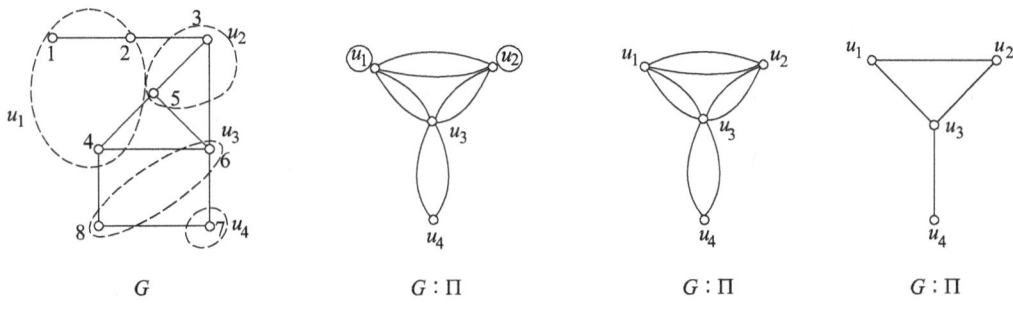

**Fig. 1.43**

Let $G = (V, E)$ be a general graph and $f : V \to W$ be a map of $V$ onto $W = \{w_1, w_2, \ldots, w_r\}$. Let $U_i = f^{-1}(w)$ and $V = \bigcup\limits_{i=1}^{r} U_i$ be the partition of $V$ induced by $f$. Then the general/multiple/ simple coalescences of $G$ are denoted by $f_g(G), f_m(G), f_s(G)$ or $G * f, G : f, G.f$, respectively, and are referred to as a *general/multiple/simple homomorph* or homomorphic image of $G$. We will use $f(G)$ as a common symbol for these and may often abbreviate this to $fG$.

Now, let $M = (W', E')$ be a general/multiple/simple graph with $W \subseteq W'$ such that $f(G)$ is a general/multiple/simple subgraph of $M$. Then $f$ is said to be a *homomorphism* of $G$ into $M$. If $W' = W$, then $f$ is said to be an *onto (surjective) homomorphism*. If $F(G) = < f(V) >$, then $f$ is called a *full homomorphism*. In case $f(G) = M$, then $f$ is called a *full onto homomorphism*.

**Remarks**

1. An injective homomorphism (i.e., $f$ is injective) is a *monomorphism*. A full onto monomorphism is an *isomorphism*.

2. If every $U_i$ is an independent set in $G$, then $f$ is a discrete homomorphism. If $U_1 = \{u, v\}$ and other $U_i$'s are singleton sets of $V$, then $f$ is called an *elementary homomorphism*, and if $uv \notin E$, then $f$ is a *discrete elementary homomorphism*.

3. If every $< U_i >$ is a connected subgraph of $G$, then $f$ is called *connected homomorphism*. An elementary homomorphism with $uv \in E$ is called a *connected elementary homomorphism*.

4. A mapping $f : V \to W'$, where $G = (V, E)$ and $M = (W', E')$ are simple graphs, is a homomorphism of $G$ into $M$ if and only if $uv \in E$ implies that $f(u)f(v) \in E'$.

The following result is due to Ore [178].

**Theorem 1.17** Any homomorphism is the product of a connected and a discrete homomorphism.

**Proof** Let $f$ be a homomorphism of $G = (V, E)$ into $M = (W', E')$ with $f(V) = W$ (i.e., $f(V_i) = W_i$). So, $V = f^{-1}(W)$ and $V_i = f^{-1}(W_i)$. Then each $< f^{-1}(W_i) >$ may consist of a number of components (which are, in fact, connected) $C_{i_1}, C_{i_2}, \ldots, C_{i_j}, \ldots, C_{i_n}$ with vertex sets $V_{i_1}, V_{i_2}, \ldots, V_{i_j}, \ldots, V_{i_{ni}}$.

The coalescence of $V_i = f^{-1}(W_i)$ can be done in two stages. First, we perform a multiple coalescence of $V_i = V_{i_1} \cup V_{i_2} \cup \ldots \cup V_{i_j} \cup \ldots \cup V_{i_{ni}}$ sending vertex set $V_{i_j}$ to a single vertex $W_{i_j}$. This, when done for all $V_i = f^{-1}(W_i)$'s, corresponds to a connected homomorphism $\Phi$ of $G$. Next, identify the vertices $W_{i_j}$, $1 \le j \le n_i$ of $\Phi(G)$ into a single vertex $W_i$ for $1 \le i \le |W|$. This corresponds to a discrete homomorphism $\theta$ of $\Phi(G)$ into $M$. Clearly, $f(G) = \theta(\Phi(G))$.   ❑

**Contractions:**   Let $G = (V, E)$ be a general graph and $F$ be a subset of E such that the edge-induced subgraph $< F >$ is connected on the vertex subset $U = V(< F >)$. Then the general/multiple/simple graphs obtained by the identification of the vertex set $U'$ in the general graph $G - F$ are called the *general/multiple/simple contractions* of $F$ in $G$ and are denoted by $G|\, þ F$, $G||F$ and $G|F$ respectively. Consider the graph of Figure 1.44. The vertices of $U = \{v_1, v_2, v_3, v_4\}$ in $G - F$ have been identified.

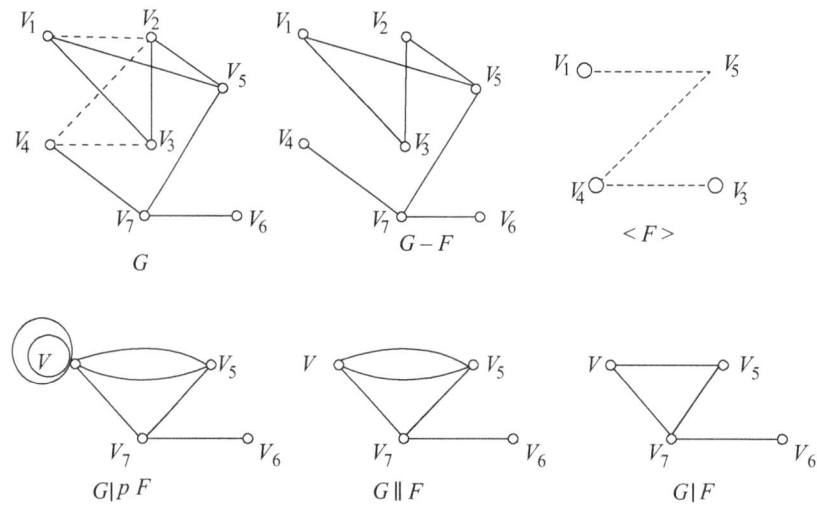

**Fig. 1.44**

Let $G = (V, E)$ be a general graph and let $F$ be a subset of $E$ such that the edge-induced subgraph $< F >$ has components $< F_i >$ on vertex sets $V(< F_i >) = U_i$, $1 \le i \le r$. Let $\pi$ be the partition of $V$ induced by the $U_i$, vertices in $U_i$ being taken as singleton sets of the partition.

The general/multiple/simple coalescences in the general graph $G - F$ are called the *general/multiple/simple contractions* of $F$ in $G$ and are denoted by $G|\, þ F$, $G||F$ and $G|F$ respectively.

Let $G = (V, E)$ be a general graph and let $W = \{w_1, w_2, \ldots, w_r\}$. Suppose $f : V \to W$ is a map such that $< U_i >$ is a connected general subgraph of $G$, where $U_i = f^{-1}(w_i)$, $1 \le i \le r(f(U_i) = w_i)$. Let $F_i \subseteq E(U_i)$ be such that $F_i$ induces a spanning connected subgraph of $< U_i >$ and let $F = \bigcup_{i=1}^{r} F_i$. Then $G|\, þ F$, $G||F$ and $G|F$ are respectively called  general/

multiple/simple contraction induced by $f$ and $F$ and are denoted by $f_F(G)$. Here $f$ is called a *contraction mapping* of $G$.

If $f_F(G) = H$, then $G$ is said to be $g/m/s$- contractible to $H$ and $H$ is called a $g/m/s$-*contraction* of $G$, according as $H$ is a $g/m/s$-graph. We call *simple contractible* and *simple contraction* as *contractible* and *contraction*.

If a subgraph of $G$ is contractible to $H$ (equivalently, if $H$ is a subgraph of a contraction of $G$), then $H$ is called a *subcontraction* of $G$, and we also say that $G$ is *subcontractible* to $H$. We denote this by $G\}H$.

**Remarks**

1.  It can be observed that corresponding to every general contraction there exists a unique contraction mapping $f$ but given a mapping $f$ from $V(G)$ to $V(H)$, with $G$ and $H$ being general graphs, there can be various contractions of $G$ into $H$ defined by suitable edge subsets of $\bigcup_{i=1}^{k} < f^{-1}(w_i) >$. Any two such contractions of the same graph $G$ corresponding to the same function $f$ differ only in the number of loops at $w_i \in W = V(H)$. Thus all such contractions correspond to a unique simple contraction induced by $f$.

2.  We note the distinction between a contraction and a connected homomorphism. A connected homomorphism $f$ is a particular case of a contraction induced by $f$ when the edges $F_i$ are the null subsets of $E < f^{-1}(w_i) >$. This is called the *null contraction* corresponding to $f$ and is denoted by $f_0(G)$. Also, the connected homomorphism induced by $f$ differs from any proper contraction induced by $f$ only in the number of loops at $w_i \in W$. The contraction induced by $f$ in which each $F_i = E < f^{-1}(w_i) >$ is called the *full contraction* induced by $f$.

3.  If $e = uv \in E$ and $F = \{e\}$, then $G| þ e$ is called an *elementary contraction* of $G$ and differs from the corresponding connected elementary homomorphism $G : \{v, u\}$ by the absence of a loop at the common vertex $w = \{u, v\}$ in $H$.

The following result is due to Behzad and Chartrand [15].

**Theorem 1.18**    If a graph $H$ is homeomorphic from a graph $G$, then $G$ is a contraction of $H$.

**Proof**    If $G = H$, then obviously $G$ is the null contraction of $H$ corresponding to the identity mapping $f : V(H) \to V(G)$.

Now, let $G \neq H$. Then $H$ is obtained from $G$ by a sequence of elementary subdivisions, say $\in_1, \in_2, \ldots, \in_k$, through the intermediate graphs $G_i = \in_i (G_{i-1})$, $G_0 = G$, $G_k = H$. Let $G_i$ be obtained from $G_{i-1}$ by subdividing the edge $uv$, introducing a new vertex $w$. Then $G_{i-1}$ is obtainable from $G_i$ by the elementary contraction with $f = \{uw\}$. That is, $G_{i-1} = G_i|uw = \Phi(G_i)$, say. Then, clearly we have $G = \Phi_1(\Phi_2\ldots(\Phi_k(H)\ldots)$. Therefore, $G$ can be obtained from $H$ by a sequence of elementary contractions and is thus a contraction of $H$.    ❏

The converse of the above theorem is not true in general. To see this, consider the following example (Fig. 1.45). Here, the graph $H$ is a contraction of the graph $G$ by using the mapping $f(a) = f(b) = f(c) = a'$, $f(g) = g'$, $f(h) = h'$, $f(d) = f(e) = f(f) = d'$. But $G$ is not homeomorphic from $H$.

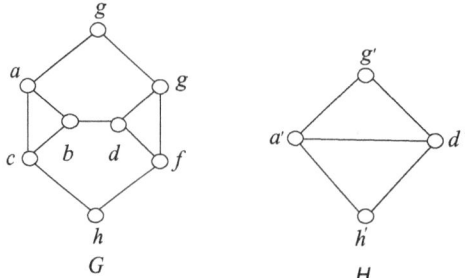

**Fig. 1.45**

**Corollary 1.1**   If a graph $H$ contains a subgraph $K$ homeomorphic from a non-trivial connected graph $G$, then $G$ has a sub contraction of $H$.

The next result due to Halin can be found in Ore [177].

**Theorem 1.19**   If a graph $G$ is contractible to a graph $H$ and $\Delta(H) \leq 3$, then $G$ has a subgraph homeomorphic from $H$.

**Proof**   Let $G$ be a graph contractible to the graph $H$. Hence there is a mapping $F : V(G) \to V(H)$ inducing the contraction $H$. Therefore, $G_i = < f^{-1}(v_i) >$ for $v_i \in V(H)$ are connected induced subgraphs of $G$ which have been contracted to the vertices $v_i$ of $H$. Let $U_i = f^{-1}(vi)$.

As $\Delta(H) \leq 3$, for given $U_1$ there are at most three $u_i$, say $u_2$, $u_3$, $u_4$ such that there is an edge from $U_1$ to $U_i$. Let the end vertices in $U_1$ of three such edges from $U_1$ to $U_2$, $U_3$, $U_4$, respectively be $u_2$, $u_3$, $u_4$. Since $G_1$ is connected, there is $u_2 - u_3$ path $P$ and a $u_4 - u_5$ path $Q$ in $G_1$, where $u_5$ is the first vertex that $Q$ has in common with $P$. (If there are only two vertices, we consider only the path $P$.) For each edge $v_i v_j$ in $H$ we choose one edge between $U_i$ and $U_j$ in $G$ and then the set of subgraphs $PUQ$ in $G_i$. The graph $H_1$ so formed is a subgraph of $G$ and clearly from the construction it is homeomorphic from $H$ (Fig. 1.46).

❑

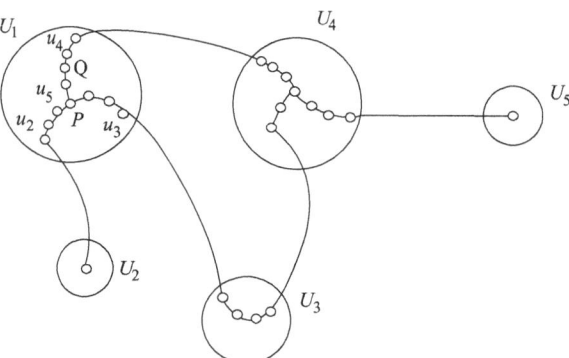

**Fig. 1.46**

**Corollary 1.2**   If the graph $G$ has a sub contraction $H$ with $\Delta(H) \leq 3$, then $G$ has a sub-graph homeomorphic from $H$.

**Theorem 1.20**   If $G$ is subcontractible to $K_5$, then either $G$ has a subgraph homeomorphic from $K_5$, or $G$ has a subcontraction to the graph $L$.

**Proof**   We assume without loss of generality, that the given graph $G$ is contractible to $K_5$. Let $V(K_5) = \{v_1, v_2, v_3, v_4, v5\}$ and let $U_i = f^{-}1(vi)$, $1 \leq i \leq 5$, where $f : V(G) \to V(K_5)$ is the contraction mapping. Then $G_i = <U_i>$ are connected subgraphs of $G$ and there is an edge in $G$ between $U_i$ and $U_j$ for each pair $(i, j)$. Let $u_2, u_3, u_4, u_5$ be the end vertices in $U_1$, of four such edges selected from $U_1$ to the other $U_i(2 \leq i \leq 5)$. Since $G_1$ is connected, there is a $u_2 - u_5$ path $P$ in $G$, and there are $u_4 - u'_4$, $u_5 - u'_5$ paths $Q$ and $R$, where $u'_4$ and $u'_5$ are the first vertices that these paths have in common with $P$.

If $u'_4 \neq u'_5$, we can contract each $U_i$, $2 \leq i \leq 5$ to a vertex to get a $K_4$ and this together with the selected edges from $U_1$ to the $U_i$, and the path $P$, $Q$, $R$ is a graph homeomorphic to $L$, so that $G$ has a subcontraction to $L$.

If $u'_4 = u'_5$, for any choice of $P$ and for every set $U_i$, $G$ has a subgraph homeomorphic from $K_5$. The same is the case when in each $U_i$, the four end vertices $u_j$ coincide (Fig. 1.47).   ❑

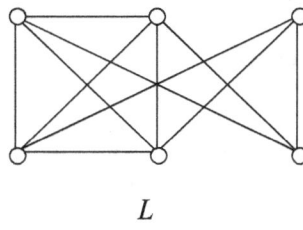

$L$

**Fig. 1.47**

## 1.9  Distance and eccentricity

**Definition:**   Distance between two vertices $u$ and $v$ of a connected graph $G$ is the length of the shortest path between $u$ and $v$. It is denoted by $d(u, v)$. Clearly, by convection, $d(u, u) = 0$. The shortest path is called a $u - v$ *geodesic*, or a $u - v$ distance path.

**Theorem 1.21**   The distance function for a connected graph is a metric defined on its vertex set.

**Proof**   Let $G$ be a connected graph and $d(u, v)$ be the distance function, with $u$ and $v$ being vertices of $G$.

1. Clearly, $d(u, v) \geq 0$, with $d(u, u) = 0$.

2. Also, $d(u, v) = d(v, u)$, because the length of the shortest path from $u$ to $v$ is same as the length of the shortest path from $v$ to $u$.

3. For any vertices $u$, $v$, $w$ in $G$, we have $d(u, v) \leq d(u, w) + d(w, v)$

Thus $d$ is a metric.                                                                    ❏

**Note**    The distance between two vertices $u$ and $v$ in different components of a disconnected graph is infinite and this distance function does not induce a metric on the vertex set.

**Neighbourhood:**    Let $v$ be any vertex of a connected graph $G$. The $i$th neighbourhood of $v$ is $N_i(v) = \{u \in V : d(v, u) = i\}$. We set $N_0(v) = \{v\}$ and denote $N_1(v)$ simply by $N(v)$, and call $N(v)$ as the neighbourhood of $v$, or the neighbours of $v$.

The $s$-ball at $v$ is $B_s(v) = \bigcup\limits_{j=0}^{s} N_j(v)$.

**Eccentricity of a vertex:**    Let $G$ be a connected graph. The eccentricity of a vertex $v$ in $G$ is the distance of the vertex $u$ farthest from $v$. It is denoted by $e(v)$. That is, $e(v) = \max\{d(u, v) : u \in V\}$.

The minimum eccentricity is called the *radius* of $G$ and the maximum eccentricity is called the *diameter* of $G$. The radius is denoted by $r$ and diameter by $d$.

Therefore, $r = \min\{e(v) : v \in V\}$ and $d = \max\{e(v) : v \in V\}$.

When the graph is to be mentioned, we use the notations, $e(v|G)$, $r(G)$ and $d(G)$ for eccentricity, radius and diameter respectively.

The vertices of minimum eccentricity in $G$ are called the *centres* of $G$ and the vertices of maximum eccentricity are called the *periphery* of $G$. That is, the centre of $G$ is $C(G) = \{v \in V : e(v) = r\}$ and periphery of $G$ is $P(G) = \{v \in V; e(v) = d\}$.

A vertex in $C(G)$ is called a *central vertex* and a vertex in $P(G)$ is called a *peripheral vertex*. A graph having $C(G) = V(G)$ is called a *self-centered graph*.

A longest geodesic (maximum among the shortest paths between any two vertices) of a graph is called a *diametral path* of $G$. Note that this is not the same thing as a longest path in $G$. The diameter is the length of any diametral path. A graph $G$ is said to be *geodetic* if any two of its vertices are joined by a unique geodesic.

**Theorem 1.22**    The radius and diameter of a graph are related as $r \leq d \leq 2r$.

**Proof**    Clearly, $r \leq d$ follows from the definition of $r$ and $d$. Now, let $u$ and $v$ be the ends of a diametral path and $w$ be a central vertex. Then,

$d = d(u, v) \leq d(u, w) + d(w, v) \leq r + r = 2r$,

by using the triangle inequality of the metric and definition of $r$.                     ❏

**Example**    Consider the graph shown in Figure 1.48. The eccentricities are 5, 4, 3, 3, 3, 4, 5, 4, 5; $r(G) = 3$, $d(G) = 5$, $C(G) = \{v_3, v_4, v_5\}$, $d(G) = \{v_1, v_7, v_9\}$. Diametral paths are $v_1 v_2 v_4 v_5 v_6 v_7$, $v_1 v_2 v_3 v_5 v_6 v_7$, $v_1 v_2 v_4 v_5 v_8 v_9$ and $v_1 v_2 v_3 v_5 v_8 v_9$.

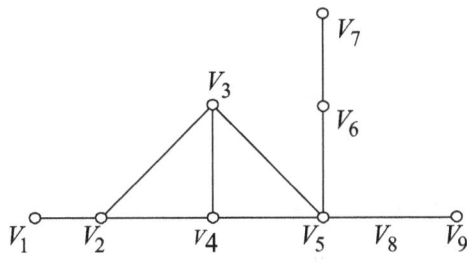

**Fig. 1.48**

**Detours:** A *detour path* between the vertices $u$ and $v$ in a graph $G$ is the path of maximum length between $u$ and $v$. The length of detour path is called *detour distance* and is denoted by $\partial(u, v)$.

The *detour eccentricity* of a vertex $v$ is defined by $\partial(v) = \max\{\partial(v, u) : u \in V\}$.

The *detour diameter*, or detour number of $G$ is defined as $\partial(G) = \max\{\partial(v) : v \in V\}$ and the *detour radius* is defined by $r_\partial(G) = \min\{\partial(v) : v \in V\}$.

A longest path in $G$ is called a *detour path* in $G$ and its length is $\partial(G)$. The *detour center* of $G$ is $C_\partial(G) = \{v \in V : \partial(v) = r_\partial(G)\}$.

**Theorem 1.23** If a connected graph $G$ of order $n$ has minimum degree $\delta$, then

$$\partial(G) = \min(n-1, 2\delta). \tag{1.23.1}$$

**Proof** Clearly, $\partial(G) \leq n-1$. If $\partial(G) = n-1$, then (1.23.1) is obviously true, as min $(n-1, 2\delta) = 2$, or $n-1$ according as $\delta = 1$, or $\delta = n-1$. Now, let $\partial(G) < n-1$. Assume the contrary,

$$\partial(G) < \min(n-1, 2\delta). \tag{1.23.2}$$

Let $P = v_0 v_1 \ldots v_\partial$ be a path of length $\partial = \partial(G)$ and $H$ be the subgraph induced by the vertex set of $P$. Then $H$ does not contain a spanning cycle, because otherwise some vertex $v \in V(G) - V(H)$ will be adjacent to some vertex $v_i$ of $V(H)$ giving a path of length $\partial + 1$. Thus, $v_0 v_\partial \notin E$, and for similar reasons $v_0$ and $v_\partial$ are adjacent only to vertices in $H$.

Let $S = \{v_i \in V(H) : v_0 v_i \in E\}$ and $T = \{v_i \in V(H) : v_{i-1} v_\partial \in E\}$.

Since $H$ does not contain a spanning cycle, $S \cap T = \Phi$. Also, $v_0 \notin S \cup T$, so that $|S \cup T| \leq \partial$. Also, $|S| \geq \delta$ and $|T| \geq \delta$, therefore $\partial \geq |S \cup T| = |S| + |T| \geq \delta + \delta = 2\delta$. That is,

$$\partial \geq 2\delta. \tag{1.23.3}$$

Therefore, $2\delta \leq \partial < n-1$, so that $2\delta < n-1$.

Thus, (1.23.2) gives $\partial < \min(n-1, 2\delta)$ implying $\partial < 2\delta$ (because min $(n-1, 2\delta) = 2\delta$ as $2\delta < n-1$), which contradicts (1.23.2) and so our assumption (1.23.2) is wrong.

Hence, $\partial(G) \geq \min(n-1, 2\delta)$. ❑

## 1.9  Exercises

1. If $G$ is a graph with $n$ vertices and $m$ edges and if $k$ is the smallest positive integer such that $k \geq 2m/n$, then prove $G$ is a vertex of degree at least $k$.

2. If $G$ is a graph with $n$ vertices and if $r$ vertices out of n have degree $k$ and the others have degree $k+1$, then prove $r = (k+1)n - 2m$, where $m$ is the number of edges in $G$.

3. If $G$ is a $k$-regular graph ($k$ being odd), prove that the number of edges in $G$ is a multiple of $k$.

4. If $G$ is a graph with $n$ vertices and exactly $n-1$ edges, then prove that $G$ has either a vertex of degree 1 or a vertex of degree zero.

5. If in a graph $G$, $\delta \geq \frac{n-1}{2}$, then show that $G$ is connected.

6. If a graph $G$ is not connected, then show that $\overline{G}$ is connected.

7. Show that if a self-complementary graph contains pendent vertex, then it must have at least another pendent vertex.

8. Prove that any two connected graphs with $n$ vertices, all of degree 2, are isomorphic.

9. Prove that if a connected graph $G$ is decomposed into two subgraphs $G_1$ and $G_2$, there must be at least one vertex common between $G_1$ and $G_2$.

10. Prove that a connected graph $G$ remains connected after removing an edge e from $G$ if and only if $e$ is in some cycle of $G$.

11. If the intersection of two paths is a disconnected graph, show that the union of the two paths has at least one cycle.

12. Show that the order of a self-complementary graph is of the form $4n$ or $4n+1$, where $n$ is a positive integer.

13. Draw all the non-isomorphic self complementary graphs on four vertices.

14. Show that the bipartition of a connected graph is unique. Prove that a graph with $n(n > 1)$ vertices has at least $(n-1)([n/2])$ cycles.

15. Prove that a graph with $n$ vertices must be connected if it has more than $(n-1)(n-2)/2$ edges.

16. Prove that a graph with $n$ vertices ($n > 2$) cannot be bipartite if it has more than $n^2/4$ edges.

17. Prove that every graph with $n$ vertices is isomorphic to a subgraph of $K_n$.

18. Prove that if a graph has more edges than vertices then it must possesses at least one cycle.

# 2. Degree Sequences

The concept of degrees in graphs has provided a framework for the study of various structural properties of graphs and has therefore attracted the attention of many graph theorists. Here, we deliberate on the various criteria for a non-decreasing sequence of non-negative integers to be a degree sequence of some graph.

## 2.1 Degree Sequences

Let $d_i$, $1 \leq i \leq n$, be the degrees of the vertices $v_i$ of a graph in any order. The sequence $[d_i]_1^n$ is called the degree sequence of the graph. The non-negative sequence $[d_i]_1^n$ is called the degree sequence of the graph if it is the degree sequence of some graph, and the graph is said to realise the sequence.

The set of distinct non-negative integers occurring in a degree sequence of a graph is called its *degree set*. A set of non-negative integers is called a *degree set* if it is the degree set of some graph, and the graph is said to realise the degree set.

Two graphs with the same degree sequence are said to be *degree equivalent*. In the graph of Figure 2.1(a), the degree sequence is $D = [1, 2, 3, 3, 3, 4]$ or $D = [1\ 2\ 3^3\ 4]$ and its degree set is $\{1, 2, 3, 4\}$, while the degree sequence of the graph in Figure 2.1(b) is $[1, 1, 2, 3, 3]$ and its degree set is $\{1, 2, 3\}$.

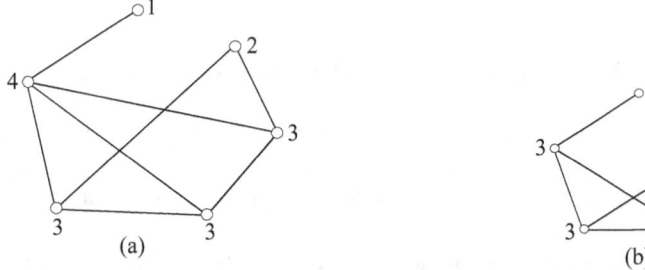

**Fig. 2.1**

If the degree sequence is arranged as the non-decreasing positive sequence $d_1^{n_1}$, $d_2^{n_2}$, ... $d_k^{n_k}$, $(d_1 < d_2 < \ldots < d_k)$, the sequence $n_1, n_2, \ldots, n_k$ is called the *frequency sequence* of the graph.

The two necessary conditions implied by Theorem 1.1 and Theorem 1.12 are not sufficient to ensure that a non-negative sequence is a degree sequence of a graph. To see this, consider the sequence $[1, 2, 3, 4, \ldots, 4, n-1, n-1]$. The sum of the degrees is clearly even and $\Delta = n-1$. However, this is not a degree sequence, since there are two vertices with degree $n-1$, and this requires that each of the two vertices is joined to all the other vertices, and therefore $\delta \geq 2$. But the minimum number in the sequence is 1.

A degree sequence is *perfect* if no two of its elements are equal that is, if the frequency sequence is $1, 1, \ldots, 1$. A degree sequence is *quasi-perfect* if exactly two of its elements are same.

**Definition:** Let $D = [d_i]_1^n$ be a non-negative sequence and $k$ be any integer $1 \leq k \leq n$. Let $D' = [d_i']_1^n$ be the sequence obtained from $D$ by setting $d_k = 0$ and $d_i' = d_i - 1$ for the $d_k$ largest elements of $D$ other than $d_k$. Let $H_k$ be the graph obtained on the vertex set $V = \{v_1, v_2, \ldots, v_n\}$ by joining $v_k$ to the $d_k$ vertices corresponding to the $d_k$ elements used to obtain $D'$. This operation of getting $D'$ and $H_k$ is called *laying off* $d_k$ and $D'$ is called the *residual sequence*, and $H_k$ the subgraph obtained by laying off $d_k$.

**Example** Let $D = [2, 2, 3, 3, 4, 4]$. We have $d_3 = 0$. Then, $D' = [2, 2, 0, 2, 3, 3]$. The subgraph $H_k$ in this case is shown in Figure 2.2.

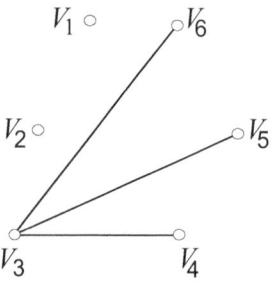

**Fig. 2.2**

## 2.2  Criteria for Degree Sequences

Havel [112] and Hakimi [99] independently obtained a recursive necessary and sufficient condition for a degree sequence, in terms of laying off a largest integer in the sequence. Wang and Kleitman [261] proved the necessary and sufficient condition for arbitrary layoffs.

**Theorem 2.1**  A non-negative sequence is a degree sequence if and only if the residual sequence obtained by laying off any non-zero element of the sequence is a degree sequence.

### Proof

*Sufficiency*  Let the non-negative sequence be $[d_i]_1^n$. Suppose $d_k$ is the non-zero element laid off and the residual sequence $[d_i']_1^n$ is a degree sequence. Then there exists a graph $G'$

realising $[d_i']_1^n$ in which $v_k$ has degree zero and some $d_k$ vertices, say $v_{i_j}$, $1 \leq j \leq d_k$ have degree $d_{i_j} - 1$. Now, by joining $v_k$ to these vertices we get a graph $G$ with degree sequence $[d_i]_1^n$. (Observe that the subgraph obtained by such joining is precisely the subgraph $H_k$ obtained by laying off $d_k$).

*Necessity*  We are given that there is a graph realising $D = [d_i]_1^n$. Let $d_k$ be the element to be laid off. First, we claim that there is a graph realising $D$ in which $v_k$ is adjacent to all the vertices in the set $S$ of $d_k$ largest elements of $D - \{d_k\}$. If not, let $G$ be a graph realising $D$ such that $v_k$ is adjacent to the maximum possible number of vertices in $S$. Then there is a vertex $v_i$ in $S$ to which $v_k$ is not adjacent and hence a vertex $v_j$ outside $S$ to which $v_k$ is adjacent (since, $d(v_k) = |S|$). By definition of $S$, $d_j \leq d_i$. Therefore, there is a vertex $v_h$ in $V - \{v_k\}$ adjacent to $v_i$, but not adjacent to $v_j$. Note that $v_h$ may be in $S$ (Fig. 2.3).

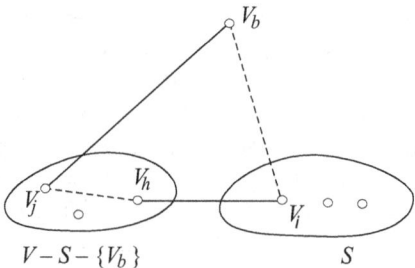

**Fig. 2.3**

Now, construct a graph $H$ from $G$ by deleting the edges $v_j v_k$ and $v_h v_i$ and adding the edges $v_j v_h$ and $v_i v_k$. This operation does not change the degree sequence. Thus $H$ is a graph realising the given sequence, in which one more vertex, namely $v_i$ of $S$ is adjacent to $v_k$, than in $G$. This contradicts the choice of $G$ and establishes the claim.

To complete the proof, if $G$ is a graph realising the given sequence and in which $v_k$ is adjacent to all vertices of $S$, let $G' = G - v_k$. Then $G'$ has the residual degree sequence obtained by laying off $d_k$. ❑

**Definition:** Let the subgraph $H$ on the vertices $v_i$, $v_j$, $v_r$, $v_s$ of a multigraph $G$ contain the edges $v_i v_j$ and $v_r v_s$. The operation of deleting these edges and introducing a pair of new edges $v_i v_s$ and $v_j v_r$, or $v_i v_r$ and $v_j v_s$ is called an elementary degree preserving transformation (EDT), or simple exchange, or 2-switching, or elementary degree-invariant transformation.

**Remarks**

1. The result of an EDT is clearly a degree equivalent multigraph.

2. If an EDT is applied to a graph, the result will be a graph only if the latter pair of edges ($v_i v_s$ and $v_j v_r$), or ($v_i v_r$ and $v_j v_s$) does not exist in $G$.

**Theorem 2.2 (Havel[112], Hakimi[99])**  The non-negative integer sequence $D = [d_i]_1^n$ is graphic if and only if $D'$ is graphic, where $D'$ is the sequence (having $n - 1$ elements) obtained from $D$ by deleting its largest element $\Delta$ and subtracting 1 from its $\Delta$ next largest elements.

## Proof

*Sufficiency* Let $D = [d_i]_1^n$ be the non-negative sequence with $d_1 \geq d_2 \geq \ldots \geq d_n$. Let $G'$ be the graph realising the sequence $D'$. We add a new vertex adjacent to vertices in $G'$ having degrees $d_2 - 1, \ldots, d_{\Delta+1} - 1$. Those $d_i$ are the $\Delta$ largest elements of $D$ after $\Delta$ itself. (But the number $d_2 - 1, \ldots, d_{\Delta+1} - 1$ need not be the $\Delta$ largest elements in $D'$).

*Necessity*   Let $G$ be a graph realising $D = [d_i]_1^n$, $d_1 \geq d_2 \geq \ldots \geq d_n$. We produce a graph $G'$ realising $D'$, where $D'$ is the sequence obtained from $D$ by deleting the largest entry $d_1$ and subtracting 1 from $d_1$ next largest entries.

Let $w$ be a vertex of degree $d_1$ in $G$ and $N(w)$ be the set of vertices which are adjacent to $w$. Let $S$ be the set of $d_1$ number of vertices in $G$ having the desired degrees $d_2, \ldots, d_{d_1+1}$.

If $N(w) = S$, we can delete $w$ to obtain $G'$. Otherwise, some vertex of $S$ is missing from $N(w)$. In this case, we modify $G$ to increase $|N(w) \cap S|$ without changing the degree of any vertex. Since $|N(w) \cap S|$ can increase at most $d_1$ times, repeating this procedure converts an arbitrary graph $G$ that realises $D$, into a graph $G^*$ that realises $D$, and has $N(w) = S$. From $G^*$, we then delete $w$ to obtain the desired graph $G'$ realising $D'$.

If $N(w) \neq S$, let $x \in S$ and $z \in S$, so that $wz$ is an edge and $wx$ is not an edge, since $d(w) = d_1 = |S|$. By this choice of $S$, $d(x) \geq d(z)$ (Fig. 2.4).

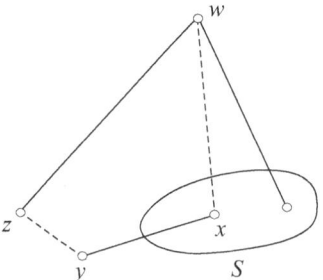

**Fig. 2.4**

We would like to add $wx$ and delete $wz$ without changing their respective degrees. It suffices to find a vertex $y$ outside $T = \{x, y, w\}$ such that $yx$ is an edge, while $yz$ is not. If such a $y$ exists, then we also delete $xy$ and add $zy$. Let $q$ be the number of copies of the edge $xz$ (0 or 1). Now $x$ has $d(x) - q$ neighbours outside $T$, and $z$ has $d(z) - 1 - q$ neighbours outside $T$. Since $d(x) \geq d(z)$, the desired $y$ outside $T$ exists and we can perform the EDT (elementary degree preserving transformation or 2-switch).                                   □

**Algorithm:**   The above recursive condition gives an algorithm to check whether a non-negative sequence is a degree sequence and if so to construct a graph realising it.

The algorithm starts with an empty graph on vertex set $V = \{v_1, v_2, \ldots, v_n\}$ and at the $k$th iteration generates a subgraph $H_k$ of $G$ by deleting (laying off) a vertex of maximum degree in the residual sequence at that stage. If the given sequence is a degree sequence, we end up with a null degree sequence (i.e., for each $i$, $d_i = 0$) and the graph realising the original sequence is simply the sum of the subgraph $H_j$. If not, at some stage, one of the elements of the residual sequence becomes negative, and the algorithm reports non-realisability of the sequence.

An obvious modification of the algorithm, obtained by choosing an arbitrary vertex of positive degree, gives the *Wang-Kleitman algorithm* for generating a graph with a given degree sequence.

**Remarks**

1. There can be many non-isomorphic graphs with the same degree sequence. The smallest example is the pair shown in Figure 2.5 on five vertices with the degree sequence [2, 2, 2, 1, 1].

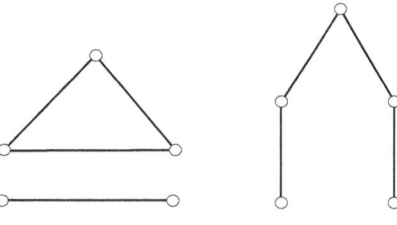

**Fig. 2.5**

    The problem of generating all non-isomorphic graphs of given order and size involves the problem of graph isomorphism for which a good algorithm is not yet known. So also is the problem of generating all non-isomorphic graphs with given degree sequence. In fact, even the problem of finding the number of non-isomorphic graphs with given order and size, or with given degree sequence (and several other problems of similar nature) has not been satisfactorily solved.

2. The Wang-Kleitman algorithm is certainly more general than the Havel-Hakimi algorithm, as it can generate more number of non-isomorphic graphs with a given degree sequence, because of the arbitrariness of the laid-off vertex. For example, not all the five non-isomorphic graphs with the degree sequence [3, 3, 2, 2, 1, 1] can be generated by the Havel-Hakimi algorithm unlike the Wang-Kleitman algorithm.

3. Even the Wang-Kleithman algorithm cannot always generate all graphs with a given degree sequence. For example, the graph $G$ with degree sequence [3, 3, 3, 3, 2, 2, 2, 2, 1, 1, 1, 1] shown in Figure 2.6, cannot be generated by this algorithm. For

   a. if we lay off a 3, it has to be laid off against the 3's and will generate a graph in which a vertex with degree 3 is adjacent to three other vertices with degree 3,

   b. if we lay off a 2 it will generate a graph with a vertex of degree 2 adjacent to two vertices of degree 3,

c. if we lay off a one it will generate a graph in which a vertex of degree one is adjacent to a vertex of degree 3. None of these cases is realised in the given graph $G$.

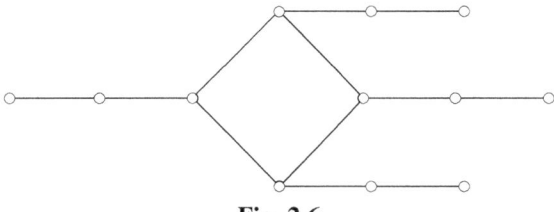

**Fig. 2.6**

However, there are other methods of generating all graphs realising a degree sequence $D$ from any one graph realising $D$ based on a Theorem by Hakimi [98], to be proved next. But those will also be inefficient unless some efficient isomorphism testing is developed.

4. The graphs in Figure 2.5 show that the same degree sequence may be realised by a connected as well as a disconnected graph. Such degree sequences are called *potentially connected*, where as a degree sequence $D$ such that every graph realising $D$ is connected is called a *forcibly connected* degree sequence.

**Definition:** If $P$ is a graph property, and $D = [d_i]_1^n$ is a degree sequence, then $D$ is said to be *potentially-P*, if at least one graph realising $D$ is a $P$-Graph, and it is said to be *forcibly-P* if every graph realising it is a $P$-graph.

**Theorem 2.3 (Hakimi [98])**    If $G_1$ and $G_2$ are degree equivalent graphs, then one can be obtained from the other by a finite sequence of EDTs.

**Proof**    Superimpose $G_1$ and $G_2$ such that each vertex of $G_2$ coincides with a vertex of $G_1$ with the same degree. Imagine the edges of $G_1$ are coloured blue and the edges of $G_2$ are coloured red. Then in the superimposed multigraph $H$, the number of blue edges incident equals the number of red edges incident at every vertex. We refer to this as blue-red parity. If there is a blue edge $v_i v_j$ and a red edge $v_i v_j$ in $H$, we call it a blue-red parallel pair.

Let $K$ be the graph obtained from $H$ by deleting all such parallel pairs. Then $K$ is the null graph if and only if $G_1$ and $G_2$ are label-isomorphic in $H$ and hence originally isomorphic. If this is not the case, we show that we can create more parallel pairs by a sequence of EDTs and delete them till the final resultant graph is null. This will prove the theorem.

Let $B$ and $R$ denote the sets of blue and red edges in $K$. If $v_i v_j \in B$, we show that we can produce a parallel pair at $v_i v_j$, so that the pair can be deleted. This would establish the claim made above.

Now, by construction, there is a blue-red degree parity at every vertex of $K$. So there are red edges $v_i v_k$, $v_j v_r$ in $K$. If $v_k \neq v_r$ (Fig. 2.7(a)) an EDT in $G_2$ switching the red edges to $v_i v_j$, $v_k v_r$ produces a blue-red parallel at $v_i v_j$.

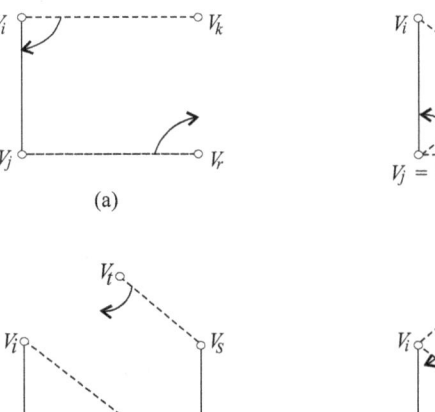

**Fig. 2.7**

If $v_k = v_r$, again by degree parity, at $v_k$ there are at least two blue edges. Let $v_k v_s$ be one such blue edge. Then $v_s$ is distinct from both $v_i$ and $v_j$, for otherwise, there is a blue-red parallel pair $v_i v_k$, or $v_j v_r$. Then there is another red edge $v_s v_t$, $v_t$ distinct from $v_i$ or $v_j$.

Let $v_t \neq v_i$. The two subcases $v_t = v_j$ and $v_t \neq v_j$ are shown in Figure 2.7(b) and (c). In the case of (b), one EDT of $G_2$ switching $v_i v_k$ and $v_s v_t$ to position $v_i v_j$ and $v_s v_k$ produces a blue-red pair at $v_i v_j$ and $v_k v_s$. In the case of (c), one EDT of $G_2$ switching $v_i v_k$ and $v_t v_s$ to positions $v_s v_k$ and $v_t v_i$ produces a blue-red parallel pair at $v_k v_s$ (which can be deleted). Another EDT of $G_2$ switching the blue-red pair $v_t v_i$ and $v_j v_k$ to positions $v_i v_j$ and $v_s v_k$ produces a blue-red pair $v_i v_j$.

Since in both cases we get a blue-red pair at $v_i v_j$ position, our claim is established and the proof of the theorem is complete. ❑

**Remarks** In the related context of a $(0, 1)$ matrix $A$ (that is, a matrix $A$ whose elements are 0's or 1's), Ryser [227] defined an interchange as a transformation of the elements of $A$ that changes a minor of type $A_1 = \begin{pmatrix} 1 & 0 \\ 0 & 1 \end{pmatrix}$ into a minor of the type $A_1 = \begin{pmatrix} 0 & 1 \\ 1 & 0 \end{pmatrix}$, or vice versa and proved an interchange theorem which can be interpreted as EDT theorem for bipartite graphs and digraphs.

Now, we give a combinatorial characterisation of degree sequences, due to Erdos and Gallai [73]. Several proofs of the criterion exist; the first proof given here is due to Choudam [58] and the second one is due to Tripathi et al.

**Theorem 2.4 (Erdos-Gallai [73])** A non-increasing sequence $[d_i]_1^n$ of non-negative integers is a degree sequence if and only if $D = [d_i]_1^n$ is even and the inequality

$$\sum_{i=1}^{k} d_i \le k(k-1) + \sum_{i=k+1}^{n} \min(d_i, k) \tag{2.4.1}$$

is satisfied for each integer $k$, $1 \le k \le n$.

## Proof

*Necessity*    Evidently $\sum_{i=1}^{n} d_i$ is even. Let $U$ denote the subset of vertices with the $k$ highest degrees in $D$. Then the sum $s = \sum_{i=1}^{k} d_i$ can be split as $s_1 + s_2$, where $s_1$ is the contribution to $s$ from edges joining vertices in $U$, each edge contributing 2 to the sum, and $s_2$ is the contribution to $s$ from the edges between vertices in $U$ and $\overline{U}$, each edge contributing 1 to the sum (Fig. 2.8).

$s_1$ is clearly bound by the degree sum of a complete graph on $k$-vertices, i.e., $k(k-1)$. Also, each vertex $v_i$ of $\overline{U}$ can be joined to at most $\min(d_i, k)$ vertices of $U$, so that $s_2$ is bounded above by $\sum_{i=k+1}^{m} \min(d_i, k)$. Together, we get (2.4.1).

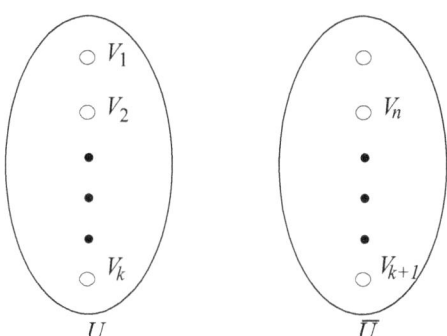

**Fig. 2.8**

*Sufficiency*    We induct on the sum $s = \sum_{i=1}^{n} d_i$ and use the obvious inequality

$$\min(a, b) - 1 \le \min(a-1, b), \tag{2.4.2}$$

for positive integers $a$ and $b$.

For $s = 2$, clearly $K_2 \cup (n-2)K_1$ realises the only sequence $[1, 1, 0, 0, \ldots 0]$ or $[1^2 0^{n-2}]$ satisfying the conditions (2.4.1).

As induction hypothesis, let all non-increasing sequences of non-negative integers with even sum at most $s-2$ and satisfying (2.4.1) be degree sequences.

Let $D = [d_i]_1^n$ be a sequence with sum $s$ and satisfying (2.4.1). We produce a new non-increasing sequence $D'$ of non-negative integers by subtracting one each from two positive terms of $D$ and verify that $D'$ satisfies the hypothesis of the theorem. Since the trailing zeros in the non-increasing sequences of non-negative integers do not essentially affect the argument, there is no loss of generality in assuming that $d_n > 0$, and we assume this to simplify the expression.

To define $D'$, let $t$ be the smallest integer ($\geq 1$) such that $d_t > d_{t+1}$. That is, let $D$ be $d_1 = d_2 = \ldots = d_t > d_{t+1} \geq d_{t+2} \geq \ldots \geq d_n > 0$.

If $D$ is regular (that is $d_i = d > 0$ for all $i$) then let $t$ be $n - 1$.

$$\text{Then, } d_i' = \begin{cases} d_i, & \text{for } 1 \leq i \leq t-1 \text{ and } t+1 \leq i \leq n-1, \\ d_t - 1, & \text{for } i = t, \\ d_n - 1, & \text{for } i = n. \end{cases}$$

Clearly, $D'$ is a non-increasing sequence of non-negative integers and $\sum_{i=1}^{n} d_i' = s - 2$ is even.

We verify that $D'$ satisfies (2.4.1) by considering several cases depending on the relative position of $k$ and the magnitudes of $d_k$ and $d_n$.

**Case I**  Let $k = n$. Therefore, $\sum_{i=1}^{k} d_i' = \sum_{i=1}^{k} d_i - 2 \leq n(n-1) - 2 < n(n-1) = $ RHS of (2.4.1) for $D'$.

**Case II**  Let $t \leq k \leq n - 1$.

$$\text{Then, } \sum_{i=1}^{k} d_i' = \sum_{i=1}^{k} d_i - 1 \leq k(k-1) + \sum_{i=k+1}^{n} \min(d_i, k) - 1 \text{ (since } D \text{ satisfies (2.4.1))}$$

$$= k(k-1) + \sum_{i=k+1}^{n-1} \min(d_i', k) + \min(d_n, k) - 1$$

$$\leq k(k-1) + \sum_{i=k+1}^{n-1} \min(d_i', k) + \min(d_n - 1, k) \qquad \text{by (2.4.2)}$$

$$= k(k-1) + \sum_{i=k+1}^{n-1} \min(d_i', k) + \min(d_n', k)$$

Therefore, $\sum_{i=1}^{k} d_i' \leq k(k-1) + \sum_{i=k+1}^{n} \min(d_i', k)$.

**Case III**  Let $k \leq t - 1$.

**Subcase III.1**  Assume $d_k \leq k - 1$.

Then, $\sum_{i=1}^{k} d_i' = kd_k \leq k(k-1) \leq k(k-1) + \sum_{i=k+1}^{n} \min(d_i', k)$, since the second term is non-negative.

**Subcase III.2**  Every $d_j = k$, $1 \leq j \leq k$. We first observe that $d_{k+2} + \ldots d_n \geq 2$.

This is obvious if $k+2 \leq n-1$, because $d_n > 0$ gives $d_n \geq 1$ and $d_{n-1} \geq 1$. When $k+2 = n$, we have $k = n-2$. As $k \leq t-1$, $t \geq k+1 = n-2+1 = n-1$. Since $t > n-1$ is not possible, $t = n-1$.

The sequence $D$ is $[n-2, n-2, \ldots, n-2, d_n]$, or $[(n-2)^{n-1} d_n]$. Then, $s = (n-1)(n-2) + d_n$. Since $s$ is even, $d_n$ is even and hence $d_n \geq 2$. Thus, $d_{k+2} + \ldots + d_n \geq 2$.

Therefore, $d_{k+2} + \ldots + d_n - 2 \geq 0$.

$$\text{Now,} \quad \sum_{i=1}^{k} d_i' \;=\; \sum_{i=1}^{k} d_i = k.k = k^2 = k^2 - k + k$$

$$= k^2 - k + d_{k+1}, \text{ because } k \leq t-1, \text{ and } d_1 = \ldots = d_{t-1} = d_t,$$

so if $d_{t-1} = k$, then $d_t = k$, and if $d_k = k$, $d_{k+1} = k$.

$$\text{Thus,} \quad \sum_{i=1}^{k} d_i' \leq k^2 - k + d_{k+1} + (d_{k+2} + \ldots + d_n - 2) = k(k-1) + \sum_{i=k+1}^{n} \min(d_i, k) - 2,$$

because $\min(d_{k+1}, k) = d_{k+1}$, $\min(d_{k+2}, k) = k = d_{k+2}$, $\ldots$, $\min(d_t, k) = k = d_t$, $\ldots$, $\min(d_{t+1}, k) = d_{t+1}$ (as $d_{t+1} < d_t = k$), $\ldots$, $\min(d_n, k) = d_n$ (as $d_n < d_t = k$).

$$\text{Hence,} \quad \sum_{i=1}^{k} d_i' \leq k(k-1) + \sum_{\substack{i=k+1 \\ i \neq t, n}}^{n} \min(d_i, k) + \min(d_t, k) + \min(d_n, k) - 2$$

$$= k(k-1) + \sum_{\substack{i=k+1 \\ i \neq t, n}}^{n} \min(d_i', k) + \min(d_t'+1, k) + \min(d_n'+1, k) - 2$$

$$\leq k(k-1) + \sum_{\substack{i=k+1 \\ i \neq t, n}}^{n} \min(d_i', k) + \min(d_t', k) + 1 + \min(d_n', k) + 1 - 2$$

$$= k(k-1) + \sum_{i=k+1}^{n} \min(d_i', k).$$

**Subcase III.3** Let $d_k \geq k+1$.

i. Let $d_n \geq k+1$.

$$\text{Then,} \quad \sum_{i=1}^{k} d_i' = \sum_{i=1}^{k} d_i \leq k(k-1) + \sum_{i=k+1}^{n} \min(d_i, k) \qquad \text{(since $D$ satisfies (2.4.1))}$$

$$= k(k-1) + \sum_{\substack{k+1 \\ i \neq t, n}}^{n} \min(d_i, k) + \min(d_t, k) + \min(d_n, k)$$

$$= k(k-1) + \sum_{\substack{k+1 \\ i \neq t, n}}^{n} \min(d_i', k) + \min(d_t - 1, k) + \min(d_n - 1, k),$$

because $\min(d_t, k) = \min(d_t - 1, k) = k$, $\min(d_n, k) = \min(d_n - 1, k) = k$, as $d_t \geq k + 1$, $d_n \geq k + 1$ implies that $d_t - 1 \geq k$, $d_n - 1 \geq k$.

So, $\displaystyle\sum_{i=1}^{k} d_i' \leq k(k-1) + \sum_{\substack{k+1 \\ i \neq t, n}}^{n} \min(d_i', k) + \min(d_t', k) + \min(d_n', k)$

$$= k(k-1) + \sum_{i=k+1}^{n} \min(d_i', k).$$

ii. Let $d_n \leq k$ and let $r$ be the smallest integer such that $d_{t+r+1} \leq k$. We verify that in (2.4.1), $D$ can not attain equality for such a choice of $k$. For, with equality, we have

$$\sum_{i=1}^{k} d_i = kd_k = k(k-1) + \sum_{k+1}^{t+r} \min(d_i, k) + \sum_{t+r+1}^{n} \min(d_i, k)$$

$$= k(k-1) + (t+r-k)k + \sum_{t+r+1}^{n} d_i,$$

because $\min(d_i, k) = \begin{cases} k, & for\ i = k+1, ..., t+r\ as\ d_i \geq k+1, \\ \\ d_i, & for\ i = t+r+1, ..., n\ as\ d_i \leq k. \end{cases}$

So, $kd_k = k(t+r-1) + \displaystyle\sum_{t+r+1}^{k} d_i.$

Then, $\displaystyle\sum_{i=1}^{k+1} d_i = (k+1)d_k = (k+1)\left\{ (t+r-1) + \frac{1}{k} \sum_{t+r+1}^{n} d_i \right\}$, (using $d_k$ from above)

$$= (k+1)(t+r-1) + \frac{k+1}{k} \sum_{t+r+1}^{n} d_i > (k+1)(t+r-1) + \sum_{t+r+1}^{n} d_i$$

$$= (k+1)k - (k+1)k + (k+1)(t+r-1) + \sum_{t+r+1}^{n} d_i$$

$$= (k+1)k + (k+1)(t+r-k-1) + \sum_{t+r+1}^{n} d_i = (k+1)k + \sum_{k+1}^{t+r} (k+1) + \sum_{t+r+1}^{n} d_i$$

$$= (k+1)k + \sum_{t+r+1}^{n} \min(d_i, k+1),$$

because    $\min(d_i, k+1) = k+1$ for $i = k+1, \ldots, t+r$, and

$$\min(d_i, k+1) = d_i, \quad \text{for } i = t+r+1, \ldots, n.$$

So,    $\sum_{i=1}^{k+1} d_i > k(k+1) + \sum_{k+1}^{n} \min(d_i, k+1).$

Therefore,    $\sum_{i=1}^{k+1} d_i > k(k+1) + (k+1) + \sum_{k+2}^{n} \min(d_i, k+1),$

which is a contradiction to (2.4.1), for $D$ for $k+1$. Hence, $D$ has strict inequality for $k$.

Therefore, $\sum_{i=1}^{k} d_i' = \sum_{i=1}^{k} d_i < k(k-1) + \sum_{k+1}^{n} \min(d_i, k).$

Thus, $\sum_{i=1}^{k} d_i' = \sum_{i=1}^{k} d_i \leq k(k-1) + \sum_{k+1}^{n} \min(d_i, k) - 1$

$$= k(k-1) + \sum_{\substack{i=k+1 \\ i \neq t}}^{n-1} \min(d_i, k) + \min(d_t, k) + \min(d_n, k) - 1$$

$$\leq k(k-1) + \sum_{\substack{i=k+1 \\ i \neq t}}^{n-1} \min(d_i', k) + \min(d_t - 1, k) + \min(d_n - 1, k),$$

as $\min(d_n, k) - 1 \leq \min(d_n - 1, k)$, $\min(d_t, k) = k$ (since, $d_t \geq k+1$), $\min(d_t - 1, k) = k$ (since $d_t - 1 \geq k$).

Therefore, $\sum_{i=1}^{k} d_i' \leq k(k-1) + \sum_{k+1}^{n} \min(d_i', k).$

Hence in all cases $D'$ satisfies (2.4.1).

Now, by induction hypothesis, there is a graph $G'$ realising $D'$. If $v_t v_n \notin E(G')$, then $G' + v_t v_n$ gives a realisation $G$ of $D$. If $v_t v_n \in E(G')$, since $d(v_t | G') = d_t - 1 \leq n-2$, there is a vertex $v_r$ such that $v_r v_t \notin E(G')$. Also, since $d(v_r | G') > d(v_n | G')$, there is a vertex $v_s$ such that $v_s v_n \notin E(G')$. Making an EDT exchanging the edge pair $v_t v_n$, $v_r v_s$ for the edge pair $v_t v_r$, $v_s v_n$, we get a realisation $G''$ of $D'$ with $v_t v_n \notin E(G'')$. Then $G'' + v_t v_n$ realises $D$.

**Second Proof of Sufficiency (Tripathi et al.)**    Let a subrealisation of a non-increasing sequence $[d_1, d_1 \ldots, d_n]$ be a graph with vertices $v_1, v_1 \ldots, v_n$ such that $d(v_i) \leq d_i$ for $1 \leq i \leq n$, where $d(v_i)$ denotes the degree of $v_i$. Given a sequence $[d_1, d_1, \ldots, d_n]$ with an even sum that satisfies (2.4.1), we construct a realisation through successive subrealisations. The initial subrealisation has $n$ vertices and no edges.

In a subrealisation, the *critical index* $r$ is the largest index such that $d(v_i) = d_i$ for $1 \leq i < r$. Initially, $r = 1$ unless the sequence is all 0, in which case the process is complete. While $r \leq n$, we obtain a new subrealisation with smaller deficiency $d_r - d(v_r)$ at vertex $v_r$ while not changing the degree of any vertex $v_i$ with $i < r$ (the degree sequence increases lexicograpically). The process can only stop when the subrelisation of $d$. ❑

Let $S = \{v_{r+1}, \ldots, v_n\}$. We maintain the condition that $S$ is an independent set, which certainly holds initially. Write $u_i \leftrightarrow v_j$ when $v_i v_j \in E(G)$; otherwise, $v_i \not\leftrightarrow v_j$

**Case 0** $v_r \not\leftrightarrow v_i$ for some vertex $v_i$ such that $d(v_i) < d_i$. Add the edge $u_r v_i$.

**Case 1** $v_r \not\leftrightarrow v_i$ for some $i$ with $i < r$. Since $d(v_i) = d_i \geq d_r > (v_r)$, there exists $u \in N(u_i) - (N(v_r) \cup \{v_r\})$, where $N(z) = \{y : z \leftrightarrow y\}$. If $d_r - d(v_r) \geq 2$, then replace $uv_i$ with $\{uv_r, v_i v_r\}$. If $d_r - d(v_r) = 1$, then since $\sum d_i - \sum d(v_i)$ is even there is an index $k$ with $k > r$ such that $d(v_k) < d_k$. Case 0 applies unless $v_r \leftrightarrow v_k$; replace $\{v_r v_k, uv_i\}$ with $\{uv_r, v_i u_r\}$.

**Case 2** $v_1 \ldots, v_{r-1} \in N(v_r)$ and $d(v_k) \neq \min\{r, d_k\}$ for some $k$ with $k > r$. In a a subrealisation, $d(v_k) \leq d_k$. Since $S$ is independent, $d(v_k) \leq r$. Hence $d(v_k) < \min\{r, d_k\}$, and case 0 applies unless $u_k \leftrightarrow v_r$. Since $d(v_k) < r$, there exists $i$ with $i < r$ such that $u_k \not\leftrightarrow v_i$. Since $d(v_i) > d(v_r)$, there exists $u \in N(v_i) - (N(v_r) \cup \{u_r\})$. Replace $uv_i$ with $\{uv_r, v_i v_k\}$.

**Case 3** $v_1, \ldots, v_{r-1} \notin N(v_r)$ and $v_i \leftrightarrow v_i$ for some $i$ and $j$ with $i < j < r$. Case 1 applies unless $v_i, v_j \in N(v_r)$. Since $d(v_i) \geq d(v_i) > d(v_r)$, there exists $u \in N(v_i) - (N(v_r) \cup \{v_r\})$ and $w \in N(v_j) - (N(v_r) \cup \{v_r\})$ (possibly $u = w$). Since $u, w \notin N(v_r)$, Case 1 applies unless $u, w \in S$. Replace $\{v_i v_j, uv_r\}$ with $\{uv_r, v, v_r\}$.

If none of these case apply, then $v_1 \ldots, v_r$ are pairwise adjacent, and $d(v_k) = \min\{r, d_k\}$ for $k > r$. Since $S$ is independent, $\sum_{i=1}^{r} d(v_i) = r(r-1) + \sum_{k=r+1}^{n} \min\{r, d_k\}$. By (2.4.1), $\sum_{i=1}^{r} d_1$ is bounded by the right side. Hence we have already eliminated the deficiency at vertex $r$. Increase $r$ by 1 and continue. ❑

Tripathi and Vijay [245] have shown that the Erdos-Gallai condition characterising graphical degree sequences of length $n$ needs to be checked only for as many $k$ as there are distinct terms in the sequence and not for all $k$, $1 \leq k \leq n$.

## 2.3 Degree Set of a Graph

The set of distinct non-negative integers occurring in a degree sequence of a graph is called its *degree set*. For example, let the degree sequence be $D = [2, 2, 3, 3, 4, 4]$, then degree set is $\{2, 3, 4\}$. A set of distinct non-negative integers is called a degree set if it is the degree set of some graph and the graph is said to *realise* the degree set.

Let $S = \{d_1, d_2, \ldots, d_k\}$ be the set of distinct non-negative integers. Clearly, $S$ is the degree set as the graph

$$G = K_{d_1+1} \cup K_{d_2+1} \cup \ldots \cup K_{d_k+1},$$

realises $S$. This graph has $d_1 + d_2 + \ldots + d_k + k$ vertices.

**Example** Let $S = \{1, 3, 4\}$. Then, $G = K_2 \cup K_4 \cup K_5$ (Fig. 2.9).

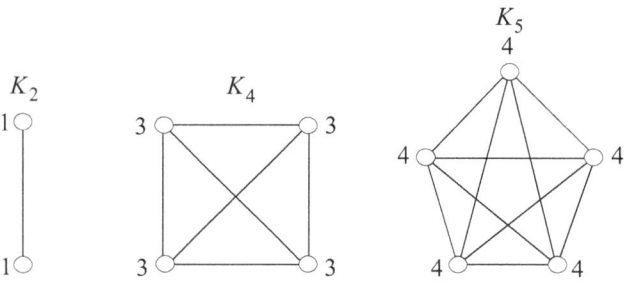

**Fig. 2.9**

The following result is due to Kapoor, Polimeni and Wall [126].

**Theorem 2.5 (Kapoor, Polimeni and Wall [126]).** Any set $S$ of distinct positive integers is the degree set of a connected graph and the minimum order of such a graph is $M + 1$, where $M$ is the maximum integer in the set $S$.

**Proof** Let $S$ be a degree set and $n_0(S)$ denote the minimum order of a graph $G$ realising $S$. As $M$ is the maximum integer in $S$, therefore in $G$ there is a vertex adjacent to $M$ other vertices, i.e., $n_0(S) \geq M + 1$. Now, if there exists a graph of order $M + 1$ with $S$ as degree set, then $n_0(S) = M + 1$. The existence of such a graph is established by induction on the number of elements $p$ of $S$.

Let $S = \{a_1, a_2, \ldots, a_p\}$ with $a_1 < a_2 < \ldots < a_p$.

For $p = 1$, the complete graph $K_{a_1+1}$ realises $\{a_1\}$ as degree set.

For $p = 2$, we have $S = \{a_1, a_2\}$. Let $G = K_{a_1} V \overline{K}_{a_2-a_1+1}$ (join of two graphs). Here every vertex of $K_{a_1}$ has degree $a_2$ and every other vertex has degree $a_1$ and therefore $G$ realises $\{a_1, a_2\}$ (Fig. 2.10(a)).

For $p = 3$, we have $S = \{a_1, a_2, a_3\}$. Then, $G = K_{a_1} V (\overline{K}_{a_3-a_2} \cup H)$, where $H$ is the graph realising the degree set $\{a_2 - a_1\}$ with $a_2 - a_1 + 1$ vertices, realises $\{a_1, a_2, a_3\}$ (Fig. 2.10 (b)).

(Note that $d(u) = a_1 - 1 + a_3 - a_2 + a_2 - a_1 + 1 = a_3$, $d(v) = a_1$, $d(w) = a_2 - a_2 + a_1 = a_2$).

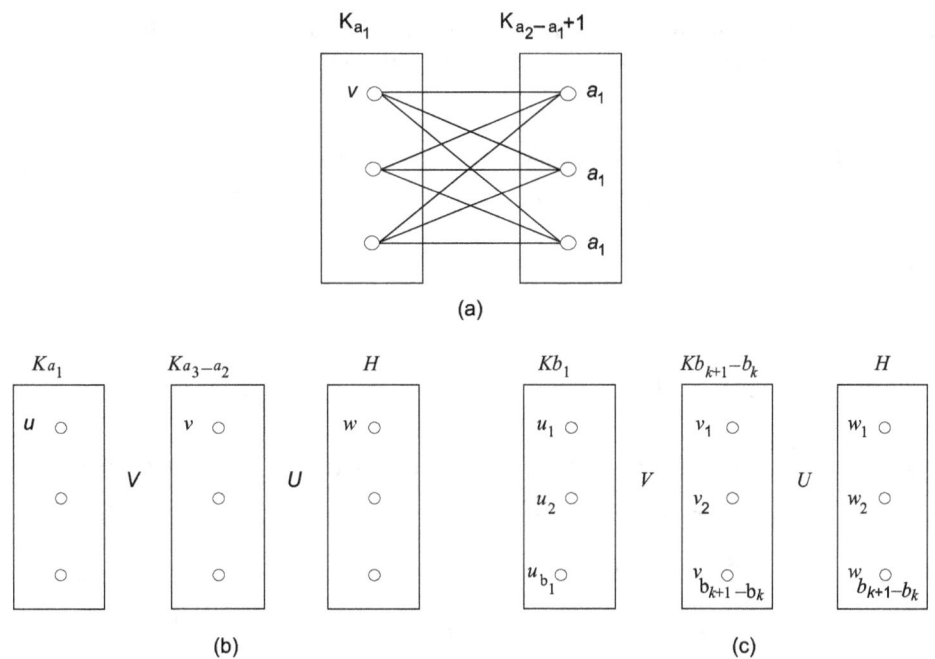

**Fig. 2.10**

Now, let every set with $h$ positive integers, $1 \leq h \leq k$ be the degree set. Let $S_1 = \{b_1, b_2, \ldots, b_{k+1}\}$ be a $(k+1)$ set of positive integers arranged in increasing order. By induction hypothesis, there is a graph $H$ realising the degree set $\{b_2 - b_1, b_3 - b_1, \ldots, b_k - b_1\}$ with order $b_k - b_1 + 1$. The graph $G = K_{b_1} V (K_{b_{k+1} - b_k} \cup H)$, with order $b_{k+1} + 1$ realises $S_1$ (Fig 2.10 (c)). Clearly by construction, all these graphs are connected.

Hence the result follows by induction.

(Note that $d(u_i) = b_1 - 1 + b_{k+1} - b_k + b_k - b_1 + 1 = b_k + 1$, $d(v_i) = b_1$, $d(w_i) = b_{i+1} - b_1 + b_1 = b_{i+1}$, that is $d(w_1) = b_2$, $d(w_2) = b_3$, $\ldots$, $d\left(w_{b_k - b_1 + 1}\right) = b_k - b_1 + b_1 = b_k$. Some results on degree sets in bipartite and tripartite graphs can be seen in [262]. ❑

## 2.4 New Criterion

We have the following notations. Let $D = [d_i]_1^n$ be a non-decreasing sequence of non-negative integers with $0 \leq d_i \leq n - 1$ for all $i$. Let $n - p_1$ be the greatest integer, $n - p_1 - p_2$, the second greatest integer and $n - \sum_{r=1}^{k} p_r$, the $k$th greatest integer in $D$, $1 \leq p_r \leq n - (r - 1)$. Let the number of times the $k$th greatest integer appears in $D$ be denoted by $a_k$. Also, we take

$$t_k = n - \left(n - \sum_{r=1}^{k} p_r\right) = \sum_{r=1}^{k} p_r, \quad 1 \leq p_r \leq n - (r - 1) \text{ and } j_k = 1, 2, \ldots, p_{k+1}.$$

The following result due to S. Pirzada and YinJian Hu [208] gives the criterion for a non-negative sequence of integers in non-decreasing order to be the degree sequence of some graph.

**Theorem 2.6** A non-decreasing sequence $[d_i]_1^n$ of non-negative integers, where $\sum_{i=1}^{n} d_i$ is even and $0 \le d_i \le n-1$ for all $i$, is a degree sequence of a graph if and only if

$$\sum_{i=1}^{t_k+j_k-1} d_i \ge \sum_{m=1}^{k} \{j_k + (k-m)\} a_m \tag{2.6.1}$$

for all $t_k + j_k - 1 + \sum_{m=1}^{k} a_m \le n$.

**Proof**   Let in $D$ the greatest integer $n - p_1$ appear $a_1$ times, the second greatest integer $n - (p_1 + p_2)$ appear $a_2$ times and the $k$th greatest integer $n - \sum_{r=1}^{k} p_r$ appear $a_k$ times, $1 \le p_r \le n - (r-1)$, so that $D$ is defined as

$$d_i = \begin{cases} d_i, & for\ i = 1, 2, ..., \left(n - \sum_{m=1}^{k} a_m\right), \\ n - p_1, & for\ i = n,\ n-1, ..., a_1, \\ n - (p_1 + p_2), & for\ i = a_1 + 1, ..., a_2, \\ n - \left(\sum_{r=1}^{k} p_r\right), & for\ i = a_{k-1} + 1, ..., a_k. \end{cases}$$

Thus,

$$d_1 \le d_2 \le ... \le d_{n - \sum_{m=1}^{k} a_m} < d_{a_k} = ... = d_{a_{k-1}+1} < d_{a_{k-1}} = ... = d_{a_{k-2}+1} < ... < d_{a_2}$$

$$= ... = d_{a_1 + 1} < d_{a_1} = ... = d_{n-1} = d_n.$$

*Necessity*   Let $D = [d_i]_1^n$ be the degree sequence of a graph $G$ whose vertex set is $V = \{v_1, v_2, ..., v_n\}$.

So, $d(v_n) = d(v_{n-1}) = ... = d(v_{a_1}) = n - p_1, \quad 1 \le p_1 \le n$.

Or briefly, $d_n = d_{n-1} = ... = d_{a_1} = n - p_1, \quad 1 \le p_1 \le n$, so that the number of times $n - p_1$ appears in $D$ is $a_1$.

Therefore, $t_1 = n - (n - p_1) = p_1$.

As $d(v_n) = n - p_1$, vertex $v_n$ has no edges with $n - 1 - (n - p_1) = p_1 - 1$ other vertices of $G$. Similarly, the vertices $v_{n-1}, ..., v_{a_1}$ each has no edges with $p_1 - 1$ other vertices of $G$. We assume without loss of generality that these $p_1 - 1$ vertices are among $v_1, v_2, ..., v_{p_1}$. This means that the contribution of edges between $v_1, v_2, ..., v_{a_1}$ towards the degrees of vertices $v_1, v_2, ..., v_{p_1}$ is $a_1$ so that

$$\sum_{i=1}^{p_1} d_i \geq a_1.$$

Since all the vertices $v_n, \ldots, v_{a_1}$ must have an edge with each of the vertices $v_{p_1+1}$, $v_{p_1+2} \ldots, v_{p_1+p_2-1}$,

$$\sum_{i=1}^{p_1+1} d_i \geq a_1 + a_1 = 2a_1, \quad \sum_{i=1}^{p_1+2} d_i \geq a_1 + a_1 + a_1 = 3a_1, \ldots, \quad \sum_{i=1}^{p_1+p_2-1} d_i \geq p_2 a_1.$$

Thus, $\sum_{i=1}^{p_1+j_1-1} d_i \geq j_1 a_1$, where $j_1 = 1, 2, \ldots, p_2$,

or $\sum_{i=1}^{t_1+j_1-1} d_i \geq j_1 a_1$, where $j_1 = 1, 2, \ldots, p_2$,

or $\sum_{i=1}^{t_1+j_1-1} d_i \geq \sum_{m=1}^{1} \{j_1 + (1-m)\} a_m$, where $j_1 = 1, 2, \ldots, p_2$,

and thus the inequalities (2.6.1) hold.

We now take

$$d(v_{a_1-1}) = d(v_{a_1-2}) = \ldots = d,(v_{a_2}) = n - (p_1 + p_2),$$

where $1 \leq p_2 \leq n-1$.

Here $t_2 = n - (n - (p_1 + p_2)) = p_1 + p_2$ and $j_2 = 1, 2, \ldots, p_3$.

As each vertex $v_{a_1-1}, \ldots, v_{a_2}$ has no edge with $p_1 + p_2 - 1$ vertices, assume these $p_1 + p_2 - 1$ vertices are among the first $v_1, v_2, \ldots, v_{p_1+p_2}$ vertices. In case we choose these $p_1 + p_2 - 1$ vertices as the first vertices $v_1, v_2, \ldots, v_{p_1+p_2-1}$, then each $v_n, \ldots, v_{a_1}, \ldots, v_{a_2}$ has an edge with $v_{p_1+p_2}$ so that

$$d(v_{p_1+p_2}) \geq a_1 + a_2.$$

Therefore, $\sum_{i=1}^{p_1+p_2} d_i = \sum_{i=1}^{p_1+p_2-1} d_i + d_{p_1+p_2} \geq a_1 + (a_1 + a_2) = 2a_1 + a_2.$

In the same way, we see that

$$\sum_{i=1}^{p_1+p_1+1} d_i \geq 3a_1 + 2a_2, \ldots, \quad \sum_{i=1}^{p_1+p_2+p_3-1} d_i \geq (p_3+1)a_1 + p_3 a_2,$$

i.e., $\sum_{i=1}^{t_2+1-1} d_i \geq 2a_1 + a_2, \quad \sum_{i=1}^{t_2+2-1} d_i \geq 3a_1 + 2a_2, \ldots, \quad \sum_{i=1}^{t_2+p_3-1} d_i \geq (p_3+1)a_1 + p_3 a_2,$

or $\sum\limits_{i=1}^{t_2+j_2-1} d_i \geq \sum\limits_{m=1}^{2} \{j_2 + (2-m)\} a_m$, where $j_2 = 1, 2, 3, \ldots, p_3$,

and thus inequalities (2.6.1) hold.

We continue the process in the same way till we reach first $t_k + j_k - 1$ vertices so that the sum of first $t_k + j_k - 1$ vertices and $\sum\limits_{m=1}^{k} a_m$ last vertices is $n$ and it follows by the same argument that

$$\sum_{i=1}^{t_k+j_k-1} d_i \geq \sum_{m=1}^{k} \{j_k + (k-m)\} a_m.$$

*Sufficiency*   We prove this by induction on $S = \sum\limits_{i=1}^{n} d_i$. When $S = 2$, clearly $K_2 \cup (n-2)K_1$ realises the only sequence $[\underbrace{0, 0, \ldots, 0}_{n-2\,times}, 1, 1]$ satisfying (2.6.1).

Let all non-decreasing sequences of non-negative integers with even sum atmost $S - 2$ and satisfying (2.6.1) be degree sequences. Let $D = [d_i]_1^n$ be a sequence with sum $S$ satisfying (2.6.1). We form a new non-decreasing sequence $D'$ of non-negative integers by subtracting one each from two positive terms of $D$ and verify $D'$ satisfies the hypothesis of the theorem. As the zeros in the non-decreasing sequence of non-negative integers do not essentially affect the argument, we can assume without loss of generality that $d_1 > 0$.
To define $D' = [d_i']_1^n$, we take

$$d_i' = \begin{cases} d_i - 1, & \text{for } i = 1,\ i = a_1, \\ d_i, & \text{otherwise.} \end{cases}$$

In case $D$ is regular, $d_i = d$ for all $i$, then we take

$$d_i' = \begin{cases} d_i - 1, & \text{for } i = 1, 2 \\ d_i, & \text{otherwise.} \end{cases}$$

Evidently $D'$ is a non-decreasing sequence of non-negative integers and $\sum\limits_{i=1}^{n} d_i' = S - 2$ is even. We now verify $D'$ satisfies inequalities (2.6.1).

We know in $D$, $a_k$ is the number of times the $k$th greatest integer $n - \sum\limits_{r=1}^{k} p_r$ appears, $1 \leq p_r \leq n - (r-1)$. Accordingly $t_k = \sum\limits_{r=1}^{k} p_r$, $1 \leq p_r \leq n - (r-1)$ and $j_k = 1, 2, \ldots, p_{k+1}$. Now in $D'$, let $a_1'$ be the number of times the greatest integer $n - p_1$ appears so that $a_1' = a_1 - 1$. Let $a_2'$ be the number of times the second greatest integer $n - (p_1 + p_2)$ appears in $D'$. We have two cases to discuss.

**Case 1**  When $p_2 = 1$.

Then, $a_2' = a_2 + 1$ and $a_m' = a_m$ for all $m = 3, 4, \ldots, k$. We observe here that $t_1' = n - (n - p_1) = p_1$, $t_2' = n - (n - (p_1 + p_2)) = p_1 + 1, \ldots, t_k' = \sum\limits_{r=1}^{k} p_r = t_k$ and $j_k' = 1, 2, \ldots, p_{k+1}$, that is, $j_k' = j_k$. Therefore,

$$\sum_{i=1}^{t_k' + j_k' - 1} d_i' \geq \sum_{m=1}^{t_k + j_k - 1} d_i - 1$$

$$\geq [j_k + (k-1)] a_1 + [j_k + (k-2)] a_2 + \ldots + [j_k + (k-k)] a_k - 1$$

$$= [j_k' + (k-1)] (a_1' + 1) + [j_k' + (k-2)] (a_2' - 1) + \ldots + [j_k' + (k-k)] a_k' - 1$$

$$= [j_k' + (k-1)] a_1' + [j_k' + (k-2)] a_2' + \ldots + [j_k' + (k-k)] a_k' + j_k' + (k-1) - j_k' - (k-2) - 1$$

$$= \sum_{m=1}^{k} [j_k' + (k-m)] a_m'.$$

Thus, the inequalities (2.6.1) hold for $D'$.

**Case 2**  Let $p_2 > 1$, then $a_1' = a_1 - 1$, $a_2' = 1$ as $a_m' = a_{m-1}$ for all $m = 3, 4, \ldots, k$. Here $t_1' = p_1 = t_1$, $t_2' = n - (n - (p_1 + 1)) = p_1 + 1 = t_1 + 1$, $t_3' = t_2 + 1, \ldots, t_k' = t_{k-1} + 1$ and $j_1' = 1$, $j_2' = j_1, \ldots, j_k' = j_{k-1}$.

Also,  $p_1' = p_1$, $p_2' = 1$, $p_3' = p_2, \ldots, p_k' = p_{k-1}$.

Now,  $\displaystyle\sum_{i=1}^{t_1' + j_1' - 1} d_i' = \sum_{i=1}^{t_1 + 1 - 1} d_i - 1$  (by taking $j_1 = 1$ for $D$)

$$= \left[ \sum_{m=1}^{1} [1 + (1-m)](a_1' + 1) \right] - 1 = \sum_{m=1}^{1} [1 + (1-m)] a_1' + 1 - 1$$

$$= \sum_{m=1}^{1} [j_i' + (1-m)] a_1',$$

and the inequalities (2.6.1) hold for $D'$.

We have,  $\displaystyle\sum_{i=1}^{t_2' + j_2' - 1} d_i' = \sum_{i=1}^{t_1 + 1 + j_i - 1} d_i' = \sum_{i=1}^{t_1 + (j_i + 1) - 1} d_i - 1 \geq \sum_{m=1}^{1} [(j_i' + 1) + (1 - m)] a_m - 1$

$$= (j_1 + 1) a_1 - 1 = (j_2' + 1)(a_1' + 1) - 1 = j_2' a_1' + a_1' + j_2' + 1 - 1$$

$$= (j_2' + 1) a_1' + j_2' a_2' \qquad \text{(because } a_2' = 1\text{)}$$

$$= \sum_{m=1}^{2} \left[ j'_2 + (2-m) \right] a'_m,$$

and the inequalities (2.6.1) hold.

$$\text{Finally,} \quad \sum_{i=1}^{t'_k + j'_K - 1} d'_i = \sum_{i=1}^{t_{k-1}+1+j_{K-1}-1} d'_i = \left( \sum_{i=1}^{t_{k-1}+(j_{k-1}+1)-1} d_i \right) - 1$$

$$\geq \sum_{m=1}^{K-1} \left[ (j_{k-1}+1) + (k-1-m) \right] a_m - 1 = \sum_{m=1}^{K-1} \left[ j_{k-1} + (k-m) \right] a_m - 1$$

$$= \left[ j_{k-1} + (k-1) \right] a_1 + \left[ j_{k-1} + (k-2) \right] a_2 + \ldots + \left[ j_{k-1} + (k(k-1)) \right] a_{k-1} - 1$$

$$= \left[ j'_k + (k-1) \right] (a'_1 + 1) + \left[ j'_k + (k-2) \right] a'_3 + \ldots + \left[ j'_k + 1 \right] a'_k - 1$$

$$= \left[ j'_k + (k-1) \right] a'_1 + \left[ j'_2 + (k-1) \right] + \left[ j'_k + (k-3) \right] a'_3 + a'_3 +$$

$$\ldots + \left[ j'_k + (k-k) \right] a'_k + a'_k - 1$$

$$= \left[ j'_k + (k-1) \right] a'_1 + \left[ j'_k + (k-2) \right] a'_2 + 1 + \left[ j'_k + (k-3) \right] a'_3 + a'_3 +$$

$$\ldots + \left[ j'_k + (k-k) \right] a'_k + a'_k - 1$$

$$= \sum_{m=1}^{k} \left[ j'_k + (k-m) \right] a'_m + (a'_3 + \ldots + a'_k)$$

$$\geq \sum_{m=1}^{k} \left[ j'_k + (k-m) \right] a'_m,$$

which shows that the inequalities (2.6.1) hold for $D'$.

Therefore, by induction hypothesis there is a graph $G'$ realising $D'$.

If $v_1 v_{a_1} \in E(G')$, then obviously $G' + v_1 v_{a_1}$ gives a realisation $G$ of $D$. If $v_1 v_{a_1} \in E(G')$, then $d(v_{a_1} \text{ in } G') = d_{a_1} - 1 \leq n - 1 - 1 = n - 2$, there is a vertex $v_q$ such that $v_q v_{a_1} \notin E(G)$. Also, $d(v_q \text{in} G') > d(v_1 \text{ in } G')$, therefore there is a vertex $v_s$ such that $v_q v_s \in E(G')$ and $v_s v_1 \notin E(G')$. Now making elementary degree preserving transformation, on exchanging the edge pair $v_{a_1} v_1, v_q v_s$ by the edges $v_q v_{a_1}, v_s v_1$, we get a realisation $G'$ of $D'$ with $v_1 v_{a_1} \notin E(G')$. Then $G' + v_1 v_{a_1}$ realises $D$.                                                                    ❑

**Note**   In the above criterion, the inequalities (2.6.1) are to be checked only for $t_k + j_k - 1 + \sum_{m=1}^{k} a_m \leq n$ (but not for greater than $n$).

We now illustrate the theorem with the help of the following examples.

**Example 1**    Let $D = [1, 2, 2, 4, 6, 6, 6, 7, 8, 8]$.

Here, $n = 10$, $a_1 = 2$, $a_2 = 1$, $a_3 = 3$, $a_4 = 1$, $p_1 = 2$, $p_2 = 1$, $p_3 = 1$, $p_4 = 2$, so $t_1 = 2$, $t_2 = 3$, $t_3 = 4$, $t_4 = 6$.

Also, $j_1 = 1$, $j_2 = 1$, $j_3 = 1, 2$.

Now, for $j_1 = 1$, $\displaystyle\sum_{i=1}^{t_1+j_1-1} d_i = \sum_{i=1}^{2+1-1} d_i = \sum_{i=1}^{2} d_i = 1 + 2 = 3,$

and $\displaystyle\sum_{m=1}^{k} [j_k + (k-m)] a_m = \sum_{m=1}^{1} [j_1 + (1-m)] a_m = j_1 a_1 = 2.$

So inequalities (2.6.1) hold.

For $j_2 = 1$, $\displaystyle\sum_{i=1}^{t_2+j_2-1} d_i = \sum_{i=1}^{3+1-1} d_i = \sum_{i=1}^{3} d_i = 5$

and $\displaystyle\sum_{m=1}^{k} [j_k + (k-m)] a_m = \sum_{m=1}^{2} [j_2 + (2-m)] a_m = 2a_1 + a_2 = 4 + 1 = 5.$

So inequalities (2.6.1) hold.

For $j_3 = 1$, $\displaystyle\sum_{i=1}^{t_3+j_3-1} d_i = \sum_{i=1}^{4+1-1} d_i = \sum_{i=1}^{4} d_i = 9$

and $\displaystyle\sum_{m=1}^{3} [j_3 + (3-m)] a_m = \sum_{m=1}^{3} [1 + (3-m)] a_m = 3a_1 + 2a_2 + a_3 = 6 + 2 + 3 = 11.$

Since the inequalities (2.6.1) do not hold (as $9 > 11$ is not true), $D$ is not the degree sequence.

**Example 2**    Let $D = [1, 2, 3, 4, 5, 6, 6, 7, 8, 8]$.

Here, $n = 10$, $a_1 = 2$, $a_2 = 1$, $a_3 = 2$, $a_4 = 1$, $p_1 = 2$, $p_2 = 1$, $p_3 = 1$, $p_4 = 1$, $p_5 = 1$. So $t_1 = 2$, $t_2 = 3$, $t_3 = 4$, $t_4 = 5$.

Also, $j_1 = 1$, $j_2 = 1$, $j_3 = 1$, $j_4 = 1$.

For $j_1 = 1$, $\displaystyle\sum_{i=1}^{t_1+j_1-1} d_i = \sum_{i=1}^{2+1-1} d_i = \sum_{i=1}^{2} d_i = 3,$

and $\displaystyle\sum_{m=1}^{1} [j_1 + (1-m)] a_m = a_1 = 2.$

Obviously the inequalities (2.6.1) hold.

For $j_2 = 1$, $\displaystyle\sum_{i=1}^{t_2+j_2-1} d_i = \sum_{i=1}^{3+1-1} d_i = \sum_{i=1}^{3} d_i = 6$

and $\displaystyle\sum_{m=1}^{2} [j_2 + (2-m)]\, a_m = \sum_{m=1}^{2} [1 + (2-m)]\, a_m = 2a_1 + a_2 = 4 + 1 = 5$ .

Here again the inequalities (2.6.1) hold.

For $j_3 = 1$, $\displaystyle\sum_{i=1}^{t_3+j_3-1} d_i = \sum_{i=1}^{4+1-1} d_i = \sum_{i=1}^{4} d_i = 10$

and $\displaystyle\sum_{m=1}^{3} [j_3 + (3-m)]\, a_m = \sum_{m=1}^{3} [1 + (3-m)]\, a_m = 3a_1 + 2a_2 + a_3 = 6 + 2 + 2 = 10$ .

Therefore the inequalities (2.6.1) hold.

For $j_4 = 1$, $t_4 + j_4 - 1 = 5 + 1 - 1 = 5$ and $a_1 + a_2 + a_3 + a_4 = 2 + 1 + 2 + 1 = 6$, therefore $t_4 + j_4 - 1 + \displaystyle\sum_{m=1}^{4} a_m = 5 + 6 = 11 > 10$ and no further verification of the inequalities is to be done.

Hence $D$ is the degree sequence.

## 2.5 Equivalence of Seven Criteria

Now, we list the seven criteria for integer sequences to be graphic.

A. **The Ryser Criterion (Bondy and Murty [36] and Ryser [227])** A sequence $[a_1, \ldots, a_p; b_1, \ldots, b_n]$ is called bipartite-graphic if and only if there is a simple bipartite graph such that one component has degree sequence $[a_1, \ldots, a_p]$ and the other one has $[b_1, \ldots, b_n]$. Define $f = \max\{i : d_i \geq i\}$ and $\tilde{d}_1 = d_i + 1$ if $i \in \langle f \rangle (= \{1, \ldots, f\})$ and $\tilde{d}_1 = d_i$ otherwise. The criterion can be stated as follows.

The integer sequence $[\tilde{d}_1, ..., \tilde{d}_n; \tilde{d}_1, ..., \tilde{d}_n]$ is bipartite-graphic. (A)

B. **The Berge Criterion (Berge [23])** Define $[\bar{d}_1, \ldots, \bar{d}_n]$ as follows: For $i \in \langle n \rangle$, $\bar{d}_i$ is the $i$th column sum of the $(0, 1)$ matrix, which has for each $k$ and $d_k$ leading terms in row $k$ equal to 1 except for the $(k, k)$th term that is 0 and also the remaining entries are 0. If $d_1 = 3$, $d_2 = 2$, $d_3 = 2$, $d_4 = 2$, $d_5 = 1$, then $\bar{d}_1 = 4$, $\bar{d}_2 = 3$, $\bar{d}_3 = 2$, $\bar{d}_4 = 1$, $\bar{d}_5 = 0$, and the $(0, 1)$ matrix becomes

$$\begin{bmatrix} 0 & 1 & 1 & 1 & 0 \\ 1 & 0 & 1 & 0 & 0 \\ 1 & 1 & 0 & 0 & 0 \\ 1 & 1 & 0 & 0 & 0 \\ 1 & 0 & 0 & 0 & 0 \end{bmatrix}$$

The criterion is

$$\sum_{i=1}^{k} \overline{d_i} \leq \sum_{i=1}^{k} d_i \text{ for each } k \in \langle n \rangle. \tag{B}$$

## C. The Erdos-Gallai Criterion. (Bondy and Murty [36])

$$\sum_{i=1}^{k} d_i \leq (k)(k-1) + \sum_{j=k+1}^{n} \min\{k, d_j\} \text{ for each } k \in \langle n \rangle. \tag{C}$$

## D. The Fulkerson-Hoffman-McAndrew Criterion (Fulkerson[83] and Grunbaum [92])

$$\sum_{i=1}^{k} d_i \leq (k)(n-m-1) + \sum_{i=n-m+1}^{n} d_i \text{ for each } k \in \langle n \rangle, m \geq 0 \text{ and } k+m \leq n. \tag{D}$$

## E. The Bollobas Criterion (Bollabas[29]))

$$\sum_{i=1}^{k} d_i \leq \sum_{j=k+1}^{n} d_i + \sum_{i=1}^{k} \min\{d_j, k-1\} \text{ for each } k \in \langle n \rangle. \tag{E}$$

## F. The Grunbaum Criterion (Grunbaum [92]).

$$\sum_{i=1}^{k} \max\{k-1, d_i\} \leq (k)(k-1) + \sum_{i=k+1}^{n} d_i \text{ for each } k \in \langle n \rangle. \tag{F}$$

## G. The Hasselbarth Criterion (Hasselbarth [111]) Define $[d_i', \ldots, d_n']$ as follows. For $i \in \langle n \rangle$, $d_i'$ is the *i*th column sum of the (0, 1)-matrix in which the $d_i$ leading terms in row *i* are 1's and the remaining entries are 0's. The criterion is

$$\sum_{i=1}^{k} d_i \leq \sum_{i=1}^{k} (d_i^* - 1) \text{ for each } k \in \langle f \rangle, \tag{G}$$

with $f = \max\{i : d_i \geq i\}$.

The following result due to Sierksma and Hoogeveen [235] gives the equivalence among the above seven criteria.

**Theorem 2.7  (Sierksma and Hoogeveen [235])** Let $[d_1, \ldots, d_n]$ be a positive integer sequence with even sum. Then each of the criteria $(A) - (G)$ is equivalent to the statement that $[d_1, \ldots, d_n]$ is graphic.

**Proof**    Refer to Ryser [227].                                                                               □

## 2.6  Signed Graphs

A signed graph is a graph in which every edge is labelled with $a$'+' or $a$'−'. An edge $uv$ labelled with $a$'+' is called a *positive edge*, and is denoted by $uv^+$. An edge $uv$ labelled with $a$'−' is called a *negative edge*, and is denoted by $uv^-$. In a signed graph $G = (V, E)$, the *positive degree* of a vertex $u$ is $\deg^+(u) = |\{uv : uv^+ \in E\}|$, the *negative degree* of a vertex $u$ is $\deg^-(u) = |\{uv : uv^- \in E\}|$, the *signed degree* of $u$ is $\mathrm{sdeg}(u) = \deg^+(u) - \deg^-(u)$ and the *degree* of $u$ is $\deg(u) = \deg^+(u) + \deg^-(u)$. An edge $uv$ labelled with $a$'+' is called a *positive edge*, and is denoted by $uv^+$. An edge $uv$ labelled with $a$'−' is called a *negative edge*, and is denoted by $uv^-$. In a signed graph $G = (V, E)$, the *positive degree* of a vertex $u$ is $\deg^+(u) = |\{uv : uv^+ \in E\}|$, the *negative degree* of a vertex $u$ is $\deg^-(u) = |\{uv : uv^- \in E\}|$, the *signed degree* of $u$ is $\mathrm{sdeg}(u) = \deg^+(u) - \deg^-(u)$ and the *degree* of $u$ is $\deg(u) = \deg^+(u) + \deg^-(u)$.

An integral sequence $[d_i]_1^n$ is the *signed degree sequence* of a signed graph $G = (V, E)$ with $V = \{v_1, v_2, \ldots, v_n\}$ if $\mathrm{s}\deg(v_i) = d_i$, for $1 \le i \le n$.

Chartrand et al. [50] have given the characterisation of signed degree sequences of signed paths, signed stars, signed double stars and complete signed graphs. An integral sequence is *s-graphical* if it is the signed degree sequence of a signed graph. An integral sequence $[d_i]_1^n$ is *standard* if $n - 1 \ge d_1 \ge d_2 \ge \ldots \ge d_n$ and $d_1 \ge |d_n|$.

The following lemma shows that a signed degree sequence can be modified and rearranged into an equivalent standard form.

**Lemma 2.1**    If $[d_i]_1^n$ is the signed degree sequence of a signed graph $G$, then $[-d_i]_1^n$ is the signed degree sequence of the signed graph $G'$ obtained from $G$ by interchanging positive edges with negative edges.

The following necessary and sufficient condition under which an integral sequence is *s*-graphical is due to Chartrand et al. [50].

**Theorem 2.8**    A standard integral sequence $[d_i]_1^n$ is *s*-graphical if and only if the sequence $[d_2 - 1, d_{d_1+s+1} - 1, d_{d_1+s+2}, \ldots, d_{n-s}, d_{n-s+1} + 1, \ldots, d_n + 1]$ is *s*-graphical for some $0 \le s \le (n - 1 - d_1)/2$.

**Remark**    We note that Hakimi's theorem for degree sequences is a case of Theorem 2.8 by taking $s = 0$. This leads to an efficient algorithm for recognising the degree sequences of a graph. But the wide degree of latitude for choosing $s$ in Theorem 2.8 makes it harder to devise an efficient algorithm implementation.

The following result due to Yan et al. [271] provides a good choice for parameter $s$ in Theorem 2.8. It leads to a polynomial time algorithm for recognising signed degree sequences.

**Theorem 2.9**   A standard sequence $D = [d_i]_1^n$ is $s$-graphical if and only if $D_m = [d_2 - 1, d_{d_1+m+1} - 1, \ldots, d_{d_1+m+2}, \ldots, d_{n-m}, d_{n-m+1} + 1 \ldots, d_n + 1]$ is $s$-graphical, where $m$ is the maximum non-negative integer such that $d_{d_1+m+1} > d_{n-m+1}$.

**Proof**   Let $D$ be the signed degree sequence of a signed graph $G = (V, E)$ with $V = \{v_1, v_2, \ldots, v_n\}$ and $\text{sdeg}(v_i) = d_i$, for $1 \leq i \leq n$. For each $s$, $0 \leq s \leq (n - 1 - d_1)/2$, consider the sequence

$$D_s = [d_2 - 1, \ldots, d_{d_1+s+1} - 1, d_{d_1+s+2}, \ldots, d_{n-s}, d_{n-s+1} + 1, \ldots, d_n + 1].$$

By Theorem 2.8, $D_s$ is $s$-graphical for some $s$. We may choose $s$ such that $|s - m|$ is minimum. Suppose $G' = (V', E')$ is a signed graph with $V' = \{v_2, v_3, \ldots, v_n\}$ whose signed degree sequence is $D_s$.

If $s < m$, then $d_a > d_b$ by the choice of $m$, where $a = d_1 + s + 2$ and $b = n - s$. Since $d_a > d_b$, there exists some vertex $v_k$ of $G'$ different from $v_a$ and $v_b$ and satisfies one of the following conditions.

i. $v_a v_k^+$ is a positive edge and $v_b v_k^-$ is a negative edge.

ii. $v_a v_k^+$ is a positive edge and $v_b$ is not adjacent to $v_k$

iii. $v_a$ is not adjacent to $v_k$ and $v_b v_k^-$ is a negative edge

For (i), remove $v_a v_k^+$ and $v_b v_k^-$ to $G'$, and for (ii), remove $v_a v_k^+$ from $G'$ and add a new positive edge $v_b v_k^+$ to $G'$ and for (iii), remove $v_b v_k^-$ from $G'$ and a new negative edge $v_a v_k^-$ to $G'$. These modifications result in a signed graph $G''$ whose signed degree sequence $D_{s+1}$. This contradicts the minimality of $|s - m|$.

If $s > m$, then $d_{d_1+s+1} = d_{n-s+1}$, and therefore, $d_{d_1+s+1} - 1 < d_{n-s+1} - 1$. An argument similar to the above leads to a contradiction in the choice of $s$. Therefore, $s = m$ and $D_m$ is $s$-graphical.

Conversely, suppose $D_m$ is the signed degree sequence of a signed graph $G' = (V', E')$ in which $V' = \{v_2, v_3, \ldots, v_n\}$. If $G$ is the signed graph obtained from $G'$ by adding a new vertex $v_1$ and new positive edges $v_1 v_i^+$ for $2 \leq i \leq d_1 + m + 1$ and new negative edges $v_1 v_j^-$ for $n - m + 1 \leq j \leq n$, then $D$ is the signed degree sequence of $G$.   $\square$

In a signed graph $G = (V, E)$ with $|V| = n$, $|E| = m$, we denote by $m^+$ and $m^-$ respectively, the numbers of positive edges and negative edges of $G$. Further, $n_+$, $n_0$ and $n_-$ denote respectively, the numbers of vertices with positive, zero and negative signed degrees.

The following result is due to Chartrand et al. [50].

**Lemma 2.2**   If $G = (V, E)$ is a signed graph with $|V| = n$, $|E| = m$, then $k = \sum_{v \in V} s \deg(v) \equiv 2m \pmod{4}$, $m^+ = \frac{1}{4}(2m + k)$ and $m^- = \frac{1}{4}(2m - k)$.

The next result is due to Yan et al [271].

**Lemma 2.3** For any signed graph $G = (V, E)$ without isolated vertices, $\sum\limits_{v \in V} |sdeg(v)| + 2n_0 \leq 2m$.

**Proof** First, each $|sdeg(v)| = |\deg^+(v) - \deg^-(v)| \leq \deg^+(v) + \deg^-(v)$. Since $G$ has no isolated vertices, $2 \leq \deg^+(v) + \deg^-(v)$ when $sdeg(v) = 0$. Thus,

$$\sum_{v \in V} |s \deg(v)| + 2n_0 \leq \sum_{v \in V} (\deg^+(v) + \deg^-(v)) = 2m^+ + 2m^- = 2m. \qquad \square$$

**Lemma 2.4** For any connected signed graph $G = (V, E)$, $\sum\limits_{v \in V} |s \deg(v)| + 2 \sum\limits_{sdeg(v) < 0} |sdeg(v)| \leq 6m + 4 - 4\alpha - 4n_+ - 4n_0$, where $\alpha = 1$ if $n_+ n_- > 0$ and $\alpha = 0$ otherwise.

**Proof** Consider the subgraph $G' = (V', E')$ of $G$ induced by those edges incident to vertices with non-negative signed degrees. We have,

$$\sum_{sdeg(v) > 0} |sdeg(v)| \leq 2 \text{ (number of positive edges in } G') -$$

$$\text{(number of negative edges in } G') \leq 3m^+ - |E'|.$$

Since $G$ is connected, each component of $G'$ contains at least one vertex of negative signed degree except for the case of $G' = G$.

Therefore, $n_+ + n_0 - 1 + \alpha \leq |E'|$. Thus,

$$\sum_{s \deg(v).0} |s \deg(v)| + n_+ + n_0 - 1 + \alpha \leq 3m^+ = 3 \left( \frac{1}{2} m + \frac{1}{4} \sum_{v \in V} s \deg(v) \right).$$

Hence, $\sum\limits_{v \in V} |s \deg(v)| + 2 \sum\limits_{s \deg(v) < 0} |s \deg(v)| \leq 6m + 4 - 4\alpha - 4n_+ - 4n_0$. $\qquad \square$

For any integer $k$, $k$ copies of $v_i v_j$ means $k$ copies of positive edges $v_i v_j^+$ if $k > 0$, no edges if $k = 0$ and $-k$ copies of negative edges $v_i v_j^-$ if $k < 0$. The next result for signed graphs with loops or multiple edges is due to Yan et al. [271].

**Theorem 2.10** An integral sequence $[d_i]_1^n$ is the signed degree sequence of a signed if and only if $\sum\limits_{i=1}^{n} d_i$ is even.

**Proof** The necessity follows from Lemma 2.2.

*Sufficiency* Let $\sum\limits_{i=1}^{n} d_i$ be even. Then the number of odd terms is even, say $d_i = 2e_i + 1$ for $1 \leq i \leq 2k$ and $d_i = 2e_i$ for $2k + 1 \leq i \leq p$. Then, $[d_1, d_2, \ldots, d_n]$ is the signed degree sequence of the signed graph with vertex set $\{v_1, v_2, \ldots, v_n\}$ and edge set $\{-d_3 = \frac{1}{2} \sum\limits_{i=1}^{n} d_i$ copies of

$v_1 v_2\} \cup \{d_2 + d_3 - \frac{1}{2} \sum\limits_{i=1}^{n} d_i$ copies of $v_2 v_3\} \cup \{d_1 + d_3 - \frac{1}{2} \sum\limits_{i=1}^{n} d_i$ copies of $v_1 v_3\} \cup \{d_i$ copies of

$v_3 v_i : 4 \leq i \leq n\}$. ❏

Various results on signed degrees in signed graphs can be found in [259], [263], [264] and [266].

## 2.7 Exercises

1. Verify whether or not the following sequences are degree sequences.
   a. $[1, 1, 1, 2, 3, 4, 5, 6, 7]$,    b. $[1, 1, 1, 2, 2, 2]$,
   c. $[4, 4, 4, 4, 4, 4]$,    d. $[2, 2, 2, 2, 4, 4]$.

2. Show that there is no perfect degree sequence.

3. What conditions on $n$ and $k$ will ensure that $k^n$ is a degree sequence?

4. Give an example of a graph that can not be generated by the Wang-Kleitman algorithm.

5. Draw the five non isomorphic graphs with degree sequence $[3, 3, 2, 2, 1, 1]$.

6. Show that a graph and its complement have the same frequency sequence.

7. Construct a graph with a degree sequence $[3, 3, 3, 3, 2, 2, 2, 2, 1, 1, 1, 1]$ by using Havel-Hakimi algorithm.

# 3. Eulerian and Hamiltonian Graphs

There are many games and puzzles which can be analysed by graph theoretic concepts. In fact, the two early discoveries which led to the existence of graphs arose from puzzles, namely, the Konigsberg Bridge Problem and Hamiltonian Game, and these puzzles also resulted in the special types of graphs, now called Eulerian graphs and Hamiltonian graphs. Due to the rich structure of these graphs, they find wide use both in research and application.

## 3.1  Euler Graphs

A closed walk in a graph $G$ containing all the edges of $G$ is called an *Euler line* in $G$. A graph containing an Euler line is called an *Euler graph*.

We know that a walk is always connected. Since the Euler line (which is a walk) contains all the edges of the graph, an Euler graph is connected except for any isolated vertices the graph may contain. As isolated vertices do not contribute anything to the understanding of an Euler graph, it is assumed now onwards that Euler graphs do not have any isolated vertices and are thus connected.

**Example**   Consider the graph shown in Figure 3.1. Clearly, $v_1\, e_1\, v_2\, e_2\, v_3\, e_3\, v_4\, e_4\, v_5\, e_5\, v_3\, v_6\, e_7\, v_1$ in (a) is an Euler line, whereas the graph shown in (b) is non-Eulerian.

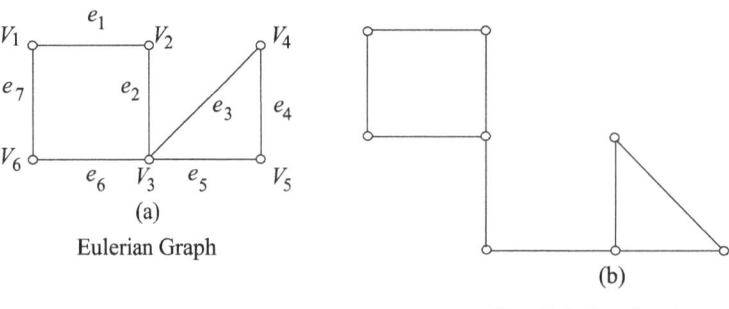

(a)

Eulerian Graph

(b)

Non-Eulerian Graph

**Fig. 3.1**

The following theorem due to Euler [74] characterises Eulerian graphs. Euler proved the necessity part and the sufficiency part was proved by Hierholzer [115].

**Theorem 3.1 (Euler [74])** A connected graph $G$ is an Euler graph if and only if all vertices of $G$ are of even degree.

## Proof

*Necessity* Let $G = (V, E)$ be an Euler graph. Thus $G$ contains an Euler line $Z$, which is a closed walk. Let this walk start and end at the vertex $u \in V$. Since each visit of $Z$ to an intermediate vertex $v$ of $Z$ contributes two to the degree of $v$ and since $Z$ traverses each edge exactly once, $d(v)$ is even for every such vertex. Each intermediate visit to $u$ contributes two to the degree of $u$, and also the initial and final edges of $Z$ contribute one each to the degree of $u$. So the degree of $u$, that is, $d(u)$ is also even.

*Sufficiency* Let $G$ be a connected graph and let degree of each vertex of $G$ be even. Assume $G$ is not Eulerian and let $G$ contain least number of edges. Since $\delta \geq 2$, $G$ has a cycle. Let $Z$ be a closed walk in $G$ of maximum length. Clearly, $G - E(Z)$ is an even degree graph. Let $C_1$ be one of the components of $G - E(Z)$. As $C_1$ has less number of edges than $G$, it is Eulerian and has a vertex $v$ in common with $Z$. Let $Z'$ be an Euler line in $C_1$. Then $Z' \cup Z$ is closed in $G$, starting and ending at $v$. Since it is longer than $Z$, the choice of $Z$ is contradicted. Hence $G$ is Eulerian.

*Second proof for sufficiency* Assume that all vertices of $G$ are of even degree. We construct a walk starting at an arbitrary vertex $v$ and going through the edges of $G$ such that no edge of $G$ is traced more than once. The tracing is continued as far as possible. Since every vertex is of even degree, we exit from the vertex we enter and the tracing clearly cannot stop at any vertex but $v$. As $v$ is also of even degree, we reach $v$ when the tracing comes to an end. If this closed walk $Z$ we just traced includes all the edges of $G$, then $G$ is an Euler graph. If not, we remove from $G$ all the edges in $Z$ and obtain a subgraph $Z'$ of $G$ formed by the remaining edges. Since both $G$ and $Z$ have all their vertices of even degree, the degrees of the vertices of $Z'$ are also even. Also, $Z'$ touches $Z$ at least at one vertex say $u$, because $G$ is connected. Starting from $u$, we again construct a new walk in $Z'$. As all the vertices of $Z'$ are of even degree, therefore this walk in $Z'$ terminates at vertex $u$. This walk in $Z'$ combined with $Z$ forms a new walk, which starts and ends at the vertex $v$ and has more edges than $Z$. This process is repeated till we obtain a closed walk that traces all the edges of $G$. Hence $G$ is an Euler graph (Fig. 3.2) ❑

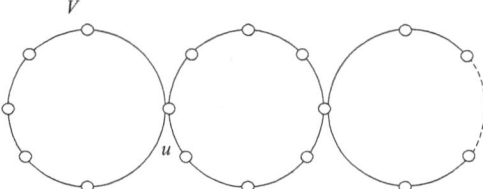

**Fig. 3.2**

## 3.2   Konigsberg Bridge Problem

Two islands *A* and *B* formed by the Pregal river (now Pregolya) in Konigsberg (then the capital of east Prussia, but now renamed Kaliningrad and in west Soviet Russia) were connected to each other and to the banks *C* and *D* with seven bridges. The problem is to start at any of the four land areas, *A*, *B*, *C*, or *D*, walk over each of the seven bridges exactly once and return to the starting point.

Euler modeled the problem representing the four land areas by four vertices, and the seven bridges by seven edges joining these vertices. This is illustrated in Figure 3.3.

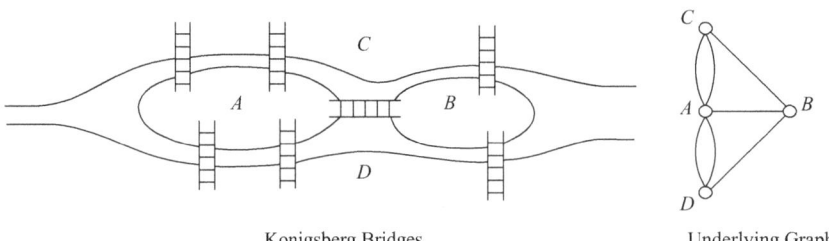

Konigsberg Bridges                                 Underlying Graph

**Fig. 3.3**

Now, we see from the graph *G* of the Konigsberg bridges that not all its vertices are of even degree. Thus, *G* is not an Euler graph, and implies that there is no closed walk in *G* containing all the edges of *G*. Hence, it is not possible to walk over each of the seven bridges exactly once and return to the starting point.

**Note**   Two additional bridges have been built since Euler's day. The first has been built between land areas *C* and *D* and the second between the land areas *A* and *B*. Now in the graph of Konigsberg bridge problem with nine bridges, every vertex is of even degree and the graph is thus Eulerian. Hence it is now possible to walk over each of the nine bridges exactly once and return to the starting point (Fig 3.4).

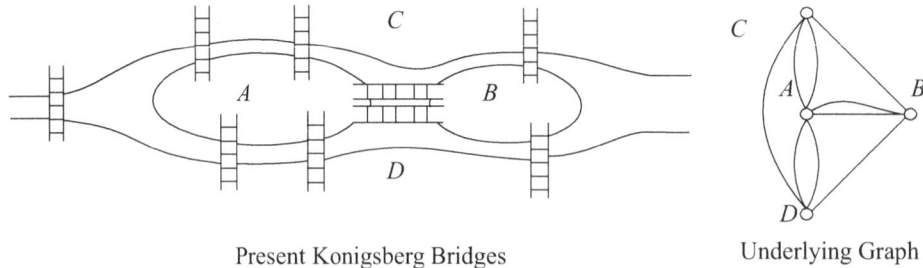

Present Konigsberg Bridges                         Underlying Graph

**Fig. 3.4**

The following characterisation of Eulerian graphs is due to Veblen [254].

**Theorem 3.2 (Veblen [254])** A connected graph $G$ is Eulerian if and only if its edge set can be decomposed into cycles.

**Proof** Let $G = (V, E)$ be a connected graph and let $G$ be decomposed into cycles. If $k$ of these cycles are incident at a particular vertex $v$, then $d(v) = 2k$. Therefore, the degree of every vertex of $G$ is even and hence $G$ is Eulerian.

Conversely, let $G$ be Eulerian. We show $G$ can be decomposed into cycles. To prove this, we use induction on the number of edges.

Since $d(v) \geq 2$ for each $v \in V$, $G$ has a cycle $C$. Then $G - E(C)$ is possibly a disconnected graph, each of whose components $C_1, C_2, \ldots, C_k$ is an even degree graph and hence Eulerian. By the induction hypothesis, each $C_i$ is a disjoint union of cycles. These together with $C$ provide a partition of $E(G)$ into cycles. $\qquad\square$

The following result is due to Toida [244].

**Theorem 3.3 (Toida [244])** If $W$ is a walk from vertex $u$ to vertex $v$, then $W$ contains an odd number of $u - v$ paths.

**Proof** Let $W$ be a walk which we consider as a graph in itself, and not as a subgraph of some other graph. Let $u$ and $v$ be initial and final vertices of the walk $W$. Clearly, $d(u_1, W)$ and $d(v_1, W)$ are odd, and $d(w_1, W)$ is even, for every $w \in V(W) - \{u, v\}$. We count the number of distinct $u - v$ walks in $W$. These walks are the subgraphs of $W$.

When we take a $u - v$ walk by successively selecting the edges $e_1, e_2, \ldots, e_s$, initial vertex of $e_1$ being $u$ and terminal vertex of $e_s$ being $v$, for each edge there are an odd number of choices. The total number of such edges is the product of these odd numbers and is therefore odd. Now from these walks, we find the $u - v$ paths. If a $u - v$ walk $W_1$ is not a path, then it contains one or more cycles. The traversal of these cycles in the two possible alternative directions (clockwise and anticlockwise) produces in all an even number of walks, all with the same edge set as $W_1$. Omitting these even number of walks which are not paths from the total odd collection of $u - v$ walks, gives an odd number of $u - v$ paths. $\qquad\square$

Toida [244] proved the necessity part and McKee [157] the sufficiency part of the next characterisation. The second proof of this result can be found in Fleischner [79], [80].

**Theorem 3.4** A connected graph is Eulerian if and only if each of its edges lies on an odd number of cycles.

### Proof

*Necessity* Let $G$ be a connected Eulerian graph and let $e = uv$ be any edge of $G$. Then $G - e$ is a $u - v$ walk $W$, and so $G - e = W$ contains an odd number of $u - v$ paths. Thus, each of the odd number of $u - v$ paths in $W$ together with $e$ gives a cycle in $G$ containing $e$ and these are the only such cycles. Therefore, there are an odd number of cycles in $G$ containing $e$.

*Sufficiency*   Let $G$ be a connected graph so that each of its edges lies on an odd number of cycles. Let $v$ be any vertex of $G$, and $E_v = \{e_1, \ldots, e_d\}$ be the set of edges of $G$ incident on $v$, then $|Ev| = d(v) = d$. For each $i$, $1 \le i \le d$, let $k_i$ be the number of cycles of $G$ containing $e_i$. By hypothesis, each $k_i$ is odd. Let $c(v)$ be the number of cycles of $G$ containing $v$. Then, clearly $c(v) = \frac{1}{2} \sum_{i=1}^{d} k_i$ implying that $2c(v) = \sum_{i=1}^{d} k_i$. Since $2c(v)$ is even, and each $k_i$ is odd, $d$ is even. Hence $G$ is Eulerian.                                                                          ❏

**Corollary 3.1**   The number of edge–disjoint paths between any two vertices of an Euler graph is even.

A consequence of Theorem 3.4 is the result of Bondy and Halberstam [37], which gives yet another characterisation of Eulerian graphs.

**Corollary 3.2 (Bondy and Halberstam [37])**   A graph is Eulerian if and only if it has an odd number of cycle decompositions.

**Proof**   In one direction, the proof is trivial. If $G$ has an odd number of cycle decompositions, then it has at least one, and hence $G$ is Eulerian.

Conversely, assume that $G$ is Eulerian. Let $e \in E(G)$ and let $C_1, \ldots, C_r$ be the cycles containing $e$. By Theorem 3.4, $r$ is odd. We proceed by induction on $m = |E(G)|$, with $G$ being Eulerian.

If $G$ is just a cycle, then the result is true. Now assume that $G$ is not a cycle. This means that for each $i$, $1 \le i \le r$, by the induction assumption, $G_i = G - E(C_i)$ has an odd number, say $s_i$, of cycle decompositions. (If $G_i$ is disconnected, apply the induction assumption to each of the nontrivial components of $G_i$). The union of each of these cycle decompositions of $G_i$ and $C_i$ yields a cycle decomposition of $G$. Hence the number of cycle decompositions of $G$ containing $C_i$ is $s_i$, $1 \le i \le r$. Let $s(G)$ denote the number of cycle decompositions of $G$. Then

$$s(G) \equiv \sum_{i=1}^{r} s_i \equiv r(\text{mod } 2) \qquad (\text{since } s_i \equiv 1(\text{mod } 2))$$

$$\equiv 1(\text{mod } 2).$$                                                                          ❏

Two examples of Euler graphs are shown in Figure 3.5.

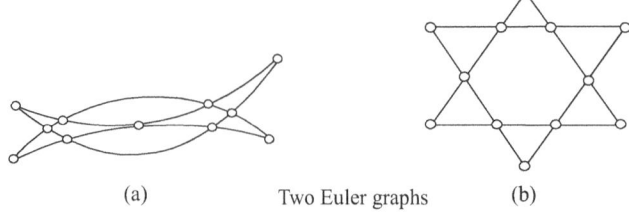

(a)                          Two Euler graphs                          (b)

**Fig. 3.5**

## 3.3 Unicursal Graphs

An open walk that includes (or traces) all edges of a graph without retracing any edge is called a unicursal line or open Euler line. A connected graph that has a unicursal line is called a unicursal graph. Figure 3.6 shows a unicursal graph.

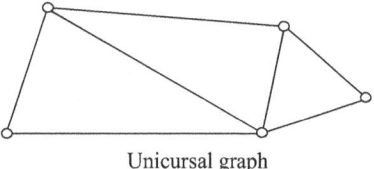

Unicursal graph

**Fig. 3.6**

Clearly by adding an edge between the initial and final vertices of a unicursal line, we get an Euler line.

The following characterisation of unicursal graphs can be easily derived from Theorem 3.1.

**Theorem 3.5**    A connected graph is unicursal if and only if it has exactly two vertices of odd degree.

**Proof**    Let $G$ be a connected graph and let $G$ be unicursal. Then $G$ has a unicursal line, say from $u$ to $v$, where $u$ and $v$ are vertices of $G$. Join $u$ and $v$ to a new vertex $w$ of $G$ to get a graph $H$. Then $H$ has an Euler line and therefore each vertex of $H$ is of even degree. Now, by deleting the vertex $w$, the degree of vertices $u$ and $v$ each get reduced by one, so that $u$ and $v$ are of odd degree.

Conversely, let $u$ and $v$ be the only vertices of $G$ with odd degree. Join $u$ and $v$ to a new vertex $w$ to get the graph $H$. So, every vertex of $H$ is of even degree and thus $H$ is Eulerian. Therefore, $G = H - w$ has a $u - v$ unicursal line so that $G$ is unicursal.    ❑

The following result is the generalisation of Theorem 3.5.

**Theorem 3.6**    In a connected graph $G$ with exactly $2k$ odd vertices, there exists $k$ edge disjoint subgraphs such that they together contain all edges of $G$ and that each is a unicursal graph.

**Proof**    Let $G$ be a connected graph with exactly $2k$ odd vertices. Let these odd vertices be named $v_1, v_2, \ldots, v_k$; $w_1, w_2, \ldots, w_k$ in any arbitrary order. Add $k$ edges to $G$ between the vertex pairs $(v_1, w_1), (v_2, w_2), \ldots, (v_k, w_k)$ to form a new graph $H$, so that every vertex of $H$ is of even degree. Therefore, $H$ contains an Euler line $Z$.

Now, if we remove from $Z$ the $k$ edges we just added (no two of these edges are incident on the same vertex), then $Z$ is divided into $k$ walks, each of which is a unicursal line. The first removal gives a single unicursal line, the second removal divides that into two unicursal lines, and each successive removal divides a unicursal line into two unicursal lines, until there are $k$ of them. Hence the result.    ❑

## 3.4 Arbitrarily Traceable Graphs

An Eulerian graph $G$ is said to be arbitrarily traceable (or randomly Eulerian) from a vertex $v$ if every walk with initial vertex $v$ can be extended to an Euler line of $G$. A graph is said to be arbitrarily traceable if it is arbitrarily traceable from every vertex (Fig. 3.7).

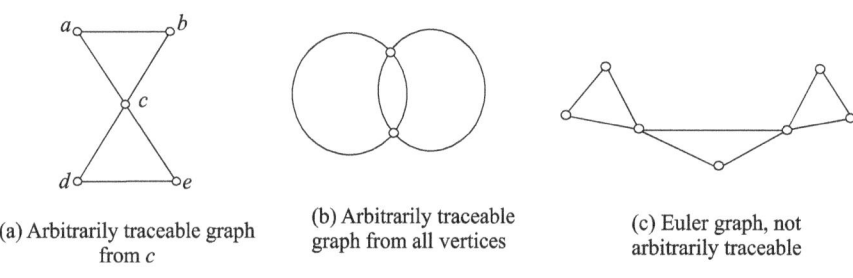

(a) Arbitrarily traceable graph from $c$  (b) Arbitrarily traceable graph from all vertices  (c) Euler graph, not arbitrarily traceable

**Fig. 3.7**

The following characterisation of arbitrarily traceable graphs is due to Ore [174]. Such graphs were also characterised by Chartrand and White [56].

**Theorem 3.7 (Ore [174])** An Eulerian graph $G$ is arbitrarily traceable from a vertex $v$ if only if every cycle of $G$ passes through $v$.

### Proof

*Necessity*  Let the Eulerian graph $G$ be arbitrarily traceable from a vertex $v$. Assume there is a cycle $C$ not passing through $v$. Let $H = G - E(C)$. Then every vertex of $H$ has an even degree and the component of $H$ containing $v$ is Eulerian. This component of $H$ can be traversed as an Euler line $Z$, starting and ending with $v$ and contains all those edges of $G$ which are incident at $v$. Clearly, this $v - v$ walk cannot be extended to contain the edges of $C$ also, contradicting that $G$ contains $v$. Thus every cycle in $G$ contains $v$.

*Sufficiency*  Let every cycle of the Eulerian graph $G$ pass through the vertex $v$ of $G$. We show that $G$ is arbitrarily traceable from $v$. Assume, on the contrary, that $G$ is not arbitrarily traceable from $v$. Then there is a $v - v$ closed walk $W$ of $G$ containing all the edges of $G$ incident with $v$, and yet not containing all the edges of $G$. Let one such edge be incident at a vertex $u$ on $W$. So every vertex of $H = G - E(W)$ is of even degree, and $v$ is an isolated vertex of $H$ and $u$ is not. The component of $H$ containing $u$ is therefore Eulerian subgraph of $G$ not passing through $v$, contradicting the assumption. Hence the result follows.  ❏

**Corollary 3.3**  Cycles are the only arbitrarily traceable graphs.

# 3.5 Sub-Eulerian Graphs

A graph $G$ is said to be *sub-Eulerian* if it is a spanning subgraph of some Eulerian graph.

The following characterisation of sub-Eulerian graphs is due to Boesch, Suffel and Tindell [28].

**Theorem 3.8 (Boesch, Suffel and Tindell [28])** A connected graph $G$ is sub-Eulerian if and only if $G$ is not spanned by a complete bipartite graph.

### Proof

*Necessity* We prove that no spanning supergraph $H$ of an odd complete bipartite graph $G$ is Eulerian. Let $V_1 \cup V_2$ be the bipartition of the vertex set of $G$. Since degree of each vertex of $G$ is odd, and $G$ is complete bipartite, therefore $|V_1|$ and $|V_2|$ are odd. If $H_1$ is the induced subgraph of $H$ on $V_1$, then at least one vertex, say $v$, of $V_1$ has even degree in $H_1$, since $|V_1|$ is odd. But then $d(v/H) = d(v/H) + |V_2|$, which is odd. Therefore, $H$ is not Eulerian.

*Sufficiency* Refer Boesch et. al., [28]. ❑

## Super-Eulerian graphs

A non-Eulerian graph $G$ is said to be *super-Eulerian* if it has a spanning Eulerian subgraph.

The following sufficient conditions for super-Eulerian graphs are due to Lesniak, Foster and Williams [148].

**Theorem 3.9 (Lesniak-Foster and Williams [148])** If a graph $G$ is such that $n \geq 6$, $\delta \geq 2$ and $d(u) + d(v) \geq n - 1$, for every pair of non-adjacent vertices $u$ and $v$, then $G$ is super-Eulerian.

The following result is due to Balakrishnan and Paulraja [12].

**Theorem 3.10 (Balakrishnan and Paulraja [12])** Let $G$ be any connected graph. If each edge of $G$ belongs to a triangle in $G$, then $G$ has a spanning Eulerian subgraph.

**Proof** Since $G$ has a triangle, $G$ has a closed walk. Let $W$ be the longest closed walk in $G$. Then $W$ must be a spanning Eulerian subgraph of $G$. If not, there exists a vertex $v \notin W$ and $v$ is adjacent to a vertex $u$ of $W$. By hypothesis, $uv$ belongs to a triangle, say $uvw$. If none of the edges of this triangle is in $W$, then $W \cup \{uv, vw, wu\}$ yields a closed walk longer than $W$ (Fig. 3.8). If $uw \in W$, then $(W - uw) \cup \{uv, vw\}$ would be a closed walk longer than $W$. This contradiction proves that $W$ is a spanning closed walk in $G$. ❑

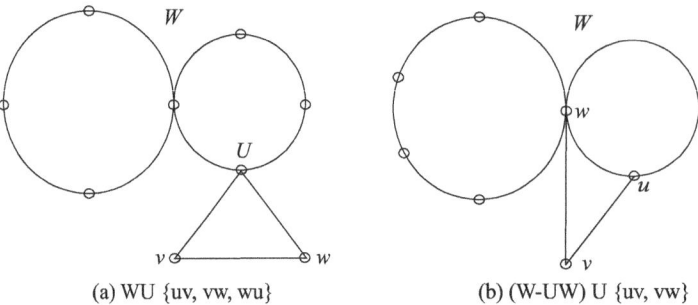

(a) WU {uv, vw, wu}                         (b) (W-UW) U {uv, vw}

**Fig. 3.8**

## 3.6   Hamiltonian Graphs

A cycle passing through all the vertices of a graph is called a *Hamiltonian cycle*. A graph
containing a Hamiltonian cycle is called a *Hamiltonian graph*. A path passing through all
the vertices of a graph is called a *Hamiltonian path* and a graph containing a Hamiltonian
path is said to be *traceable*. Examples of Hamiltonian graphs are given in Figure 3.9.

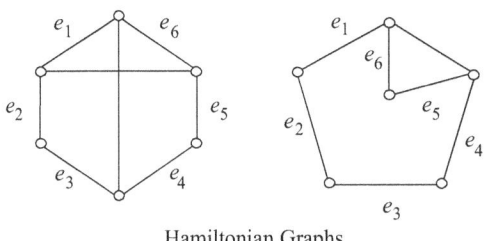

Hamiltonian Graphs

**Fig. 3.9**

If the last edge of a Hamiltonian cycle is dropped, we get a Hamiltonian path. However,
a non-Hamiltonian graph can have a Hamiltonian path, that is, Hamiltonian paths cannot
always be used to form Hamiltonian cycles. For example, in Figure 3.10, $G_1$ has no Hamiltonian path, and so no Hamiltonian cycle; $G_2$ has the Hamiltonian path $v_1v_2v_3v_4$, but has no
Hamiltonian cycle, while $G_3$ has the Hamiltonian cycle $v_1v_2v_4v_3v_1$.

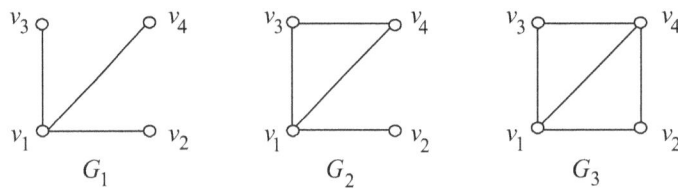

**Fig. 3.10**

Hamiltonian graphs are named after Sir William Hamilton, an Irish Mathematician (1805–1865), who invented a puzzle, called the Icosian game, which he sold for 25 guineas to a game manufacturer in Dublin. The puzzle involved a dodecahedron on which each of the 20 vertices was labelled by the name of some capital city in the world. The aim of the game was to construct, using the edges of the dodecahedron a closed walk of all the cities which traversed each city exactly once, beginning and ending at the same city. In other words, one had essentially to form a Hamiltonian cycle in the graph corresponding to the dodecahedron. Figure 3.11 shows such a cycle.

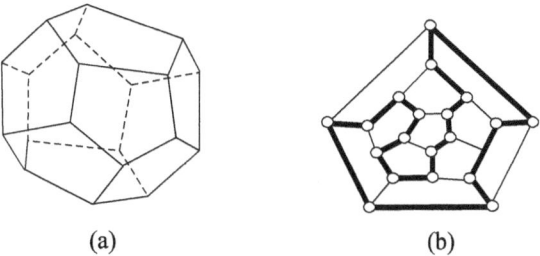

(a)                                                   (b)

Dedecahedron and its graph shown with the Hamiltonian circuit

**Fig. 3.11**

Clearly, the $n$-cycle $C_n$ with $n$ distinct vertices (and $n$ edges) is Hamiltonian. Now, given any Hamiltonian graph $G$, the supergraph $G'$ (obtained by adding in new edges between non-adjacent vertices of $G$) is also Hamiltonian. This is because any Hamiltonian cycle in $G$ is also a Hamiltonian cycle of $G'$. For instance, $K_n$ is a supergraph of an $n$-cycle and so $K_n$ is Hamiltonian.

A multigraph or general graph is Hamiltonian if and only if its underlying graph is Hamiltonian, because if $G$ is Hamiltonian, then any Hamiltonian cycle in $G$ remains a Hamiltonian cycle in the underlying graph of $G$. Conversely, if the underlying graph of a graph $G$ is Hamiltonian, then $G$ is also Hamiltonian.

Let $G$ be a graph with $n$ vertices. Clearly, $G$ is a subgraph of the complete graph $K_n$. From $G$, we construct step by step supergraphs of $G$ to get $K_n$, by adding an edge at each step between two vertices that are not already adjacent (Fig. 3.12).

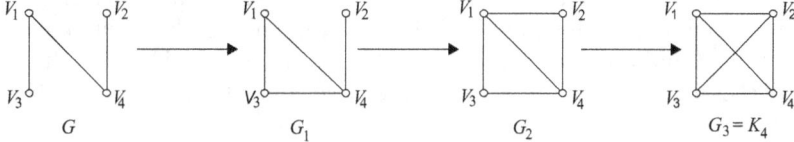

**Fig. 3.12**

Now, let us start with a graph $G$ which is not Hamiltonian. Since the final outcome of the procedure is the Hamiltonian graph $K_n$, we change from a non-Hamiltonian graph to a Hamiltonian graph at some stage of the procedure. For example, the non-Hamiltonian graph $G_1$ above is followed by the Hamiltonian graph $G_2$. Since supergraphs of Hamiltonian graphs are Hamiltonian, once a Hamiltonian graph is reached in the procedure, all the subsequent supergraphs are Hamiltonian.

**Definition:**   A simple graph $G$ is called *maximal non-Hamiltonian* if it is not Hamiltonian and the addition of an edge between any two non-adjacent vertices of it forms a Hamiltonian graph. For example, $G_1$ above is maximal non-Hamiltonian. Figure 3.13 shows a maximal non-Hamiltonian graph.

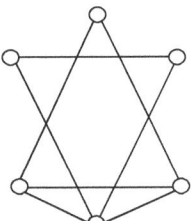

**Fig. 3.13**

Now, it follows from the above procedure that any non-Hamiltonian graph with $n$-vertices is a subgraph of a maximal non-Hamiltonian graph with $n$ vertices.

The above procedure is used to prove the following sufficient conditions due to Dirac [68].

**Theorem 3.11 (Dirac [68])**   If $G$ is a graph with $n$ vertices, where $n \geq 3$ and $d(v) \geq n/2$, for every vertex $v$ of $G$, then $G$ is Hamiltonian.

**Proof**   Assume that the result is not true. Then for some value $n \geq 3$, there is a non-Hamiltonian graph $H$ in which $d(v) \geq n/2$, for every vertex of $H$. Now in any spanning super graph $K$ (i.e., with the same vertex set) of $H$, $d(v) \geq n/2$ for every vertex of $K$, since any proper supergraph of this form is obtained by adding more edges. Thus there is a maximal non-Hamiltonian graph $G$ with $n$ vertices and $d(v) \geq n/2$ for every $v$ in $G$. Using this $G$, we obtain a contradiction.

Clearly, $G \neq K_n$, as $K_n$ is Hamiltonian. Therefore, there are non-adjacent vertices $u$ and $v$ in $G$. Let $G + uv$ be the supergraph of $G$ by adding an edge between $u$ and $v$. Since $G$ is maximal non-Hamiltonian, $G + uv$ is Hamiltonian. Also, if $C$ is a Hamiltonian cycle of $G + uv$, then $C$ contains the edge $uv$, since otherwise $C$ is a Hamiltonian cycle of $G$, which is not possible.

Let this Hamiltonian cycle $C$ be $u = v_1, v_2, \ldots, v_n = v, u$.

Now, let $S = \{v_i \in C : \text{there is an edge from } u \text{ to } v_{i+1} \text{ in } G\}$
and $T = \{v_j \in C : \text{there is an edge from } v \text{ to } v_j \text{ in } G\}$.

Then $v_n \notin T$, since otherwise there is an edge from $v$ to $v_n = v$, that is a loop, which is impossible.

Also $v_n \notin S$, (taking $v_{n+1}$ as $v_1$), since otherwise we again get a loop from $u$ to $v_1 = u$. Therefore, $v_n \in S \cup T$ (Fig. 3.14).

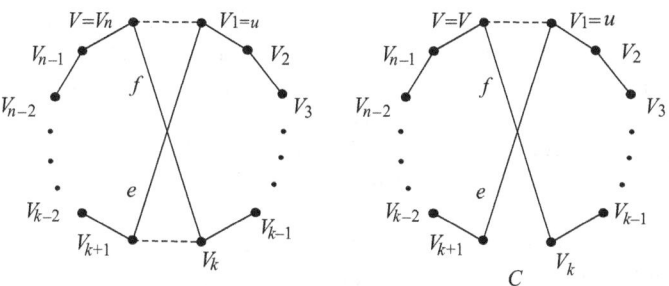

**Fig. 3.14**

Let $|S|$, $|T|$ and $|S \cup T|$ be the number of elements in $S$, $T$ and $S \cup T$ respectively. So $|S \cup T| < n$. Also, for every edge incident with $u$, there corresponds one vertex $v_i$ in $S$. Therefore, $|S| = d(u)$. Similarly, $|T| = d(v)$.

Now, if $v_k$ is a vertex belonging to both $S$ and $T$, then there is an edge $e$ joining $u$ to $v_{k+1}$ and an edge $f$ joining $v$ to $v_k$. This implies that $C' = v_1, v_{k+1}, v_{k+2}, \ldots, v_n, v_k, v_{k-1}, \ldots, v_2, v_1$ is a Hamiltonian cycle in $G$, which is a contradiction as $G$ is non-Hamiltonian. This shows that there is no vertex $v_k$ in $S \cap T$, so that $S \cap T = \Phi$.

Thus, $|S \cup T| = |S| + |T| - |S \cap T|$ gives $|S| + |T| = |S \cup T|$, so that $d(u) + d(v) < n$. This is a contradiction, because $d(u) \geq n/2$ for all $u$ in $G$, and so $d(u) + d(v) \geq n/2 + n/2$ giving $d(u) + d(v) \geq n$. Hence the theorem follows. ❑

The following result is due to Ore [176].

**Theorem 3.12 (Ore [176])** Let $G$ be a graph with $n$ vertices and let $u$ and $v$ be non-adjacent vertices in $G$ such that $d(u) + d(v) \geq n$. Let $G + uv$ denote the super graph of $G$ obtained by joining $u$ and $v$ by an edge. Then $G$ is Hamiltonian if and only if $G + uv$ is Hamiltonian.

**Proof** Let $G$ be a graph with $n$ vertices and suppose $u$ and $v$ are non-adjacent vertices in $G$ such that $d(u) + d(v) \geq n$. Let $G + uv$ be the super graph of $G$ obtained by adding the edge $uv$. Let $G$ be Hamiltonian. Then obviously $G + uv$ is Hamiltonian. Conversely, let $G + uv$ be Hamiltonian. We have to show that $G$ is Hamiltonian.

Then, as in Theorem 3.11, we get $d(u) + d(v) < n$, which contradicts the hypothesis that $d(u) + d(v) \geq n$. Hence $G$ is Hamiltonian. ❑

Now, we give the proof of Bondy [35] of Theorem 3.12, and this proof bears a close resemblance to the proof of Dirac's theorem given by Newman [170], but is more direct.

**Proof (Bondy [35])** Consider the complete graph $K$ on the vertex set of $G$ in which the edges of $G$ are coloured blue and the remaining edges of $K$ are coloured red. Let $C$ be a

Hamiltonian cycle of $K$ with as many blue edges as possible. We show that every edge of $C$ is blue, in other words, that $C$ is Hamiltonian cycle of $G$.

Suppose to the contrary, $C$ has a red edge $uu^-$ (where $u^-$ is the successor of $u$ on $C$). Consider the set $S$ of vertices joined to $u$ by blue edges (that is, the set of neighbours of $u$ in $G$). The successor $u^-$ of $u$ on $C$ must be joined by a blue edge to some vertex $v^-$ of $S^-$, because if $u^-$ is adjacent in $C$ only to vertices $V - (S^- \cup \{u^-\})$, $d_G(u) + d_G(u^-) = |N_G(u)| + |N_G(u^-)| \leq |S| + (|V| - |S^-| - 1) = |V(G)| - 1$, contradicting the hypothesis that $d_G(u) + d_G(u^-) \geq |V(G)|$, $u$ and $u^-$ being non-adjacent in $G$. But now the cycle $C$ obtained from $C$ by exchanging the edges $uu^-$ and $vv^-$ has more blue edges than $C$, which is a contradiction.                          ❏

**Definition:** Let $G$ be a graph with $n$ vertices. If there are two non-adjacent vertices $u_1$ and $v_1$ in $G$ such that $d(u_1) + d(v_1) \geq n$, join $u_1$ and $v_1$ by an edge to form the super graph $G_1$. Now, if there are two non-adjacent vertices $u_2$ and $v_2$ in $G_1$ such that $d(u_2) + d(v_2) \geq n$, join $u_2$ and $v_2$ by an edge to form supergraph $G_2$. Continue in this way, recursively joining pairs of non-adjacent vertices whose degree sum is at least $n$ until no such pair remains. The final supergraph thus obtained is called the *closure* of $G$ and is denoted by $c(G)$.

The example in Figure 3.15 illustrates the closure operation.

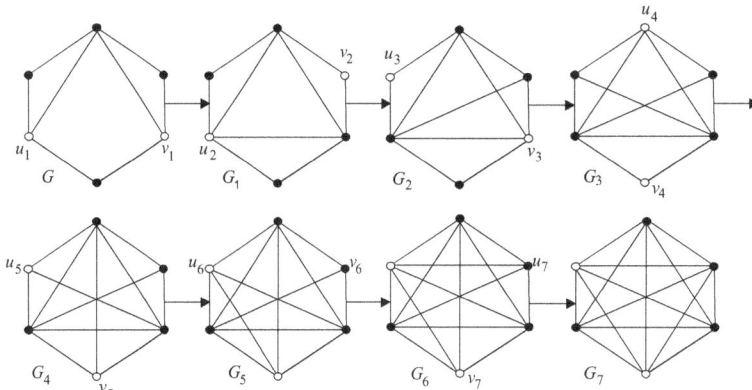

**Fig. 3.15**

We observe in this example that there are different choices of pairs of non-adjacent vertices $u$ and $v$ with $d(u) + d(v) \geq n$. Therefore, the closure procedure can be carried out in several different ways and each different way gives the same result.

Now, in the graph shown in Figure 3.16, $n = 7$ and $d(u) + d(v) < 7$, for any pair $u$, $v$ of adjacent vertices. Therefore, $c(G) = G$.

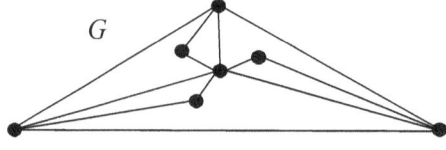

**Fig. 3.16**

The importance of $c(G)$ is given in the following result due to Bondy and Chvatal [36].

**Theorem 3.13 (Bondy and Chvatal [36])** A graph $G$ is Hamiltonian if and only if its closure $c(G)$ is Hamiltonian.

**Proof** Let $c(G)$ be the closure of the graph $G$. Since $c(G)$ is a supergraph of $G$, therefore, if $G$ is Hamiltonian, then $c(G)$ is also Hamiltonian.

Conversely, let $c(G)$ be Hamiltonian. Let $G$, $G_1$, $G_2$, ..., $G_{k-1}$, $G_k = c(G)$ be the sequence of graphs obtained by performing the closure procedure on $G$. Since $c(G) = G_k$ is obtained from $G_{k-1}$ by setting $G_k = G_{k-1} + uv$, where $u$, $v$ is a pair of non adjacent vertices in $G_{k-1}$ with $d(u) + d(v) \geq n$, therefore, it follows that $G_{k-1}$ is Hamiltonian. Similarly $G_{k-2}$, so $G_{k-3}$, ..., $G_1$ and thus $G$ is Hamiltonian. ❏

**Corollary 3.4** Let $G$ be a graph with $n$ vertices with $n \geq 3$. If $c(G)$ is complete, then $G$ is Hamiltonian.

There can be more than one Hamiltonian cycle in a given graph, but the interest lies in the edge-disjoint Hamiltonian cycles. The following result gives the number of edge-disjoint Hamiltonian cycles in a complete graph with odd number of vertices.

**Theorem 3.14** In a complete graph with $n$ vertices there are $(n-1)/2$ edge-disjoint Hamiltonian cycles, if $n$ is an odd number, $n \geq 3$.

**Proof** A complete graph $G$ of $n$ vertices has $n(n-1)/2$ edges and a Hamiltonian cycle in $G$ contains $n$ edges. Therefore the number of edge-disjoint Hamiltonian cycles in $G$ cannot exceed $(n-1)/2$. When $n$ is odd, we show there are $(n-1)/2$ edge-disjoint Hamiltonian cycles. ❏

The subgraph of a complete graph with $n$ vertices shown in Figure 3.17 is a Hamiltonian cycle. Keeping the vertices fixed on a circle, rotate the polygonal pattern clockwise by $\frac{360}{n-1}$, $2 \cdot \frac{360}{n-1}$, ..., $\frac{n-3}{2} \cdot \frac{360}{n-1}$ degrees. We see that each rotation produces a Hamiltonian cycle that has no edge in common with any of the previous ones. Therefore, there are $(n-3)/2$ new Hamiltonian cycles, all disjoint from the one in Figure 3.17, and also edge-disjoint among themselves. Thus there are $(n-1)/2$ edge disjoint Hamiltonian cycles.

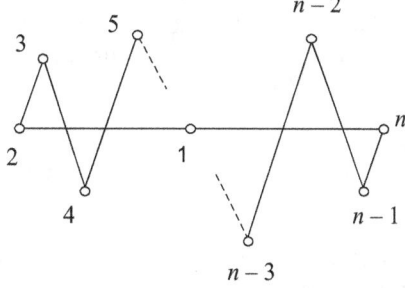

**Fig. 3.17**

The next result involving degrees give the sufficient conditions for a graph to be Hamiltonian.

**Theorem 3.15**   Let $D = [d_i]_1^n$ be a degree sequence of a graph $G = (V, E)$, $d_1 \leq d_2 \leq \ldots \leq d_n$. Each of the following gives the sufficient conditions for $G$ to be Hamiltonian.

A. $1 \leq k \leq n \Rightarrow d_k \geq \dfrac{n}{2}$, (Dirac [68])

B. $uv \notin E \Rightarrow d(u) + d(v) \geq n$, (Ore [176])

C. $1 \leq k \leq \frac{n}{2} \Rightarrow d_k > k$, (Posa [210]).

D. $j < k$, $d_j \leq j$ and $d_k \leq k - 1 \Rightarrow d_j + d_k \geq n$, (Bondy [33])

E. $d_k \leq k < \frac{n}{2} \Rightarrow d_{n-k} \geq n - k$, (Chvatal [59])

F. For every $i$ and $j$ with $1 \leq i \leq n$, $1 \leq j \leq n$, $i + j \geq n$, $v_i v_j \notin E$, $d(v_i) \leq i$ and $d(v_j) \leq j - 1 \Rightarrow d(v_i) + d(v_j) \geq n$, (Las Vergnas [256].

G. $c(G)$ is complete. (Bondy and Chvatal [36]).

**Proof**   We first prove that

$$A \overset{(i)}{\Rightarrow} B \overset{(ii)}{\Rightarrow} C \overset{(iii)}{\Rightarrow} D \overset{(iv)}{\Rightarrow} E \overset{(v)}{\Rightarrow} F \overset{(vi)}{\Rightarrow} G.$$

i. This can be easily established.

ii. Assume that (C) is not true, so that there exists $a$ $k$ with $1 \leq k < \frac{n}{2}$ and $d_k \leq k$. Then the induced subgraph on the vertices $v_1$, $v_2$, ..., $v_k$ is a complete graph. For, if there are vertices $i$ and $j$ with $1 \leq i < j \leq k$ and $v_i v_j \notin E$, then $d_i + d_j \leq 2d_k < n$, contradicting (B). Since $d_k \leq k$, each $v_i$, $1 \leq i \leq k$, is adjacent to at most one $v_j$, $k + 1 \leq j \leq n$. Also, $n - k > k$, because $k < \frac{n}{2}$. Therefore, there is a vertex $v_j$, $k + 1 \leq j \leq n$ not adjacent to any of the vertices $v_1$, $v_2$, ..., $v_k$. For this $v_j$, we have $d_j \leq n - k - 1$. But then $d_j + d_k \leq (n - k - 1) + k = n - 1$. Thus there is a $v_j v_k \notin E$ with $d_j + d_k \leq n - 1$, contradicting (B). Hence proving (ii).

iii. Assume that (D) is not true, so that there exist $j$ and $k$ with $j < k$, $d_j \leq j$, $d_k \leq k - 1$ and $d_j + d_k < n$. This gives $i = d_j < \frac{n}{2}$. But then $d_j \leq j$ gives $d_{d_j} \leq d_j$, since the sequence is non-decreasing. Therefore, $d_i \leq d_j = i$. Thus there is an $i$, $1 \leq i \leq \frac{n}{2}$ with $d_i \leq i$, contradicting (C). This proves (iii).

iv. If (E) is not true, there is a $k$ with $d_k \leq k < \frac{n}{2}$ and $d_{n-k} \leq n - k - 1$. Then $d_k + d_{n-k} \leq n - 1$. Setting $n - k = j$, we have $k < j$, $d_k \leq k$, $d_j \leq j - 1$ and $d_j + d_k \leq n - 1$. This contradicts (D) and so (iv) is proved.

v. Assume that (F) is not true, so that there is a pair of vertices $v_i$ and $v_j$, $i < j$ with $v_i v_j \notin E$ and violating (F). Choose $i$ to be the least such possible integer. Then by minimality of $i$, $d_{i-1} > i - 1$. Thus $d_i \geq d_{i-1} \geq i$ and since $d_i \leq i$, we obtain $d_i = i$. If

$i \geq \frac{n}{2}$, we get $d_i + d_j \geq 2d_i \geq n$, contradicting the violation of (F). Therefore, $i < \frac{n}{2}$. Thus there is an $i$, $1 \leq i < \frac{n}{2}$ with $d_i = i$. Now, if (E) is satisfied, we have $d_{n-i} \geq n-i$ and since $j \geq n-i$, we obtain $d_j \geq d_{n-i} \geq n-i$. By minimality of $i$, $d_i = i$ and we have $d_j + d_i \geq (n-i) + i = n$, again contradicting the violation of (F). Thus negation of (F) implies negation of (E) and (V) is established.

vi. Assume that $c(G) = H$ is not complete. Let $v_i$ and $v_j$ be non-adjacent vertices in $H$ such that (a) $j$ is as large as possible and (b) $i$ is as large as possible subject to (a). Then $i < j$, and since $H$ is the closure of $G$, therefore

$$d(v_i|H) + d(v_j|H) \leq n-1, \qquad (3.15.1)$$

$$d(v_i|G) + d(v_j|G) \leq n-1.$$

Now, by the choice of $j$, $v_i$ is adjacent in $H$ to all $v_k$ with $k > j$, so that

$$d(v_i|H) \geq n-j. \qquad (3.15.2)$$

Again, by the choice of $i$, $v_j$ is adjacent in $H$ to all $v_k$ with $k > i$, $k \neq j$, so that

$$d(v_j|H) \geq n-i-1. \qquad (3.15.3)$$

From (3.15.1) and (3.15.2), we have

$$d(v_j|G) \leq d(v_j|H) \leq (n-1) - (n-j) = j-1.$$

From (3.15.1) and (3.15.3), we have

$$d(v_j|G) \leq d(v_i|H) \leq (n-1) - (n-i-1) = i.$$

From (3.15.2) and (3.15.3), we have

$$i + j \geq (2n-1) - d(v_i|H) - d(v_j|H) \geq n. \qquad \text{(using (1))}$$

Therefore, $i$ and $j$ contradict the given conditions. Thus $H = c(G)$ is complete. This proves (vi). $\qquad \qquad \square$

Now, by Theorem 3.13 it follows that if (G) holds, then $G$ is Hamiltonian.

The next result is due to Nash-Williams [168].

**Theorem 3.16 (Nash-Williams [168])** Every $k$-regular graph on $2k+1$ vertices is Hamiltonian.

**Proof** Let $G$ be a $k$-regular graph on $2k+1$ vertices. Add a new vertex $w$ and join it by an edge to each vertex of $G$. The resulting graph $H$ on $2k+2$ vertices has $\delta = k+1$. Thus by Theorem 3.15 (A), $H$ is Hamiltonian. Removing $w$ from $H$, we get a Hamiltonian path, say $v_0 v_1 \ldots v_{2k}$.

Assume that $G$ is not Hamiltonian, so that (a) if $v_0v_i \in E$, then $v_{i-1}v_{2k} \notin E$, (b) if $v_0v_i \notin E$, then $v_{i-1}v_{2k} \in E$, since $d(v_0) = d(v_{2k}) = k$.

Now, the following cases arise.

**Case (i)** $v_0$ is adjacent to $v_1, v_2, \ldots, v_k$ and $v_{2k}$ is adjacent to $v_k, v_{k+1}, \ldots, v_{2k-1}$. Then there is an $i$ with $1 \leq i \leq k$ such that $v_i$ is not adjacent to some $v_j$ for $0 \leq j \leq k (j \neq i)$. But $d(v_i) = k$. So, $v_i$ is adjacent to $v_j$ for some $j$ with $k+1 \leq j \leq 2k-1$. Then the cycle $C$ given by $v_iv_{i-1}\ldots v_0v_{i+1}\ldots v_{j-1}v_{2k}v_{2k+1}\ldots v_j$ is a Hamiltonian cycle of $G$ (Fig 3.18).

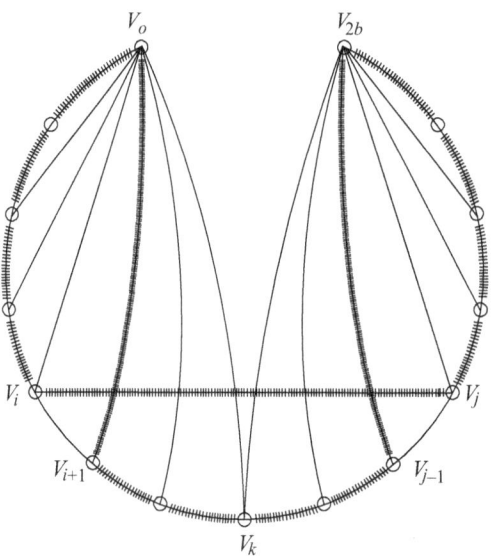

**Fig. 3.18**

**Case (ii)** There is an $i$ with $1 \leq i \leq 2k-1$ such that $v_{i+1}v_0 \in E$, but $v_iv_0 \notin E$. Then by (b), $v_{i-1}v_{2k} \in E$. Thus, $G$ contains the $2k$-cycle $v_{i-1}v_{i-2}\ldots v_0v_{i+1}$. Renaming the $2k$-cycle $C$ as $u_1u_2\ldots u_{2k}$ and let $u_0$ be the vertex of $G$ not on $C$. Then $u_0$ cannot be adjacent to two consecutive vertices on $C$ and hence $u_0$ is adjacent to every second vertex on $C$, say $u_1, u_3, \ldots, u_{2k-1}$. Replacing $u_{2i}$ by $u_0$, we obtain another maximum cycle $C'$ of $G$ and hence $u_{2i}$ must be adjacent to $u_1, u_3, \ldots, u_{2k-1}$. But then $u_1$ is adjacent to $u_0, u_2, \ldots, u_{2k}$, implying $d(u_1) \geq k+1$. This is a contradiction and hence $G$ is Hamiltonian. ❑

## 3.7 Pancyclic Graphs

**Definition:** A graph $G$ of order $n (\geq 3)$ is *pancyclic* if $G$ contains all cycles of lengths from 3 to $n$. $G$ is called *vertex-pancyclic* if each vertex $v$ of $G$ belongs to a cycle of every length $\ell$, $3 \leq \ell \leq n$.

**Example** Clearly, a vertex-pancyclic graph is pancyclic. However, the converse is not true. Figure 3.19 displays a pancyclic graph that is not vertex-pancyclic.

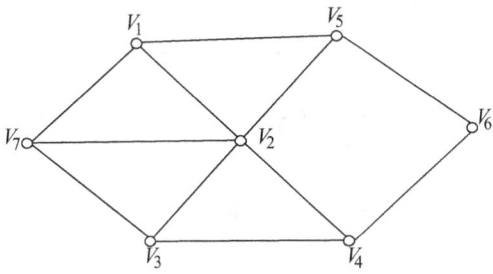

**Fig. 3.19**

The result of pancyclic graphs was initiated by Bondy [34], who showed that Ore's sufficient condition for a graph $G$ to be Hamiltonian (Theorem 6.2.5) actually implies much more. Note that if $\delta \geq \frac{n}{2}$, then $m \geq \frac{n^2}{2}$. The proof of the following result due to Thomassen can be found in Bollobas [29].

**Theorem 3.17** Let $G$ be a simple Hamiltonian graph on $n$ vertices with at least $\left\lceil \frac{n^2}{2} \right\rceil$ edges. Then $G$ is either pancyclic or else is the complete bipartite graph $K_{\frac{n}{2}, \frac{n}{2}}$. In particular, if $G$ is Hamiltonian and $m > \frac{n^2}{4}$, then $G$ is pancyclic.

**Proof** The result can easily be verified for $n = 3$. We may therefore assume that $n \geq 4$. We apply induction on $n$. Suppose the result is true for all graphs of order at most $n - 1$ ($n \geq 4$), and let $G$ be a graph of order $n$.

First, assume that $G$ has a cycle $C = v_0 v_1 \dots v_{n-2} v_0$ of length $n - 1$. Let $v$ be the (unique) vertex of $G$ not belonging to $C$. If $d(v) \geq \frac{n}{2}$, $v$ is adjacent to two consecutive vertices on $C$ and hence $G$ has a cycle of length 3. Suppose for some $r$, $2 \leq r \leq \frac{n-1}{2}$, $C$ has no pair of vertices $u$ and $w$ on $C$ adjacent to $v$ in $G$ with $d_C(u, w) = r$. Then if $v_{i_1}, v_{i_2}, \dots v_{i_{d(v)}}$ are the vertices of $C$ that are adjacent to $v$ in $G$ (recall that $C$ contains all the vertices of $G$ except $v$), then $v_{i_1+r}, v_{i_2+r}, \dots, v_{i_{d(v)}+r}$ are nonadjacent to $v$ in $G$, where the suffixes are taken modulo $(n-1)$. Thus, $2d(v) \leq n - 1$, a contradiction. Hence, for each $r$, $2 \leq r \leq \frac{n-1}{2}$, $C$ has a pair of vertices $u$ and $w$ on $C$ adjacent to $v$ in $G$ with $d_C(u, w) = r$. Thus for each $r$, $2 \leq r \leq \frac{n-1}{2}$, $G$ has a cycle of length $r + 2$ as well as a cycle of length $n - 1 - r + 2 = n - r + 1$ (Fig. 3.20). Thus $G$ is pancyclic.

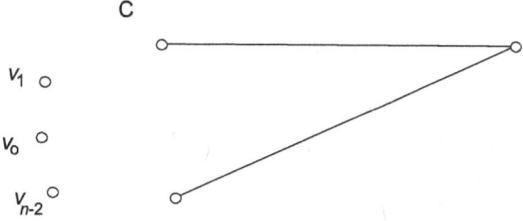

**Fig. 3.20**

If $d(v) \leq \frac{n-1}{2}$, then $G[V(C)]$, the subgraph of $G$ induced by $V(C)$ has at least $\frac{n^2}{4} - \frac{n-1}{2} > \frac{(n-1)^2}{4}$ edges. So, by the induction assumption, $G[V(C)]$ is pancyclic and hence $G$ is pancyclic. (By hypothesis, $G$ is Hamiltonian).

Next, assume that $G$ has no cycle of length $n - 1$. Then $G$ is not pancyclic. In this case, we show that $G$ is $K_{\frac{n}{2}, \frac{n}{2}}$.

Let $C = v_0 v_1 v_2 \ldots v_{n-1} v_0$ be a Hamilton cycle of $G$. We claim that of the two pairs $v_i v_k$ and $v_{i+1} v_{k+2}$ (where suffixes are taken modulo $n$), at most only one of them can be an edge of $G$. Otherwise, $v_k v_{k-1} v_{k-2} \ldots v_{i+1} v_{k+2} v_{k+3} v_{k+4} \ldots v_i v_k$ is an $(n-1)$-cycle in $G$, a contradiction. Hence, if $d(v_i) = r$, then there are $r$ vertices adjacent to $v_i$ in $G$ and hence at least $r$ vertices (including $v_{i+1}$ since $v_i v_{i-1} \in E(G)$) that are nonadjacent to $v_{i+1}$. Thus, $d(v_{i+1}) \leq n - r$ and $d(v_i) + d(v_{i+1}) \leq n$.

Summing the last inequality over $i$ from 0 to $n - 1$, we get $4m \leq n^2$. But by hypothesis, $4m \geq n^2$. Hence, $m = \frac{n^2}{4}$ and so $n$ must be even.

This gives $d(v_i) + d(v_{i+1}) = n$ for each $i$, and thus, for each $i$ and $k$, exactly one of $v_i v_k$ and $v_{i+1} v_{k+2}$ is an edge of $G$.                 (3.17.1)

Thus, if $G \neq K_{\frac{n}{2}, \frac{n}{2}}$, then certainly there exist $i$ and $j$ such that $v_i v_j \in E$ and $i \equiv j \pmod 2$. Hence for some $j$, there exists an even positive integer $s$ such that $v_{j+1} v_{j+1+s} \in E$. Choose $s$ to be the least even positive integer with the above property. Then $v_j v_{j+1+s} \notin E$. Hence, $s \geq 4$ (as $s = 2$ would mean that $v_j v_{j+1} \notin E$). Again, by (3.17.1), $v_{j-1} v_{j+s-3} = v_{j-1} v_{j-1+(s-2)} \in E(G)$ contradicting the choice of $s$. Thus, $G = K_{\frac{n}{2}, \frac{n}{2}}$. The last part follows from the fact that $|E(K_{\frac{n}{2}, \frac{n}{2}},)| = \frac{n^2}{4}$.          ❑

**Theorem 3.18**    Let $G \neq K_{\frac{n}{2}, \frac{n}{2}}$, be a simple graph with $n \geq 3$ vertices and let $d(u) + d(v) \geq n$ for every pair of non-adjacent vertices of $G$. Then $G$ is pancyclic.

**Proof**    By Ore's Theorem (Theorem 3.12), $G$ is Hamiltonian. We show that $G$ is pancyclic by first proving that $m \geq \frac{n^2}{4}$ and then invoking Theorem 3.17. This is true if $\delta \geq \frac{n}{2}$ (as $2m = \sum_{i=1}^{n} d_i \geq \delta n \geq n^2/2$). So assume that $\delta < \frac{n}{2}$.

Let $S$ be the set of vertices of degree $\delta$ in $G$. For every pair $(u, v)$ of vertices of degree $\delta$, $d(u) + d(v) < \frac{n}{2} + \frac{n}{2} = n$. Hence by hypothesis, $S$ induces a clique of $G$ and $|S| \leq \delta + 1$. If $|S| = \delta + 1$, then $G$ is disconnected with $G[S]$ as a component, which is impossible (as $G$ is Hamiltonian). Thus, $|S| \leq \delta$. Further, if $v \in S$, $v$ is nonadjacent to $n - 1 - \delta$ vertices of $G$. If $u$ is such a vertex, $d(v) + d(u) \geq n$ implies that $d(u) \geq n - \delta$. Further, $v$ is adjacent to at least one vertex $w \notin S$ and $d(w) \geq \delta + 1$, by the choice of $S$. These facts give that

$$2m = \sum_{i=1}^{n} d_i \geq (n - \delta - 1)(n - \delta) + \delta^2 + (\delta + 1),$$

where the last $(\delta + 1)$ comes out of the degree of $w$. Thus,

$$2m \geq n^2 - n(2\delta + 1) + 2\delta^2 + 2\delta + 1,$$

which implies that

$$4m \geq 2n^2 - 2n(2\delta + 1) + 4\delta^2 + 4\delta + 2$$

$$= (n - (2\delta + 1)^2 + n^2 + 1$$

$$\geq n^2 + 1, \text{ since } n > 2\delta.$$

Consequently, $m > \dfrac{n^2}{4}$, and by Theorem 3.17, $G$ is pancyclic.  ❏

## 3.8  Exercises

1. Prove that the wheel $W_n$ is Hamiltonian for every $n \geq 2$, and $n$-cube $Q_n$ is Hamiltonian for each $n \geq 2$.

2. If $G$ is a $k$-regular graph with $2k - 1$ vertices, then prove that $G$ is Hamiltonian.

3. Show that if a cubic graph $G$ has a spanning closed walk, then $G$ is Hamiltonian.

4. If $G = G(X, Y)$ is a bipartite Hamiltonian graph, then show that $|X| = |Y|$.

5. Prove that for each $n \geq 1$, the complete tripartite graph $K_{n, 2n, 3n}$ is Hamiltonian, but $K_{n, 2n, 3n+1}$ is not Hamiltonian.

6. How many spanning cycles are there in the complete bipartite graphs $K_{3, 3}$ and $K_{4, 3}$?

7. Prove that a graph $G$ with $n \geq 3$ vertices is arbitrarily traceable if and only if it is one of the graphs $C_n$, $K_n$ or $K_{n, n}$ with $n = 2p$.

8. Prove that a graph $G$ with $n \geq 3$ vertices is randomly traceable if and only if it is randomly Hamiltonian.

9. Find the closure of the graph given in Figure 3.2. Is it Hamiltonian?

10. Does there exist an Eulerian graph with

    i. an even number of vertices and an odd number of edges,
    ii. and odd number of vertices and an even number of edges.

    Draw such a graph if it exists.

11. Characterise graphs which are both Eulerian and Hamiltonian.

12. Characterise graphs which possess Hamiltonian paths but not Hamiltonian cycles.

13. Characterise graphs which are unicursal but not Eulerian.

14. Give an example of a graph which is neither pancyclic nor bipartite, but whose $n$-closure is complete.

# 4. Trees

One of the important classes of graphs is the trees. The importance of trees is evident from their applications in various areas, especially theoretical computer science and molecular evolution.

## 4.1 Basics

**Definition:** A graph having no cycles is said to be *acyclic*. A *forest* is an acyclic graph.

**Definition:** A *tree* is a connected graph without any cycles, or a tree is a connected acyclic graph. The edges of a tree are called *branches*. It follows immediately from the definition that a tree has to be a simple graph (because self-loops and parallel edges both form cycles). Figure 4.1(a) displays all trees with fewer than six vertices.

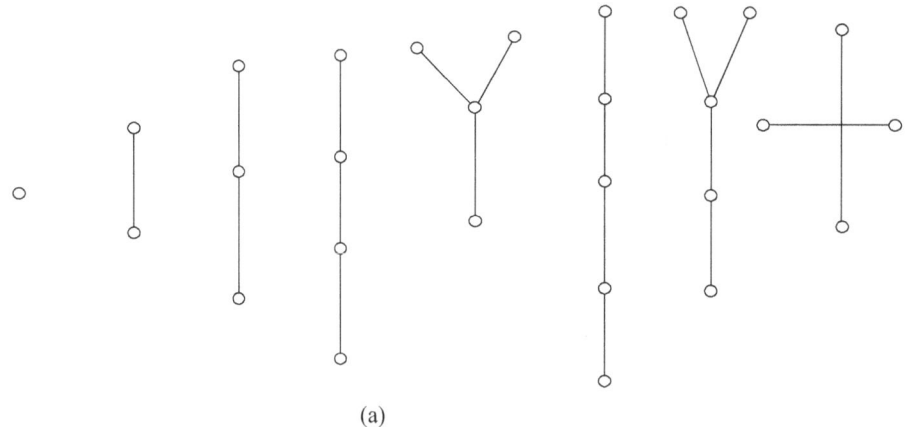

(a)

**Fig. 4.1(a)**

The following result characterises trees.

**Theorem 4.1**   A graph is a tree if and only if there is exactly one path between every pair of its vertices.

**Proof** Let $G$ be a graph and let there be exactly one path between every pair of vertices in $G$. So $G$ is connected. Now $G$ has no cycles, because if $G$ contains a cycle, say between vertices $u$ and $v$, then there are two distinct paths between $u$ and $v$, which is a contradiction. Thus $G$ is connected and is without cycles; therefore it is a tree.

Conversely, let $G$ be a tree. Since $G$ is connected, there is at least one path between every pair of vertices in $G$. Now, let there are two distinct paths between two vertices $u$ and $v$ of $G$. The union of these two paths contains a cycle which contradicts the fact that $G$ is a tree. Hence, there is exactly one path between every pair of vertices of a tree. ❑

The next two results give alternative methods for defining trees.

**Theorem 4.2** A tree with n vertices has $n - 1$ edges.

**Proof** We prove the result by using induction on $n$, the number of vertices. The result is obviously true for $n = 1$, 2 and 3. Let the result be true for all trees with fewer than $n$ vertices. Now, let $T$ be a tree with $n$ vertices and let $e$ be an edge with end vertices $u$ and $v$. So, the only path between $u$ and $v$ is $e$. Therefore, deletion of $e$ from $T$ disconnects $T$. Now, $T - e$ consists of exactly two components $T_1$ and $T_2$ say, and as there were no cycles to begin with, each component is a tree. Let $n_1$ and $n_2$ be the number of vertices in $T_1$ and $T_2$ respectively, so that $n_1 + n_2 = n$. Also, $n_1 < n$ and $n_2 < n$. Thus, by induction hypothesis, number of edges in $T_1$ and $T_2$ are respectively $n_1 - 1$ and $n_2 - 1$. Hence, number of edges in $T = n_1 - 1 + n_2 - 1 + 1 = n_1 + n_2 - 1 = n - 1$. ❑

**Theorem 4.3** Any connected graph with $n$ vertices and $n - 1$ edges is a tree.

**Proof** Let $G$ be a connected graph with $n$ vertices and $n - 1$ edges. We show that $G$ contains no cycles. Assume to the contrary that $G$ contains cycles.

Remove an edge from a cycle so that the resulting graph is again connected. Continue this process of removing one edge from one cycle at a time till the resulting graph $H$ is a tree. As $H$ has $n$ vertices, so number of edges in $H$ is $n - 1$. Now, the number of edges in $G$ is greater than the number of edges in $H$. So $n - 1 > n - 1$, which is not possible. Hence, $G$ has no cycles and therefore is a tree. ❑

**Definition:** A graph is said to be *minimally connected* if removal of any one edge from it disconnects the graph. Clearly, a minimally connected graph has no cycles.

Here is the next characterisation of trees.

**Theorem 4.4** A graph is a tree if and only if it is minimally connected.

**Proof** Let the graph $G$ be minimally connected. Then $G$ has no cycles and therefore is a tree.

Conversely, let $G$ be a tree. Then $G$ contains no cycles and deletion of any edge from $G$ disconnects the graph. Hence, $G$ is minimally connected. ❑

The following results give some more properties of trees.

**Theorem 4.5**   A graph $G$ with $n$ vertices, $n-1$ edges and no cycles is connected.

**Proof**   Let $G$ be a graph without cycles with $n$ vertices and $n-1$ edges. We have to prove that $G$ is connected. Assume that $G$ is disconnected. So $G$ consists of two or more components and each component is also without cycles. We assume without loss of generality that $G$ has two components, say $G_1$ and $G_2$ (Fig. 4.1(b)). Add an edge $e$ between a vertex $u$ in $G_1$ and a vertex $v$ in $G_2$. Since there is no path between $u$ and $v$ in $G$, adding $e$ did not create a cycle. Thus $G \cup e$ is a connected graph (tree) of $n$ vertices, having $n$ edges and no cycles. This contradicts the fact that a tree with $n$ vertices has $n-1$ edges. Hence $G$ is connected.

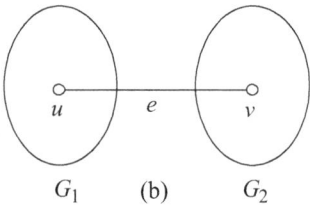

$G_1$        (b)        $G_2$

**Fig. 4.1(b)**

**Theorem 4.6**   Any tree with at least two vertices has at least two pendant vertices.

**Proof**   Let the number of vertices in a given tree $T$ be $n(n > 1)$. So the number of edges in $T$ is $n-1$. Therefore the degree sum of the tree is $2(n-1)$. This degree sum is to be divided among the $n$ vertices. Since a tree is connected it cannot have a vertex of 0 degree. Each vertex contributes at least 1 to the above sum. Thus there must be at least two vertices of degree exactly 1.

**Second proof**   We use induction on $n$. The result is obviously true for all trees having fewer than $n$ vertices. We know that $T$ has $n-1$ edges, and if every edge of $T$ is incident with a pendant vertex, then $T$ has at least two pendant vertices, and the proof is complete. So let there be some edge of $T$ that is not incident with a pendant vertex and let this edge be $e = uv$ (Fig. 4.2). Removing the edge $e$, we see that the graph $T - e$ consists of a pair of trees say $T_1$ and $T_2$ with each having fewer than $n$-vertices. Let $u \in V(T_1)$, $v \in V(T_2)$, and $|V(T_1)| = n_1$, $|V(T_2)| = n_2$. Applying induction hypothesis on both $T_1$ and $T_2$, we observe that each of $T_1$ and $T_2$ has two pendant vertices. This shows that each of $T_1$ and $T_2$ has at least one pendant vertex that is not incident with the edge $e$. Thus the graph $T - e + e = T$ has at least two pendant vertices.

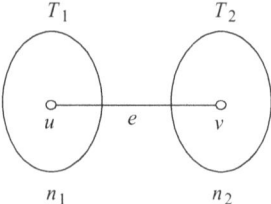

$T_1$                $T_2$

$n_1$                $n_2$

**Fig. 4.2**

**Third proof**    Let $T$ be a tree with $n(n > 1)$ vertices. The number of edges in $T$ is $n - 1$ and the sum of degrees in $T$ is $2(n - 1)$, that is, $\sum d_i = 2(n - 1)$. Assume $T$ has exactly one vertex $v_1$ of degree one, while all the other $n - 1$ vertices have degree $\geq 2$. Then sum of degrees is $d(v_1) + d(v_2) + \ldots + d(v_n) \geq 1 + 2 + 2 + \ldots + 2 = 1 + 2(n - 1)$. So, $2(n - 1) \geq 1 + 2(n - 1)$, implying $0 \geq 1$, which is absurd. Hence $T$ has at least two vertices of degree one.    ❏

The following result characterises tree degree sequences.

**Theorem 4.7**    The sequence $[d_i]_1^n$ of positive integers is a degree sequence of a tree if and only if

$$\text{(i) } d_i \geq 1 \text{ for all } i, 1 \leq i \leq n \text{ and (ii) } \sum_{i=1}^{n} d_i = 2n - 2.$$

**Proof**

*Necessity*    Since a tree has no isolated vertex, therefore $d_i \geq 1$ for all $i$. Also, $\sum_{i=1}^{n} d_i = 2(n - 1)$, as a tree with $n$ vertices has $n - 1$ edges.

*Sufficiency*    We use induction on $n$. For $n = 2$, the sequence is $[1, 1]$ and is obviously the degree sequence of $K_2$. Suppose the claim is true for all positive sequences of length less than $n$.

Let $[d_i]_1^n$ be the non-decreasing positive sequence of $n$ terms, satisfying conditions (i) and (ii). Then, $d_1 = 1$ and $d_n > 1$ (by Theorem 4.5).

Now, consider the sequence $D' = [d_2, d_3, \ldots, d_{n-1}, d_n - 1]$, which is a sequence of length $n - 1$. Obviously in $D'$, $d_i \geq 1$ and $\sum d_i = d_2 + d_3 + \ldots + d_{n-1} + d_n - 1 = d_1 + d_2 + d_3 + \ldots + d_{n-1} + d_n - 1 - 1 = 2n - 2 - 2 = 2(n - 1) - 2$ (because $d_1 = 1$). So $D'$ satisfies conditions (i) and (ii), and by induction hypothesis there is a tree $T'$ realising $D'$. Now in $T'$, add a new vertex and join it to the vertex having degree $d_n - 1$ to get a tree $T$. Therefore the degree sequence of $T$ is $[d_1, d_2, \ldots, d_n]$.

**Theorem 4.8**    A forest of $k$ trees which have a total of $n$ vertices has $n - k$ edges.

**Proof**    Let $G$ be a forest and $T_1, T_2, \ldots, T_k$ be the $k$ trees of $G$. Let $G$ have $n$ vertices and $T_1, T_2, \ldots, T_k$ have respectively $n_1, n_2, \ldots, n_k$ vertices. Then, $n_1 + n_2 + \ldots + n_k = n$. Also, the number of edges in $T_1, T_2, \ldots, T_k$ are respectively $n_1 - 1, n_2 - 1, \ldots, n_k - 1$. Thus number of edges in $G = n_1 - 1 + n_2 - 1 + \ldots + n_k - 1 = n_1 + n_2 + \ldots + n_k - k = n - k$.    ❏

Now we give an interesting result about trees as subgraphs.

**Theorem 4.9**    Let $T$ be a tree with $k$ edges. If $G$ is a graph whose minimum degree satisfies $\delta(G) \geq k$, then $G$ contains $T$ as a subgraph. Alternatively, $G$ contains every tree of order atmost $\delta(G) + 1$ as a subgraph.

**Proof**    We use induction on $k$. If $k = 0$, then $T = K_1$ and it is clear that $K_1$ is a subgraph of any graph. Further, if $k = 1$, then $T = K_2$ and $K_2$ is a subgraph of any graph whose minimum degree is one. Assume that the result is true for all trees with $k-1$ edges $(k \geq 2)$ and consider a tree $T$ with exactly $k$ edges. We know that $T$ contains at least two pendant vertices. Let $v$ be one of them and let $w$ be the vertex that is adjacent to $v$. Consider the graph $T-v$. Since $T-v$ has $k-1$ edges, the induction hypothesis applies, so $T-v$ is a subgraph of $G$. We think of $T-v$ as actually sitting inside $G$ (meaning $w$ is a vertex of $G$, too). Now, since $G$ contains at least $k+1$ vertices, and $T-v$ contains $k$ vertices, there exist vertices of $G$ that are not a part of the subgraph $T-v$. Further, since the degree in $G$ of $w$ is at least $k$, there must be a vertex $u$ not in $T-v$ that is adjacent to $w$. The subgraph $T-v$ together with $u$ forms the tree $T$ as a subgraph of $G$ (Fig. 4.3).                    ❑

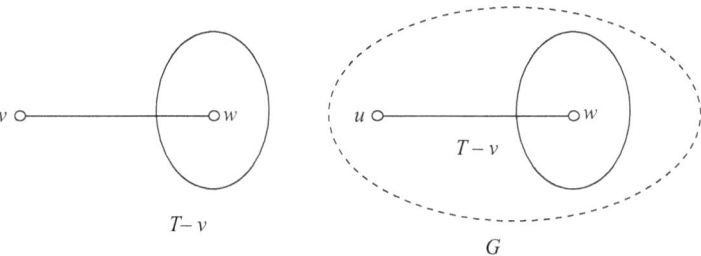

**Fig. 4.3**

**Distance and centers in a tree:** The distance $d(u, v)$ between any two vertices $u$ and $v$ in a connected graph is the length of the shortest path between them. For example, consider the graph of Figure 4.4(a). The paths between $u$ and $v$ are $uw_1 w_2 v$, $u w_4 w_5 w_6 v$ and $uw_7 w_8 v$, which is the shortest. Here, $d(u, v) = 3$.

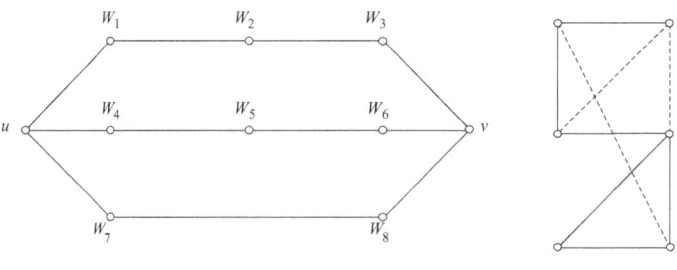

**Fig. 4.4(a)**

A function $f(x, y)$ is said to be a *metric* if it satisfies the following.

1. Non-negativity: $f(x, y) \geq 0$ and $f(x, y) = 0$ if and only if $x = y$.

2. Symmetry: $f(x, y) = f(y, x)$.

3. Triangle inequality: $f(x, y) \leq f(x, z) + f(z, y)$, for any $z$.

Clearly, the distance between vertices of a connected graph is a metric.

**Definition:** The *eccentricity* $e(v)$ of a vertex in a graph $G$ is the distance from $v$ to the vertex farthest from $v$ in $G$.

i.e., $e(v) = \max\limits_{u \in G} d(u, v)$.

A vertex with minimum eccentricity in a graph $G$ is called the *center* of $G$ and the minimum eccentricity is the *radius* of $G$, while a vertex with maximum eccentricity is called the peripheral vertex of $G$ and the maximum eccentricity is the *diameter* of $G$.

## 4.2 Rooted and Binary Trees

A tree in which one vertex (called the *root*) is distinguished from all the others is called a *rooted tree*.

A *binary tree* is defined as a tree in which there is exactly one vertex of degree two and each of the remaining vertices is of degree one or three. Obviously, a binary tree has three or more vertices. Since the vertex of degree two is distinct from all other vertices, it serves as a root, and so every binary tree is a rooted tree. ❑

Below are given some properties of binary trees.

**Theorem 4.10** Every binary tree has an odd number of vertices.

**Proof** Apart from the root, every vertex in a binary tree is of odd degree. We know that there are even number of such odd vertices. Therefore when the root (which is of even degree) is added to this number, the total number of vertices is odd.

**Corollary 4.1** There are $\frac{1}{2}(n+1)$ pendant vertices in any binary tree with $n$ vertices.

**Proof** Let $T$ be a binary tree with $n$ vertices. Let $q$ be the number of pendant vertices in $T$. Therefore there are $n-q$ internal vertices in $T$ and so $n-q-1$ vertices of degree 3. Thus, the number of edges in $T = \frac{1}{2}[3(n-q-1)+2+q]$. But the number of edges in $T$ is $n-1$. Hence, $\frac{1}{2}[3(n-q-1)+2+q] = n-1$, so that $q = \frac{1}{2}(n+1)$. ❑

The following result is due to Jordan [122].

**Theorem 4.11 (Jordan [122])** Every tree has either one or two centers.

**Proof** The maximum distance, $\max d(v, v_i)$ from a given vertex $v$ to any other vertex occurs only when $v_i$ is a pendant vertex. With this observation, let $T$ be a tree having more than two vertices. Tree $T$ has two or more pendant vertices. Deleting all the pendant vertices from $T$, the resulting graph $T'$ is again a tree. The removal of all pendant vertices from $T$ uniformly reduces the eccentricities of the remaining vertices (vertices in $T'$) by one. Therefore the centers of $T$ are also the centers of $T'$. From $T'$ we remove all pendant vertices and get another tree $T''$. Continuing this process, we either get a vertex, which is a center of $T$, or an edge whose end vertices are the two centers of $T$.

**Definition:** Trees with center $K_1$ are called *unicentral* and trees with center $K_2$ are called *bicentral trees*.

## Spanning trees

A tree is said to be a spanning tree of a connected graph $G$, if $T$ is a subgraph of $G$ and $T$ contains all vertices of $G$.

**Example**   Consider the graph of Fig. 4.4(b), where the bold lines represent a spanning tree.

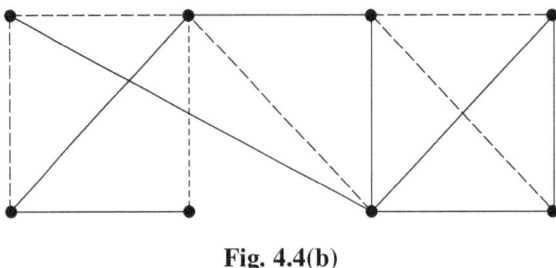

**Fig. 4.4(b)**

❏

The following result shows the existence of spanning trees.

**Theorem 4.12**   Every connected graph has at least one spanning tree.

**Proof**   Let $G$ be a connected graph. If $G$ has no cycles, then it is its own spanning tree. If $G$ has cycles, then on deleting one edge from each of the cycles, the graph remains connected and cycle free containing all the vertices of $G$.

**Definition:**   An edge in a spanning tree $T$ is called a *branch* of $T$. An edge of $G$ that is not in a given spanning tree $T$ is called a *chord*. It may be noted that branches and chords are defined only with respect to a given spanning tree. An edge that is a branch of one spanning tree $T_1$ (in a graph $G$) may be a chord with respect to another spanning tree $T_2$. In Figure 4.5, $u_1 u_2 u_3 u_4 u_5 u_6$ is a spanning tree, $u_2 u_4$ and $u_4 u_6$ are chords.

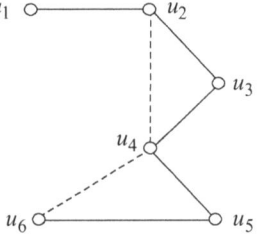

**Fig. 4.5**

A connected graph $G$ can be considered as a union of two subgraphs $T$ and $\overline{T}$, that is $G = T \cup \overline{T}$, where $T$ is a spanning tree, $\overline{T}$ is the complement of $T$ in $G$. $\overline{T}$ being the set of chords is called the *co tree*, or chord set. ❏

The following result provides the number of chords in any graph with a spanning tree.

**Theorem 4.13** With respect to any of its spanning trees, a connected graph of $n$ vertices and $m$ edges has $n - 1$ tree branches and $m - n + 1$ chords.

**Proof** Let $G$ be a connected graph with $n$ vertices and $m$ edges. Let $T$ be the spanning tree. Since $T$ contains all $n$ vertices of $G$, $T$ has $n - 1$ edges and thus the number of chords in $G = m - (n - 1) = m - n + 1$.

**Definition:** Let $G$ be a graph with $n$ vertices, $m$ edges and $k$ components. The *rank r* and *nullity* $\mu$ of $G$ are defined as $r = n - k$ and $\mu = m - n + k$.

Clearly, the rank of a connected graph is $n - 1$ and the nullity is $m - n + 1$.

It can be seen that rank of $G =$ number of branches in any spanning tree (or forest) of $G$. Also, nullity of $G =$ number of chords in $G$. So, rank + nullity = number of edges in $G$.

The nullity of a graph is also called its *cyclomatic number*, or first Betti number.

**Theorem 4.14** If $T$ is a tree with $2k \geq 0$ vertices of odd degree, then $E(T)$ is the union of $k$ pair-wise edge-disjoint paths.

**Proof** We prove the result for every forest $G$, using induction on $k$. If $k = 0$, then $G$ has no pendant vertex and therefore no edge. Let $k > 0$ and let each forest with $2k - 2$ vertices of odd degree has decomposition into $k - 1$ paths. Since $k > 0$, therefore some component of $G$ is a tree with at least two vertices. This component has at least two pendant vertices. Let $P$ be the path connecting two pendant vertices. Deleting $E(P)$ changes the parity of the vertex degree only for the end vertices of $P$ and it makes them even. Thus $G - E(P)$ is a forest with $2k - 2$ vertices of odd degree. So by the induction hypothesis, $G - E(P)$ is the union of $k - 1$ pair wise edge-disjoint paths. These $k - 1$ edge-disjoint paths together with $P$ partition $E(G)$ into $k$ pair wise edge-disjoint paths (Fig. 4.6).

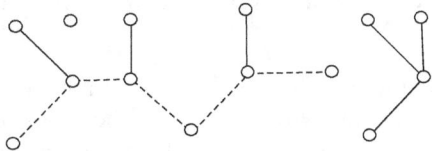

Dotted lines represent path $P$

**Fig. 4.6**

**Theorem 4.15** Let $T$ be a non-trivial tree with the vertex set $S$ and $|S| = 2k$, $k \geq 1$. Then there exists a set of $k$ pairwise edge-disjoint paths whose end vertices are all the vertices of $S$.

**Proof**   Obviously, there exists a set of $k$ paths in $T$ whose end vertices are all the vertices of $S$. Let $P = \{P_1, P_2, \ldots, P_k\}$ be such a set of $k$ paths and let the sum of their lengths be the minimum.

Now we show that the paths of $P$ are pairwise edge-disjoint. Assume to the contrary, and let $P_i$ and $P_j$, $i \neq j$, be paths having an edge in common. Then $P_i$ and $P_j$ have path $P_{ij}$ of length $\geq 1$ in common. Therefore, $P_i \Delta P_j$ the symmetric difference of $P_i$ and $P_j$ is a disjoint union of two paths, say $Q_i$ and $Q_j$, with their end vertices being disjoint pairs of vertices belonging to $S$ (Fig. 4.7).

If $P_i$ and $P_j$ are replaced by $Q_i$ and $Q_j$ in $P$, then the resulting set of paths has the property that their end vertices are all the vertices of $S$ and that the sum of their lengths is less than the sum of the lengths of the paths in $P$. This is a contradiction to the choice of $P$.

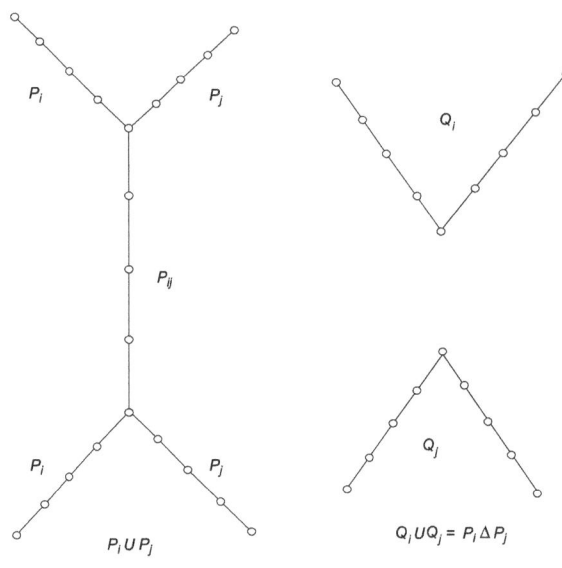

**Fig. 4.7**

**Theorem 4.16**   If $u$ is a vertex of an $n$-vertex tree $T$, then $\displaystyle\sum_{v \in V(T)} d(u, v) \leq \binom{n}{2}$.

**Proof**   Let $T = (V, E)$ be a tree with $|V| = n$. Let $u$ be any vertex of $T$. We use induction on $n$. If $n = 2$, the result is trivial. Now, let $n > 2$. The graph $T - u$ is a forest and let the components of $T - u$ be $T_1, T_2, \ldots, T_k$, where $k \geq 1$. Since $T$ is connected, $u$ has a neighbour in each $T_i$. Also, since $T$ has no cycles, $u$ has exactly one neighbour $v_i$ in each $T_i$. For any $v \in V(T_i)$, the unique $u - v$ path in $T$ passes through $v_i$ and we have $d_T(u, v) = 1 + d_{T_i}(v_i, v)$. Let $n_i = n(T_i)$ (Fig. 4.8). Then, we have

$$\sum_{v \varepsilon V(T_i)} d_T(u, v) = n_i + \sum_{v \varepsilon V(T_i)} d_{T_i}(v_i, v). \tag{4.16.1}$$

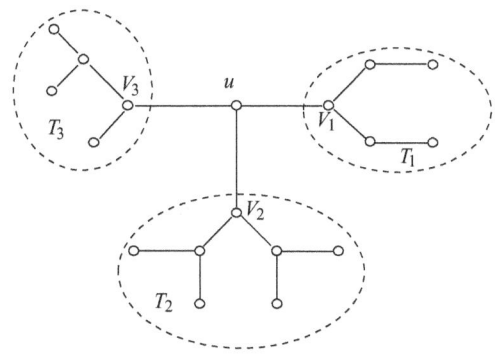

**Fig. 4.8**

By the induction hypothesis, we have

$$\sum_{v \varepsilon V(T_i)} d_{T_i}(v_i, v) \le \binom{n_i}{2}.$$

We now sum the formula (4.16.1) for distances from $u$ over all the components of $T - u$ and we get

$$\sum_{v \varepsilon V(T_i)} d_T(u, v) \le (n-1) + \sum_i \binom{n_i}{2}.$$

Now, we have $\sum_i n_i = n - 1$. Clearly, $\sum_i \binom{n_i}{2} \le \binom{\sum n_i}{2}$, because the right side counts the edges in $K_{\sum n_i}$ or $K_{n-1}$, and the left side counts the edges in a subgraph of $K_{\sum n_i}$, the subgraph being union of disjoint cliques $K_{n_1}, K_{n_1}, \ldots, K_{n_k}$.

Thus, $\displaystyle\sum_{v \varepsilon V(T)} d_T(u, v) \le (n-1) + \binom{n-1}{2} = \binom{n}{2}$.

**Corollary 4.2**    The sum of the distances from a pendant vertex of the path $P_n$ to all other vertices is $\displaystyle\sum_{i=0}^{n-1} i = \binom{n}{2}$.

**Corollary 4.3**    If $H$ is a subgraph of a graph $G$, then $d_G(u, v) \le d_H(u, v)$.

**Proof**    Every $u - v$ path in $H$ appears also in $G$, and $G$ may have additional $u - v$ paths that are shorter than any $u - v$ path in $H$.

**Corollary 4.4**   If $u$ is a vertex of a connected graph $G$, then

$$\sum_{v \varepsilon V(G)} d(u, v) \leq \binom{n(G)}{2}.$$

**Proof**   Let $T$ be a spanning tree of $G$. Then, $d_G(u, v) \leq d_T(u, v)$, so that

$$\sum_{v \varepsilon V(G)} d_G(u, v) \leq \sum_{v \varepsilon V(G)} d_T(u, v) \leq \binom{n(G)}{2}.$$

Thus, $\sum_{v \varepsilon V(G)} d_G(u, v) \leq \binom{n(G)}{2}.$

The sum of the distances over all pairs of distinct vertices in a graph G is the Wiener index $W(G) = \sum_{u, v \varepsilon V(G)} d(u, v)$. On assigning vertices for the atoms and edges for the atomic bonds, we can use graphs to study molecules. Wiener [268] originally used this to study the boiling point of paraffin.

**Theorem 4.17**   Let $v$ be any vertex of a connected graph $G$. Then $G$ has a spanning tree preserving the distances from $v$.

**Proof**   Let $G$ be a connected graph. We find a spanning tree $T$ of $G$ such that for each $u \in V = V(G) = V(T)$, $d_G(v, u) = d_T(v, u)$.
Consider the neighbourhoods of $v$,

$$N_i(v) = \{u \in V : d_G(v, u) = i\}, \ 1 \leq i \leq e, \text{ where } e = e(v).$$

Let $H$ be the graph obtained from $G$ by removing all edges in each $< N_i(v) >$. Clearly, $H$ is connected. Let $< B_i(v) >_H$ denote the induced subgraph of $H$, induced by the ball $B_i(v)$. Clearly, $< B_1(v) >_H$ does not contain any cycle. If $< B_2(v) >_H$ contains cycles, remove edges from $[N_1(v), N_2(v)]$ sequentially, one edge from each cycle, till it becomes acyclic. Proceeding successively by removing edges from $[N_i(v), N_{i+1}(v)]$ to make $< B_{i+1}(v) >_H$ acyclic for $1 \leq i \leq e-1$, we get a spanning tree of $H$ and hence of $G$.
Since in this procedure one distance path from $v$ to each of the other vertices remains intact, we have $d_G(v, u) = d_T(v, u)$ for each $u \in V$.

**Remarks**   The above result implies that for any vertex $v$ of a connected graph $G$, there exists an image $\Phi_v(G)$ which is a spanning tree of $G$ preserving distances from $v$. This is called an *isometric tree* of $G$ at $v$. If there is only one such tree (upto isomorphism) at $v$, we say that $G$ has a *unique isometric tree* at $v$. If $G$ has the same unique isometric tree at each vertex $v$, then $G$ is said to have a *unique isometric tree* (or unique distance tree). $K_{2,2}$ and the Peterson graph are examples of graphs having unique isometric trees, while $K_{3,3}$ does not have a unique isometric tree at any vertex. Every tree has a unique isometric tree. ❏

The next result due to Chartrand and Stewart [52] gives the necessary condition for a graph to have a unique isometric tree.

**Theorem 4.18 (Chartrand and Stewart [52])** Let $G$ be a connected graph with $d = 2r$, which has a unique isometric tree. Then the end vertices of every diametral path of $G$ has degree 1.

**Proof** Let $G$ be a connected graph with $d = 2r$ and let $P$ be a diametral path with end vertices $u$ and $v$. If possible let $d(u|G) > 1$. Let $T_u$ be the isometric tree at $u$. It is easy to see that $T_u$ can be chosen to contain $P$.

Since $u$ has degree at least 2 in $G$, there is a vertex $u_i$ adjacent to $u$ and not lying in $P$. Clearly, $D_{T_u}(u_i, v) = 1 + d$.

Let $c$ be a central vertex of $G$. Then for any two vertices $w_1$ and $w_2$ of $G$, we have

$$d_G(w_1, c) \le r = \tfrac{1}{2}d \text{ and } d_G(w_2, c) \le r = \tfrac{1}{2}d.$$

Therefore, $d_G(w_1, w_2) \le d(w_1, c) + d(c, w_2) \le d$.

Since $T_c$ is isometric with $G$ at $c$, we also have $d_{T_c}(w_1, w_2) \le d$.

Thus no path $T_c$ has length greater than $d$, whereas there is a path in $T_u$ of length $1 + d$. Therefore, $T_c \ne T_u$ and $G$ does not have a unique isometric tree. This contradicts the hypothesis. Hence the result follows.

**Remark** The above condition is necessary but not sufficient. To see this, consider the graph given in Figure 4.9.

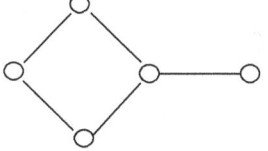

**Fig. 4.9**

Chartrand and Schuster [54], and Kundu [142] have given some more results on the graphs with unique isometric trees.

**Definition:** The *complexity* $\tau(G)$ of a graph $G$ is the number of different spanning trees of $G$. ❏

The following result gives a recursive formula for $\tau(G)$.

**Theorem 4.19** For any cyclic edge $e$ of a graph $G$, $\tau(G) = \tau(G - e) + \tau(G|pe)$.

**Proof**   Let $S$ be the set of spanning trees of $G$ and let $S$ be partitioned as $S_1 \cup S_2$, where $S_1$ is the set of spanning trees of $G$ not containing $e$ and $S_2$ is the set of the spanning trees of $G$ containing $e$.

Since $e$ is a cyclic edge, $G - e$ is connected and there is a one-one correspondence between the elements of $S_1$ and the spanning trees of $G - e$. Also, there is a one-one correspondence between the spanning trees of $G|\text{þ}e$ and the elements of $S_2$.

Thus, $\tau(G) = |S_1| + |S_2| = \tau(G-e) + \tau(G|\text{þ}e)$.

**Remarks**

1. The above recurrence relation is valid even if $e$ is a cut edge. This is because $\tau(G - e) = 0$ and every spanning tree of $G$ contains every cut edge.

2. The recurrence relation is valid even if $G$ is a general graph and $e$ is a multiple edge, but not when $e$ is a loop.

3. The complexity of any graph $G$ is computed by repeatedly applying the above recurrence. We observe that on applying the elementary contraction to a multiple edge, the resulting graph can have a loop and by remark (2) the procedure can be still continued. At each stage of the algorithm, only an edge belonging to the proper cycle is chosen. The algorithm starts with a given graph and produces two graphs (possibly general) at the end of the first stage. At each subsequent stage one proper cyclic edge from each graph is chosen (if it exists) for applying the recurrence. On termination of the algorithm, we get a set of graphs (or general graphs) none of which have a proper cycle. Then $\tau(G)$ is the sum of the number of these graphs. If $H$ is any of these graphs, then $\tau(H)$ is the product of its edges, ignoring the loops.

**Example**   Consider the graph $G$ given in Figure 4.10.

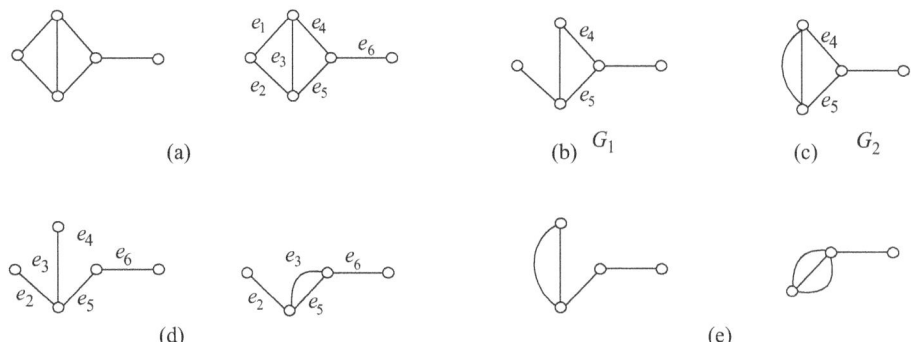

**Fig. 4.10**

Label the edges of $G$ arbitrarily. Choose $e_1$ as the first cyclic edge. Then $\tau(G)$ is the sum of the complexities of the graphs given in Figure 4.10(b) and (c). Now, choose $e_4$ in both $G_1$ and $G_2$ as the next cyclic edge. Then $\tau(G)$ is the sum of the complexities of the graphs in Figure 4.10(d) and (e). Since there are no more cyclic edges, the algorithm terminates, and we have $\tau(G) = 1 + 2 + 2 + 3 = 8$.

## 4.3 Number of Labelled Trees

Let us consider the problem of constructing all simple graphs with $n$ vertices and $m$ edges. There are $n(n-1)/2$ unordered pairs of vertices. If the vertices are distinguishable from each other (i.e., labelled graphs), then the number of ways of selecting $m$ edges to form the graph is $\binom{\frac{n(n-1)}{2}}{m}$.

Thus, the number of simple labelled graphs with $n$ vertices and $m$ edges is

$$\binom{\frac{n(n-1)}{2}}{m}. \tag{A}$$

Clearly, many of these graphs can be isomorphic (that is they are same except for the labels of their vertices). Thus, the number of simple, unlabelled graphs of $n$ vertices and $m$ edges is much smaller than that given by (A) above.

**Theorem 4.20** The number of simple, labelled graphs of $n$ vertices is $2^{\frac{n(n-1)}{2}}$.

**Proof** The number of simple graphs of $n$ vertices and $0, 1, 2, \ldots, n(n-1)/2$ edges are obtained by substituting $0, 1, 2, \ldots, n(n-1)/2$ for $m$ in (A). The sum of all such numbers is the number of all simple graphs with $n$ vertices.

Therefore the total number of simple, labelled graphs of $n$ vertices is

$$\binom{\frac{n(n-1)}{2}}{0} + \binom{\frac{n(n-1)}{2}}{1} + \binom{\frac{n(n-1)}{2}}{2} + \ldots + \binom{\frac{n(n-1)}{2}}{\frac{n(n-1)}{2}} = 2^{\frac{n(n-1)}{2}},$$

by using the identity $\binom{k}{0} + \binom{k}{1} + \binom{k}{2} + \ldots + \binom{k}{k} = 2^k$.

The following result was given independently by Tutte [252] and Nash-Williams [167]. We prove the necessity and for sufficiency the reader is referred to the original papers of Tutte and Nash-Williams.

**Theorem 4.21** A simple connected graph $G$ contains $k$ pairwise edge-disjoint spanning trees if and only if, for each partition $\pi$ of $V(G)$ into $p$ parts, the number $m(\pi)$ of edges of $G$ joining distinct parts is at least $k(p-1)$.

### Proof

*Necessity* Let $G$ has $k$ pairwise edge-disjoint spanning trees. If $T$ is one of them, and if $\pi = \{V_1, V_2, \ldots, V_p\}$ is a partition of $V(G)$ into $p$ parts, then identification of each part $V_i$ into a single vertex $v_i$, $1 \leq i \leq p$, results in a connected graph $G_0$ (possibly with multiple edges) on $\{V_1, V_2, \ldots, V_p\}$. Clearly, $G_0$ contains a spanning tree with $p-1$ edges, and each such edge belongs to $T$, and joins distinct partite sets of $\pi$. Since this is true for each of the $k$ edge −disjoint spanning trees of $G$, the number of edges joining distinct parts of $\pi$ is at least $k(p-1)$.

Cayley [46] in 1889 discovered the formula $\tau(K_n) = n^{n-2}$. Clearly, the number of spanning trees of $K_n$ is same as the number of non-label-isomorphic trees on $n$ vertices. Several proofs of this result have appeared since Cayley's discovery. Moon [164] has outlined ten such proofs, and a complete presentation of some of these can also be found in Lovasz [152]. Here, we give two proofs, and the first is due to Prufer [212].

**Theorem 4.22 (Cayley [46])**   There are $n^{n-2}$ labelled trees with $n$ vertices, $n \geq 2$.

**Proof**   Let $T$ be a tree with $n$ vertices and let the vertices be labelled $1, 2, \ldots, n$. Remove the pendant vertex (and the edge incident to it) having the smallest label, say $u_1$. Let $v_1$ be the vertex adjacent to $u_1$. Now, from the remaining $n-1$ vertices, let $u_2$ be the pendant vertex with the smallest label and let $v_2$ be the vertex adjacent to $u_2$. We remove $u_2$ and the edge incident on it. We repeat this operation on the remaining $n-2$ vertices, then on $n-3$ vertices, and so on. This process completes after $n-2$ steps, when only two vertices are left.

Let the vertices after each removal have labels $v_1, v_2, \ldots, v_{n-2}$. Clearly, the tree $T$ uniquely defines the sequence

$$(v_1, v_2, \ldots, v_{n-2}). \tag{4.22.1}$$

Conversely, given a sequence of $n-2$ labels, an $n$-vertex tree is constructed uniquely as follows. Determine the first number in the sequence

$$1, 2, 3, \ldots, n, \tag{4.22.2}$$

that does not appear in (4.22.1). Let this number be $u_1$. Thus the edge $(u_1, v_1)$ is defined. Remove $v_1$ from sequence (4.22.1) and $u_1$ from (4.22.2). In the remaining sequence of (4.22.2), find the first number which does not appear in the remaining sequence of (4.22.1). Let this be $u_2$ and thus the edge $(u_2, v_2)$ is defined. The construction is continued till the sequence (4.22.1) has no element left. Finally, the last two vertices remaining in (4.22.2) are joined.

For each of the $n-2$ elements in sequence (4.22.1), we choose any one of the $n$ numbers, thus forming $n^{n-2}$ $(n-2)$-tuples, each defining a distinct labelled tree of $n$ vertices. Since each tree defines one of these sequences uniquely, there is a one−one correspondence between the trees and the $n^{n-2}$ sequences.                                                        ❑

**Example**   Consider the tree shown in Figure 4.11. Pendant vertex with smallest label is $u_1$. Remove $u_1$. Let $v_1$ be adjacent to $u_1$ (label of $v_1$ is 1). Pendant vertex with smallest label is 4. Remove 4. Here, 4 is adjacent to 1. Pendant vertex with smallest label is 1. Remove 1. Here, 1 is adjacent to 3. Remove 3. Then 3 is adjacent to 5. Remove 6. So 6 is adjacent to 5. Remove 5. So 5 is adjacent to 9. Sequence $(v_1, v_2, \ldots, v_{n-2})$ is $(1, 1, 3, 5, 5, 5, 9)$.

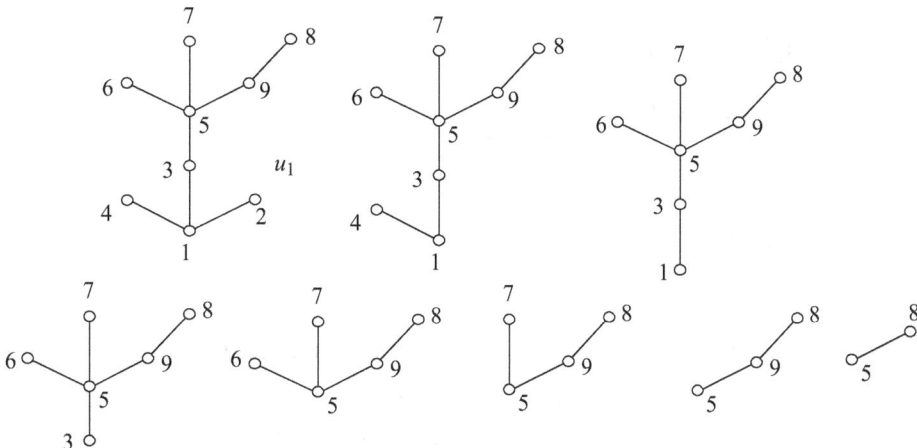

**Fig. 4.11**

**Theorem 4.23**  If $D = [d_i]_1^n$ is the degree sequence of a tree, then the number of labelled trees with this degree sequence is

$$\frac{(n-2)!}{(d_1-1)!(d_2-1)!\ldots(d_n-1)!}.$$

**Proof**  We first observe that, when asking for all possible trees with the vertex label set $V = \{v_1, v_2, \ldots, v_n\}$ with degree sequence $D = [d_i]_1^n$, it is not necessary that $d_i = d(v_i)$ and it is not necessary that the sequence be monotonic non-decreasing.

Therefore we assume that $D = [d_i]_1^n$ is an integer sequence satisfying the conditions $\sum d_i = 2(n-1)$ and $d_i \geq 1$. We use induction on $n$. The result is obvious for $n = 1, 2$. For $n = 2$, the sequence is $[d_1, d_2]$ and the only degree sequence in this case is $[1, 1]$. Clearly, there is only one labelled tree with this degree sequence.

Also, $\dfrac{(n-2)!}{(d_1-1)!\ldots(d_n-1)} = \dfrac{(2-2)!}{(1-1)!(1-1)!} = 1.$

Now, assume that the result is true for all sequences of length $n-1$. Let $D = [d_i]_1^n$ be an $n$ length sequence. By assumption there is a $d_i = 1$ and let it be $d_n = 1$. Let $T_n$ be a tree realising $D = [d_i]_1^n$. Now, removing $v_n$, we get a tree $T_{n-1}$ on the vertex set $\{v_1, v_2, \ldots, v_{n-1}\}$ with degrees $d_1, \ldots, d_{j-1}, d_{j-1}, d_{j+1}, \ldots, d_{n-1}$, where $v_j$ is the vertex to which $v_n$ is adjacent in $T_n$. Clearly, the converse is also true. Therefore, by induction hypothesis, the number of trees $T_{n-1}$ is

$$\frac{(n-3)!}{(d_1-1)!\ldots(d_{j-1}-1)!(d_j-1-1)!(d_{j+1}-1)!\ldots(d_{n-1}-1)!}$$

$$= \frac{(n-3)!(d_j-1)}{(d_1-1)!...(d_{j-1}-1)!\,[(d_j-1)\,(d_j-2)!]\,(d_{j+1}-1)!...(d_{n-1}-1)!}$$

$$= \frac{(n-3)!(d_j-1)}{(d_1-1)!...(d_j-1)!...(d_{n-1}-1)!(d_n-1)!}$$

$$= \frac{(n-3)!(d_j-1)}{\prod\limits_{j=1}^{n}(d_j-1)!}.$$

Since $v_j$ is any one of the vertices $v_1, \ldots, v_{n-1}$, the number of trees $T_n$ is

$$\sum_{j=1}^{n-1} \frac{(n-3)!(d_j-1)}{\prod\limits_{j=1}^{n}(d_j-1)!} = \frac{(n-3)!}{\prod\limits_{j=1}^{n}(d_j-1)!} \sum_{j=1}^{n}(d_j-1), \text{ as } d_n=1, \text{ and } d_n-1=1-1=0$$

$$= \frac{(n-3)!}{\prod\limits_{j=1}^{n}(d_j-1)!}(n-2), \text{ since } \sum_{j=1}^{n}(d_j-1)=2(n-1)-n=n-2$$

$$= \frac{(n-2)!}{\prod\limits_{j=1}^{n}(d_j-1)!}. \qquad \qquad \square$$

Now, we use Theorem 4.22 to obtain $\tau(K_n)=n^{n-2}$, which forms the second proof of Cayley's Theorem.

**Second Proof of Theorem 4.22**   We know the number of labelled trees with a given degree sequence $[d_i]_1^n$ is

$$\frac{(n-2)!}{\prod\limits_{j=1}^{n}(d_j-1)!}.$$

The total number of labelled trees with $n$ vertices is obtained by adding the number of labelled trees with all possible degree sequences.

Therefore, $\tau(K_n) = \sum\limits_{\substack{d_i \geq 1 \\ \sum\limits_{i=1}^{n} d_i = 2n-2}} \left[ \frac{(n-2)!}{\prod\limits_{j=1}^{n}(d_j-1)!} \right].$

Let $d_i-1=k_i$. So $d_i \geq 1$ gives $d_i-1 \geq 0$, or $k_i \geq 0$.

Also, $\sum\limits_{i=1}^{n} k_i = \sum\limits_{i=1}^{n}(d_i-1) = \sum\limits_{i=1}^{n} d_i - n = 2n-2-n = n-2.$

Thus, $\tau(K_n) = \sum_{\substack{k_i \geq 0 \\ \sum_1^n k_i = n-2}} \frac{(n-2)!}{k_1! k_2! \ldots k_n!} = \sum_{\substack{k_i \geq 0 \\ \sum_1^n k_i = n-2}} \frac{(n-2)!}{k_1! k_2! k_n!} 1^{k_1} 1^{k_2} \ldots 1^{kn}$

$$= (1 + 1 + \ldots + 1)^{n-2}, \text{ by multinomial theorem.}$$

Thus, $\tau(K_n) == n^{n-2}$. ❏

**Note** The multinomial distribution is given by

$$\frac{n!}{x_1! x_2! \ldots x_k!} p_1^{x_1} p_2^{x_2} \ldots p_k^{x_k} = (p_1 + p_2 + \ldots + p_k)^n, \text{ where } \sum_{i=1}^n x_i = n.$$

# 4.4 The Fundamental Cycles

**Definition:** Let $T$ be a spanning tree of a connected graph $G$. Let $\overline{T}$ be the spanning subgraph of $G$ containing only the edges of $G$ which are not in $T$ (i.e., $\overline{T}$ is the relative complement of $T$ in $G$). Then $\overline{T}$ is called the co tree of $T$ in $G$. The edges of $T$ are called branches and the edges of $\overline{T}$ are called chords of $G$ relative to the spanning tree $T$.

**Theorem 4.24** If $T$ is a spanning tree of a connected graph $G$ and $f$ is a chord of $G$ relative to $T$, then $T + f$ contains a unique cycle of $G$.

**Proof** Let $f = uv$. Then there is a unique $u - v$ path $P$ in $T$. Clearly, $P + f$ is a cycle of $G$, since $T$ is acyclic, any cycle $C$ of $T + e$ should contain $e$, and $C - e$ is a $u - v$ path in $T$. Since there is a unique path in $T$, $T + e$ contains a unique cycle of $G$. ❏

**Remarks**

1. If $f_1$ and $f_2$ are two distinct chords of the connected graph $G$ relative to a spanning tree $T$, then there are two unique distinct cycles $C_1$ and $C_2$ of $G$ containing respectively $f_1$ and $f_2$.

2. If $e \in E(\overline{G})$ and $T$ is a spanning tree of $G$, then $T + e$ contains a unique cycle of $K_n$.

**Definition:** Let $G$ be a connected graph with $n$ vertices and $m$ edges. The number of chords of $G$ relative to a spanning tree $T$ of $G$ is $m - n + 1 = \mu$. The $\mu$ distinct cycles of a connected graph $G$ corresponding to the distinct chords of $G$ relative to a spanning tree $T$ of $G$ are said to form a set of fundamental cycles of $G$.

If $G$ is a disconnected graph with $k$ components $G_1, G_2, \ldots, G_k$ and $T_i$, $1 \leq i \leq k$, are a set of $k$ spanning trees of $G_i$, then the union of the set of fundamental cycles of $G_i$ with respect to $T_i$ is a set of fundamental cycles for $G$. It is to be noted that different spanning trees give different sets of fundamental cycles. ❏

The following result characterises cycles in terms of the set of all spanning trees.

**Theorem 4.25**   Any cycle of a connected graph $G$ contains at least one chord of every spanning tree of $G$.

**Proof**   Let $C$ be a cycle and assume that the result is not true. So there exists a spanning tree $T$ of $G$ such that $C$ is contained in the edge set $E(G) - E(\overline{T})$, where $\overline{T}$ is the cotree of $G$ corresponding to $T$. This means that the tree $T$ contains the cycle $C$, which is a contradiction.                                                                                           ❏

**Theorem 4.26**   A set of edges $C$ of a connected graph $G$ is a cycle of $G$ if and only if it is a minimal set of edges containing at least one chord of every spanning tree of $G$.

**Proof**   Let $C$ be a cycle of $G$. Then it contains at least one chord of every spanning tree of $G$. If $C'$ is any proper subset of $C$, then $C'$ does not contain a cycle and is a forest. A spanning tree $T$ of $G$ can therefore be constructed containing $C'$. Clearly, $C'$ does not contain any chord of $T$. Thus no proper subset of $C$ has the stated property, proving that $C$ is minimal with respect to the property.

To prove sufficiency, let $C$ be minimal set with the stated property. Then $C$ is not acyclic. Therefore $C$ contains at least a cycle $C'$. But by the necessary part, $C'$ is minimal with respect to the property and hence $C' = C$, that is, $C$ is a cycle.                                     ❏

## 4.5   Generation of Trees

**Definition:**   Let $T_1$ and $T_2$ be two spanning trees of a connected graph $G$ and let there be edges $e_1 \in T_1$ and $e_2 \in T_2$ such that $T_1 - e_1 + e_2 = T_2$ (and hence $T_2 - e_2 + e_1 = T_1$). The transformation $T_1 \leftrightarrow T_2$ is called an *elementary tree transformation* (ETT), or a fundamental exchange. If $e_1$ and $e_2$ are adjacent in $G$, then the ETT is called a *neighbour transformation* (NT). If $e$ is a pendant edge of $T_1$ (and hence $e_2$ is a pendant edge of $T_2$) the ETT is called a *pendant-edge transformation* (PET) or an end-line transformation.

**Definition:**   Let $I$ be the collection of all spanning trees of a connected graph $G$. Let $Tr(G)$ be the graph whose vertices $t_i$ correspond to the elements $T_i$ of $I$, and in which $t_i$ and $t_j$ are adjacent if and only if there is an ETT between $T_i$ and $T_j$, that is, if and only if $E(T_i) \Delta E(T_j) = \{e_i, e_j\}$. Then $Tr(G)$ is called the *tree graph* of $G$. The distance $d(T_i, T_j)$ between the spanning trees $T_i$ and $T_j$ of $G$ is defined to be the distance between $t_i$ and $t_j$ in $Tr(G)$.

**Theorem 4.27**   The tree graph $Tr(G)$ of a connected graph is connected.

**Proof**   Let $G$ be a connected graph with $n$ vertices and let $Tr(G)$ be its tree graph. To prove that $Tr(G)$ is connected, it is enough to prove that any two spanning trees of $G$ can be obtained from each other by a finite sequence of ETT's.

Let $T$ and $T'$ be two distinct spanning trees of $G$. Then there is a set $S = \{e_1, e_2, \ldots, e_k\}$ of some $k$ edges of $T$ which are not in $T'$. Since a spanning tree has $n-1$ edges, there is a corresponding set $S' = \{e'_1, e'_2, \ldots, e'_k\}$ of edges of $T'$ which are not in $T$. Thus, $T + e'_1$ contains a unique fundamental cycle $T\, e'_1$. As $T'$ is a tree, at least one edge of $T\, e'_1$ (which is a branch of $T$) will not be in $T'$ and thus a member of $S$. Without loss of generality, let this edge be $e_1$. Define $T_1 = T - e_1 + e'_1$. Then $T_1$ can be obtained from $T$ by an ETT and therefore $T_1$ and $T'_1$ have one more edge in common.

Repeating this process $k-1$ more times, we get a sequence of spanning trees $T_0 = T, T_1, T_2, \ldots, T_{k-1}, T_k = T'$ such that there is an ETT $T_i \longleftrightarrow T_{i+1}$, $0 \leq i \leq k-1$. ❏

**Theorem 4.28** An elementary tree transformation can be obtained by a sequence of neighbor transformations.

**Proof** Let $T$ and $T' = T - x + y$ be spanning trees of the graph $G$, where $x$ and $y$ are non-adjacent edges of $G$. Then we can choose a set of edges $e_1, e_2, \ldots, e_k$ such that $x, e_1, e_2, \ldots, e_k, y$ is a path in $T + y$. Define $T_1 = T = x + e_1$ and $T_i = T_{i-1} + e_{i-1} + e_i$, $2 \leq i \leq k$ and $T_{k+1} = T_k - e_h + y$. Then, $T_{k+1} = T'$, and is obtained from $T$ by a sequence of $k+1$ neighbour transformations through the intermediate trees $T_i$, $1 \leq i \leq k-1$. ❏

**Definition:** A spanning tree of a graph $G$ corresponding to a central vertex of the tree $Tr(G)$ is called a *central tree*.

The set of diameters of the spanning trees of a connected graph $G$ is the *tree diameter set* of $G$. A set of positive integers is a *feasible tree diameter set* if it is the tree diameter set of some graph. For example, the graph in Figure 4.12 has one spanning tree of diameter seven and all others of diameter five.

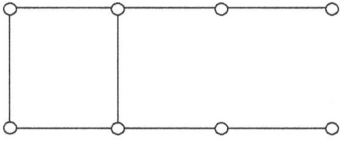

**Fig. 4.12**

The *girth* $g(G)$ of a graph $G$ is the length of a smallest cycle of $G$. A cycle of smallest length is called a *girdle* of $G$. The *circumference* $c(G)$ of a graph $G$ is the length of the longest cycle of $G$. A cycle of maximum length is called a *hem* of $G$.

Let $\underline{n}(\delta, g)$ denote the minimum order (minimum vertices) of a graph with minimum degree at least $\delta\ (\geq 3)$ and girth at least $g\ (\geq 2)$.

Let $\bar{n}(\Delta, g)$ denote the maximum order of a graph with degree at most $\Delta$ and girth at most $g$. ❏

The following upper bound for $\underline{n}(\delta, g)$ can be found in Bollobas [29].

**Theorem 4.29 (Bollobas [29])** $\underline{n}(\delta, g) \leq (2\delta)^g$.

**Proof** Clearly, $\underline{n}\,(\delta,\,g)$ denotes the minimum order of a graph with minimum degree at least $\delta\,(\geq 3)$ and girth at least $g\,(\geq 2)$. Therefore, we construct a graph with atmost $(2\delta)^g$ vertices with these properties. Let $n = (2\delta)^g$.

Consider all graphs with vertex set $V = \{1,\,2,\,\ldots,\,n\}$ and having exactly $\delta n$ edges.

Since there are $\binom{n}{2}$ possible positions to accommodate these $\delta n$ edges, the number of such graphs

$$= \left( \begin{array}{c} \binom{n}{2} \\ \delta n \end{array} \right).$$

Among the $n$ available vertices, the number of ways an $h$-cycle can be formed is

$$= \frac{1}{2} \binom{n}{h} (h-1)!$$

Obviously, $\frac{1}{2} \binom{n}{h} (h-1)! < \frac{1}{2h} n^h$.

The number of graphs in the set which contain a given $h$-cycle is

$$= \left( \begin{array}{c} \binom{n}{2} - h \\ \delta n - h \end{array} \right).$$

Hence, the average number of cycles of length at most $g - 1$ in these graphs is

$$< \sum_{h=3}^{g-1} \frac{1}{2h} n^h \left( \begin{array}{c} \binom{n}{2} - h \\ \delta n - h \end{array} \right) \Big/ \left( \begin{array}{c} \binom{n}{2} \\ \delta n \end{array} \right)$$

$$< \sum_{h=3}^{g-1} (2\delta)^h < (2\delta)^g = n.$$

Since the average is less than $n$, there is an element in the set with value less than or equal to $n - 1$. Thus, there is a graph $G$ on $n$ vertices with $\delta n$ edges and at most $n - 1$ cycles of length at most $g - 1$. Removing one edge from each of these cycles, we get a graph $G_0$ with girth at least $g$. The number of edges removed is atmost $n - 1$, so that $m(G_o) \geq n\delta - (n-1) \geq n\,(\delta - 1) + 1$ and $n\,(G_0) = n$. Thus, $G_0 \in G_{\delta - 1}$, and hence $G_0$ contains a subgraph $H$ with $\delta(H) \geq \delta$. By construction, $g(H) \geq g$ and $n(H) \leq n = (2\delta)^g$. Thus, we have constructed a graph $H$ with the desired properties.  $\square$

**Note**  If $G$ is a graph with at least $n_0$ vertices and at least $n_0' n(G) - \binom{n_0 + 1}{2} + 1$ edges, then $G$ contains a subgraph $H$ with $\delta(H) \geq n_0 + 1$.

We denote by $G_{n_o} = \left\{ G : n(G) > n_0,\ m(G) \geq n_0.n(G) - \binom{n_0 + 1}{2} + 1 \right\}$.

Now, we give a lower bound for $\underline{n}(\delta, g)$ which is due to Tutte [248].

**Theorem 4.30 (Tutte [248])**

$$\underline{n}(\delta, g) \geq \begin{cases} \dfrac{\delta(\delta-1)^{\frac{g-1}{2}}-2}{\delta-2}, & \text{if } g \text{ is odd,} \\[2ex] \dfrac{2(\delta-1)^{\frac{g}{2}}-1}{\delta-2}, & \text{if } g \text{ is even.} \end{cases}$$

**Proof**

i. Let $g$ be odd, say $g = 2d+1$. Then clearly the diameter of $G$ is at least $d$. Let $v$ be a vertex with eccentricity at least $d$. Consider the neighbourhoods

$$N_i = N_i(V),\ 1 \leq i \leq d = (g-1)/2.$$

Now, no vertex of $N_i$ is adjacent to more than one vertex of $N_{i-1}$, because otherwise, there will be a cycle of length $1 \leq 2i < g$. Similarly, there is no edge in $< N_i >$.

Therefore, for every $u \in N_i$, we have

$$|N(u) \cap N_{i-1}| = 1,\ |N(u) \cap N_i + 1| = d(u) - 1 \text{ and}$$

$$|N_{i+1}| = \sum_{ui N_i} \{d(\mu) - 1\} \geq (\delta - 1)|N_i|. \tag{4.30.1}$$

As, $V \supseteq \{v\} \cup \bigcup_{u \varepsilon N_i} N_i(v)$, therefore

$$n \geq 1 + \sum_{i=1}^{d} |Ni| \geq 1 + \delta + \delta(\delta-1) + \ldots + \delta(\delta-1)^{d-1}$$

$$= 1 + \frac{\delta}{\delta-2}\left\{(\delta-1)^d - 1\right\} = \frac{\left\{\delta(\delta-1)^{\frac{g-1}{2}}-2\right\}}{\delta-2}.$$

ii. Let $g$ be even, say $g = 2d$. Then again the diameter is at least $d$. Let $xy$ be an edge of $G$ and let

$$S_i = \{v \in V : d(x, v) = I, \text{ or } d(y, v) = i\}, \text{ for } 1 \leq I \leq d-1, \text{ and } S_0 = \{x, y\}.$$

The girth requirement forces that there are no edges in $< S_i >$, for $1 \leq i \leq d-2$, and that each vertex of $S_i$ be adjacent to at most one vertex of $S_{i-1}$ for $1 \leq i \leq d-1$.

Thus, for each $u \in S_i$, we have $|N(u) \cap S_{i+i}| = d(u) - 1$ and

$$|S_{i+1}| = \sum_{u \varepsilon S_i} (d(u) - 1) \geq (\delta - 1) |S_i|. \tag{4.30.2}$$

Since, $V \supseteq \{x, y\} \cup \bigcup_{i=1}^{d-1} S_i$,

$$n \geq \sum_{i=0}^{d-1} |S_i| = 2 \sum_{i=0}^{d-1} (\delta - 1)^i = \frac{2}{\delta - 2} \left[ (\delta - 1)^{\frac{g}{2}} - 1 \right]. \qquad \Box$$

By using arguments as in Theorem 4.30 and by replacing $\delta$ by $\Delta$, we obtain the following result.

**Theorem 4.31**

$$\bar{n}(\Delta, g) \leq \begin{cases} \dfrac{\Delta(\Delta - 1)^{\frac{g-1}{2}} - 2}{\Delta - 2}, & \text{if } g \text{ is odd}, \\ \dfrac{2 \left[ (\Delta - 1)^{\frac{g}{2}} - 1 \right]}{\Delta - 2}, & \text{if } g \text{ is even}. \end{cases}$$

**Definition:** A $k$-regular graph with girth $g$ and with minimum order $\underline{n}(k, g)$ is called a *(k, g)-cage*.

$$\text{The integer } n_0 = \begin{cases} \dfrac{k(k - 1)^{\frac{g-1}{2}} - 2}{k - 2}, & \text{if } g \text{ is odd}, \\ \dfrac{2 \left[ (k - 1)^{\frac{g}{2}} - 1 \right]}{k - 2}, & \text{if } g \text{ is even}, \end{cases}$$

is called the *Moore bound* for a $k$-regular graph with $g$.

# 4.6  Helly Property

**Definition:** A family $\{A_i : i \in I\}$ of subsets of a set $A$ is said to satisfy the Helly property if $J \subseteq I$, and $A_i \cap A_j \neq \phi$, for every $i, j \in J$, then $\bigcap_{j \in J} A_j \neq \phi$.

The following result is reported by Balakrishnan and Ranganathan [13].

**Theorem 4.32 (Balakrishnan and Ranganathan [13])**    A family of subtrees of a tree satisfies the Helly property.

**Proof** Let $\tau = \{T_i : i \in I\}$ be a family of subtrees of a tree $T$. Suppose that for all $i$, $j \in J \subseteq I$, $T_i \cap T_j \neq \phi$. We have to prove $\bigcap_{j \in J} T_j \neq \phi$. If some tree $T_i \in \tau$, $i \in J$, is a single vertex tree $\{v\}$ (that is, $K_1$), then clearly, $\bigcap_{j \in J} T_j = \{v\}$. So assume that each tree $T_i \in T$ with $i \in J$ has at least two vertices.

We induct on the number of vertices of $T$. Suppose the result is true for all trees with at most $n$ vertices and let $T$ be a tree with $(n+1)$ vertices. Let $v_0$ be an end vertex of $T$ and $u_0$ its unique neighbour in $T$. Let $T_i' = T_i - v_0$, $i \in J$ and $T' = T - v_0$. By induction hypothesis, the result is true for the tree $T'$. Also, $T_i' \cap T_j' \neq \phi$, for any $i$, $j \in J$. In fact, if $T_i$ and $T_j$ have a vertex $u$ ($\neq v_0$) in common then $T_i'$ and $T_j'$ also have $u$ in common, whereas if $T_i$ and $T_j$ have $v_0$ in common, then $T_i$ and $T_j$ have $u_0$ also in common, and so do $T_i'$ and $T_j'$. Hence by induction hypothesis, $\bigcap_{j \in J} T_j' \neq \phi$ and therefore $\bigcap_{j \in J} T_j \neq \phi$. $\qquad\square$

## 4.7 Signed Trees

The following result by Yan et al. [271] characterises signed degree sequences in signed trees.

**Theorem 4.33 (Yan et al. [271])** Let $D = [d_i]_1^n$ be an integral sequence of $n \geq 2$ terms and let $D$ has $n_+$ positive terms, $n_0$ zero and $n_-$ negative terms. Let $\alpha = 1$ if $n_+ n_- > 0$, and $\alpha = 0$, otherwise. Then $D$ is the signed degree sequence of a signed tree if and only if (i) to (iv) hold.

i. $\sum_{i=1}^n d_i \equiv 2n - 2 \pmod 4$.

ii. $\sum_{i=1}^n |d_i| \leq 2n - 2 - 2n_0$.

iii. $\sum_{i=1}^n |d_i| + 2 \sum_{d_i > 0} |d_i| \leq 2n - 2 - 4\alpha + 4p_-$.

iv. $\sum_{i=1}^n |d_i| + 2 \sum_{d_i > 0} |d_i| \leq 2n - 2 - 4\alpha + 4p_+$.

**Proof** Note that condition (iv) for $D$ is same as condition (iii) for $-D$. The necessity of the theorem follows from the fact that $m = n - 1$ and Lemmas 2.2, 2.3 and 2.4.

We prove the sufficiency by induction on $n$. For $n = 2$, by (i) and (iii), $d_1 = d_2 = 1$ or $-1$. Therefore $D$ is the signed degree sequence of $K_2$ with positive edge or a negative edge. Assume that the theorem is true for $n - 1$. Let $n \geq 3$.

By (ii), $D$ has at least two terms in which $|d_i| = 1$. After rearranging the terms in $D$ or taking $-D$, we may assume without loss of generality that $d_n = 1$ and one of the following holds.

1. $|d_i| = 1$, for $1 \leq i \leq n$, $d_1 \geq 0$ and $d_1 = 0$, if $n_0 > 0$.

2. $d_1 \geq 2$.

3. $d_i \leq 1$ but $d_i \neq -1$ for $1 \leq i \leq n$ and $d_1 = 0$ and $\alpha = 1$.

4. $d_i = 1$ or $d_i \leq -2$, for $1 \leq i \leq n$ and $d_1 = \alpha = 1$.

For any of the above, consider the sequence $D' = [d_i']_1^{n'}$, where $n' = n - 1$ and $d_i' = d_1 - 1$ and $d_i' = d_i$, for $2 \leq i \leq n - 1$.

Note that $\sum_{i=1}^{n'} d_i' = \left( \sum_{i=1}^{n} d_i \right) - 2 \equiv (2n - 2) - 2 \equiv 2n' - 2 (\mathrm{mod}\, 4)$, that is, (i) holds for $D'$. We check conditions (ii) to (iv) for $D'$ according to the four cases above.

**Case 1**   In this case, $|d_i'| \leq 1$, for $1 \leq i \leq n - 1$, we have

$$\sum_{i=1}^{n'} |d_i'| = n_+' + n_-', \quad \sum_{d_i' > 0} |d'| = n_+', \quad \sum_{d_i' < 0} |d_{i,}'| = n_-'.$$

Thus, (ii) to (iv) holds for $D'$ as $n_+' + n_-' \geq 2$.

**Case 2**   In this case, since $d_1 \geq 2$ and $d_n = 1$, we have

$$n' = n - 1, \ n_+' = n_+ - 1, \ n_0' = n_0, \ n_-' = n_-, \ \alpha' = \alpha,$$

$$\sum_{i=1}^{n'} |d_i'| = \sum_{i=1}^{n} |d_i'| - 2, \quad \sum_{d_i' > 0} |d_i'| = \sum_{d_i' > 0} |d_{i,}'| - 2, \quad \sum_{d_i' < 0} |d_i'| = \sum_{d_i' < 0} |d_i'|.$$

Therefore (ii) to (iv) holding for D imply that (ii) to (iv) hold for $D'$.

**Case 3**   In this case, since $d_1 = 0$ and $d_n = 1$, we have

$$n' = n - 1, \ n_+' = n_+ - 1, \ n_0' = n_0 - 1, \ n_-' = n_- + 1, \ \alpha' \leq \alpha,$$
$$\sum_{i=1}^{n'} |d_i'| = \sum_{i=1}^{n} |d_i|, \quad \sum_{d_i > 0} |d_i'| = \sum_{d_i > 0} |d_{i,}'| - 1, \quad \sum_{d_i' < 0} |d_i'| = \sum_{d_i' < 0} |d_i'| + 1.$$

So (ii) and (iii) holding for $D$ imply that (ii) and (iii) hold for $D'$. Since $d_i' \leq 1$ for $1 \leq i \leq n - 1$, $\sum_{d_i'}^{n'} |d_i'| = n_+'$. By (iii) for $D$ and the fact that $d_i \leq -2$ when $d_i < 0$,

$$n_+ + 6n_- \leq \sum_{i=1}^{n'} |d_i| + 2 \sum_{d_i < 0} |d_i| \leq 2n - 2 - 4\alpha + 4n_- = 2n_+ + 2n_0 + 6n_- 6,$$

and so $6 \leq n_+ + 2n_0$. Therefore, $3 \leq n'_+ + 2n'_0$, and then $4 \leq 2n'_+ + 2n'_0$.

This together with (ii) for $D'$ and $\sum_{d'_i>0} |d'_i| = n'_+$ implies (iv) for $D'$.

**Case 4**  In this case, since $d_1 = d_n = 1$, therefore

$$n' + n - 1, \ n'_+ = n_+ - 2, \ n'_0 = n_0 + 1 = 1, \ n'_- = n_-, \ \alpha' \leq \alpha,$$

$$\sum_{i=1}^{n'} |d'_i| = \sum_{i=1}^{n} |d_i| - 2, \ \sum_{d'_i>0} |d'_i| = \sum_{d'_i>0} |d'_i| - 2, \ \sum_{d'_i<0} |d'_i| = \sum_{d_i<0} |d'_i|.$$

(iii) for $D$ implies that (iii) holds for $D'$. As in the argument for Case 3, we have $\sum_{d'_i>0} |d_i| = n'_+$ and $6 \leq n_+ + 2n_0$. Therefore, $4 \leq n'_+$. Adding $2 \sum_{d'_i>0} |d'_i| = 2n'_+$ to the equality in (iii) for $D'$ and dividing the resulting equality by 3, we get (ii) for $D'$ as $2n'_0 \leq 2n'_+$. Adding $2 \sum_{d'_i>0} |d'_i| = 2n'_+$ to the equality in (ii) for $D'$, we get (iv) for $D'$ as $4\alpha' \leq 2n'_0 + 2n'_+$.

From the above discussion, $D'$ satisfies (i) to (iv). By the induction hypothesis, there exists a signed tree $T'$ with the vertex set $\{v_1, v_2, \ldots, v_{n-1}\}$ and signed degree $T'(v_i) = d'_i$, for $1 \leq i \leq n-1$. Suppose $T$ is the signed tree obtained from $T'$ by adding a new vertex $v_n$ and a new positive edge $v_1 v_n^+$, then $T$ has a signed degree sequence $D$. □

**Corollary**  Let $D = [d_i]_1^n$ be an integral sequence of $n \geq 3$ terms. Let $D$ has at least two terms in which $|d_i| = 1$, $|d_n| = 1$ and one of the following condition holds.

1. $|d_i| \leq 1$, for $1 \leq i \leq n$, $d_i \geq 0$, and $d_1 = 0$ if $n_o > 0$.

2. $d_1 \geq 2$.

3. $d_i \leq 1$ but $d_i \neq -1$ for $1 \leq i \leq n$, and $d_1 = 0$ and $\delta = 1$

4. $d_i = 1$ or $d_i \leq -2$ for $1 \leq i \leq n$, and $d_1 = \delta = 1$.

Then $D$ is the signed degree sequence of a signed tree if and only if $D' = [d_1 - 1, d_2, \ldots, d_{n-1}]$ is the signed degree sequence of a signed tree.

## 4.8  Exercises

1. Draw all unlabelled trees with seven and eight vertices.

2. Draw a tree which has radius five and diameter ten.

3. If a tree has an even number of edges, then show that it contains at least one vertex of even degree.

4. If the maximum degree of a vertex in a tree is $\Delta$, then show that it has $\Delta$ pendant vertices.

5. If $T$ is a tree such that every vertex adjacent to a pendant vertex has degree at least three, then prove that some pair of pendant vertices in $T$ has a common neighbour.

6. Show that a path is its own spanning tree.

7. Prove that every tree is a bipartite graph.

8. If for a simple graph $G$, $m(G) \geq n(G)$, prove that $G$ contains a cycle.

9. Show that, for a unicentral tree, $d = 2r$, and for a bicentral tree, $d = 2r - 1$.

10. Prove that if $K_{r,s}$ is a tree, then it must be a star.

11. How many spanning trees does $K_4$ have?

12. Prove that each spanning tree of a connected graph $G$ contains all the pendant edges of $G$.

13. Prove that each edge of a connected graph $G$ belongs to at least one spanning tree of $G$.

# 5. Connectivity

In a connected graph there is at least one path between every pair of its vertices. If in a graph, it happens that by deleting a vertex, or by removing an edge, or performing both, the graph becomes disconnected, we can say that such vertices or edges hold the whole graph, or in other words have the property of destroying the connectedness of a graph. For example, consider a communication network which is modelled as the graph $G$ shown in Figure 5.1, where vertices correspond to communication centers and the edges represent communication channels. Clearly, deletion of vertex $v$ results in the breakdown of the communication. This implies that in the above communication network, the center represented by vertex $v$ has the property of destroying the communication system and thus communication network depends on the connectivity. We start this chapter with the following definitions.

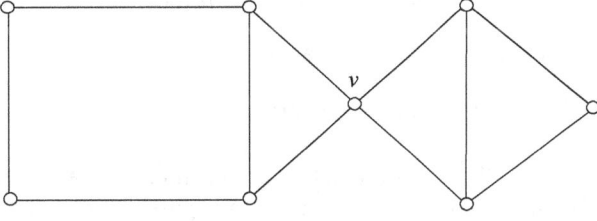

**Fig. 5.1**

## 5.1 Basic Concepts

**Cut vertex:** Let $G$ be a graph with $k(G)$ components. A vertex $v$ of $G$ is called a cut vertex of $G$ if $k(G-v) > k(G)$. For example, in the graph of Figure 5.2, the vertices $u$ and $v$ are cut vertices.

**Cut edge:**    An edge $e$ of a graph $G$ is said to be a cut edge if $k(G-e) > k(G)$. In the graph of Figure 5.2, $e$ and $f$ are cut edges.

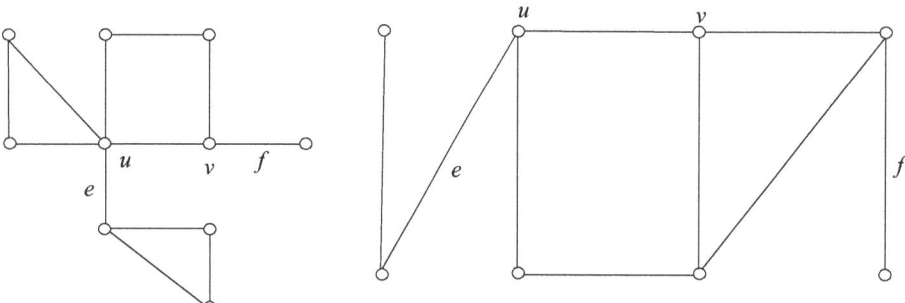

**Fig. 5.2**

The following observations are the immediate consequences of the definitions introduced above.

1. Removal of a vertex may increase the number of components in a graph by at least one, while removal of an edge may increase the number of components by at most one.

2. The end vertices of a cut edge are cut vertices if their degree is more than one.

3. Every non-pendant vertex of a tree is a cut vertex.

We now give the first result which characterises cut vertices.

**Theorem 5.1**    If $G = (V, E)$ is a connected graph, then $v$ is a cut vertex if there exist vertices $u, w \in V - \{v\}$ such that every $u - w$ path in $G$ passes through $v$.

**Proof**    Let $G = (V, E)$ be a connected graph and let $v$ be the cut vertex of $G$. Then $G - v$ is disconnected. Let $G_1, G_2, \ldots, G_k$ be the components of $G - v$. Now, let $U = V(G_1)$ and $W = \bigcup_{i=2}^{k} V(G_i)$. Also, let $u \in U$ and $w \in W$, and to be definite, let $w \in V(G_i)$, $i \neq 1$. If there is a $u - w$ path $P$ in $G$ not passing through $v$, then $P$ connects $u$ and $w$ in $G - v$ also. Therefore $G_1 \cup G_i$ is a single component in $G - v$, contradicting our assumption. Thus every $u - w$ path in $G$ passes through $v$.

Conversely, let there be vertices $u, w \in V - \{v\}$ such that every $u - w$ path in $G$ passes through $v$. Then there is no $u - w$ path in $G - v$. Therefore $u$ and $w$ belong to different components of $G - v$. Thus $G - v$ is disconnected and $v$ is a cut vertex of $G$.    ❑

The following result characterises cut edges.

**Theorem 5.2**  For a connected graph $G$, the following statements are equivalent.

    i. $e$ is a cut edge of $G$.

    ii. If $e = ab$, there is a partition of the edge subset $E - \{e\}$ as $E_1 \cup E_2$ with $a \in V(< E_1 >)$ and $b \in V(< E_2 >)$ such that for any $u \in V(< E_1 >)$ and any $w \in V(< E_2 >)$, every $u - w$ path contains $e$.

    iii. There exists vertices $u$ and $w$ such that every $u - w$ path in $G$ contains $e$.

    iv. $e$ is not a cycle edge of $G$.

**Proof**  (i) $\Rightarrow$ (ii). Let $e$ be a cut edge of $G$. So $G - e$ is disconnected. Let $G_1$ and $G_2$ be two components of $G - e$ and $E_1 = E(G_1)$ and $E_2 = E(G_2)$. If $u \in V(G_1)$ and $w \in v(G_2)$ exist such that there is a $u - w$ path $P$ in $G$ which does not contain $e$, then $u$ and $w$ are connected in $G - e$ by the path $P$. This implies that $G_1 \cup G_2$, that is, $G - e$ is connected, contradicting the hypothesis. This proves (ii).

(ii) $\Rightarrow$ (iii). Obvious.

(iii) $\Rightarrow$ (iv). Suppose $e$ lies on a cycle $C$. Then $C - e$ gives an $a - b$ path $Q$ not containing $e$. With vertices $u$ and $w$ following the condition given in (iii), let $P$ be any $u - w$ path. Without loss of generality, assume that $a$ and $b$ occur in that order in $P$. Let $u_0$ and $w_0$ be the first and last vertices that $P$ has in common with $C$ (the possibility of these coinciding with $a$, $b$, $u$ or $w$ is not ruled out). Then $P_{u, u_0} \cup Q_{u_0, w_0} \cup P_{w_0, w}$ is a $u - w$ path $P'$ of $G$ which does not contain $e$, contradicting (iii). (See Figure 5.3(a), where broken curves represent path $Q$ and thick curves represent path $P$.)

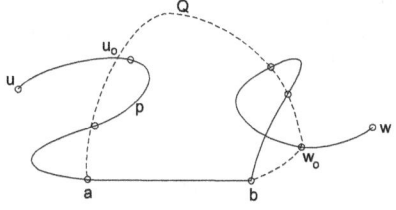

**Fig. 5.3(a)**

(iv) $\Rightarrow$ (i). Let $e$ be not a cyclic edge of $G$. We have to prove that $e$ is a cut edge of $G$, that is, $G - e$ is disconnected. Assume $G - e$ is disconnected. Then there is an $a - b$ path $P$ in $G - e$. But then $P \cup e$ is a cycle containing $e$, which contradicts (iv).  ❑

**Block:**  A block is a connected graph which does not have any cut edge. We observe that a block does not have any cut vertex. The graph $K_2 = (\{a, b\}, e)$ does not have a cut vertex and hence is a block. However, $e$ is a cut edge in this case. We call $K_2$ a *trivial block*. All other blocks are *non-trivial*.

**Separable graph:**   A connected graph with at least one cut vertex is called a separable graph. A block of a graph $G$ is a maximal graph $H$ of $G$ such that $H$ is a block. That is, $H$ has no cut vertex, but for any $v \in V(G) - V(H)$, $\langle V(H) \cup \{v\} \rangle$ is either a disconnected graph or a separable graph.

The next result characterises blocks.

**Theorem 5.3**   For a connected graph $G$, the following are equivalent.

    i. $G$ is a non-trivial block.

    ii. Any two vertices of $G$ lie on a cycle.

    iii. Given any vertex $u$ and any edge $vw$, there is a cycle of $G$ containing both.

    iv. Given any pair of edges $e = uv$ and $e' = u'v'$, there is a cycle of $G$ containing both.

    v. Given any pair of vertices $u$ and $u'$ and any edge $e = vw$, there is a $u - u'$ path of $G$ containing $e$.

**Proof**

    a. (i) $\Rightarrow$ (ii). The proof is by induction on the distance between the vertices. If $d(u, v) = 1$, then $uv$ is an edge, and since $G$ is a block, $uv$ is not a cut edge. Hence, $uv$ is a cyclic edge, and so $u$ and $v$ lie on a cycle. Now, for the induction hypothesis, we assume that if $u$ is any vertex, then any vertex $v'$ at a distance at most $k - 1$ from $u$ lies on a cycle with $u$.

        Let $v$ be a vertex at a distance $k$ from $u$. We prove that $u$ and $v$ lie on a cycle. Let $P$ be a shortest $u - v$ path and $v'$ the nearest vertex on $P$ from $u$. By induction hypothesis there is a cycle $C$ containing $u$ and $v'$. Since $v'$ is not a cut vertex of $G$, there is a $u - v$ path $Q$ not passing through $v'$. Let $z$ be the last vertex from $u$ that $Q$ has in common with $C$. Then $C_{uv'} \cup \{v'v\} \, Q_{vz} \cup C_{zu}$ is a cycle of $G$ containing $u$ and $v$. (Here $C_{uv'}$ is the $uv'$ segment of $C$ not containing $z$, and $C_{zu}$ is the $zu$ segment of $C$ not containing $v'$) (Fig. 5.3(b)).

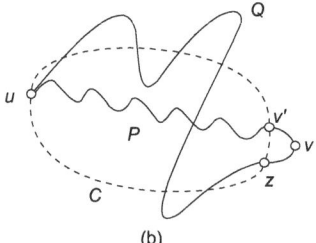

(b)

**Fig. 5.3(b)**

    b. (ii) $\Rightarrow$ (i). Let any two vertices of $G$ lie on a cycle. We prove that $G$ is a non-trivial block, that is, $G$ has no cut vertex. Assume to the contrary that $G$ has a cut vertex *u*. Then there are vertices *v* and *w* such that every *v* − *w* path passes through *u*. But then there is no cycle containing *v* and *w*, which is a contradiction. Thus *G* has no cut vertex.

c. (ii) ⇒ (iii). Let any two vertices of *G* lie on a cycle. Let vertex *u* and edge *vw* be given. So by (b), *G* is a block and therefore *vw* is not a cut edge.

    Let *C* be a cycle containing *vw*. If *C* contains *u*, the proof is complete. If not, by (ii), there is a cycle *Z* containing *u* and *v*. Taking any orientation of *Z*, let *x* and *y* be the first and last vertices from *u* that *Z* has in common with *C*. Then the *u* − *x* segment of *Z*, the *x* − *y* segment of *C* containing *vw* and the *y* − *u* segment of *Z* constitute a cycle of *G* containing *u* and *vw*.

d. (iii) ⇒ (ii). Let *u* and *v* be any two vertices. Since *v* cannot be an isolated vertex, there is an edge *vw*. By (iii) there is a cycle containing *u* and *vw* (Fig. 5.4).

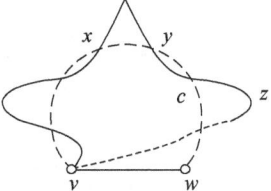

**Fig. 5.4**

e. (iii) ⇒ (iv). Let *uv* and *u'v'* be the given edges. By (iii) there is a cycle *C* through *u* containing *u'v'*. If it passes through *v*, then the *u* − *v* segment of *C* containing *u'v'* and the edge *vu* constitute a cycle as required. If not, then as the earlier implications show that *G* is a block, *u* is not a cut vertex, and hence there is a *v* − *v'* path *P* in *G* not passing through *u* (Fig. 5.5).

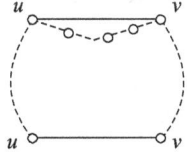

**Fig. 5.5**

Let *w* be the first vertex from *v* that *P* has in common with *C*. Then *v* − *w* segment of *P*, the *w* − *u* segment of *C* containing *u'v'* and the edge *uv* constitute a cycle of *G* as desired (Fig. 5.6).

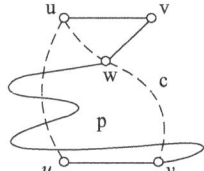

**Fig. 5.6**

f. (iv) ⇒ (iii). This can be proved as in (d).

g. (iv) ⇒ (v). Let $u$ and $u'$ be the vertices and $vw$ the given edge. If $uv'$ is also an edge, then there is nothing to prove. If not, since $u$ is not an isolated vertex, there is an edge $uv_1$ and by (iv) there is a cycle $C$ containing $uv_1$ and $vw$. By previous implications, $G$ is a block and hence $u$ is not a cut vertex. Therefore there is a $u'w$ path $P$ not passing through $u$. Let $x$ be the first vertex from $u\prime$ that $P$ has in common with $C$. Then the $u'x$ segment of $P$ and the $xu$ segment of $C$ containing $vw$ constitute a path as desired (Fig. 5.7).

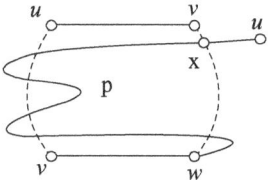

**Fig. 5.7**

h. (v) ⇒ (iv). Obvious.                                                                ❏

**Remark**    Property (iv) of the above theorem can be used to define an equivalence relation on the edge set $E$ of a graph $G$ by $e \sim f$, if and only if $e$ and $f$ lie on a common cycle in $G$. The equivalence classes are simply the blocks of $G$ and the edge set $E$ is partitioned into blocks. These blocks are joined at cut vertices, two blocks having at most one vertex in common.

## 5.2  Block–Cut Vertex Tree

Let $B$ be the set of blocks and $C$ be the set of cut vertices of a separable graph $G$. Construct a graph $H$ with vertex set $B \cup C$ in which adjacencies are defined as follows. $c_i \in C$ is adjacent to $b_j \in B$ if and only if the block $b_j$ of $G$ contains the cut vertex $c_i$ of $G$. The bipartite graph $H$ constructed above is called the *block-cut vertex tree* of $G$.

**Example**   Consider the graph in Figure 5.8. The blocks are $b_1 = <1, 2>$, $b_2 = <2, 3, 4>$, $b_3 = <2, 5, 6, 7>$, $b_4 = <7, 8, 9, 10, 11>$, $b_5 = <8, 12, 13, 14, 15>$, $b_6 = <10, 16>$, $b_7 = <10, 17, 18>$ and cut vertices are $c_1 = 2$, $c_2 = 7$, $c_3 = 8$, $c_4 = 10$.

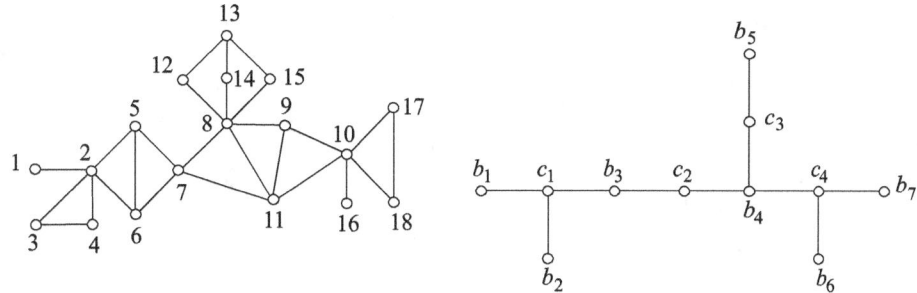

**Fig. 5.8**

**End block:**   A block of a graph $G$ containing only one cut vertex is called an *end-block* of $G$.

We have the following result on blocks.

**Theorem 5.4**   Every separable graph has at least one cut vertex and therefore has at least two end blocks.

**Proof**   A separable graph $G$ has at least one cut vertex and therefore has at least two blocks. Thus its block-cut vertex tree $T$ has at least three vertices. Now, for any separable graph the end blocks correspond to the pendant vertices of its block-cut vertex tree. Also, any tree with at least two vertices has two vertices of degree one. Thus the block-cut vertex tree $T$ has at least two pendant vertices. Hence $G$ has at least two end-blocks.   ❑

The next result is due to Harary and Norman [109].

**Theorem 5.5 (Harary and Norman [109])** The center of any connected graph $G$ lies on a block of $G$.

**Proof**   If not, let $B_1$, $B_2$ be blocks of $G$ containing central vertices. If $b_1$, $b_2$ are the vertices of the block-cut vertex tree $T$ of $G$ corresponding to $B_1$ and $B_2$, then there is at least one vertex $c$ in the unique $b_1 - b_2$ path of $T$, corresponding to a cut-vertex $c$ of $G$. So there are two components $G_1$ and $G_2$ of $G - c$ such that $B_1 - c \subseteq G_1$ and $B_2 - c \subseteq G_2$. Let $\bar{c}$ be an eccentric vertex of $c$ in $G$ and $P$ be a $c - \bar{c}$ path of $G$ having length $e(c)$. Then at least one of the components $G_1$ and $G_2$, say $G_2$, contains no vertex of $P$. Let $s$ be a central vertex in $G_2$ and $Q$ be a shortest $s - c$ path in $G$. Then $Q \cup P$ is clearly an $s - \bar{c}$ path in $G$ and thus $d(s, \bar{c}) = d(s, c) + e(c)$. Therefore, $e(s) > e(c)$ contradicting the fact that $s$ is a central vertex. Thus the center of $G$ lies in a single block. Hence the center of any connected graph $G$ lies on a block of $G$.   ❑

**Definition:** If $G$ is a separable graph and $c$ a cut-vertex of $G$, then a maximal connected subgraph of $G$ containing $c$ in which $c$ is not a cut-vertex is called a *branch* of $G$ at $c$. The induced subgraph $\langle C \rangle$ on the central vertices of $G$ is called the *central graph* of $G$. If $G$ has a unique central vertex $c$, then $G$ is said to be a *unicentric graph*. The unique block $B$ of $G$ to which the center $c$ of $G$ belongs is called the *central block* of $G$. This is unambiguously defined except when $G$ is unicentric and the unique central vertex is a cut-vertex of $G$. When $B = \langle C \rangle = G$, then $G$ is called a *self-centered graph*. If the unique central vertex $c$ of $G$ is a cut-vertex of $G$, the unique block of any of the branches of $G$ at $c$ in which $c$ has an eccentric vertex $\bar{c}$ may be taken as the chosen central block of $G$.

We note that the central graph of a tree is either $K_1$ or $K_2$. Buckley, Miller and Slater [53] have studied graphs with specified central graphs. The following result is atributed to Hedetniemi and is reported in Parthasarathy [180].

**Theorem 5.6** For any graph $H$ there exists a graph $G$ with 4 more vertices such that $H$ is the central graph of $G$.

**Proof** Take two new vertices $v$ and $w$, and join each to every vertex of $H$. Take two other vertices $x$ and $y$, join $x$ to $v$, and $y$ to $w$. Then in the resulting graph $G$, $e(x) = e(y) = 4$, $e(v) = e(w) = 3$ and $e(u) = 2$, for every vertex $u \in V(H)$. Thus $H$ is an induced subgraph of $G$ and the central graph of $G$. ❏

## 5.3 Connectivity Parameters

Now, assume that a graph does not get disconnected by deleting a single vertex, or by removing a single edge. A natural question then arises: what is the minimum number of vertices or edges required to disconnect a graph? This and other related questions are answered in this section. Before proceeding, we have the following definitions.

**Definition:** Let $G = (V, E)$ be a graph. A subset $S$ of $V \cup E$ is called a *disconnecting set* of the graph $G$ if $k(G - S) > k(G)$, or $G - S$ is the trivial graph.

If a disconnecting set $S$ is a subset of $V$, it is called a *vertex cut* of $G$, and if it is a subset of $E$ it is called an *edge cut* of $G$. If a disconnecting set $S$ contains vertices and edges it is called a *mixed cut*.

**Example** For the graph shown in Figure 5.9, $S = \{3, e_3\}$ is mixed cut $S = \{3\}$ is vertex cut, and $S = \{e_1, e_3\}$ is edge cut.

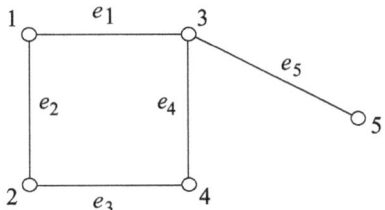

**Fig. 5.9**

A mixed cut/vertex cut/edge cut $S$ is *minimal* if no proper subset of $S$ has the same property as $S$. A mixed cut/vertex cut/edge cut $S$ is *minimum* if it has least cardinality among all such minimal sets. A minimal vertex cut is called a *knot* and a minimum vertex cut is called a *clot*. The cardinality of a clot is called the *vertex-connectivity* number, or clot number of the graph $G$ and is denoted by $\kappa(G)$.

A minimal edge cut is called a *bond* and a minimum edge cut is called a *band*. The cardinality of a band is called the *edge-connectivity* number, or band number of the graph $G$, and is denoted by $\lambda(G)$.

The minimum cardinality of a mixed set is denoted by $\sigma(G)$.

Let $S$ be a disconnected set of the graph $G = (V, E)$. Let vertices $s$ and $t$ be in the same component of $G$, but in different components of $G - S$. Then $S$ is called an $s - t$ *separating set* in $G$. Minimal $s - t$ separating vertex cut is called an $s - t$ *knot*, and the minimum $s - t$ separating vertex cut is called an $s - t$ *clot*. Minimal $s - t$ separating edge cut is called an $s - t$ *bond*, and the minimum $s - t$ separating edge cut is called an $s - t$ *band*. The cardinality of an $s - t$ clot is called the $s - t$ *clot number* and is denoted by $\kappa(s, t)$, and the cardinality of an $s - t$ band, called the $s - t$ *band number*, is denoted by $\lambda(s, t)$. The cardinality of a minimum $s - t$ separating mixed cut is denoted by $\sigma(s, t)$.

The following result gives vertex connectivity of complete graphs and an upper bound for non-complete graphs.

**Theorem 5.7**    $\kappa(K_n) = n - 1$. If $G$ is incomplete, then $\kappa(G) \leq n - 2$.

**Proof**

i. Clearly, $K_n$ is a connected graph with $n$ vertices. Now, deleting of a vertex $v_1$ keeps the graph $G - v_1$ connected. Clearly, $G - v_1$ has $n - 1$ vertices. Deleting one more vertex, say $v_2$ from $G - v_1$, gives a graph $G - \{v_1, v_2\}$, which is again connected. Continuing this process, we observe that deleting any number of vertices $i$, $1 \leq i \leq n - 1$ does not disconnect the graph, but deleting exactly $n - 1$ vertices gives a trivial graph with one vertex. Thus, $\kappa(K_n) = n - 1$.

ii. Let $G$ be an incomplete graph with $n$ vertices. Then there are at least two vertices, say $v_i$ and $v_j$ which are not adjacent. If there is exactly one edge $v_iv_j$ missing, then deleting the $n - 2$ vertices other than $v_i$ and $v_j$ disconnects the graph. So in this case $\kappa(G) = n - 2$. If there are more edges missing, then clearly $\kappa(G) < n - 2$ (Fig. 5.10). Hence, $\kappa(G) \leq n - 2$.  ❏

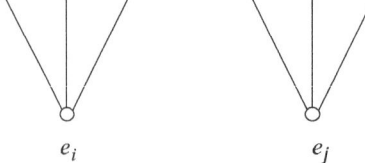

**Fig. 5.10**

The following result is obvious.

**Theorem 5.8**  $\kappa(G) = \min_{\substack{s,\,t\in v \\ st\notin E}} \kappa(s,\,t),\ \lambda(G) = \min_{s,\,t\in v} \lambda(s,\,t),\ \sigma(G) = \min_{s,\,t\in v} \sigma(s,\,t).$

**Cut of a graph:**   Let $G = (V,\,E)$ be a graph and let $A$ be any non-empty sub-set of the vertex set $V$. Let $\bar{A} = V - A$. The set of all edges with one end in $A$ and the other end in $\bar{A}$, denoted by $[A,\bar{A}]$ is called a *cut* of $G$. The concept of a cut of a graph is intermediate between that of an edge cut and a bond.

We note that every cut is an edge cut, but the converse is not true. Consider the graph in Figure 5.11. Here, $F = \{e_1,\,e_2,\,e_3,\,e_4,\,e_5\}$ is an edge cut, but $F$ is not a cut.

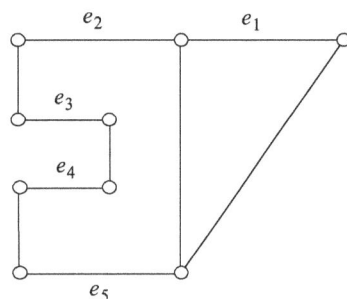

**Fig. 5.11**

Also, every bond is a cut, but the converse is not true. This is illustrated by the graph in Figure 5.12. Let $A = \{1,\,2,\,3,\,4\}$. Then $\bar{A} = \{5,\,6,\,7,\,8\}$. So, $[A,\,\bar{A}] = \{e_1,\,e_2,\,e_3,\,e_4\}$. Here, $[A,\,\bar{A}]$ is a cut but not a bond.

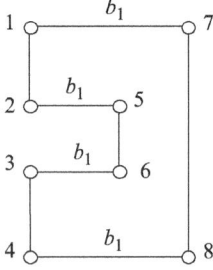

**Fig. 5.12**

**Theorem 5.9**    Every minimal cut is a bond and every bond is a minimal cut.

**Proof**    Let $G(V, E)$ be a graph. Let $[A, \bar{A}]$ be a cut of $G$. Assume $C = [A, \bar{A}]$ to be a minimal cut. Then no subset of the edges of $C$ is a cut and this implies that $G - C$ has only two components $\langle A \rangle$ and $\langle \bar{A} \rangle$. Therefore $C$ is a bond.

Conversely, let $F$ be a bond. Then $G - F$ has only two components, say $C_1$ and $C_2$. Then $F = [V_1, V_2]$, with $V_2 = \bar{V}_1$. Thus $F$ is a cut and hence a minimal cut.    ❏

**Theorem 5.10**    Every cut is a disjoint union of minimal cuts.

**Proof**    Let $G = (V, E)$ be a graph and let $C = [A, \bar{A}]$ be a cut of $G$. Let $C$ be not a minimal cut. Then at least one of $\langle A \rangle$, or $\langle \bar{A} \rangle$ has more than one component.

Assume $C_1, C_2, \ldots, C_r$ to be the components of $\langle A \rangle$ and $C_1', C_2', \ldots, C_s'$ be the components of $\langle \bar{A} \rangle$. (Clearly, at least one of $r$ and $s$ is greater than one.) Let $C_i$ be coalesced to vertices $c_i$, $1 \le i \le r$ and $C_i'$ be coalesced to vertices $c_i'$, $1 \le i \le s$, and let $H$ be the simple coalescence thus obtained. Obviously in $H$, there are no edges of the form $c_i c_j$ and $c_i' c_j'$, $i \ne j$. Thus $H$ is a bipartite graph (because there are edges $c_i c_j'$ in $H$) (Fig. 5.13).

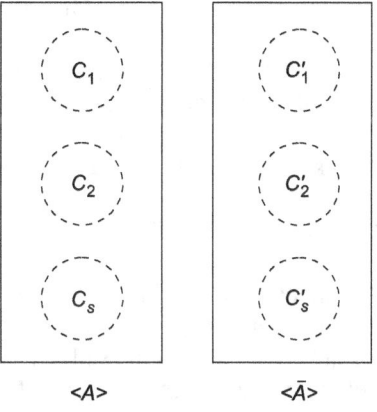

**Fig. 5.13**

If we can partition the edge set of $H$ into a disjoint union of bonds of $H$, the edges of $G$ corresponding to these bonds will be disjoint bonds of $G$ whose union is $C$. To achieve such a partition of $E(H)$, we first take the cut edges of $H$ as members of the partition and let $F$ be the set of such cut edges. For the remaining members of the partition, we take the stars at the remaining (non-isolated) vertices $c_i$ (or $c_i'$). This gives the required partition and hence the result follows.    ❏

**Illustration**    Consider the graph of Figure 5.14. Partition of $E(H)$ is $\{e_1\} \cup \{e_2\} \cup \{e_3, e_4\}$ $\cup \{e_5, e_6\}$. $e_1$ is a cut edge, $e_2$ is a cut edge, $\{e_3, e_4\}$ form the star $K_{1,2}$ and $\{e_5, e_6\}$ form the star $K_{1,2}$.

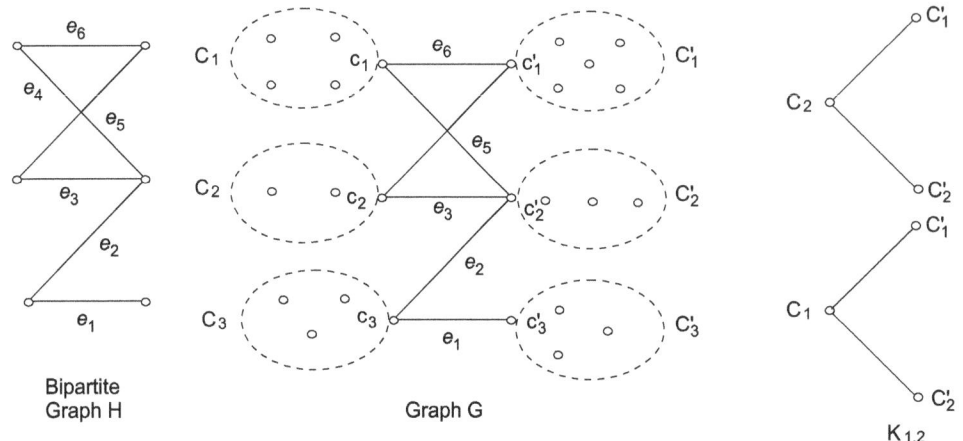

**Fig. 5.14**

**Remark**   Though the equivalence of minimal edge cuts and minimal cuts is brought out by Theorem 5.10, there is an essential difference between edge cuts and cuts as already mentioned. To emphasise this, we observe that Theorem 5.10 cannot be generalised to state that every edge cut is a disjoint union of bonds. The example in Figure 5.11 illustrates this point.

Since it is enough to consider connected graphs for discussing connectivity concepts, in what follows, we shall assume that graphs are connected, unless stated otherwise.

The following results are reported by Harary and Frisck [105].

**Theorem 5.11** In a connected graph $G = (V, E)$, if $st \notin E$, then $\kappa(s, t) \leq \sigma(s, t)$.

**Proof**   Let $G = (V, E)$ be a connected graph, and $s, t$ be vertices in $V$ such that $st \notin E$. Let $\kappa(s, t)$ be the cardinality of the minimum $s - t$ separating vertex cut ($s - t$ clot). Let $\sigma(s, t)$ be the cardinality of a minimum $s - t$ separating mixed cut. We prove that from any mixed $s - t$ separating set, we can get an $s - t$ separating vertex cut with no more elements.

Let $S$ be a minimum mixed $s - t$ separating set. If $ij$ is an edge in $S$, then both $i$ and $j$ cannot coincide with $s$ and $t$, since $st \notin E$.

If $i = s$, add $i$ to $S$, and remove from $S$ all edges with $i$ as an end vertex. If $i \neq s$, add $j$ to $S$ and remove from $S$ all edges with $j$ as an end vertex. The resulting, possibly mixed set is clearly an $s - t$ separating set with no more elements than $S$.

We repeat this process and remove all edges from $S$, and obtain a vertex cut $S'$ with atmost $|S|$ elements.

Since, $\kappa(s, t) \leq |S'| \leq |S| = \sigma(s, t)$, we have

$$\kappa(s, t) \leq \sigma(s, t). \qquad \qquad \square$$

**Corollary** In a connected graph $G = (V, E)$, if $st \notin E$, then $\kappa(s, t) \leq \lambda(s, t)$.

**Note** If $st \in E$, then $\kappa(s, t)$ is not defined.

**Theorem 5.12** For any graph $G$, $\sigma(G) = \kappa(G)$.

### Proof

**Case (i)** When $G = K_n$, then $\kappa(G) = n - 1$ and $\lambda(G) = n - 1$.

Let $S$ be a minimum mixed disconnecting set of $G$ and let $S = T \cup F$, where $T \subseteq V$, $F \subseteq E$, and $|T| = n_1$, $|F| = m_1$. Then $G - T$ is $K_{n-n_1}$. Therefore, $|F| \geq \lambda(K_{n-n_1-1}) = n - n_1 - 1$. Thus, $m_1 \geq n - n_1 - 1$. So, $\sigma = |S| = m_1 + n_1 \geq n - n_1 - 1 + n_1 = n - 1$. Therefore, $\sigma \geq n - 1 = \kappa$.

Also, $\sigma = \kappa$. Hence, $\sigma = \kappa$.

**Case (ii)** When $G$ is incomplete, then clearly $\sigma = \kappa$. We have to prove that $\sigma = \kappa$ when $G$ is complete. If possible, let there be a minimum $s - t$ separating mixed set $S = M \cup \{st\}$ with $\sigma = |S| < \kappa$. Now, $M$ can be replaced by a set of vertices $T$ (a subset of the vertex set of the induced subgraph $\langle M \rangle$) to provide a vertex cut of $G^* = G - st$ with cardinality at most $|M|$.

Let $C_1$ and $C_2$ be the components of $G - S$ to which $s$ and $t$ respectively belong. Let there be another component $C_3$ of $G - S$ and let $v$ be a vertex of $C_3$. Then $T \cup \{s\}$ is a $v - t$ separating vertex cut of $G$. But then $|T \cup \{s\}| \leq |S| < \kappa$, a contradiction (Fig. 5.15).

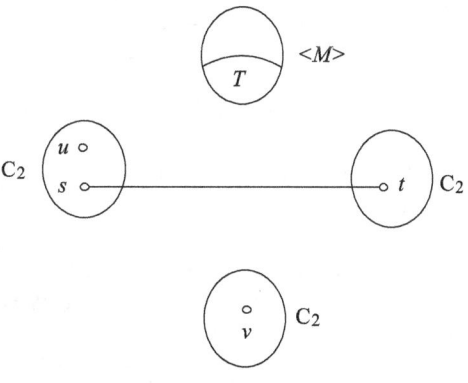

**Fig. 5.15**

Thus $C_1$ and $C_2$ are the only components of $G - S$. Also, if $u \in V(C_1)$, and $u \neq s$, then $T \cup \{s\}$ is a $u - t$ separating vertex cut of $G$, again leading to a contradiction. Thus, $C_1 = \{s\}$, and similarly $C_2 = \{t\}$. So $G$ has $|V(M)| + 2$ vertices, and is incomplete. Therefore, $\kappa(G) \leq n - 2 = |V(M)| + 2 - 2 = |V(M)| < \kappa$ implying $\kappa(G) < \kappa$, a contradiction. Thus, $\sigma \not< \kappa$. Hence, $\sigma = \kappa$. ☐

The following inequalities are due to Whitney [265].

**Theorem 5.13 (Whitney [265])** For any graph $G$, $\kappa(G) \leq \lambda(G) \leq \delta(G)$.

**Proof**   We first prove $\lambda(G) \leq \delta(G)$.

If $G$ has no edges, then $\lambda = 0$ and $\delta = 0$. If $G$ has edges, then we get a disconnected graph, when all edges incident with a vertex of minimum degree are removed. Thus, in either case, $\lambda(G) \leq \delta(G)$.

We now prove $\kappa(G) \leq \lambda(G)$. For this, we consider the various cases. If $G = K_n$, then $\kappa(G) = \lambda(G) = n - 1$. Now let $G$ be an incomplete graph. In case $G$ is disconnected or trivial, then obviously $\kappa = \lambda = 0$.

If $G$ is disconnected and has cut edge (bridge) $x$, then $\lambda = 1$. In this case, $\kappa = 1$, since either $G$ has a cut vertex incident with $x$, or $G$ is $K_2$.

Finally, let $G$ have $\lambda \geq 2$ edges whose removal disconnects it. Clearly, the removal of $\lambda - 1$ of these edges produces a graph with a cut edge (bridge) $x = uv$. For each of these $\lambda - 1$ edges, select an incident vertex different from $u$ or $v$. The removal of these vertices also removes the $\lambda - 1$ edges and quite possibly more. If the resulting graph is disconnected, then $\kappa < \lambda$. If not, $x$ is a cut edge (bridge) and hence the removal of $u$ or $v$ will result in either a disconnected or a trivial graph, so that $\kappa \leq \lambda$ in every case.   ❑

**Illustration**   We illustrate this by the graph shown in Figure 5.16. Here, $\kappa = 2$, $\lambda = 3$ and $\delta = 4$.

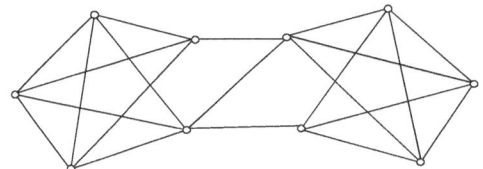

**Fig. 5.16**

**Theorem 5.14**   For any $v \in V$ and any $e \in E$ of a graph $G = (V, E)$, $\kappa(G) - 1 \leq \kappa(G - v)$ and $\lambda(G) - 1 < \lambda(G - e) \leq \lambda(G)$.

**Proof**   We observe that the removal of a vertex or an edge from a graph can bring down $\kappa$ or $\lambda$ by at most one, and that while $\kappa$ may be increased by the removal of a vertex, $\lambda$ cannot be increased by the removal of an edge.   ❑

**Theorem 5.15**   For any three integers $r$, $s$, $t$ with $0 < r \leq s \leq t$, there is a graph $G$ with $\kappa = r$, $\lambda = s$ and $\delta = t$.

**Proof**   Take two disjoint copies of $K_{t+1}$. Let $A$ be a set of $r$ vertices in one of them and $B$ be a set of $s$ vertices in the other. Join the vertices of $A$ and $B$ by $s$ edges utilising all the vertices of $B$ and all the vertices of $A$. Since $A$ is a vertex cut and the set of these $s$ edges is an edge cut of the resulting graph $G$, it is clear that $\kappa(G) = r$ and $\lambda(G) = s$. Also, there is at least one vertex which is not in $A \cup B$, and it has degree $t$, so that $\delta(G) = t$.   ❑

**Illustration**   Let $r = 1$, $s = 2$, $t = 3$. Take two copies of $K_4$. Here, $\kappa(G) = 1$, $\lambda(G) = 2$, $\delta(G) = 3$ (Fig. 5.17).

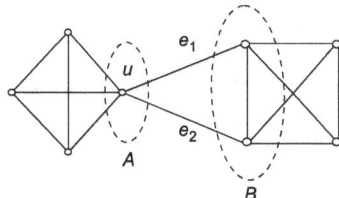

**Fig. 5.17**

**Theorem 5.16**   For a graph, $\delta \geq \dfrac{n}{2}$ ensures $\lambda = \delta$.

**Proof**   Let $G$ be a graph with $\delta \geq \dfrac{n}{2}$. Let $\lambda < \delta$. Let $F$ be a set of $\lambda$ edges disconnecting $G$. Let $C_1$ and $C_2$ be the components of $G - F$, and $A_1$ and $A_2$ be the end vertices of $F$ in $C_1$ and $C_2$, respectively.

Suppose $|A_1| = r$, $|A_2| = s$ and also $V(C_1) = A_1$. Then each vertex of $C_1$ is adjacent with at least one edge of $F$. So the number $m_1$ of edges in $C_1$ satisfies the inequality

$$m_1 \geq \frac{1}{2}(r\delta - \lambda) > \frac{1}{2}(r\delta - \delta), \text{ since } \lambda < \delta \text{ by assumption.}$$

Therefore, $m_1 > \dfrac{1}{2}(r-1)\delta > \dfrac{1}{2}(r-1)r$, since $r \leq |F| = \lambda < \delta$.

But a graph on $n$ vertices cannot have more than $\dfrac{1}{2}r(r-1)$ edges. Thus, $|V(C_1)| > |A_2|$. Similarly, $V(C_2) > A_2$. Thus, each of $C_1$ and $C_2$ contains at least $\delta + 1$ vertices.

Therefore, $n = |V(G)| \geq 2(\delta + 1) \geq 2(\dfrac{n}{2} + 1) = n + 2$ or $n \geq n + 2$, which is a contradiction. Hence $\lambda < \delta$ is not possible. So, $\lambda = \delta$.                                                              ❑

## 5.4  Menger's Theorem

Harary [104] has listed eighteen variations of Menger's theorem including those for digraphs. Clearly, all these are equivalent and one can be obtained from the other. Several proofs of the various forms of Menger's theorems have appeared, for example, in Dirac [67], Ford and Fulkerson [81], Lovasz [150], McCuaig [156], Menger [158], Nash-Williams and Tutte [169], O'Neil [173], Pym [213] and Wilson [269].

Let $u$ and $v$ be two distinct vertices of a connected graph $G$. Two paths joining $u$ and $v$ are called disjoint (vertex disjoint) if they have no vertices other than $u$ and $v$ (and hence no edges) in common. The maximum number of such paths between $u$ and $v$ is denoted by $p(u, v)$. If the graph $G$ is to be specified, it is denoted by $p(u, v|G)$.

Now, we give the vertex form of Menger's theorem. The proof is due to Nash-Williams [9] and Tutte [169].

**Theorem 5.17 (Menger-vertex form)**   The minimum number of vertices separating two non-adjacent vertices $s$ and $t$ is equal to the maximum number of disjoint $s-t$ paths, that is, for any pair of non-adjacent vertices $s$ and $t$, the clot number equals the maximum number of disjoint $s-t$ paths. That is, $\kappa(s, t) = p(s, t)$, for every pair $s, t \in V$ with $st \notin E$.

**Proof**   Let $G = (V, E)$ be a graph with $|E| = m$. We use induction on $m$, the number of edges. The result is obvious for a graph with $m = 1$ or $m = 2$. Assume that the result is true for all graphs with less than $m$ edges. Let the result be not true for the graph $G$ with $m$ edges. Then we have

$$p(s, t|G) < \kappa(s, t|G) = q \text{ (say)}, \tag{5.17.1}$$

as for any graph, we obviously have $p(s, t) \leq \kappa(s, t)$.

Let $e = uv$ be an edge of $G$. The deletion graph $G_1 = G - e$, and the contraction graph $G_2 = G|e$ have less number of edges than $G$. Therefore, by induction hypothesis, we have

$$p(s, t|G_1) = \kappa(s, t|G_1), \text{ and } p(s, t|G_2) = \kappa(s, t|G_2). \tag{5.17.2}$$

Let $I$ be an $(s, t)-$ clot in $G_1$ and $J'$ be an $(s, t)-$ clot in $G_2$. Then we have

$$|I| = \kappa(s, t|G_1) = p(s, t|G_1) \leq p(s, t|G) < q, \text{ and}$$

$|J'| = \kappa(s, t|G_2) = p(s, t|G_2) \leq p(s, t|G) < q$, by using (5.17.2) and (5.17.1).

So, $|J'| < q$ and therefore $|J'| \leq q - 1$.

Now to $J'$ there corresponds an $(s-t)$ vertex cut $J$ of $G$ such that $|J| \leq |J'| + 1$, since, by elementary contraction, $\kappa(s, t)$ can be decreased by at most one, and this decrease actually occurs when $e \in E(\langle J \rangle)$.

Thus, $|J| \leq |J'| + 1 \leq q - 1 + 1 = q$, that is,

$$|J| \leq q. \tag{5.17.3}$$

Since $J$ is an $(s, t)$ vertex cut in $G$, $\kappa(s, t) \leq |J|$, $q \leq |J|$.

Thus, $q \leq |J| \leq q$, so that $|J| = q$.

Therefore, $|I| < q$ and $|J| = q$ and                                                                       (5.17.4)

$u, v \in J$ by (5.17.3).

Now, let

$$H_s = \{w \in I \cup J : \text{there exists an } s - w \text{ path in } G, \text{ vertex-disjoint from } I \cup J - \{w\}\}, \text{ and}$$

$H_t = \{w \in I \cup J : \text{there exists a } t - w \text{ path in } G, \text{ vertex-disjoint from } I \cup J - \{w\}\}.$

Clearly, $H_s$ and $H_t$ are $(s - t)$ separating vertex cuts in $G$. Therefore,

$$|H_s| \geq q \text{ and } |H_t| \geq q. \tag{5.17.5}$$

Obviously, $H_s \cup H_t \subseteq I \cup J$.

We claim that $H_s \cap H_t \subseteq I \cup J$. For this, let $w \in H_s \cap H_t$. Then there exists an $s - w$ path $P_1$ and $w - t$ path $P_2$ in $G$ vertex disjoint from $I \cup J - \{w\}$. So $P_1 \cup P_2$ contains a path, say $P$. If $e \in P$ then we have $u, v \in V(P) \cap J \subseteq \{w\}$, which is impossible. Therefore $e \notin P$ and so $P \subseteq G - e$. Since $I$ is an $(s, t)$ separator in $G - e$ and $J$ is an separator in $G$, $P$ has a vertex common with $I$ and also with $J$. So $w \in I \cap J$. Thus, $H_s \cap H_t \subseteq I \cap J$.

Combining (5.17.4) and (5.17.5), and the above observation, we have

$$q + q \leq |H_s| + |H_t| = |H_s \cup H_t| + |H_s \cap H_t| \leq |I \cup J| + |I \cap J|$$

$$= |I| + |J| < q + q,$$

which is a contradiction.

Thus, (5.17.1) is not true, and therefore, we have

$$\kappa(s, t|G) = p(s, t|G). \qquad \square$$

**Definition:** Two paths joining $u$ and $v$ are said to be *edge-disjoint* if they have no edges in common. The maximum number of edge-disjoint paths between $u$ and $v$ is denoted by $l(u, v)$.

We now give the edge form of Menger's theorem and the proof is adopted from Wilson [196].

**Theorem 5.18 (Menger-edge form)** For any pair of vertices $s$ and $t$ of a graph $G$, the minimum number of edges separating $s$ and $t$ equals the maximum number of edge-disjoint paths joining $s$ and $t$, that is, $\lambda(s, t) = l(s, t)$ for every pair $s, t \in V$.

**Proof** Let $G = (V, E)$ be a graph and let $|E| = m$. We use induction on the number of edges $m$ of $G$. For $m = 1, 2$, the result is obvious. Assume the result to be true for all graphs with fewer than $m$ edges. Let $\lambda(s, t) = k$. We have two cases to consider.

**Case (i)** Suppose $G$ has an $(s - t)$ band $F$ such that not all edges of $F$ are incident with $s$, nor all edges of $F$ are incident with $t$. Then $G - F$ consists of two non-trivial components $C_1$ and $C_2$ with $s \in C_1$ and $t \in C_2$. Let $G_1$ be the graph obtained from $G$ by contracting the edges of $C_1$ and $G_2$ be a graph obtained from $G$ by contracting the edges of $C_2$. Therefore,

$$G_1 = G||E(C_1) \text{ and } G_2 = G||E(C_2).$$

Since $G_1$ and $G_2$ have less edges than $G$, the induction hypothesis applies to them. Also, the edges corresponding to $F$ provide an $(s - t)$ band in $G_1$ and $G_2$, so that $(s, t|G_1) = k$

and $\lambda(s, t|G_2) = k$. Thus, by induction hypothesis, there are $k$ edge-disjoint paths joining $s$ and $t$ in $G_1$, there are $k$ edge-disjoint paths joining $s$ and $t$ in $G_2$. Thus, $l(s, t|G_1) = k$ and $l(s, t|G_2) = k$.

The section of the path of the $k$ edge-disjoint paths joining $s$ and $t$ in $G_2$ which are in $C_1$ and the section of the paths of the $k$ edge-disjoint paths joining $s$ and $t$ in $G_1$ which are in $C_2$ can now be combined to get $k-$ edge disjoint paths between $s$ and $t$ in $G$. Hence, $l(s, t|G) = k$.

**Case (ii)**    Every $(s-t)$ band of $G$ is such that either all its edges are incident with $s$, or all its edges are incident with $t$.

If $G$ has an edge $e$ which is not in any $(s-t)$ band of $G$, then $\lambda(s, t|G-e) = \lambda(s, t|G) = k$. Since the induction hypothesis is applicable to $G-e$, there are $k$ edge-disjoint paths between $s$ and $t$ in $G-e$ and thus in $G$. Hence, $l(s, t|G) = k$.

Now, assume that every edge of $G$ is in at least one $(s-t)$ band of $G$. Then every $s-t$ path $P$ of $G$ is either a single edge or a pair of edges. Any such path $P$ can therefore contain at most one edge of any $(s-t)$ band. Then $G - E(P) = G_1$ is a graph with $\lambda(s, t|G_1) = \kappa - 1$.

Appling induction hypothesis, we have $l(s, t|G_1) = \kappa - 1$. Together with $P$, we get $l(s, t|G) = K$.                                                                                       ❏

**Definition:**    A graph $G$ is said to be *n-(vertex)* connected if $\kappa(G) = n$ and *n-(edge)* connected if $\lambda(G) = n$. Thus a separable graph ($\kappa = 1$) is 1-connected and not 2-connected. A separable graph without cut edges is only 1-edge connected.

# 5.5  Some Properties of a Bond

We give some properties of a bond (bond is also called a cut-set). The first property follows.

**Theorem 5.19**    Every bond in a connected graph $G$ connects at least one branch of every spanning tree of $G$.

**Proof**    Let $G$ be a connected graph and $T$ be a spanning tree of $G$. Let $S$ be an arbitrary bond in $G$. Clearly, there are edges which are common in $S$ and $T$. For, if there is no edge of $S$ which is also in $T$, then removal of the bond $S$ from $G$ will not disconnect the graph, as $G - S$ contains $T$ and is therefore connected. Thus, $S$ and $T$ have at least one common edge.                                                                                       ❏

**Theorem 5.20**    In a connected graph $G$, any minimal set of edges containing at least one branch of every spanning tree of $G$ is a bond.

**Proof**    Let $G$ be a connected graph and let $Q$ be a minimal set of edges containing at least one branch of every spanning tree of $G$.

Consider $G - Q$, the subgraph that remains after removing the edges $Q$ from $G$. Since $G - Q$ contains no spanning tree of $G$, therefore $G - Q$ is disconnected (one component of which may just consist of an isolated vertex). Also, since $Q$ is a minimal set of edges with this property, therefore any edge $e$ from $Q$ returned to $G-Q$ creates at least one spanning tree. Thus, the subgraph $G-Q+e$ is a connected graph. Therefore, $Q$ is a minimal set of edges whose removal from $G$ disconnects $G$. This, by definition, is a bond. ❑

**Theorem 5.21** Every cycle has an even number of edges in common with any bond.

**Proof** Let $G$ be a graph and let $S$ be a bond of $G$. Let the removal of $S$ partition the vertices of $G$ into two mutually disjoint subsets $V_1$ and $V_2$. Consider a cycle $C$ in $G$ (Fig. 5.18).

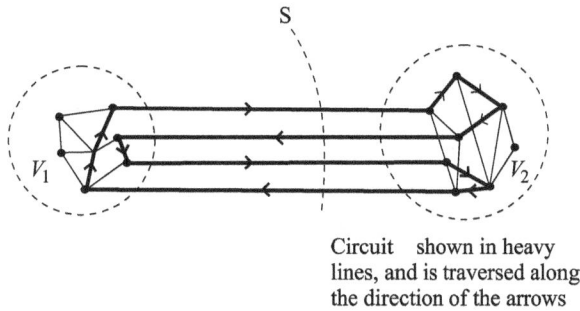

Circuit shown in heavy
lines, and is traversed along
the direction of the arrows

**Fig. 5.18**

If all the vertices in $C$ are entirely within vertex set $V_1$ (or $V_2$), then the number of edges common to $S$ and $C$ is zero, which is an even number. If, on the other hand, some vertices in $C$ are in $V_1$ and some in $V_2$, we traverse back and forth between the sets $V_1$ and $V_2$ as we traverse the cycle. Because of the closed nature of a cycle, the number of edges between $V_1$ and $V_2$ must be even. And, since every edge in $S$ has one end in $V_1$ and other in $V_2$, and no other edge in $G$ has the property of separating sets $V_1$ and $V_2$, the number of edges common to $S$ and $C$ is even. ❑

## 5.6 Fundamental Bonds

Consider a spanning tree $T$ of a connected graph $G$. Take any branch $b$ in $T$. Since $\{b\}$ is a bond in $T$, therefore $\{b\}$ partitions all vertices of $T$ into two disjoint sets, one at each end of $b$. Consider the same partition of vertices in $G$ and the bond $S$ in $G$ that corresponds to this partition. Bond $S$ will contain only one branch $b$ of $T$ and the rest (if any) of the edges in $S$ are chords with respect to $T$. Such a bond $S$ containing exactly one branch of a tree $T$ is called a *fundamental bond* with respect to $T$.

**Theorem 5.22** The ring sum of any two bonds is either a third bond, or an edge-disjoint union of bonds.

**Proof** Let $G$ be a connected graph, and $S_1$ and $S_2$ be two bonds. Let $V_1$ and $V_2$ be the unique and disjoint partitioning of the vertex set $V$ of $G$ corresponding to $S_1$. Let $V_3$ and $V_4$ be the partitioning corresponding to $S_2$.

Clearly, $V_1 \cup V_2 = V$, $V_1 \cap V_2 = \varphi$, $V_3 \cup V_4 = V$ and $V_3 \cap V_4 = \varphi$ (Fig. 5.19(a) and (b)).

Now, let $(V_1 \cap V_4) \cup (V_2 \cap V_3) = V_5$ and $(V_1 \cap V_3) \cup (V_2 \cap V_4) = V_6$.

Clearly, $V_5 = V_1 \oplus V_3$ and $V_6 = V_2 \oplus V_3$ (Fig. 5.19(c)).

Now, the ring sum of two bonds $S_1 \oplus S_2$ consists only of edges that join vertices in $V_5$ to those in $V_6$. Also, there are no edges outside $S_1 \oplus S_2$ that joins vertices in $V_5$ to those in $V_6$. Thus, the set of edges $S_1 \oplus S_2$ produces a partitioning of $V$ into $V_5$ and $V_6$ such that $V_5 \cup V_6 = V$ and $V_5 \cap V_6 = \varphi$. Hence, $S_1 \oplus S_2$ is a bond if the subgraphs containing $V_5$ and $V_6$ each remain connected after $S_1 \oplus S_2$ is removed from $G$. Otherwise, $S_1 \oplus S_2$ is an edge disjoint union of bonds.                                                                          ❑

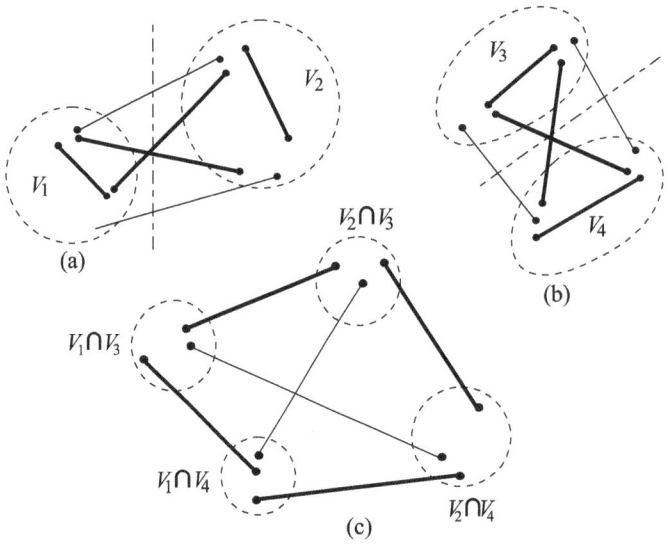

**Fig. 5.19**

**Example** Consider the graph in Figure 5.20. Here, $\{d, e, f\} \oplus \{f, g, h\} = \{d, e, g, h\}$ is a bond, $\{a, b\} \oplus \{b, c, e, f\} = \{a, c, e, f\}$ is another bond and $\{d, e, g, h\} \oplus \{f, g, k\} = \{d, e, f, h, k\} = \{d, e, f\} \cup \{h, k\}$ an edge disjoint union of bonds.

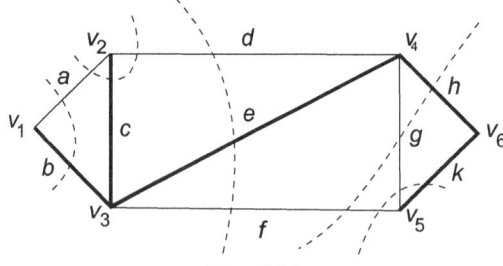

**Fig. 5.20**

**Theorem 5.23** With respect to a given spanning tree $T$, a chord $c_i$ that determines a fundamental cycle $C$ occurs in every fundamental bond associated with the branches in $C$ and in no other.

**Proof** Let $G$ be a connected graph and $T$ be a spanning tree of $G$. Let $c_i$ be a chord with respect to $T$ and let the fundamental cycle made by $c_i$ be called $C$, consisting of $k$ branches $b_1, b_2, \ldots, b_k$ in addition to the chord $c_i$. So $C = \{c_i, b_1, b_2, \ldots, c_k\}$ is a fundamental cycle with respect to $T$.

Now every branch of any spanning tree has a fundamental bond associated with it. So let $S_1$ be the fundamental bond associated with $b_1$, consisting of $q$ chords in addition to the branch $b_1$. Thus, $S_1 = \{b_1, c_1, c_2, \ldots, c_q\}$ is a fundamental bond with respect to $T$.

We know that there are even number of edges common to $C$ and $S_1$. Clearly, $b_1$ is in both $C$ and $S_1$. So there is exactly one more edge which is in both $C$ and $S_1$. Obviously, the edge $c_i$ in $C$ can possibly be in $S_1$. Thus, $c_i$ is one of the chords $c_1, c_2, \ldots, c_q$.

The same argument holds for fundamental bonds associated with $b_2, b_3, \ldots, b_k$. Thus the chord $c_i$ is contained in every fundamental bond associated with branches in $C$.

Now we show that the chord $c_i$ is not in any other fundamental bond $S'$ with respect to $T$, besides those associated with $b_1, b_2, \ldots, b_k$. Let this be possible. Then since none of the branches in $C$ are in $S'$, there is only one edge $c_i$ common to $S'$ and $C$, which gives a contradiction to the fact that there are even number of edges common to a fundamental bond and a cycle. $\qquad\Box$

**Example** In the graph of Figure 5.20, consider the spanning tree $\{b, c, e, h, k\}$. The fundamental cycle made by the chord is $C = \{f, e, h, k\}$. The three fundamental bonds determined by the three branches $e$, $h$ and $k$ are as follows: (i) determined by $e$ is $\{d, e, f\}$, (ii) determined by $h$ is $\{f, g, h\}$ and (iii) determined by $k$ is $\{f, g, k\}$. Clearly, chord $f$ occurs in each of these three fundamental bonds and there is no other fundamental bond that contains $f$.

**Theorem 5.24** With respect to a given spanning tree $T$, a branch $b_i$ that determines a fundamental bond $S$ is contained in every fundamental cycle associated with the chords in $S$, and in no others.

**Proof** Let $G$ be a connected graph and $T$ be a spanning tree in $G$. Let the fundamental bond determined by a branch $b_i$ be $S = \{b_i, c_1, c_2, \ldots, c_p\}$.

Let $C_1$ be the fundamental cycle determined by chord $c_1$, so that

$$C_1 = \{c_1, b_1, b_2, \ldots, b_q\}.$$

We know that $S$ and $C_1$ have even number of edges in common. One common edge is obviously $c_1$. Thus, the second common edge should be $b_i$, so that $b_i$ is also in $C_1$. Therefore, $b_i$ is one of the branches $b_1, b_2, \ldots, b_q$.

The same is true for the fundamental cycles made by the chords $c_2, c_3, \ldots, c_p$.

Now assume that $b_i$ occurs in a fundamental cycle $C_{p+1}$ made by a chord other than $c_1, c_2, \ldots, c_p$. Since none of the chords $c_1, c_2, \ldots, c_p$ is in $C_{p+1}$, there is only one edge $b_i$ common to a cycle $C_{p+1}$ and the bond $S$, which is not possible. Hence the result follows.

❑

**Example** Consider the graph of Figure 5.20. Consider the branch $e$ of spanning tree $\{b, c, e, h, k\}$. The fundamental bond determined by $e$ is $\{e, d, f\}$. The two fundamental cycles determined by chords $d$ and $f$ are respectively $\{d, c, e\}$ and $\{f, e, h, k\}$. Clearly, branch $e$ is contained in both these fundamental cycles and none of the remaining three fundamental cycles contains branch $e$.

**Theorem 5.25** Let $A$, $B$ be two disjoint vertex subsets of a graph $G$ and let any vertex subset of $G$ which meets every $A - B$ path in $G$ have at least $k$ vertices. Then there are $k$ vertex disjoint $A - B$ paths in $G$.

**Proof** Let $G$ be a graph and let $A$ and $B$ be two disjoint vertex subsets of $G$. Let $S$ be any vertex subset of $G$ which meets every $A - B$ path in $G$ and let $|S| \geq k$.

Take two new vertices $s$ and $t$, and join $s$ by an edge to each vertex of $A$, and join $t$ by an edge to each vertex of $B$. Let $G'$ be the resulting graph, and in $G'$ we have $\kappa(s, t) \geq k$.

Hence, by Menger's theorem, there are $k$ vertex disjoint paths between $s$ and $t$ in $G'$. Omitting the edge incident with $s$ and $t$ in these paths, we get $k$ vertex-disjoint $A - B$ path in $G$.

❑

**Definition:** A graph $G$ is *k-connected* if $\kappa(G) = k$, and $G$ is *k-edge connected* if $\lambda(G) = k$. A $k$-connected ($k$-edge connected) graph is $r$-connected ($r$-edge 1-connected) for each $r$, $0 \leq r \leq k - 1$. Clearly, a separable graph ($\kappa = 1$) is connected and not 2-connected. A separable graph without cut edge is 2-edge connected. A separable graph with cut edges is only 1-connected.

The following result is due to Whitney [265].

**Theorem 5.26 (Whitney [265])** A graph $G$ with at least three vertices is 2-connected if and only if any two vertices of $G$ are connected by at least two internally disjoint paths.

**Proof** Let $G$ be 2-connected so that $G$ contains no cut vertex. Let $u$ and $v$ be two distinct vertices of $G$. To prove the result, we induct on $d(u, v)$.

If $d(u, v) = 1$, let $e = uv$. Since $G$ is 2-connected and $n(G) \geq 3$, therefore $e$ cannot be a cut edge of $G$. For, if $e$ is a cut edge, then at least one of $u$ and $v$ is a cut vertex. Now, by Theorem 5.2, $e$ belongs to a cycle $C$ in $G$. Then $C - e$ is a $u - v$ path in $G$, internally disjoint from the path $uv$.

Assume that any two vertices $x$ and $y$ of $G$, such that $d(x, y) = t - 1$, $t \geq 2$, are joined by two internally disjoint $x - y$ paths in $G$. Let $d(u, v) = t$ and let $P$ be a $u - v$ path of length $t$, and $w$ be the vertex before $v$ on $P$. Then $d(u, w) = t - 1$. Therefore, by induction hypothesis, there are two internally disjoint $u - w$ paths, say $P_1$ and $P_2$, in $G$. Since $G$ has no cut vertex, $G - w$ is connected and therefore there exists a $u - v$ path $Q$ in $G - w$. Clearly, $Q$ is a $u - v$

path in $G$ not containing $w$. Suppose $x$ is the vertex of $Q$ such that $x-v$ section of $Q$ contains only the vertex $x$ in common with $P_1 \cup P_2$ (Fig. 5.21). Assume that $x$ belongs to $P_1$. Then the union of the $u-x$ section of $P_1$ and $x-v$ section of $Q$ together with $P_2 \cup \{wv\}$ are two internally disjoint $u-v$ paths in $G$.

Conversely, assume that any two distinct vertices of $G$ are connected by at least two internally disjoint paths. Then $G$ is connected. Also, $G$ has no cut vertex. For, if $v$ is a cut vertex of $G$, then there exist vertices $u$ and $w$ such that every $u-w$ path contains $v$, contradicting the hypothesis. Thus $G$ is 2-connected. ❏

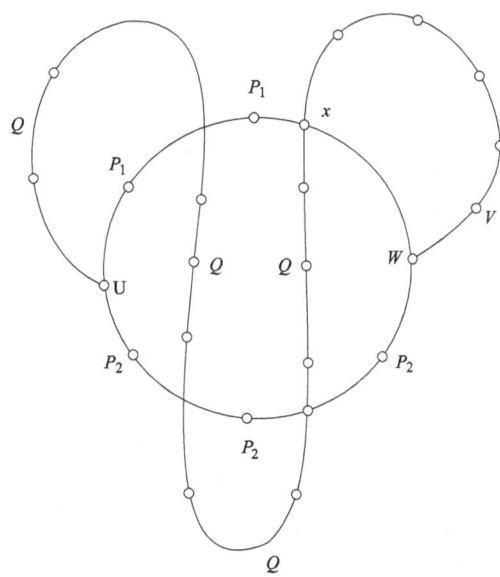

**Fig. 5.21**

The following property of 3-connected graphs is given in Thomassen [241] and is attributed to Barnette and Grunbaum [14] and Titov [243].

**Theorem 5.27** If $G$ is a 3-connected graph with at least five vertices, then $G$ has an edge $e$ such that $G-e$ is a subdivision of a 3-connected graph.

**Proof** Since $G$ is 3-connected, $\delta \geq 3$, and so, by Menger's theorem, $G$ has a subdivision of $K_4$.

Let $H$ be a proper subgraph of $G$ which is a subdivision of a 3-connected graph, and let $H$ have maximum possible number of edges. If $H$ is a 3-connected spanning subgraph of $G$, then by the maximality of $H$, $G$ has an edge $e$ such that $H = G-e$ is a subgraph of $G$ with the desired property.

Now, let $H$ be 3-connected but not spanning (Fig. 5.22). Then there is a vertex $v \in V(G) - V(H)$, so that there are three $v - V(H)$ paths, say $P_1$, $P_2$ and $P_3$ which have only vertex $v$ in common. Let the other end vertices of these paths in $V(H)$ be $v_1$, $v_2$ and $v_3$. If $v_1 v_2 = e \in E(H)$, then $H + P_1 + P_2 - e$ is a subdivision of a 3-connected graph. Otherwise,

$H + P_1 + P_2$ is a subdivision of a 3-connected graph. In both cases the maximality of $H$ is contradicted.

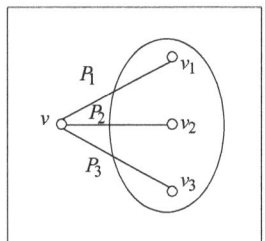

**Fig. 5.22**

Now, let $H$ be not 3-connected. Then $H$ has a suspended path $P$ of length at least 2. Let $u$ and $v$ be the end vertices of $P$. Since $G$ is 3-connected, $G - \{u, v\}$ has a path $P'$ joining an internal vertex $w$ of $P$ to a vertex in $V(H) - V(P)$. But then, $H \cup P'$ is a subdivision of a 3-connected graph. By the choice of $H$, $H \cup P' = G$ and $P'$ consists of a single edge $e'$.   ❏

The following property of 3-connected graphs is attributed to Thomassan [241].

**Theorem 5.28 (Thomassan [241])**   If $G$ is a 3-connected graph with at least five vertices, then $G$ has an edge $e$ such that $G|e$ is 3-connected.

**Proof**   Let $G$ be a 3-connected graph with at least five vertices. Let $e = uv$ be an edge of $G$ such that $G|e$ is not 3-connected. Then $G|e$ is 2-connected.

Let $\{x, y\}$ be a vertex cut of $G|e$ and let $z$ be the vertex into which $x$ and $y$ have been coalesced. Assume both $x$ and $y$ are different from $z$. Then $G|e - \{x, y\}$ is a graph obtained by contracting an edge of the connected graph $G - \{x, y\}$. This implies that $G|e - \{x, y\}$ is connected, which is a contradiction. Thus, one of $x$ and $y$ coincides with $z$. Renaming the other as $w$, we see that $G$ has a vertex cut $\{u, v, w\}$.

Let $G_1$ be the smallest component of $G - \{u, v, w\}$. Since $G$ is 3-connected, $G_1$ is joined to $w$ by an edge $e_1 = wx_1$. If $G|e_1$ is not 3-connected, by a similar argument, there is a vertex $y_1$ such that $G - \{w, x, y\} = G_2$ is disconnected. But then, the smallest component of $G_2$ is a proper subgraph of $G_1$.

Continuing in this way, we reach a stage when the smallest sub-graph is a single vertex, and the edge $f$ joining it to the previous vertex cut is such that $G|f$ is a 3-connected graph.
                                                                                              ❏

**Illustration**   We illustrate this in Figure 5.23, where graph $G$ is 3-connected having vertex cut $\{u, v, w\}$. $G|e$ is 2-connected with vertex cut $\{z, w\}$ and $G|e_1$ is 3-connected. In $G - \{u, v, w\}$, we observe that the smallest subgraph is a single vertex.

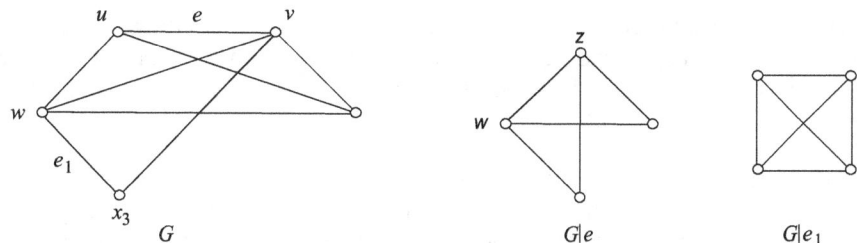

**Fig. 5.23**

## 5.7 Block Graphs and Cut Vertex Graphs

The block graph $B(G)$ of a graph $G$ is a graph whose vertices are the blocks of $G$ and two of these vertices are adjacent whenever the corresponding blocks contain a common cut vertex of $G$. The cut vertex graph $C(G)$ of a graph $G$ has vertices as cut vertices of $G$ and two such vertices are adjacent if the cut vertices of $G$ to which they belong lie on a common block (Fig. 5.24).

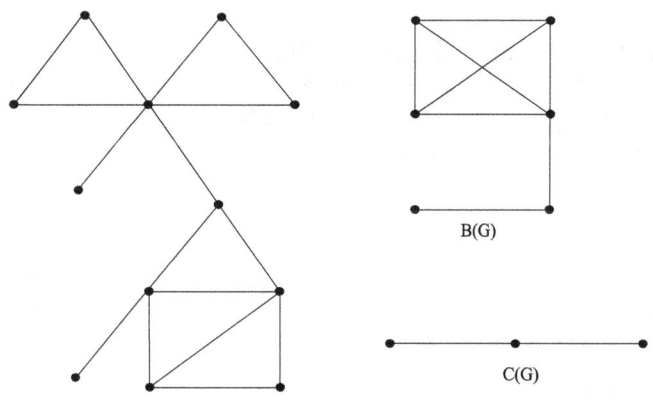

**Fig. 5.24**

The following characterisation of block graphs is due to Harary [103].

**Theorem 5.29 (Harary [103])** A graph $H$ is the block graph of some graph if and only if every block of $H$ is complete.

**Proof** Let $H = B(G)$ and assume there is a block $H_i$ of $H$ which is not complete. Then there are two vertices in $H_i$ which are non-adjacent and lie on a shortest common cycle $Z$ of length at least 4. But the union of the blocks of $G$ corresponding to the vertices of $H_i$ which lie on $Z$ is then connected and has no cut vertex, so it itself is contained in a block, contradicting the maximality property of a block of a graph.

Conversely, let $H$ be a given graph in which every block is complete. Form $B(H)$, and then form a new graph $G$ by adding to each vertex $H_i$ of $B(H)$ a number of end edges equal to the number of vertices of the block $H_i$ which are not cut vertices of $H$. Then it is easy to see that $B(G)$ is isomorphic to $H$.                                                              ❑

**1-Isomorphism**  A separable graph consists of two or more non-separable subgraphs, and each of the largest non-separable subgraph is a block. The graph in Figure 5.25 has five blocks and three cut vertices $u$, $v$ and $w$. We note that a non-separable connected graph consists of just one block.

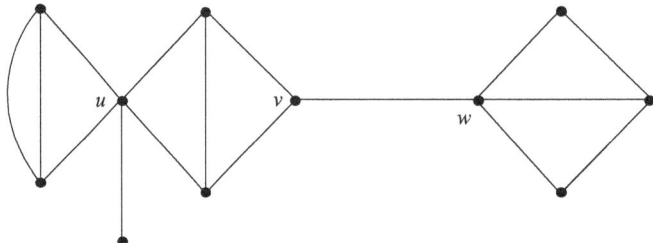

**Fig. 5.25**

Now, compare the disconnected graphs of Figure 5.26 with the graph of Figure 5.25. Clearly, these two graphs are not isomorphic, as they do not have the same number of vertices. Evidently, the blocks of the graph of Figure 5.25 are isomorphic to the components of the graph of Figure 5.26. We call such graphs 1-isomorphic.

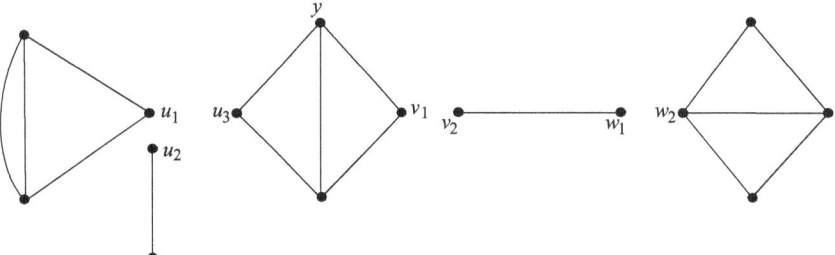

**Fig. 5.26**

These observations lead to the following definition.

**Definition:**  Two graphs $G_1$ and $G_2$ are said to be *1-isomorphic* if they become isomorphic to each other under repeated application of the following operation.

**Operation 1**  Split a cut vertex into two vertices to produce two disjoint subgraphs.
This definition implies that two non-separable graphs are 1-isomorphic if and only if they are isomorphic.

We now have the following result. Two 1-isomorphic graphs have the following property.

**Theorem 5.30** If $G_1$ and $G_2$ are 1-isomorphic graphs, then rank $G_1$ = rank $G_2$ and nullity $G_1$ = nullity $G_2$.

**Proof** Under operation 1, whenever a cut vertex in a graph $G$ is split into two vertices, the number of components in $G$ increases by one. Therefore, rank $G$ = number of vertices in $G$-number of components in $G$ remains invariant under operation 1.

Since no edges are destroyed or new edges created by operation 1, two 1-isomorphic graphs have the same number of edges. Two graphs with same rank, and same number of edges have the same nullity, since nullity = number of edges−rank. ❏

Suppose the two vertices $x$ and $y$ belonging to different components of the graph in Figure 5.26 are superimposed, then the graph obtained is shown in Figure 5.27. Clearly, the graph in Figure 5.27 is 1-isomorphic to the graph in Figure 5.26. Also, since the blocks of the graph in Figure 5.27 are isomorphic to the blocks of the graph in Figure 5.25, these two graphs are 1-isomorphic. Hence, the three graphs in Figures 5.25, 5.26 and 5.27 are 1-isomorphic.

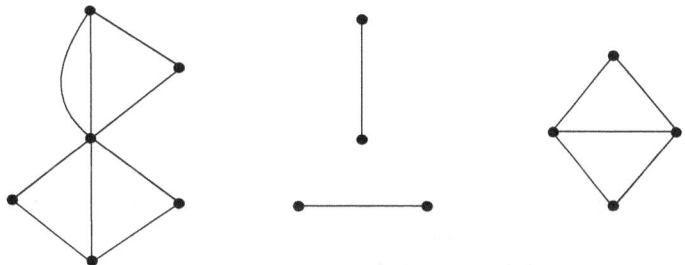

**Fig. 5.27**

We have seen that a graph $G_1$ is 1-isomorphic to a graph $G_2$ if the blocks of $G_1$ are isomorphic to the blocks of $G_2$. Since a non-separable graph is a block, 1-isomorphism for non-separable graphs is same as isomorphism. For separable graphs, obviously 1-isomorphism is different from isomorphism. In fact, graphs that are isomorphic are also 1-isomorphic, but the converse need not be true.

**2-isomorphism** In a 2-connected graph $G$, let vertices $x$ and $y$ be a pair of vertices whose removal from $G$, leaves the remaining graph disconnected. That is, $G$ consists of a subgraph $H$ and its complement $\bar{H}$ such that $H$ and $\bar{H}$ have exactly two vertices, $x$ and $y$, in common. Now, we perform the following operation on $G$.

**Operation 2** Split the vertex $x$ into $x_1$ and $x_2$, and the vertex $y$ into $y_1$ and $y_2$ such that $G$ is split into $H$ and $\bar{H}$. Let vertices $x_1$ and $y_1$ go with $H$, and $x_2$ and $y_2$ with $\bar{H}$. Now, rejoin the graphs $H$ and $\bar{H}$ by merging $x_1$ with $y_2$ and $x_2$ with $y_1$. Clearly, edges whose end vertices are $x$ and $y$ in $G$ can go with $H$ or $\bar{H}$, without affecting the final graph.

Two graphs are said to be *2-isomorphic* if they become isomorphic after undergoing operation 1, or operation 2, or both any number of times. For example, Figure 5.28 shows how the two graphs in (a) and (d) are 2-isomorphic.

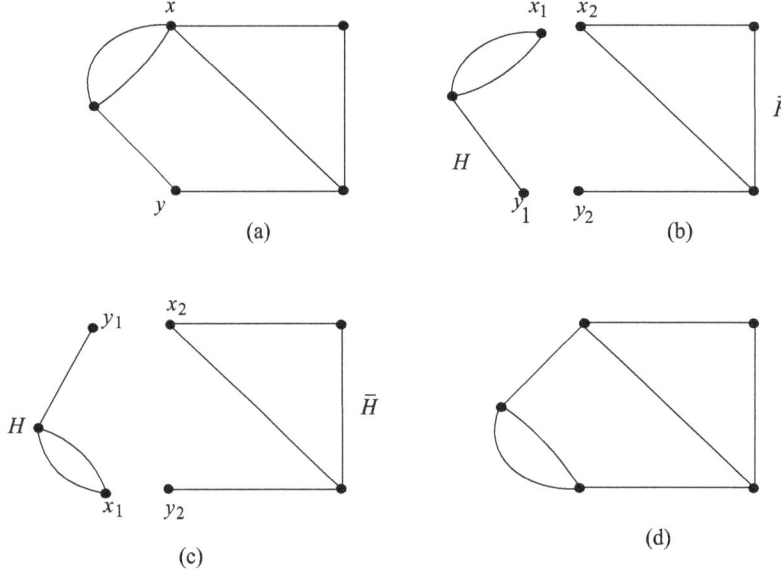

**Fig. 5.28**

It follows from the definition that isomorphic graphs are always 1-isomorphic and 1-isomorphic graphs are always 2-isomorphic. But 2-isomorphic graphs are not necessarily 1-isomorphic and 1-isomorphic graphs are not necessarily isomorphic. However, for graphs with three or more connectivity, isomorphism, 1-isomorphism and 2-isomorphism are same.

Clearly, no edges or vertices are created or destroyed under operation 2. So the rank and nullity of a graph remain unchanged under operation 2. Therefore the 2-isomorphic graphs are equal in rank and equal in nullity.

**Cycle correspondence:**   Two graphs $G_1$ and $G_2$ are said to have a cycle correspondence if there is a one-one correspondence between the edges of $G_1$ and $G_2$, and a one-one correspondence between the cycles of $G_1$ and $G_2$, such that a cycle in $G_1$ formed by certain edges of $G_1$ has a corresponding cycle in $G_2$ formed by the corresponding edges of $G_2$, and vice versa. Clearly, isomorphic graphs have cycle correspondence. Since in a separable graph $G$, every cycle is confined to a particular block, every cycle in $G$ retains its edges as $G$ undergoes operation 1. Thus 1-isomorphic graphs have cycle correspondence.

Following is the important result for 2-isomorphic graphs and is due to Whitney [266].

**Theorem 5.31 (Whitney [266])**   Two graphs are 2-isomorphic if and only if they have cycle correspondence.

## 5.8 Exercises

1. Prove that a vertex $v$ of a tree is a cut vertex if and only if $d(v) > 1$.

2. Prove that a unicentric graph need not be separable.

3. Prove that a graph $H$ is the block-cut vertex graph of some graph $G$ if and only if it is a tree in which the distance between any two end vertices is even.

4. Prove that a unicentric graph need not have $d = 2r$.

5. Prove that a non-separable graph with at least two edges has nullity greater than zero.

6. Prove that a non-separable graph of nullity one is a cycle and its converse.

7. If $v$ is a cut vertex of a simple connected graph $G$, prove that $v$ is not a cut vertex of $\bar{G}$.

8. Prove that a connected $k$-regular bipartite graph is 2-connected.

9. Show that a simple connected graph with at least three vertices is a path if and only if it has exactly two vertices that are not cut vertices.

10. If $b(v)$ denotes the number of blocks of a simple connected graph $G$ containing vertex $v$, prove that the number of blocks $b(G)$ of $G$ is given by

$$b(G) = 1 + \sum_{v \in V(G)} (b(v) - 1).$$

11. Prove that if a graph $G$ is $k$-connected or $k$-edge-connected, then $m \geq \dfrac{nk}{2}$.

12. Prove that a connected graph with at least two vertices contains at least two vertices that are not cut vertices.

13. Prove that a 3-regular connected graph has a cut vertex if and only if it has a cut edge.

14. Prove that the connectivity and edge connectivity of a cubic graph are equal.

15. Prove that a graph with at least three vertices is 2-connected if and only if any two vertices of $G$ lie on a common cycle.

16. Prove that a graph is 2-connected if and only if for every pair of disjoint connected subgraphs $G_1$ and $G_2$, there exist two internally disjoint paths $P_1$ and $P_2$ of $G$ between $G_1$ and $G_2$.

17. In a 2-connected graph $G$, prove that any two longest cycles have at least two vertices in common.

18. Prove that a connected graph $G$ is 3-connected if and only if every edge of $G$ is the exact intersection of the edge sets of two cycles of $G$.

19. Prove that a connected graph is Eulerian if and only if each of its blocks is Eulerian.

20. Prove that a connected graph is Eulerian if and only if each of its edge cuts has an even number of edges.

# 6. Planarity

Let $G = (V, E)$ be a graph with $V = \{v_1, v_2, \ldots, v_n\}$ and $E = \{e_1, e_2, \ldots, e_m\}$. Let $S$ be any surface (like the plane, sphere) and $P = \{p_1, p_2, \ldots, p_n\}$ be a set of $n$ distinct points of $S$, $p_i$ corresponding to $v_i$, $1 \le i \le n$. If $e_i = v_j v_k$, draw a Jordan arc $J_i$ on $S$ from $p_j$ to $p_k$ such that $J_i$ does not pass through any other $p_i$. Then $P \cup \{J_1, J_2, \ldots, Jm\}$ is called a *drawing* of $G$ on $S$, or a diagram representing $G$ on $S$. $p_i$ are called the *points* of the diagram and $J_i$, the *lines* of the diagram.

An *embedding* of a graph $G$ on a surface S is a diagram of $G$ drawn on the surface such that the Jordan arcs representing any two edges of $G$ do not intersect except at a point representing a vertex of $G$.

A graph is *planar* if it has an embedding on the plane. A graph which has no embedding on the plane is *nonplanar*. That is, a graph $G$ is said to be planar if there exists some geometric representation of $G$ which can be drawn on a plane such that no two of its edges intersect and a graph that cannot be drawn on a plane without a crossover between its edges is called nonplanar.

In order that a graph $G$ is nonplanar, we have to show that of all possible geometric representations of $G$, none can be embedded in a plane. Equivalently, a geometric graph $G$ is planar if there exists a graph isomorphic to $G$ that is embedded in a plane.

An embedding of a planar graph $G$ on a plane is called a *plane representation* of $G$. Figure 6.1 shows three diagrams of the same graph which is planar. The two graphs in Figure 6.2 represent the same planar graph.

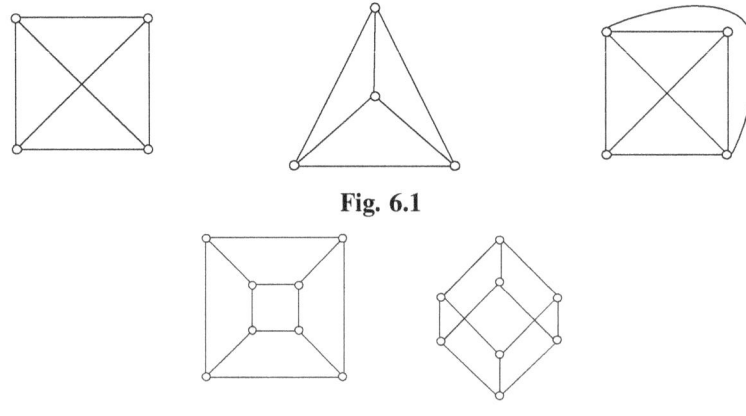

**Fig. 6.1**

**Fig. 6.2**

# 6.1 Kuratowski's Two Graphs

The complete graph $K_5$ and the complete bipartite graph $K_{3,3}$ are called Kuratowski's graphs, after the polish mathematician Kasimir Kurtatowski, who found that $K_5$ and $K_{3,3}$ are nonplanar.

**Theorem 6.1**  The complete graph $K_5$ with five vertices is nonplanar.

**Proof**  Let the five vertices in the complete graph be named $v_1$, $v_2$, $v_3$, $v_4$, $v_5$. Since in a complete graph every vertex is joined to every other vertex by means of an edge, therefore there is a cycle $v_1 v_2 v_3 v_4 v_5 v_1$ that is a pentagon. This pentagon divides the plane of the paper into two regions, one inside and the other outside, Figure 6.3(a).

Since vertex $v_1$ is to be connected to $v_3$ by means of an edge, this edge may be drawn inside or outside the pentagon (without intersecting the five edges drawn previously). Suppose that we choose to draw the line from $v_1$ to $v_3$ inside the pentagon, Figure 6.3(b). In case we choose outside, we end with the same argument. Now we have to draw an edge from $v_2$ to $v_4$ and another from $v_2$ to $v_5$. Since neither of these edges can be drawn inside the pentagon without crossing over the edge already drawn, we draw both these edges outside the pentagon, Figure 6.3(c). The edge connecting $v_3$ and $v_5$ cannot be drawn outside the pentagon without crossing the edge between $v_2$ and $v_4$. Therefore $v_3$ and $v_5$ have to be connected with an edge inside the pentagon, Figure 6.3(d). Now, we have to draw an edge between $v_1$ and $v_4$ and this cannot be placed inside or outside the pentagon without a crossover. Thus the graph cannot be embedded in a plane.  ❑

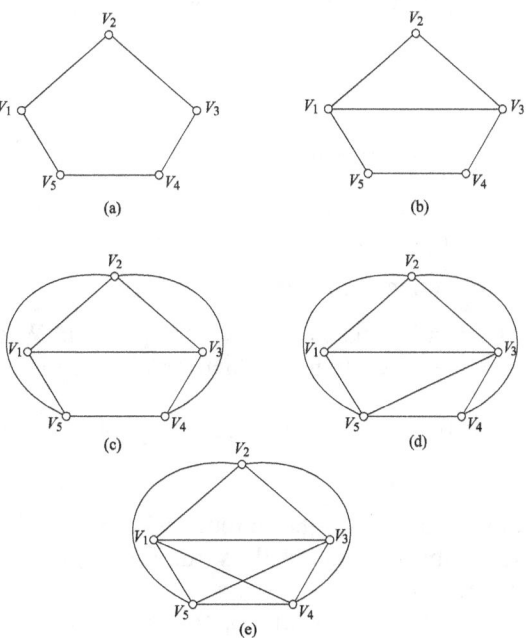

**Fig. 6.3**

**Theorem 6.2**    The complete bipartite graph $K_{3,3}$ is nonplanar.

**Proof**    The complete bipartite graph has six vertices and nine edges. Let the vertices be $u_1$, $u_2$, $u_3$, $v_1$, $v_2$, $v_3$. We have edges from every $u_i$ to each $v_i$, $1 \le i \le 3$. First we take the edges from $u_1$ to each $v_1$, $v_2$ and $v_3$. Then we take the edges between $u_2$ to each $v_1$, $v_2$ and $v_3$, Figure 6.4(a). Thus we get three regions namely I, II and III. Finally we have to draw the edges between $u_3$ to each $v_1$, $v_2$ and $v_3$. We can draw the edge between $u_3$ and $v_3$ inside the region II without any crossover, Figure 6.4(b). But the edges between $u_3$ and $v_1$, and $u_3$ and $v_2$ drawn in any region have a crossover with the previous edges. Thus the graph cannot be embedded in a plane. Hence $K_{3,3}$ is nonplanar.                    ❏

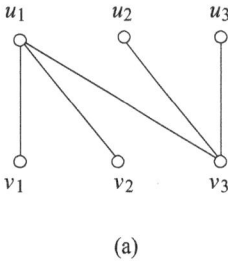

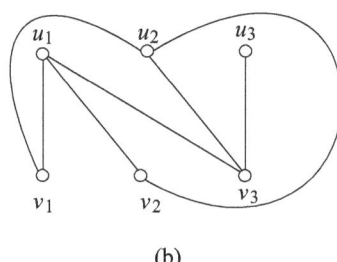

(a)

(b)

**Fig. 6.4**

We observe that the two graphs $K_5$ and $K_{3,3}$ have the following common properties.

1. Both are regular.

2. Both are nonplanar.

3. Removal of one edge or a vertex makes each a planar graph.

4. $K_5$ is a nonplanar graph with the smallest number of vertices, and $K_{3,3}$ is the nonplanar graph with smallest number of edges.

Thus both are the simplest nonplanar graphs.

The following result given independently by Fary [77] and Wagner [260] implies that there is no need to bend edges in drawing a planar graph to avoid edge intersections.

**Theorem 6.3 (Fary [77])** Every triangulated planar graph has a straight line representation.

**Proof**    The proof is by induction on the number of vertices. The result is obvious for $n = 4$. So, let $n \ge 5$ and assume that the result is true for all planar graphs with fewer than $n$ vertices. Let $G$ be a plane graph with $n$ vertices.

First, we show that $G$ has an edge $e$ belonging to just two triangles. For this, let $x$ be any vertex in the interior of a triangle $T$ and choose $x$ and $T$ such that the number of regions inside $T$ is minimal.

Let $y$ be a neighbor of $x$, and the edge $xy$ lies inside $T$, and let $xy$ belong to three triangles $xyz_1$, $xyz_2$ and $xyz_3$. Then one of these triangles lies completely inside another. Assume that $z_3$ lies inside $xyz_1$. Then $z_3$ and $xyz_1$ contradict the choice of $x$ and $T$ (Fig. 6.5).

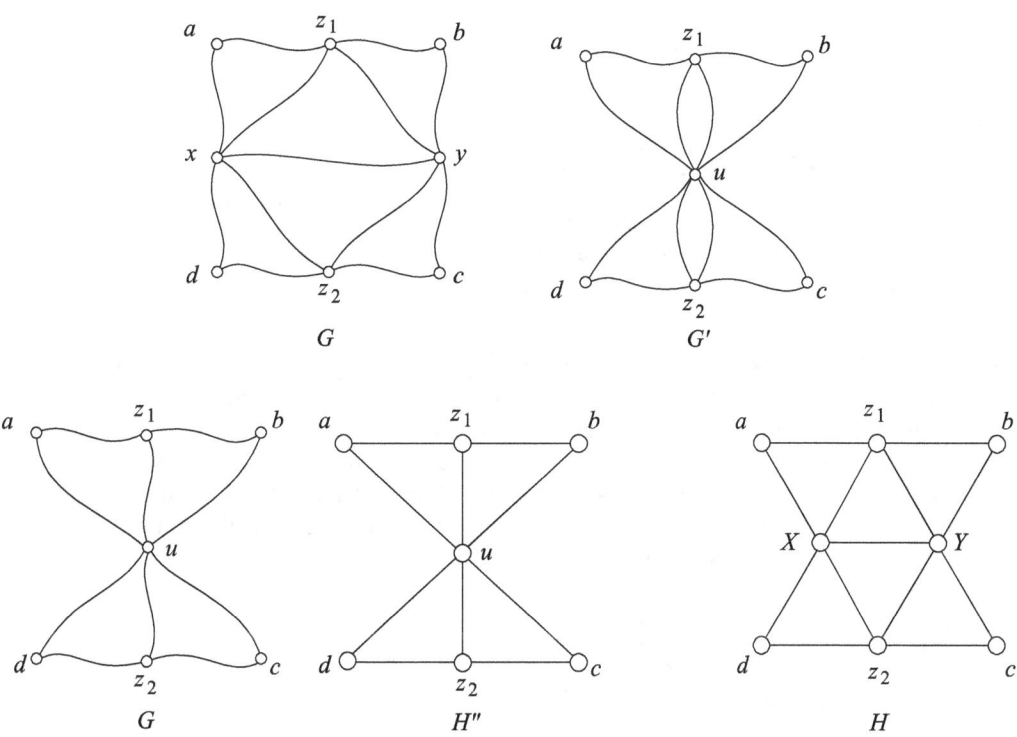

**Fig. 6.5**

Thus there is an edge $e = xy$ lying in just two triangles $xyz_1$ and $xyz_2$. Contracting $xy$ to a vertex $u$, we get a new graph $G'$ with a pair of double edges between $u$ and $z_1$, and $u$ and $z_2$. Remove one each of this pair of double edges to get a graph $G''$ which is a triangulated graph with $n-1$ vertices. By the induction hypothesis, it has a straight line representation $H''$. The edges of $G''$ correspond to $uz_1$, $uz_2$ in $H''$. Divide the angle around $u$ into two parts in one of which the pre-images of the edges adjacent to $x$ in $G$ lie, and in the other, the pre-images of the edges adjacent to $y$ in $G$. Thus $u$ can be pulled apart to $x$ and $y$, and the edge $xy$ is restored by a straight line to get a straight line representation of $G$. ❑

## 6.2 Region

A plane representation of a graph divides the plane into regions (also called windows, faces or meshes). A region is characterised by the set of edges (or the set of vertices) forming its boundary. We note that a region is not defined in a nonplanar graph, or even in a planar graph not embedded in a plane. Thus, a region is a property of the specific plane representation of a graph and not of an abstract graph.

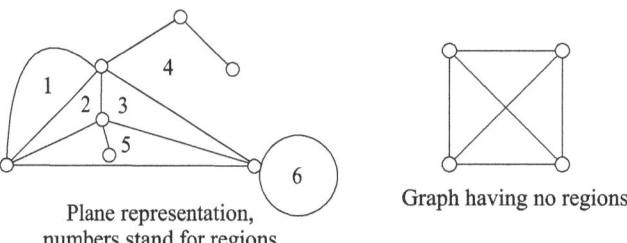

Plane representation,
numbers stand for regions

Graph having no regions

**Fig. 6.6**

**Infinite region:**    The portion of the plane lying outside a graph embedded in a plane such as region 4 in the graph given in Figure 6.6, is infinite in its extent. Such a region is called an infinite, unbounded, outer or exterior region for that particular plane representation. Like other regions, the infinite region is also characterised by a set of edges (or vertices). Clearly, by changing the embedding of a given planar graph, we can change the infinite region. Consider the graphs of Figure 6.7. In the two embeddings of the same graph, finite region $v_1 v_3 v_5$ in (a) is infinite in (b).

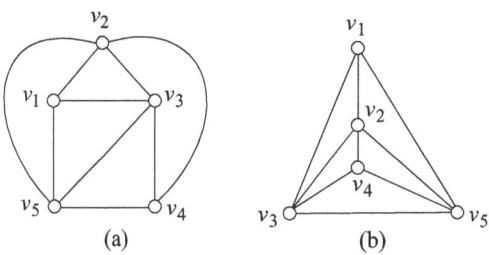

(a)                                   (b)

**Fig. 6.7**

**Embedding on a sphere:**    A graph is spherical if it can be embedded on the surface of a sphere.

The following result implies that a planar graph can often be embedded on the surface of a sphere, so that there is no distinction between finite and infinite region.

**Theorem 6.4**    A graph can be embedded on the surface of a sphere if it can be embedded in a plane.

**Proof**    Consider the stereographic projection of a sphere on the plane. Put the sphere on the plane and call the point of contact as *SP* (south-pole). At point *SP*, draw a straight line perpendicular to the plane, and let the point where this line intersects the surface of the sphere be called *NP* (north-pole).

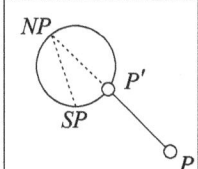

**Fig. 6.8**

Now, corresponding to any point $p$ on the plane, there exists a unique point $p'$ on the sphere, and vice versa, where $p'$ is the point where the straight line from point $p$ to point $NP$ intersects the surface of the sphere. Thus, there is a one-one correspondence between the points of the sphere and the finite points on the plane, and points at infinity in the plane corresponding to the point $NP$ on the sphere.

Therefore from this construction, it is clear that any graph that can be embedded in a plane can also be embedded on the surface of the sphere, and vice versa. ❏

We now have the following result.

**Theorem 6.5** A planar graph can be embedded in a plane such that any specified region, specified by the edges forming it, can be made the infinite region.

**Proof** A planar graph embedded on the surface of the sphere divides the surface of the sphere into different regions. Each region of the sphere is finite, the infinite region on the plane having been mapped onto the region containing the point $NP$. Now, clearly by suitably rotating the sphere, we can make any specified region map onto the infinite region on the plane. Hence the result. ❏

**Remark** Now, thinking in terms of the regions on the sphere, we observe that there is no real difference between the infinite region and the finite regions on the plane. Therefore, when we discuss the regions in a plane representation of a graph, we include the infinite region. Also, since there is no essential difference between an embedding of a planar graph on a plane, or on a sphere (a plane can be regarded as the surface of the sphere of infinitely large radius), the term plane representation of a graph is often used to include spherical as well as plane embedding.

## 6.3 Euler's Theorem

The following important result due to Euler gives a relation between the number of vertices, edges, regions and the components of a planar graph.

**Theorem 6.6** If $G$ is a planar graph with $n$ vertices, $m$ edges, $f$ regions and $k$ components, then

$$n - m + f = k + 1. \tag{6.6.1}$$

**Proof**   We construct the graph $G$ by the addition of successive edges starting from the null graph $\overline{K_n}$. For this starting graph, $k = n$, $m = 0$, $f = 1$, so that (6.6.1) is true.

Let $G_{i-1}$ be the graph at the start of $i$th stage and $G_i$ be the graph obtained from $G_{i-1}$ by addition of the $i$th edge $e$. If $e$ connects two components of $G_{i-1}$, then $f$ is not altered, $m$ is increased by 1 and $k$ is reduced by 1, so that (6.6.1) holds for $G_i$ as it holds for $G_{i-1}$. If $e$ joins two vertices of the same components of $G_{i-1}$, $k$ is unaltered, $m$ is increased by 1 and $f$ is increased by 1, so that again (6.6.1) holds for $G_i$.                                                                   ❏

The following relation between the number of vertices, edges and regions is the discovery of Euler [75] and is also called as Euler's formula for planar graphs.

**Theorem 6.7 (Euler [75])**   In a connected planar graph with $n$ vertices, $m$ edges and $f$ regions (faces), $n - m + f = 2$.

(The proof can be deduced from Theorem 6.6 by taking $k = 1$, as the connected graph has one component.)

**Proof**   Without loss of generality, assume that the planar graph is simple. Since any simple planar graph can have a plane representation such that each edge is a straight line, any planar graph can be drawn such that each region is a polygon (a polygon net). Let the polygon net representing the given graph consist of f regions. Let $k_p$ be the number of $p$-sided regions. Since each edge is on the boundary of exactly two regions,

$$3k_3 + 4k_4 + 5k_5 + \ldots + rk_r = 2m, \tag{6.7.1}$$

where $k_r$ is the number of polygons with $r$ edges.

Also, $k_3 + k_4 + k_5 + \ldots + k_r = f.$                                                               (6.7.2)

The sum of all angles subtended at each vertex in the polygon net is $2\pi n$.           (6.7.3.)

Now, the sum of all interior angles of a $p$-sided polygon is $\pi(p - 2)$ and the sum of the exterior angles is $\pi(p + 2)$. The expression in (6.7.3) is the total sum of all interior angles of $f - 1$ finite regions plus the sum of the exterior angles of the polygon defining the infinite region. This sum is

$$\pi(3 - 2)k_3 + \pi(4 - 2)k_4 + \ldots + \pi(r - 2)kr + 4\pi m$$

$$= \pi[k_3 + 4k_4 + \ldots + rk_r + 2(k_3 + k_4 + \ldots + k_r)] + 4\pi$$

$$= \pi(2m - 2f) + 4\pi = 2\pi(m - f + 2). \tag{6.7.4}$$

Equating (6.7.3) and (6.7.4) we get

$$2\pi(m - f + 2) = 2n\pi,$$

so that $f = m - n - 2.$                                                                              ❏

**Definition:** Let $\phi$ be a region of a planar graph $G$. We define the *degree* of $\phi$, denoted by $d(\phi)$, as the number of edges on the boundary of $\phi$.

**Second proof of Theorem 6.7**   We use induction on $m$, the number of edges. If $m = 0$, then $G$ is $K_1$, a graph with 1 vertex and 1 region. So, $n - m + f = 1 - 0 + 1 = 2$, and the result is true. If $m = 1$, then the number of vertices in $G$ is either one or two, the first possibility occurring when the edge is a loop. These two possibilities give rise respectively to two regions and one region.

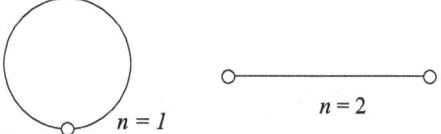

**Fig. 6.9**

Therefore, $n - m + f = \begin{cases} 1 - 1 + 2, & \text{in the loop case}, \\ 2 - 1 + 1, & \text{in the non loop case}. \end{cases}$

Thus, $n - m + f = 2$, and again the result is true.

Assume that the result is true for all connected planar graphs with fewer than $m$ edges. Let $G$ have $m$ edges.

**Case I**   Suppose $G$ is a tree. Then, $m = n - 1$ and $f = 1$, because a planar representation of a tree has only one region. Thus, $n - m + f = n - (n - 1) + 1 = 2$, and the result holds.

**Case II**   Suppose $G$ is not a tree. Then $G$ has cycles. Let $C$ be a cycle in $G$. Let $e$ be an edge of $C$. The graph $G - e$ has one edge less than the graph $G$. Also the number of vertices in $G - e$ and $G$ are same. Since removing $e$ coalesces two regions in $G$ into one in $G - e$, therefore $G - e$ has one region less than that in $G$. Thus by induction hypothesis in $G - e$, we have $n - (m - 1) + (f - 1) = 2$, so that $n - m + f = 2$. Hence the result.

The next result gives an upper bound for the edges of a simple planar graph.

**Theorem 6.8**   Let $G$ be a simple planar graph with $n$ vertices and $m$ edges, where $n \geq 3$. Then $m \leq 3n - 6$.

**Proof**   First assume that the planar graph $G$ is connected. If $n = 3$, then since $G$ is simple, therefore $G$ has at most three edges. Thus $m \leq 3$, that is, $m \leq 3 \times 3 - 6$, and the result is true.

Now, let $n = 4$. First let $G$ be a tree so that $m = n - 1$. Since $n \geq 4$, obviously we have $n - 1 \leq 3n - 6$.

Now, let $G$ be not a tree. Then $G$ has cycles. Clearly, there is a cycle in $G$, all of whose edges lie on the boundary of the exterior region of $G$. Then, since $G$ is simple, we have $d(\phi) \geq 3$, for each region $\phi$ of $G$.

Let $b = \sum\limits_{\phi \in F} d(\phi)$, where $F$ denotes the set of all regions of $G$.

Since each region has at least three edges on its boundary, therefore we have $b \geq 3f$, where $f$ is the number of regions of $G$. However, when we sum to get $b$, each edge of $G$ is counted either once or twice (twice when it occurs as a boundary edge for two regions). Therefore, $b \leq 2m$ so that $3f \leq b \leq 2m$. In particular, we have $3f \leq 2m$.

Using Euler's formula $n - m + f = 2$, we get $m \leq 3n - 6$.

Now, let $G$ be not connected and let $G_1$, $G_2$, ..., $G_t$ be its connected components. For each $i$, $1 \leq i \leq t$, let $n_i$ and $m_i$ denote the number of vertices and edges in $G_i$. Since each $G_i$ is a planar simple graph, by the above argument, we have $m_i \leq 3n_i - 6$, for each $i$, $1 \leq i \leq t$.

Also, $n = \sum\limits_{i=1}^{t} n_i$ and $m = \sum\limits_{i=1}^{t} m_i$.

Hence, $m = \sum\limits_{i=1}^{t} m_i \leq \sum\limits_{i=1}^{t} (3n_i - 6) = 3\sum\limits_{i=1}^{t} n_i - 6t \leq 3n - 6$.                    ❏

The following result gives the existence of a vertex of degree less than six in a simple planar graph.

**Theorem 6.9**    If $G$ is a simple planar graph, then $G$ has a vertex $v$ of degree less than 6.

**Proof**    If $G$ has only one vertex, then this vertex has degree zero. If $G$ has only two vertices, then both vertices have degree at most one.

Let $n \geq 3$. Assume degree of every vertex in $G$ is at least six.

Then, $\sum\limits_{v \in v(G)} d(v) \geq 6n$.

We know $\sum\limits_{v \in v(G)} d(v) \geq 2m$.

Thus, $2m \geq 6n$ so that $m \geq 3n$.

This is not possible because, by Theorem 6.8, we have $m \leq 3n - 6$. Thus we get a contradiction. Hence $G$ has at least one vertex of degree less than 6.

**Corollary 6.1** $K_5$ is nonplanar.

**Proof** Here $n = 5$ and $m = 10$. So, $3n - 6 = 15 - 6 = 9$. Thus, $m > 3n - 6$. Therefore, $K_5$ is nonplanar.

**Corollary 6.2** $K_{3,3}$ is nonplanar.

**Proof** Since $K_{3,3}$ is bipartite, it contains no odd cycles, and so no cycle of length three. It follows that every region of a plane drawing of $K_{3,3}$ if it exists, has at least four boundary edges. We have $d(\phi) \geq 4$, for each region $\phi$ of $G$.

Let $b = \sum_{\phi \in F} d(\phi)$, where $F$ denotes the set of all regions of $G$. Since each region has at least four edges on its boundary, we have $b \geq 4f$, where $f$ is the number of regions of $G$.

Now, when we sum up to get $b$, each edge of $G$ is counted either once or twice, and so $b \leq 2m$. Thus, $4f \leq b \leq 2m$, so that $2\mathrm{f} \leq m$.

For $K_{3,3}$, we have $m = 9$, and so $2f \leq 9$ giving $f \leq \dfrac{9}{2}$. But by Euler's formula, $f = m - n + 2 = 9 - 6 + 2 = 5$, a contradiction. Hence $K_{3,3}$ is nonplanar. ❏

Euler's formula gives a necessary condition for connected planar graphs.

**Theorem 6.10** For a connected simple planar graph with $n$ vertices and $m$ edges, and girth $g$, we have $m \leq \dfrac{g(n-2)}{g-2}$.

**Proof** Let $G$ be a connected simple planar graph with $n$ vertices, $m$ edges and girth $g$. Let $F$ be the set of regions and $d(\phi)$ be the degree of the region $\phi$. Then, $\sum_{\phi \in F} d(\phi) = 2m$. Let $f$ be the number of regions, that is $|F| = f$. As $g$ is the length of the smallest cycle in $G$, therefore $g$ is the smallest degree of a region. So, $\sum_{\phi \in F} d(\phi) \geq gf$. Thus, $2m \geq gf$. Now, using Euler's formula, $f = m - n + 2$, we have $2m \geq g(m - n + 2)$, so that $m \leq \dfrac{g(n-2)}{g-2}$.

**Corollary 6.3** The Peterson graph $P$ is nonplanar.

**Proof** The girth of the Perterson graph $P$ is 5, $n(P) = 10$, and $m(P) = 15$. Thus, if $P$ were planar, $15 = 5\left(\dfrac{10-2}{5-2}\right)$, which is not true. Hence $P$ is nonplanar. ❏

**Corollary 6.4** Any graph containing a homeomorph of a $K_5$ or $K_{3,3}$ is nonplanar. This follows from the fact that any homeomorph of a nonplanar graph is nonplanar.

The following result characterises nonplanar 3-connected graphs with minimum edges.

**Theorem 6.11** A nonplanar connected graph $G$ with minimum number of edges that contains no subdivision of $K_5$ or $K_{3,3}$ is simple and 3-connected.

**Proof**   Since $G$ has minimum number of edges, $G$ is minimal with respect to these properties. The minimality of $G$ ensures that it is simple. Also, a graph is planar if and only if each of its blocks is planar. This implies that a minimal nonplanar graph is a block.

Suppose that such a graph has a two vertex cut $S = \{u, v\}$. Then $G - S$ is disconnected. Let $G_1$ be one of its components and $G_2$ be the union of all of its other components.

Let $H_1 = \langle V(G_1) \cup S \rangle + e$ and $H_2 = \langle V(G_2) \cup S \rangle + e$, where $e = uv$

Clearly, these are nonempty graphs with fewer edges than $G$. Since $G$ is nonplanar, at least one of $H_1$ and $H_2$ is nonplanar. If both are planar, let $\tilde{H}_1$ be a plane embedding of $H_1$ and let $\phi$ be a region of $\tilde{H}_1$ containing the edge $e$.

A plane embedding $\tilde{H}_2$ of $H_2$ inside $\phi$ can then be obtained with the edge $e$ of both $\tilde{H}_1$ and $\tilde{H}_2$ coinciding. Then $\tilde{H}_1 \cup \tilde{H}_2 - e$ is a plane embedding of $G$, contradicting our hypothesis.

Suppose $H_1$ is nonplanar. Since $m(H_1) < m(G)$, therefore $H_1$ contains a subdivision $K$ of either $K_5$ or $K_{3,3}$ according to the hypothesis. Clearly $e \in E$, since otherwise, $K \subset G$, contradicting the assumption. Now, replacing $e$ of $K$ by a $u - v$ path in $H_2$, we get a homeomorph of $K$ contained in $G$. This is again a contradiction and shows that $G$ has no 2-vertex cut. Hence $G$ is 3-connected.                                                                    ❑

The following result is used in proving Kuratowski's theorem.

**Theorem 6.12**

   a. If $G|e$ contains a subdivision of $K_5$, then $G$ contains a subdivision of $K_5$ or $K_{3,3}$.

   b. If $G|e$ contains a subdivision of $K_{3,3}$, then $G$ contains a subdivision of $K_{3,3}$.

**Proof**   Let $G' = G|e$ be a graph obtained by contracting the edge $e = xy$ of $G$. Let $w$ be the vertex of $G'$ obtained by contracting $e = xy$.

   a. Let $G|e$ contains a subdivision of $K_5$, say $H$. If $w$ is not a branch vertex of $H$, then $G$ also contains a subdivision of $K_5$, obtained by expanding $w$ back into the edge $xy$, if necessary (Fig. 6.10).

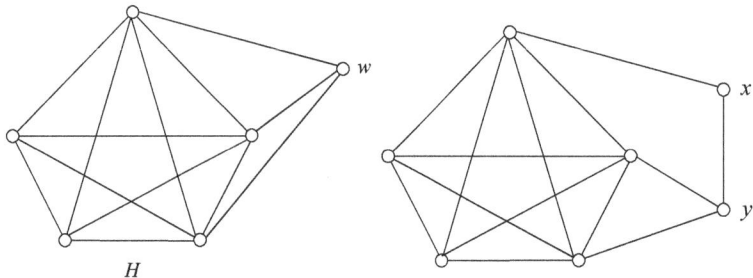

**Fig. 6.10**

Assume $w$ is a branch vertex of $H$ and each of $x$, $y$ is incident in $G$ to two of the four edges incident to $w$ in $H$. Let $u_1$ and $u_2$ be the branch vertices of $H$ that are at the other ends of the paths leaving $w$ on edges incident to $x$ in $G$. Let $v_1$, $v_2$ be the branch vertices of $H$ that are at the other ends of the paths leaving $w$ on edges incident to $y$ in $G$ (Fig. 6.11).

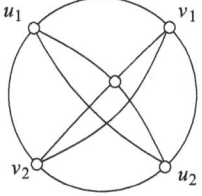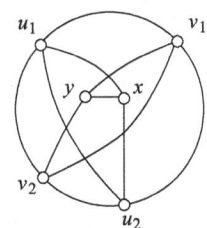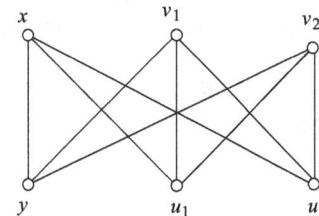

**Fig. 6.11**

By deleting the $u_1 - u_2$ path and $v_1 - v_2$ path from $H$, we obtain a subdivision of $K_{3,3}$ in $G$, in which $y$, $u_1$, $u_2$ are branch vertices for one partite set, and $x$, $v_1$, $v_2$ are branch vertices of the other.

b. Let $G|e$ contain a subdivision of $K_{3,3}$, say $H$. If $w$ is not a branch vertex of $H$, then $G$ also contains a subdivision of $K_{3,3}$, obtained by expanding $w$ back into the edge $xy$, if necessary (Fig. 6.12).

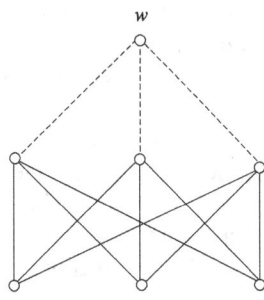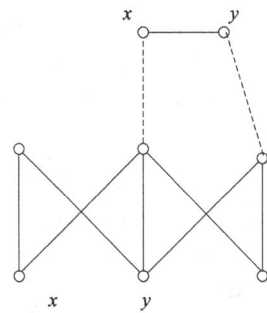

**Fig. 6.12**

Now, assume that $w$ is a branch vertex in $H$ and at most one of the edges incident to $w$ in $H$ is incident to $x$ in $G$. Then $w$ can be expanded into $xy$ to lengthen that path and $y$ becomes the corresponding branch vertex of $K_{3,3}$ in $G$ (Fig. 6.13).  ❑

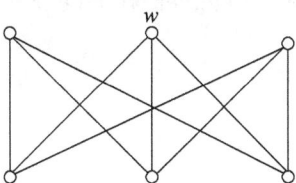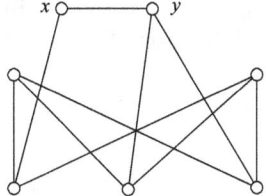

**Fig. 6.13**

## 6.4 Kuratowski's Theorem

This theorem was independently given by Kuratowski [144] and Frink and Smith. In 1954, Dirac and Schuster [69] found a proof that was slightly shorter than the original proof. The proof given here is due to Thomassen [241].

**Theorem 6.13 (Kuratowski [144])**     A graph is planar if and only if it does not have any subdivision of $K_5$ or $K_{3,3}$.

### Proof

*Necessity*     Let $G$ be a planar graph. Then any of its subgraphs is neither $K_5$ nor $K_{3,3}$ nor does it contain any subdivision of $K_5$ or $K_{3,3}$.

*Sufficiency*     It is enough to prove sufficiency for 3-connected graphs. Let $G$ be a 3-connected graph with $n$ vertices. We prove that the 3-connected graph $G$ either contains a subdivision of $K_5$ or $K_{3,3}$ or has a convex plane representation. This we prove by using induction on $n$. Since $G$ is 3-connected, therefore $n \geq 4$. For $n = 4$, $G = K_4$ and clearly has a plane representation.

Now, let $n \geq 5$. Assume the result to be true for all 3-connected graphs with fewer than $n$ vertices. Since $G$ is 3-connected, $G$ has an edge $e$ such that $G|e$ is connected. Let $e = xy$. If $G|e$ contains a subdivision of $K_5$ or $K_{3,3}$, then $G$ also contains a subdivision of $K_5$ or $K_{3,3}$. Therefore, let $G|e = H$ have a convex plane representation. Let $z$ be the vertex obtained by contraction of $e = xy$. The plane graph obtained by deleting the edges incident to $z$ has a region containing $z$ (this may be the exterior region). Let $C$ be the cycle of $H - z$ bounding this region.

Since we started with a convex plane representation of $H$, we have straight segments from $z$ to all its neighbours. Let $x_1, x_2, \ldots, x_k$ be the neighbours of $x$ in that order on $C$.

If all the neighbours of $y$ belong to a single segment from $x_i$ to $x_{i+1}$ on $C$, then we obtain a convex plane representation of $G$ by putting $x$ at $z$ in $H$, and putting $y$ at a point close to $z$ in the wedge formed by $x x_i$ and $x x_{i+1}$.

If all the neighbours of $y$ do not belong to any single segment $x_i x_{i+1}$ on $C$ ($1 \leq i \leq k$, $x_{k+1} = x_1$), then we have the following cases (Fig. 6.14).

a. $y$ shares three neighbours with $x$. In this case $C$ together with these six edges involving $x$ and $y$ form a subdivision of $K_5$.

b. $y$ has two $u$, $v$ in $C$ that are in different components of the subgraph of $C$ obtained by deleting $x_i$ and $x_{i+1}$, for some $i$. In this case, $C$ together with the paths $uyv$, $x_i x x_{i+1}$ and $xy$ form a subdivision of $K_{3,3}$.                                                                           □

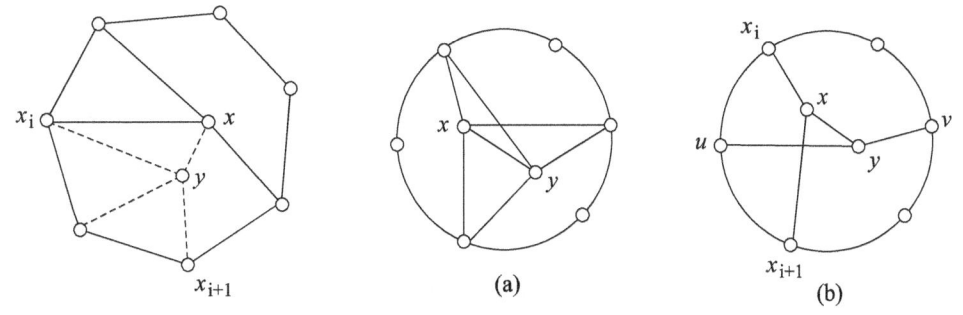

**Fig. 6.14**

## 6.5 Geometric Dual

Let $G$ be a plane graph. The dual of $G$ is defined to be the graph $G^*$ constructed as follows. To each region $f$ of $G$ there is a corresponding vertex $f^*$ of $G^*$ and to each edge $e$ of $G$ there is corresponding edge $e^*$ in $G^*$ such that if the edge $e$ occurs on the boundary of the two regions $f$ and $g$, then the edge $e^*$ joins the corresponding vertices $f^*$ and $g^*$ in $G^*$. If the edge $e$ is a bridge, i.e., the edge $e$ lies entirely in one region $f$, then the corresponding edge $e^*$ is a loop incident with the vertex $f^*$ in $G^*$. For example, consider the graph shown in Figure 6.15.

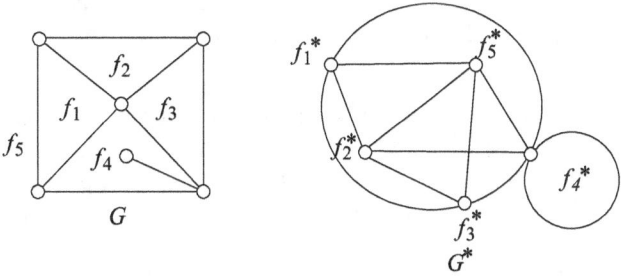

**Fig. 6.15**

**Theorem 6.14**   The dual $G^*$ of a plane graph is planar.

**Proof**   Let $G$ be a plane graph and let $G^*$ be the dual of $G$. The following construction of $G^*$ shows that $G^*$ is planar.

Place each vertex $f_k^*$ of $G^*$ inside its corresponding region $f_i$. If the edge $e_i$ lies on the boundary of two regions $f_j$ and $f_k$ of $G$, join the two vertices $f_j^*$ and $f_i^*$ by the edge $e_i^*$, drawing so that it crosses the edge $e$ exactly once and crosses no other edge of $G$ (Fig. 6.16). ❑

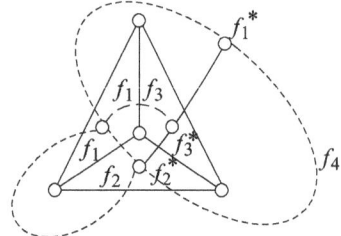

**Fig. 6.16**

**Remarks** Clearly, there is one-one correspondence between the edges of plane graph $G$ and its dual $G^*$ with one edge of $G^*$ intersecting one edge of $G$.

1. An edge forming a self-loop in $G$ gives a pendant edge in $G^*$ (An edge incident on a pendant vertex is called a pendant edge).

2. A pendant edge in $G$ gives a self loop in $G^*$.

3. Edges that are in series in $G$ produce parallel edges in $G^*$.

4. Parallel edges in G produce edges in series in $G^*$.

5. The number of edges forming the boundary of a region $f_i$ in $G$ is equal to the degree of the corresponding vertex $f_i^*$ in $G^*$.

6. Considering the process of drawing a dual $G^*$ from $G$, it is evident that $G$ is a dual of $G^*$. Therefore, instead of calling $G^*$ a dual of $G$, we usually say that $G$ and $G^*$ are dual graphs.

7. Let $n$, $m$, $f$, $r$ and $\mu$ denote the number of vertices, edges, regions, rank and nullity of a connected plane graph $G$ and let $n^*$, $m^*$, $f^*$, $r^*$ and $\mu^*$ be the corresponding numbers in $G^*$. Then, $n^* = f$, $m^* = m$, $f^* = n$. We have $r^* = n^* - 1$, $\mu^* = m^* - n^* + 1$, $r = n - 1$, $\mu = m - n + 1$. So, $r^* = f - 1$, $\mu^* = m - f + 1$, $r = n - 1$, $\mu = m - n + 1$.

Using Euler's formula, $n - m + f = 2$ or $f = m - n + 2$, we have $r^* = m - n - 2 - 1 = m - n + 1 = \mu$, and $\mu^* = m - f + 1 = n + f - 2 - f + 1 = n - 1 = r$.

We now have the following result.

**Theorem 6.15** The edge $e$ is a loop in $G$ if and only if $e^*$ is a bridge in $G^*$.

**Proof** Let the edge $e$ be a loop in a plane graph $G$. Then it is the edge on the common boundary of two regions on which, say $f$, lies within the area of the plane surrounded by $e$ with the other, say $g$, lying outside the area. Thus, from definition of $G^*$, $e^*$ is the only path from $f^*$ to $g^*$ in $G^*$. Thus, $e^*$ is a bridge in $G^*$.

Conversely, let $e^*$ be a bridge in $G^*$, joining vertices $f^*$ and $g^*$. Thus, $e^*$ is the only path in $G^*$ from $f^*$ to $g^*$. This implies, again from the definition of $G^*$, that the edge $e$ in $G$ completely encloses one of the regions $f$ and $g$. So $e$ is a loop in $G$.     ❑

**Remark** The occurrence of parallel edges in $G^*$ is easily described. Given two regions $f$ and $g$ of $G$, there are $k$ parallel edges between $f^*$ and $g^*$ if and only if $f$ and $g$ have $k$ edges on their common boundary.

**Note** We have defined the dual of a plane graph instead of a planar graph. The reason for this is that different plane drawings $G_1$ and $G_2$ of the same planar graph $G$ may result in non-isomorphic duals $G_1^*$ and $G_2^*$.

**2-isomorphism** Two graphs $G = (V, E)$ and $H = (W, F)$ are said to be 2-isomorphic if there exists a bijection $\phi: E \to F$ such that both $\phi$ and $\phi^{-1}$ preserve cycles.

Two graphs $G$ and $H$ are said to be 2-isomorphic if there exists a one-one correspondence between their edge sets such that the edges of a cycle in $G$ correspond to the edges of a cycle in $G_2$ and vice versa.

**Example** Figure 6.17(a) shows 2-isomorphic graphs.

**1-isomorphism** Graphs which become isomorphic after the splitting of all of their cut vertices are said to be 1-isomorphic.

**Example** Figure 6.17(b) shows 1-isomorphic graphs.

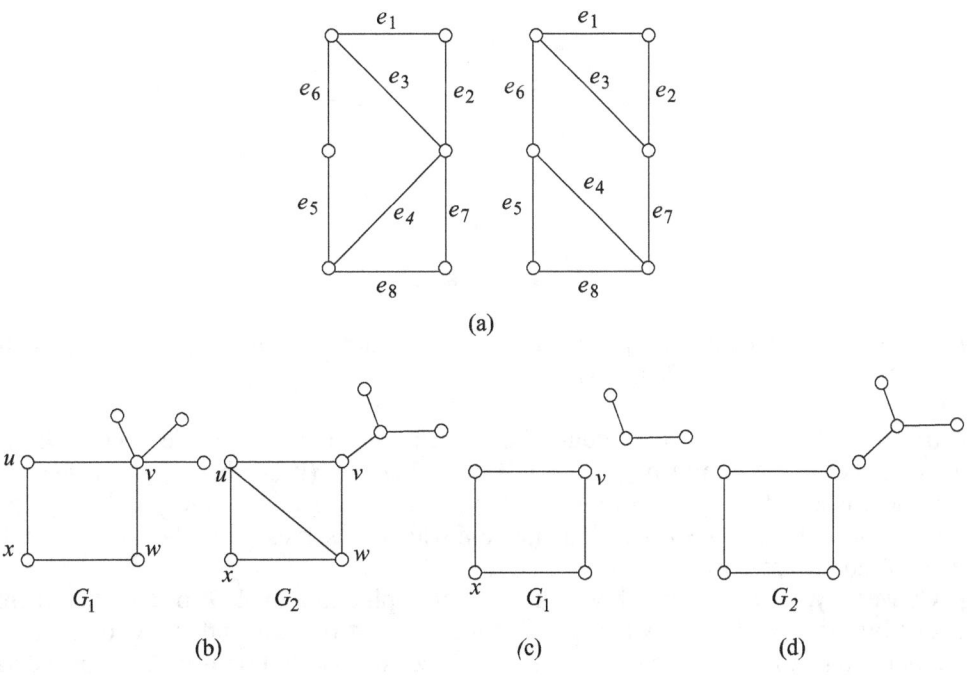

(a)

(b)  (c)  (d)

**Fig. 6.17**

Consider the two non-isomorphic graphs $G_1$ and $G_2$ shown in Fig. 6.17(a) and (b). They both have a single cut vertex, namely $v$. If this cut vertex is split into two vertices in each of $G_1$ and $G_2$, we obtain the edge disjoint graphs $G_1'$ and $G_2'$ as shown in (c) and (d). Clearly, $G_1'$ and $G_2'$ are isomorphic. Thus, $G_1$ and $G_2$ are 1-isomorphic.

The next result gives a correspondence between the edges of a graph with the edges of its dual.

**Theorem 6.16**   Edges in a plane graph $G$ form a cycle in $G$ if and only if the corresponding dual edges form a bond in $G^*$.

**Proof**   Let $D \subseteq E(G)$. It is sufficient to prove that $D$ contains a cycle if and only if the set $D^*$ of dual edges contains a bond of $G^*$. Now, let $D$ contain a cycle $C$. Then by the Jordan Curve theorem, some region of $G$ lies inside $C$ and some region lies outside $C$. These regions correspond to vertices $v^*$, $w^*$ in $G^*$, one drawn inside $C$ and one outside $C$. A $v^* - w^*$ path in $G^*$ crosses $C$ and hence uses an edge of $G^*$, that is dual to an edge of $C$. Thus $D^*$ disconnects $v^*$ from $w^*$ and hence $D^*$ contains a bond (Fig. 6.18(a)).

Conversely, let $D^*$ contain a bond. We show that $D$ contains a cycle. Assume that $D$ does not contain a cycle. Then $D$ encloses no region. It remains possible to reach each region of $G$ from every other without crossing $D$. Hence $G^* - D^*$ is connected, and so $D^*$ contains no bond. This is a contradiction. Hence $D$ contains a cycle. ❏

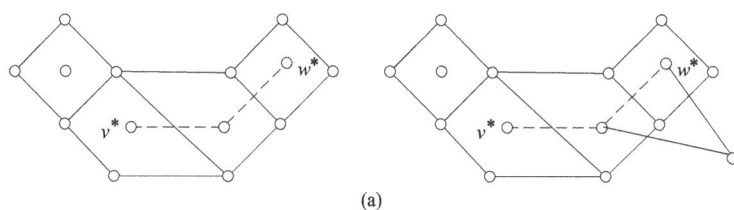

(a)

**Fig. 6.18(a)**

**Theorem 6.17**   All duals of a planar graph $G$ are 2-isomorphic and any graph $H$ 2-isomorphic to a dual $G^*$ of $G$ is itself a dual of $G$.

**Proof**   Let $G^* = (V^*, E^*)$ be a dual of $G$. Then there is an edge bijection $\phi : E \to E^*$ such that cycles in $G^*$ correspond to bonds in $G$. Let $H = (W, F)$ be any other dual of $G$. Then there is an edge bijection $\psi : E \to E^*$ such that bonds in $G^*$ correspond to cycles in $H$. Therefore, $\phi\psi : E^* \to F$ is an edge bijection which preserves cycles between $G^*$ and $H$. Hence $G^*$ and $H$ are 2-isomorphic.

Conversely, let $H$ be a graph which is 2-isomorphic to a dual $G^*$ of $G$. Then there is an edge bijection $\chi : E^* \to F$ which preserves cycles between $G^*$ and $H$. Also, there is an edge bijection $\phi : E \to E^*$ such that cycles in $G^*$ correspond to bonds in $G$. Thus we have an edge bijection $\chi\phi : F \to E$ such that cycles in $F$ correspond to bonds in $G$. Hence $H$ is a dual of $G$. ❏

The following result provides the condition for two given graphs to be duals of each other.

**Theorem 6.18** A necessary and sufficient condition for two planar graphs $G_1$ and $G_2$ to be the duals of each other is that there should be a one-one correspondence of the edges in $G_1$ with the edges in $G_2$ such that the subset of edges in $G_1$ forms a cycle if and only if the corresponding set of edges in $G_2$ forms a bond.

**Proof**

*Necessity* Consider a plane representation of a planar graph $G_1$ and let $G_1^*$ be the dual of $G_1$. Let $C$ be an arbitrary cycle in $G_1$.

Clearly, $C$ will form some closed simple curve in the plane representation of $G_1$, dividing the plane into two areas. Thus, the vertices of $G_1^*$ are partitioned into two non-empty, mutually exclusive subsets, one inside $C$ and the other outside. In other words, the set of edges $C^*$ in $G_1^*$ corresponding to the set $C$ in $G_1$ is a bond in $G_1^*$.

Similarly, corresponding to a bond $S^*$ in $G_1^*$, there is a unique cycle consisting of the corresponding edge set $S$ in $G_1$ such that $S$ is a cycle.

*Sufficiency* Let $G$ be a planar graph, and let $G'$ be a graph for which there is a one-one correspondence between the bonds of G and the cycles of $G'$ and vice versa. Let $G^*$ be the dual graph of $G$. Therefore, there is a one-one correspondence between the bonds of $G$, and cycles of $G^*$. So there is a one-one correspondence between the cycles of $G'$ and $G^*$. Thus $G'$ and $G^*$ are 2-isomorphic. Hence $G'$ is the dual of $G$. ❑

Let $G$ be a plane graph and $G^*$ be the dual of $G$. So to every vertex of $G$ we have a region of $G^*$. Let the region $\phi$ of the plane graph $G^*$ corresponding to the vertex $v$ of $G$ has $e_1^*, e_2^*, \ldots, e_k^*$ as its boundary edges. Then each of these edges $e_i^*$ crosses the corresponding edge $e_i$ of $G$, and these edges are all incident at the vertex $v$. It follows that $\phi$ contains the vertex $v$. As $G^*$ is a plane graph, we can construct the dual of $G^*$, called the double dual of $G$ and is denoted by $G^{**}$.

We now have the following result.

**Theorem 6.19** Let $G$ be a plane connected graph. Then $G$ is isomorphic to its double dual $G^{**}$.

**Proof** Let $G$ be a plane connected graph and $G^*$ be the dual of $G$. Now, any region $\phi$ of dual $G^*$ contains exactly one vertex of $G$, namely its corresponding vertex $v$ of $G$. This is because the number of the regions of $G^*$ is same as the number of vertices of $G$.

Thus in the construction of $G^{**}$, we take the vertex $v$ to be the vertex in $G^{**}$ corresponding to the region $\phi$ of $G^*$. This choice gives the required isomorphism. ❑

**Definition:** A connected graph G is called *self-dual* if it is isomorphic to its dual $G^*$. For example, $K_4$ is self-dual.

The next result is a characterisation of bipartite graphs in terms of duality.

**Theorem 6.20** A connected plane graph $G$ is bipartite if and only if its dual graph $G^*$ is Eulerian.

**Proof** Let $G$ be a bipartite graph. Then $G$ does not contain odd cycles. Since $G$ is a plane, all the regions of $G$ are of even length. So degree of every region in $G$ is even. Let $G^*$ be the dual graph of $G$. Since the vertex degrees of $G^*$ are same as the corresponding degrees of the regions of $G$, degree of every vertex in $G^*$ is even. Hence $G^*$ is Eulerian.

Conversely, assume $G^*$ is Eulerian. Then degree of every vertex in $G^*$ is even. Since the degree of a vertex in $G^*$ corresponds to the degree of the region in $G$, degree of every region in $G$ is even. This shows that $G$ contains only even cycles. Hence $G$ is bipartite.

**Dual of a subgraph:** Let $G$ be a plane graph and let $G^*$ be its dual. Let $e$ be an edge in $G$ and let $e^*$ be the corresponding edge in $G^*$. We delete the edge $e$ and find dual of $G - e$. If edge $e$ is on the boundary of two regions, then removal of $e$ merges these two regions into one.

Thus, the dual $(G - e)^*$ is obtained from $G^*$ by deleting the corresponding edge $e^*$ and then fusing the two end vertices of $e^*$ in $G^* - e^*$. On the other hand, if edge $e$ is not on the boundary, then $e^*$ forms a self-loop. In that case, $G^* - e^*$ is the same as $(G - e)^*$.

Thus, if a graph $G$ has a dual $G^*$, then the dual of any subgraph of $G$ can be obtained by successive application of this procedure.

**Dual of a homeomorphic graph:** Let $G$ be a plane graph and $G^*$ be its dual. Let $e$ be an edge in $G$ and let $e^*$ be the corresponding edge in $G^*$. We create an additional vertex in $G$ by introducing a vertex of degree two in edge $e$ ($e$ now becomes two edges in series). This simply adds an edge parallel to $e^*$ in $G^*$. Likewise, the reverse process of merging two edges in series will simply eliminate one of the corresponding parallel edges in $G^*$. Thus, by this procedure, if the graph $G$ has dual $G^*$, the dual of any graph homeomorphic to $G$ can be obtained from $G^*$. These are illustrated in Figure 6.18(b).

The following important result is due to Whitney [267].

**Theorem 6.21 (Whitney [267])** A graph has a dual if and only if it is planar.

Let $G$ be a nonplanar graph. Then, according to Kuratowski's theorem, $G$ contains $K_5$ or $K_{3,3}$, or a sub-division of $K_5$ or $K_{3,3}$.

**Proof** We have only to prove that a nonplanar graph does not have a dual.

We know that a graph $G$ has a dual only if every subgraph $H$ of $G$ and every subdivision of $H$ has a dual. Thus, to prove the result, we show that neither $K_5$ nor $K_{3,3}$ has a dual.

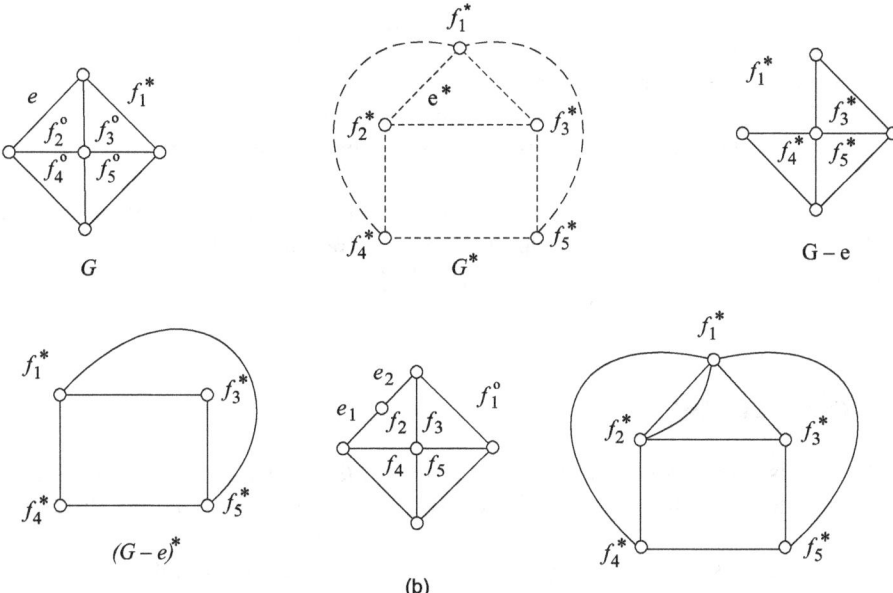

**Fig. 6.18(b)**

a. Suppose that $K_{3,3}$ has a dual $D$. Then the bonds in $K_{3,3}$ correspond to cycles in $D$, and vice-versa. Since $K_{3,3}$ has no bond consisting of two edges, $D$ has no cycle consisting of two edges. That is, $D$ contains no pair of parallel edges. Since every cycle in $K_{3,3}$ is of length four or six, $D$ has no bond with less than four edges. Therefore, the degree of every vertex in $D$ is at least four. As $D$ has no parallel edges, and the degree of every vertex is at least four, $D$ has at least five vertices each of degree four or more ($\geq 4$). That is, $D$ has at least $\frac{5 \times 4}{2} = 10$ edges. This is a contradiction to the fact that $K_{3,3}$ has nine edges. Hence $K_{3,3}$ has no dual.

b. Suppose $K_5$ has a dual $H$. We note that $K_5$ has

    i. 10 edges,

    ii. no pair of parallel edges,

    iii. no bond with two edges, and

    iv. bonds with only four or six edges.

Therefore graph $H$ has

    i. 10 edges,

    ii. no vertex with degree less than three,

    iii. no bond with two edges, and

    iv. cycles of length four or six only.

Now graph $H$ contains a hexagon (a cycle of length six) and no more than three edges can be added to a hexagon without creating a cycle of length three, or a pair of parallel edges. Since both of these are not present in $H$, and $H$ has 10 edges, there must be at least seven vertices in $H$. The degree of each of these vertices is at least three. This implies that $H$ has at least 11 edges, which is a contradiction. Hence $K_5$ has no dual.

**Definition:** Two plane graphs $G$ and $G^*$ are said to be *duals* (or combinatorial duals) of each other if there is a one-one correspondence between the edges of $G$ and $G^*$ such that if $H$ is any subgraph of $G$, and $H^*$ is the corresponding subgraphs of $G^*$, then rank $(G^* - H^*) =$ rank $G^* -$ nullity $H$.

## 6.6 Polyhedron

A polyhedron is a solid bounded by surfaces, called *faces*, each of which is a plane. That is, a solid bounded by plane surfaces is called a polyhedron. For example, brick, window frame, tetrahedron.

A polyhedron is said to be *convex* if any two of its interior points can be joined by a straight line lying entirely within the interior. For example, a brick is convex, but a window frame is not convex.

The vertices and edges of a polyhedron, which form a skeleton of the solid, give a simple graph in three dimensional space. It can be shown that for a convex polyhedron this graph is planar.

To see this, imagine the faces of the convex polyhedron to be made of rubber, with one face, say at the base, missing. Then taking hold of the edges at the missing face, we are able to stretch out the rubber to form one plane sheet. The vertices and edges in this transformation now form a plane graph. Moreover, each face of this plane graph corresponds to a face of a solid, with the exterior face of the graph corresponding to the missing face we used in stretching process.

Clearly, the plane graph is also connected, the degree of every vertex is at least 3 and the degree of every face is also at least 3. Also, the graph is simple.

A simple connected plane graph $G$ is called *polyhedral*, if $d(v) \geq 3$, for each vertex $v$ of $G$, and $d(\phi) \geq 3$, for every face $\phi$ of $G$.

We now have the following simple but useful property of polyhedral graphs.

**Theorem 6.22** Let $P$ be a convex polyhedron, and $G$ be its polyhedral graph. For each $n \geq 3$, let $n_i$ denote the number of vertices of $G$ of degree $i$ and let $f_i$ denote the number of faces of $G$ of degree $i$. Then

a. $\sum_{i \geq 3} i n_i = \sum_{i \geq 3} i f_i = 2m$, where $m$ is the number of edges of $G$.

b. The polyhedron $P$, and so the graph $G$, has at least one face bounded by a cycle of length $n$ for either $n = 3$, 4 or 5.

## Proof

a. The expression $\sum_{i \geq 3} i n_i$ is simply $\sum_{v \in V(G)} d(v)$, since $d(v) \geq 3$ for each vertex $v$.

We know, $\sum_{v \in V(G)} d(v) = 2m$, so that $\sum_{i \geq 3} i n_i = 2m$.

For each $n \geq 3$, the expression $i f_i$ is obtained by going round the boundary of each face having a cycle of length $i$ as its boundary, counting up the edges as we go. If we do this over all possible $i$, then we count each edge twice, and so

$$\sum_{i \geq 3} i f_i = 2m.$$

b. Assume to the contrary that $P$ has no faces bounded by cycles of length 3, 4 or 5. Then, $f_3 = f_4 = f_5 = 0$.

So by (a), $2m = \sum_{i \geq 6} i f_i \geq \sum_{i \geq 6} 6 f_i = 6 \sum_{i \geq 6} f_i = 6f$, where $f$ is the total number of faces of $P$. Thus, $f \leq \frac{1}{3} m$.

Also by (a), $2m = \sum_{i \geq 3} i n_i \geq \sum_{i \geq 3} 3 n_i = 3 \sum_{i \geq 3} n_i = 3n$, where $n$ is the total number of vertices of $P$. Thus, $n \leq \frac{2}{3} m$.

Now, by Euler's formula, $m = n + f - 2$ and we have

$$m \leq \frac{2}{3} m + \frac{1}{3} m - 2 = m - 2,$$ and this implies that $2 \leq 0$,

which is impossible. Hence $P$ must have a face of length 3, 4 or 5. ❑

**Regular polyhedra:** A polyhedron is called regular if it is convex and its faces are congruent regular polygons (so that the polyhedral angles are all equal). The regular polyhedra are also called Platonic bodies or Platonic solids.

A fact that has been known for at least 2000 years is that there are only five regular polyhedra, which is proved next by using graph theoretic argument.

**Theorem 6.23** The only regular polyhedra are the tetrahedron, the cube, the octahedron, the dodecahedron and the icosahedron.

**Proof** Let $P$ be a regular polyhedron and let $G$ be its corresponding polyhedral graph. Let $n$, $m$ and $f$ denote the number of vertices, edges and faces of $G$. Since the faces of $G$ are congruent to each other, each one is bounded by the same number of edges. Thus, $d(\phi) = k, k \geq 3$ for each face $\phi$ of $G$. Also, we know that there is at least one cycle of length 3, 4 or 5, therefore, $3 \leq k \leq 5$. Similarly, since the polyhedral angles are all equal to each other, the graph $G$ is regular, say $r$-regular, where $r \geq 3$. Thus, we have

$rn = 2m$, and $kf = 2m$, so that

$$2m = rn = kf. \tag{6.23.1}$$

Using this in Euler's formula $n - m + f = 2$, we have

$$8 = 4n - 4m + 4f = 4n - 2m + 4f - 2m = 4n - rn + 4f - kf.$$

Therefore, $8 = (4 - r)n + (4 - k)f.$ $\tag{6.23.2}$

Since $n$ and $f$ are both positive, and also $3 \leq k \leq 5$ and $r = 3$, there are only five possible cases (Fig. 6.19).

**Case 1**    Let $r = 3$ and $k = 3$. Then, $8 = (4 - 3)n + (4 - 3)f$, so that $8 = n + f$.

From (6.23.1.), $n = f$. So, $n + f = 8$ gives $n = 4$, $f = 4$.

Thus, $n = 4$, $m = 6$, $f = 4$, which is a tetrahedron.

**Case 2**    Let $r = 3$, $k = 4$. Then, from (6.23.2), $8 = (4 - 3)n + (4 - 4)f$.

So, $n = 8$. Then, (6.23.1) gives $3n = 4f$, so that $f = 6$. Also, $m = 12$.

So, $n = 8$, $m = 12$, $f = 6$, and thus is a cube.

**Case 3**    Let $r = 3$, $k = 5$. Then (6.23.2) gives $8 = (4 - 3)n + (4 - 5)f$, so that $8 = n - f$.

From (6.23.1), $3n = 5f$. Solving $8 = n - f$ and $3n = 5f$, we get $n = 20$, $f = 12$.

Thus, $n = 20$, $m = 30$, $f = 12$, which is a dodecahedron.

**Case 4**    Let $r = 4$, $k = 3$. Then, we get $n = 6$, $f = 8$.

So, $n = 6$, $m = 12$, $f = 8$, which is an octahedron.

**Case 5**    Let $r = 5$, $k = 3$. Then, we have $n = 12$, $f = 20$.

Thus, $n = 12$, $m = 30$, $f = 20$, which is an icosahedron.                                           ❑

Tetrahedron

Cube

Octahedron

Dodecahedron

Icosahedron

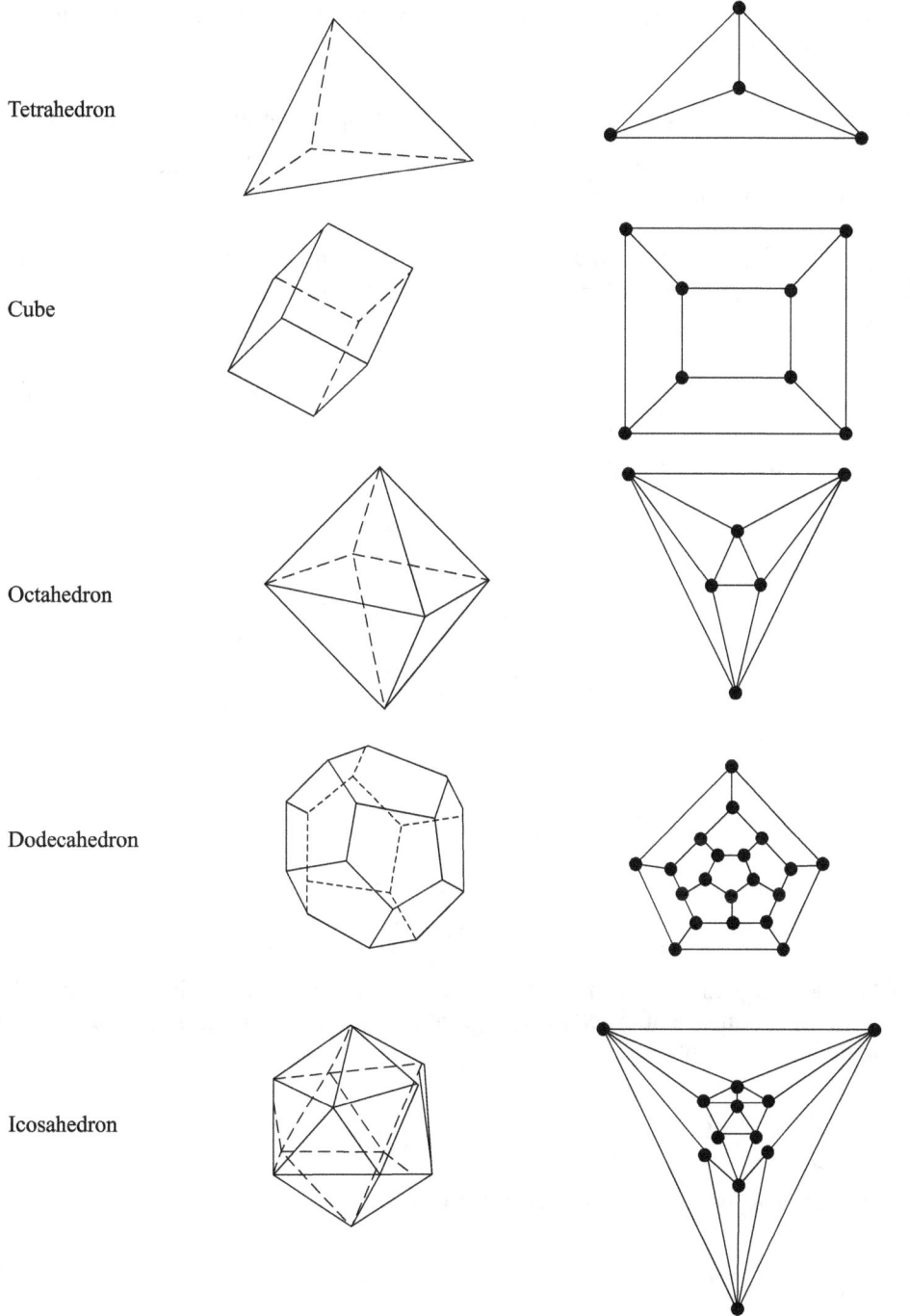

**Fig. 6.19**

The following is an elegant necessary condition for a plane graph to be Hamiltonian and is due to Grinberg [86].

**Theorem 6.24** If $G$ is a loopless plane graph having a Hamiltonian cycle $C$, then $\sum_{i=2}^{n}(i-2)(\phi_i' - \phi_i'') = 0$, where $\phi_i'$ and $\phi_i''$ are the numbers of regions of $G$ of degree $i$ contained in interior $C$ and exterior $C$ respectively.

**Proof** Let $E'$ and $E'$ denote the sets of edges of $G$ contained in int $C$ and ext $C$, respectively, and let $|E'| = m'$ and $|E''| = m''$. Then int $C$ contains exactly $m' + 1$ regions (Fig. 6.20) and so

$$\sum_{i=2}^{n}\phi_i' = m' + 1, \tag{6.24.1}$$

Since $G$ is loopless, $\phi_1' = \phi_1'' = 0$.

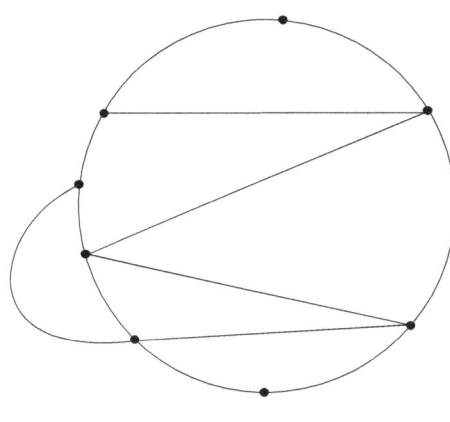

**Fig. 6.20**

Also, each edge in int $C$ is on the boundary of exactly two regions in int $C$ and each edge of $C$ is on the boundary of exactly one region in int $C$. Thus, counting the edges of all the regions in int $C$, we obtain

$$\sum_{i=2}^{n}i\phi_i'' = 2m' + n. \tag{6.24.2}$$

Eliminating $m'$ between (6.24.1) and (6.24.2), we get

$$\sum_{i=2}^{n}(i-2)\phi_i'' = n - 2. \tag{6.24.3}$$

Similarly, $\sum_{i=2}^{n}(i-2)\phi_i'' = n-2.$ (6.24.4)

The required result follows from equation (6.24.3) and (6.24.4). ❑

**Example** Consider the Herschel graph $G$ shown in Figure 6.21. Clearly, $G$ has 9 regions, and all the regions are of degree 4. Thus, if $G$ were Hamiltonian, then $2(\phi_4' - \phi_4'') = 0$. This implies $\phi_4' = \phi_4''$, which is impossible, since $\phi_4' + \phi_4'' = $ number of regions of degree 4 in $G = 9$ is odd. Hence $G$ is non-Hamiltonian.

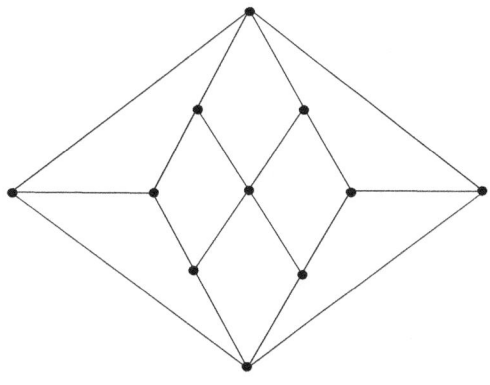

**Fig. 6.21**

Konig observed that every graph is embeddable on some orientable surface. This is seen by drawing an arbitrary graph G on the plane, possibly with edges that cross over each other and then attaching a handle to the plane at each crossing and allowing one edge to go over the handle and the other under it. Konig further showed that any embedding of a graph on an orientable surface with a minimum number of handles has all its regions simply connected.

Obviously, planar graphs can be embedded on a sphere. A toroidal graph is a graph embedded on a torus, for instance, $K_5$, $K_{3,3}$, $K_7$ and $K_{4,4}$.

**Definition:** The *genus* $\gamma(G)$ of a graph $G$ is the minimum number of handles which are added to the sphere so that $G$ can be embedded on the resulting surface. Obviously $\gamma(G) = 0$ if and only if $G$ is planar. The homeomorphic graphs have the same genus.

The genus of a polyhedron is the number of handles required on the sphere for a surface to contain the polyhedron.

**Definition:** The least number of planar subgraphs whose union is a given graph $G$ is called the *thickness* of $G$. Clearly the thickness of any planar graph is 1. Also, $K_5$ and $K_{3,3}$ have thickness 2, while the thickness of $K_9$ is 3.

The *crossing number* of a graph $G$ is the minimum number of pairwise intersections of its edges when $G$ is drawn in the plane. Clearly, crossing number of a graph is zero if and only if $G$ is planar.

The following result is due Euler and the proof can be found in Courant and Robbins [61].

**Theorem 6.25**  For a polyhedron of genus $\gamma$ with $n$ vertices, $m$ edges and $f$ regions

$$n - m + f = 2 - 2\gamma.$$

We have the following observations.
If $G$ is a connected graph of genus $\gamma$, then

$$m = \begin{cases} 3(n-2+2\gamma), & \text{if every region in } G \text{ is a triangle,} \\ 2(n-2+2\gamma), & \text{if every region in } G \text{ is a quadrilateral.} \end{cases}$$

From this, we have the following.
If $G$ is a connected graph of genus $\gamma$, then $\gamma \geq \frac{1}{6}m - \frac{1}{2}(n-2)$, and if $G$ has no triangles, then $\gamma \geq \frac{1}{4}m - \frac{1}{2}(n-2)$.

The proof of the following result is due to Ringel and Youngs [223] and some details can be found in Harary [104].

**Theorem 6.26**  For $n \geq 3$, the genus of the complete graph is

$$\gamma(K_n) = \frac{(n-3)(n-4)}{12}.$$

## 6.7  Decomposition of Some Planar Graphs

Schnyder [232] proved that each triangulated planar graph $G$ can be decomposed into three edge disjoint trees. If instead triangles all faces (or regions) of $G$ are quadrilaterals, and $G$ is without loops and multiple edges, Petrovic [183] proved that two trees are sufficient.

The following two lemmas are used in proving Theorem 6.27.

**Lemma 6.1**  If $G$ is a planar graph on $n(n \geq 4)$ vertices whose all regions are quadrilaterals, then $|E(G)| = 2n - 4$.

**Lemma 6.2**  If $G$ is a planar graph on $n(n \geq 4)$ vertices whose all regions are quadrilaterals, then $G$ contains at least three vertices of degree $\leq 3$.

The following is the result due to Petrovic [183].

**Theorem 6.27 (Petrovic [183])**  Let $G$ be a planar graph on $n(n \geq 4)$ vertices whose all faces are quadrilaterals, and $v_r$ and $v_b$ any two non-adjacent vertices of $G$ (we assume $v_r$ is coloured red and $v_b$ blue). Then the edges of $G$ can be partitioned into red and blue ones so that red ones form a spanning tree $T_r$ of $G - v_b$, and blue ones a spanning tree $T_b$ of $G - v_r$.

**Proof**  We call $T_r$ the red, and $T_b$ the blue tree. A vertex of $G$ is red (respectively blue) if it belongs to $T_r$ (respectively $T_b$). A vertex is red-blue if it belongs to both $T_r$ and $T_b$. Thus $v_r$ is the only red, and $v_b$ is only blue vertex in $G$. We proceed by induction on $n$. For $n = 4$, the assertion is obvious. Assume that it holds for each graph on less than $n(n > 4)$ vertices, and consider a graph $G$ on $n$ vertices. Let $u$ ($u \neq v_r, v_b$) be a vertex of $G$ of degree $\leq 3$ existing by Lemma 6.2.

   a. $d(u) = 2$. Let $uv_1$, $uv_2$ be the edges $s$, and $uv_1xv_2u$, $uv_1yv_2u$ the faces incident with $u$. Since $G \neq C_4$, each of vertices $v_1$ and $v_2$ has degree $\geq 3$. It implies that all regions of $G - u$ are quadrilaterals. By induction hypothesis $G - u$ can be partitioned into trees $T_r'$ and $T_b'$. Now, independent of the kind of vertices $v_1$ and $v_2$, (red, blue or red-blue) we can always colour the edges $uv_1$ and $uv_2$ so that the obtained trees $T_r$ and $T_b$ are as desired.

   b. $d(u) = 3$. Let $uv_1$, $uv_2$, $uv_3$ be the edges and $uv_1xv_2u$, $uv_2yv_3u$, $uv_3zv_1u$ be the faces incident with $u$.

First assume that all vertices $x$, $y$, $z$ are distinct. Since no triangle can be partitioned into disjoint quadrilaterals whose vertices are vertices of the triangle and some its inner points, $xy$, $yz$, $zx \notin E(G)$. Suppose that one of vertices $x$, $y$, $z$, is blue and there is no edge connecting it with the opposite vertex of the hexagon $v_1xv_2yv_3z$. Without loss of generality, we may assume that it is $x$. Then, $xv_3 \notin E(G)$. Denote by $G''$ the graph obtained from $G$ by deleting the edges $xv_1$ and $xv_2$ and identifying vertices $u$ and $x$. Since each face of $G''$ is a quadrilateral it can be partitioned into two trees $T_r''$ and $T_b''$, where $V(T_r'') = V(G'' - v_b)$ and $V(T_b'') = V(G'' - v_r)$. If edges $uv_1$ and $uv_2$ are coloured differently in $G'''$, say $uv_1$ red and $uv_2$ blue, we colour $xv_1$ red and $xv_2$ blue obtaining the required trees $T_r = T_r' + xv_1$ and $T_b = T_b'' + xv_2$ in $G$. Therefore, assume that $uv_1$ and $uv_2$ are monocoloured, say red Then one of vertices $v_1$ and $v_2$, say $v_1$, is red–blue. As $u$ is also a red-blue vertex, there is the unique blue $u - v_1$ path $P$ in $G''$. If $P$ contains $uv_3$, we colour $xv_1$ and $xv_2$ blue and red, respectively. If not, we colour both edges $xv_1$ and $xv_2$ red, and recolour $uv_1$ blue. It is routine to check that obtained trees $T_r = T_r'' + xv_1$, $T_b = T_b'' + xv_2$ in the first case, and $T_r = T_r'' - uv_1 + xv_1 + xv_2$, $T_b = T_b'' + uv_1$ in the second, are as required.

Now, assume that none of the vertices $x$, $y$, $z$ satisfy the condition above. Since at most one of edges $xv_3$, $yv_1$, $zv_2$ can exist, we may assume without loss of generality that $x$ is red-blue, $y$ is the red, $z$ is the blue vertex, $xv_3 \in E(G)$ and $yv_1 \notin E(G)$. Then each region of the graph $G'''$ obtained from $G$ by deleting edges $yv_2$ and $yv_3$ and identifying $y$ and $u$ is a quadrilateral. By induction hypothesis, $G'''$ can be composed into trees $T_r'''$ and $T_b'''$, where $V(T_r''') = V(G''' - z)$ and $V(T_b''') = V(G''' - y)$. Colouring all edges incident with $y$ red, and recolouring the edge $uv_3$ blue, we obtain the decomposition of $G$ into trees $T_r$ and $T_b$, where $V(T_r) = V(G-z)$ and $V(T_b) = V(G-y)$.

Suppose that two of vertices $x$, $y$, $z$, say $y$ and $z$, coincide. Then $d(v_3) = 2$. If $v_3$ is red-blue, we finish the proof as in case (a). If $x$ is red-blue we proceed as above. So we may assume that $\{v_3, x\} = \{v_r, v_b\}$. Let $v_3 = v_r$, $x = v_b$. By induction hypothesis the graph $G^{IV} = G - v_3$ can be decomposed into trees $T_r^{IV}$ and $T_b^{IV}$, where $V(T_r^{IV}) = V(G^{IV} - x)$ and $V(T_b^{IV}) = V(G^{IV} - u)$. Then edges $uv_1$, $uv_2$ are coloured red and edges $xv_1$, $xv_2$, blue. It implies that edges $yv_1$, $yv_2$ are coloured differently, $yv_1$, red and $yv_2$, blue. Now, colouring $v_3u$ and $v_3y$ red and recolouring $uv_1$ blue, we get from $T_r^{IV}$ and $T_b^{IV}$ the desired trees $T_r$ and $T_b$, where $V(T_r) = V(T_r) = V(G-x)$ and $V(T_b) = V(G-v_3)$.

If $x = y = z$, then $v(G) = \{u, x, v_1, v_2, v_3\}$, $E(G) = \{uv_1, uv_2, uv_3, xv_1, xv_2, xv_3\}$ and the statement holds trivially.                                                                                          ❑

## 6.8  Exercises

1. If $G$ is a connected planar graph of order less than 12, prove that $\delta(G) = 4$.

2. If $G$ is a planar graph of order 24 and is regular of degree 3, then what is the number of regions in a planar representation of $G$?

3. Prove that Euler's formula fails for disconnected graphs.

4. Show that every graph with at most three cycles is planar.

5. Find a simple graph $G$ with degree sequence $[4, 4, 3, 3, 3, 3]$ such that

   a. $G$ is planar,

   b. $G$ is nonplanar.

6. Show that every simple bipartite cubic planar graph contains a $C_4$.

7. Prove that a simple planar graph has at least 4 vertices of degree 5 at most.

8. Let $G$ be a planar, triangle free graph of order $n$. Prove that $G$ has no more than $2n - 4$ edges.

9. Let $G$ be of order $n = 11$. Show that at least one of $G$ or $\overline{G}$ is non planar.

10. Show that the average degree of a planar graph is less than 6.

11. Use Kuratowski's theorem to prove that the Peterson graph is nonplanar.

12. Show that the edges forming a spanning tree in a planar graph $G$ correspond to the edges forming a set of chords in the dual $G^*$.

13. Prove that the complete graph $K_4$ is self-dual.

# 7. Colourings

Presently, colouring is one of the important branches of graph theory and has attracted the attention of almost all graph theorists, mainly because of the four colour theorem, the details of which can be seen in Chapter 12.

## 7.1 Vertex colouring

A vertex colouring (or simply colouring) of a graph $G$ is a labelling $f : V(G) \rightarrow \{1, 2, \ldots\}$; the labels called *colours*, such that no two adjacent vertices get the same colour and each vertex gets one colour. A $k$-colouring of a graph $G$ consists of $k$ different colours and $G$ is then called *k-colourable*. A 2-colourable and a 3-colourable graph are shown in Figure 7.1. It follows from this definition that the $k$-colouring of a graph $G = (V, E)$ partitions the vertex set $V$ into $k$ independent sets $V_1, V_2, \ldots, V_k$ such that $V = V_1 \cup V_2 \cup \ldots \cup V_k$. The independent sets $V_1, V_2, \ldots, V_k$ are called the *colour classes* and the function $f : V(G) \rightarrow \{1, 2, \ldots, k\}$ such that $f(v) = i$ for $v \in V_i$, $1 \le i \le k$, is called the *colour function*.

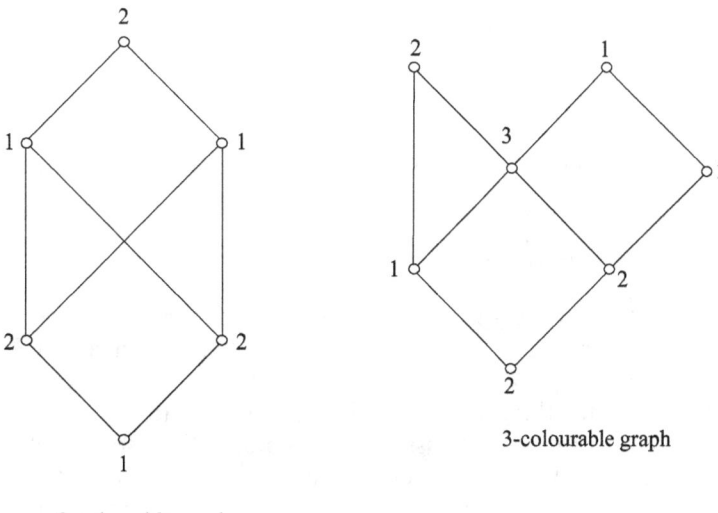

2-colourable graph

3-colourable graph

**Fig. 7.1**

The minimum number $k$ for which there is a $k$-colouring of the graph is called the *chromatic number* (chromatic index) of $G$ and is denoted by $\chi(G)$. If $\chi(G) = k$, the graph $G$ is said to be *k-chromatic*.

We observe that colouring any one of the components in a disconnected graph does not affect the colouring of its other components. Also, parallel edges can be replaced by single edges, since it does not affect the adjacencies of the vertices. Thus, for colouring considerations, we opt only for simple connected graphs.

The following observations are the immediate consequences of the definitions introduced above.

1. A graph is 1-chromatic if and only if it is totally disconnected.

2. A graph having at least one edge is at least 2-chromatic (bichromatic).

3. A graph $G$ having $n$ vertices has $\chi(G) \leq n$.

4. If $H$ is subgraph of a graph $G$, then $\chi(H) \leq \chi(G)$.

5. A complete graph with $n$ vertices is $n$-chromatic, because all its vertices are adjacent. So, $\chi(K_n) = n$ and $\chi(\overline{K}_n) = 1$. Therefore, we see that a graph containing a complete graph of $r$ vertices is at least $r$-chromatic. For example, every graph containing a triangle is at least 3-chromatic.

6. A cycle of length $n \geq 3$ is 2-chromatic if $n$ is even and 3-chromatic if $n$ is odd. To see this, let the vertices of the cycle be labelled $1, 2, \ldots, n$, and assign one colour to odd vertices and another to even. If $n$ is even, no adjacent vertices get the same colour, if $n$ is odd, the $n$th vertex and the first vertex are adjacent and have the same colour, therefore need the third colour for colouring.

7. If $G_1, G_2, \ldots, G_r$ are the components of a disconnected graph $G$, then

$$\chi(G) = \max_{1 \leq i \leq r} \chi(G_i).$$

We note that trees with greater or equal to two vertices are bichromatic as is seen in the following result.

**Theorem 7.1**    Every tree with $n \geq 2$ vertices is 2-chromatic.

**Proof**    Let $T$ be a tree with $n \geq 2$ vertices. Consider any vertex $v$ of $T$ and assume $T$ to be rooted at vertex $v$ (Fig. 7.2). Assign colour 1 to $v$. Then assign colour 2 to all vertices which are adjacent to $v$. Let $v_1, v_2, \ldots, v_r$ be the vertices which have been assigned colour 2. Now assign colour 1 to all the vertices which are adjacent to $v_1, v_2, \ldots, v_r$. Continue this process till every vertex in $T$ has been assigned the colour. We observe that in $T$ all vertices at odd distances from $v$ have colour 2, and $v$ and vertices at even distances from $v$ have colour 1. Therefore along any path in $T$, the vertices are of alternating colours. Since there is one and only one path between any two vertices in a tree, no two adjacent vertices have the same colour. Thus $T$ is coloured with two colours. Hence $T$ is 2-chromatic.    ❑

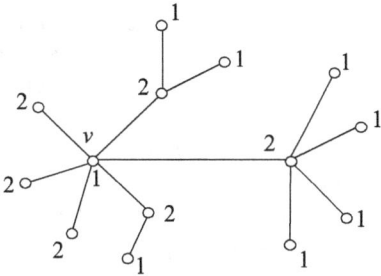

**Fig. 7.2**

The converse of the above theorem is not true, i. e., every 2-chromatic graph need not be a tree. To see this, consider the graph shown in Figure 7.3. Clearly, $G$ is 2-chromatic, but is not a tree.

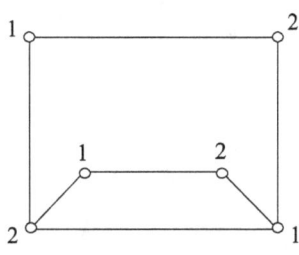

**Fig. 7.3**

The next result due to Konig [134] characterises 2-chromatic graphs.

**Theorem 7.2 (Konig [134])**    A graph is bicolourable (2-chromatic) if and only if it has no odd cycles.

**Proof**    Let $G$ be a connected graph with cycles of only even length and let $T$ be a spanning tree in $G$. Then, by Theorem 7.1, $T$ can be coloured with two colours. Now add the chords to $T$ one by one. As $G$ contains cycles of even length only, the end vertices of every chord get different colours of $T$. Thus $G$ is coloured with two colours and hence is, 2-chromatic.

Conversely, let $G$ be bicolourable, that is, 2-chromatic. We prove $G$ has even cycles only. Assume to the contrary that $G$ has an odd cycle. Then by observation (6), $G$ is 3-chromatic, a contradiction. Hence $G$ has no odd cycles.    ❏

**Corollary 7.1**    For a graph $G$, $\chi(G) \geq 3$ if and only if $G$ has an odd cycle.

The following result is yet another characterisation of 2-chromatic graphs.

**Theorem 7.3**    A nonempty graph $G$ is bicolourable if and only if $G$ is bipartite.

**Proof**  Let $G$ be a bipartite graph. Then its vertex set $V$ can be partitioned into two nonempty disjoint sets $V_1$ and $V_2$ such that $V = V_1 \cup V_2$. Now assigning colour 1 to all vertices in $V_1$ and colour 2 to all vertices in $V_2$ gives a 2-colouring of $G$. Since $G$ is nonempty, $\chi(G) = 2$.

Conversely, let $G$ be bicolourable, that is, $G$ has a 2-colouring. Denote by $V_1$ the set of all those vertices coloured 1 and by $V_2$ the set of all those vertices coloured 2. Then, no two vertices in $V_1$ are adjacent and no two vertices in $V_2$ are adjacent. Thus, any edge in $G$ joins a vertex in $V_1$ and a vertex in $V_2$. Hence $G$ is bipartite with bipartition $V = V_1 \cup V_2$.   ❑

## 7.2  Critical Graphs

If $G$ is a $k$-chromatic graph and $\chi(G - v) = k - 1$ for every vertex $v$ in $G$, then $G$ is called a $k$-critical graph. A 4-critical graph is shown in Figure 7.4. If $G$ is $k$-chromatic, but $\chi(G - e) = k - 1$ for each edge $e$ of $G$, then $G$ is called *k-edge-critical graph*, or *k-minimal*. A graph $G$ is said to be *contraction critical* or con-critical if $\chi(H) < \chi(G)$ for every proper contraction $H$ of $G$. A graph $G$ is said to be *critical* if $\chi(H) < \chi(G)$ for every proper subgraph $H$ of $G$.

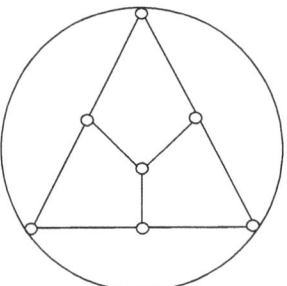

**Fig. 7.4**

Now, we have the following observations.

1. Every critical or minimal graph is connected.

2. Every connected $k$-chromatic graph contains a critical or minimal $k$-chromatic graph.

3. $\chi(G) = \max \{\chi(B) : B$ is a block of $G\}$.

4. The only 1-critical or 1-minimal graph is $K_1$, the only 2-critical or 2-minimal graph is $K_2$ and the only 3-critical or 3-minimal graphs are $C_{2n+1}, n \geq 1$, that is, odd cycles.

The following result due to Dirac [66] describes some of the important properties of a $k$-critical graph.

**Theorem 7.4 (Dirac [66])**   If $G$ is a $k$-critical graph, then

a. $G$ is connected,

b.  $\delta(G) \geq k - 1$,

c.  $G$ has no pair of subgraphs $G_1$ and $G_2$ for which $G = G_1 \cup G_2$, and $G_1 \cap G_2$ is a complete graph,

d.  $G - v$ is connected for every vertex $v$ of $G$, provided $k > 1$.

## Proof

a.  Assume that $G$ is not connected. Since $\chi(G) = k$, by observation 7, there is a component $G_1$ of $G$ such that $\chi(G_1) = k$. If $v$ is any vertex of $G$ which is not in $G_1$, then $G_1$ is a component of the subgraph $G - v$. Therefore, $\chi(G - v) = \chi(G_1) = k$. This contradicts the fact that $G$ is $k$-critical. Hence $G$ is connected.

b.  Let $v$ be a vertex of $G$ so that $d(v) < k - 1$. Since $G$ is $k$-critical, the subgraph $G - v$ has a $(k - 1)$-colouring. As $v$ has at most $k - 2$ neighbours, these neighbours use at most $k - 2$ colours in this $(k - 1)$-colouring of $G - v$. Now, colour $v$ with the unused colour and this gives a $(k - 1)$-colouring of $G$. This contradicts the given assumption that $\chi(G) = k$. Hence every vertex $v$ has degree at least $k - 1$.

c.  Let $G = G_1 \cup G_2$, where $G_1$ and $G_2$ are subgraphs with $G_1 \cap G_2 = K_t$. Since $G$ is $k$-critical, therefore $G_1$ and $G_2$ both have chromatic number at most $k - 1$. Consider a $(k - 1)$-colouring of $G_1$ and a $(k - 1)$-colouring of $G_2$. As $G_1 \cap G_2$ is complete, in the overlap, every vertex in $G_1 \cap G_2$ has a different colour (in each of the $(k - 1)$-colourings). This implies that colours in the $(k - 1)$-colouring of $G_2$ can be rearranged such that it assigns the same colour to each vertex in $G_1 \cap G_2$, as is given by the colouring of $G_1$. Combining the two colourings then produces a $(k - 1)$-colouring of all of $G$. This is impossible, since $\chi(G) = k$. Thus, no subgraphs of the type $G_1$ and $G_2$ exist.

d.  Assume that $G - v$ is disconnected, for some vertex $v$ of $G$. Then $G - v$ has a subgraph $H_1$ and $H_2$ with $H_1 \cup H_2 = G - v$ and $H_1 \cap H_2 = \Phi$. Let $G_1$ and $G_2$ be the subgraphs of $G$, where $G_1$ is induced by $H_1$ and $v$, while $G_2$ is induced by $H_2$ and $v$. Then, $G = G_1 \cup G_2$ and $G_1 \cap G_2 = K_1$ (with $K_1$ as a single vertex). This contradicts (c) and thus $G - v$ is connected.                                                    ❏

Let $S = \{u, v\}$ be a 2-vertex cut of a critical $k$-chromatic graph $G$. Since no separating set of a critical graph is a complete graph, therefore $uv$ is not an edge of $G$. Let $G_i$ be the $S$-component of $G$. $G_i$ is said to be of type 1 if every $(k - 1)$ colouring of $G_i$ assigns the same colour to $u$ and $v$, $G_i$ is of type 2 if every $(k - 1)$-colouring of $G_i$ assigns different colours to $u$ and $v$, and $G_i$ is of type 3 if some $(k - 1)$-colouring of $G_i$ assigns same colour to $u$ and $v$, while some other $(k - 1)$-colouring assigns different colours to $u$ and $v$.

The following characterisation of $k$-critical graphs with a 2-vertex cut is due to Dirac.

**Theorem 7.5**   If $G$ is a minimal $k$-chromatic graph with a 2-vertex cut $S = \{u, v\}$, then (i) $G = G_1 \cup G_2$, where $G_i$ is the $S$-component of type $i$, $i = 1, 2$ and (ii) both $G + uv$ and $G : uv$ are $k$-minimal.

**Proof** Let $G$ be a minimal $k$-chromatic graph with a 2-vertex cut $S = \{u, v\}$. So the $S$-components of $G$ are $(k-1)$-colourable. Now, there is no set of $(k-1)$-colourings of the $S$-components all of which agree on $S$, then $G$ is $(k-1)$-colourable. Therefore there is a type 1 $S$-component $G_1$ and a type 2 $S$-component $G_2$. Then $G_1 \cup G_2$ is not $(k-1)$-colourable. Since $G$ is $k$-critical, there is no third $S$-component $G_3$. Hence, $G = G_1 \cup G_2$.

Now, let $H_1 = G_1 + uv$ and $H_2 = G_2 : uv$. We prove that $H_2$ is $k$-minimal. Since $G_2$ is of type 2, therefore every $(k-1)$-colouring of $G_2$ assigns different colours to $u$ and $v$. As $u$ and $v$ are identified to a single vertex, say $w$ in $H_2$, so a $k$-colouring is necessary to colour $H_2$, that is, $H_2$ is $k$-chromatic. We further prove that $\chi(H_2 - e) = k - 1$ for any edge of $H_2$. Any such edge $e$ can be considered to belong to $G$ and in the $(k-1)$-colouring of $G - e$, $u$ and $v$ get the same colour, since they can be considered to belong to $G_1$ which is a subgraph of $G - e$. The restriction of such a colouring of $G - e$ to $H_2 - e$ (with $u$ and $v$ identified as $w$ with the common colour of $u$ and $v$) is a $(k-1)$-colouring of $H_2 - e$. This proves the result.

That $H_1$ is minimal, can be proved in a similar manner.  ❑

**Theorem 7.6** Every $k$-chromatic graph can be contracted into a con-critical chromatic graph.

**Proof** Let $G$ be a $k$-chromatic graph and let the edge $e$ of $G$ be contracted. Then a colouring of $G$ can be used to give a colouring of $G|e$ except that, possibly the vertex formed by the contraction may be assigned an extra colour. Thus, $\chi(G|e) \le \chi(G) + 1$. On the other hand, a colouring of $G|e$ can be used to get a colouring for $G$ by using an extra colour for one of the end vertices of $e$. Therefore, $\chi(G) \le \chi(G|e) + 1$. Thus the contraction of an edge changes the chromatic number by at most one. Sometimes contraction of an edge may increase the chromatic number, but by repeated contractions, the number of edges and therefore the chromatic number gets reduced. Clearly, the connected graph can be contracted to a single vertex whose chromatic number is one. In between, a stage arises where the chromatic number of the graph is the same as the original, but the contraction of any edge reduces the chromatic number by one.  ❑

As noted earlier, every connected $k-$ chromatic graph contains a critical or minimal $k$-chromatic graph. To see this, we observe that if $G$ is not $k$-critical, then $\chi(G - v) = k$, for some vertex $v$ of $G$. If $G - v$ is $k-$critical, then this is the required subgraph. If not, then $G - \{v, w\} = (G - v) - w$ has chromatic number $k$, for some vertex $w$ in $G - v$. If this new subgraph is $k$-critical, then again this is the required subgraph. If not, we continue this vertex deletion procedure, and we will clearly get a $k$-critical subgraph.

We have the following immediate observation.

**Theorem 7.7** Any $k$-chromatic graph has at least $k$ vertices of degree at least $k-1$ each.

**Proof** Let $G$ be a $k$-chromatic graph and let $H$ be a $k$-critical subgraph of $G$. Then, by Theorem 7.4 (b), every vertex of $H$ has degree at least $k-1$ in $H$ and hence in $G$. Since $H$ is $k$-chromatic, $H$ has at least $k$ vertices. This completes the proof. ❑

We note that there is no easy characterisation of graphs with chromatic number greater or equal to three. The graph vertex - colouring problem in a standard NP-complete problem and no good algorithm for finding $\chi(G)$ has been discovered for the class of all graphs, though for some special classes of graphs polynomial time algorithms have been found. There are various results which give upper bounds for the chromatic number of an arbitrary graph $G$, provided the degrees of all the vertices of $G$ are known. The first of these is due to Szekeres and Wilf [237].

**Theorem 7.8 (Szekeres and Wilf [237])** Let $G$ be a graph and $k = \max\{\delta(G') : G'$ is a subgraph of $G\}$. Then, $\chi(G) = k-1$.

**Proof** Let $H$ be a $k$-minimal subgraph of $G$. Then $H$ is a subgraph of $G$ and therefore $\delta(H) \leq k$. Using Theorem 7.4, we have, $\delta(H) \geq \chi(H) - 1 = \chi(G) - 1$. Thus, $\chi(G) \leq \delta(H) + 1 = k+1$. ❑

The next result is due to Welsh and Powell [262] and its proof is due to Bondy [32].

**Theorem 7.9** Let $G$ b a graph with degree sequence $[d_i]_1^n$ such that $d_1 \geq d_2 \geq \ldots \geq d_n$. Then, $\chi(G) \leq \max\{\min\{i, d_i + 1\}\}$.

**Proof** Let $G$ be $k$-chromatic. Then, by Theorem 7.8, $G$ has at least $k$ vertices of degree at least $k-1$. Therefore, $d_k \geq k-1$ and $\max\{\min\{i, d_i + 1\}\} = \min\{k, d_k + 1\} = k = \chi(G)$. ❑

Now we have the following upper bounds for chromatic number.

**Theorem 7.10** For any graph $G$, $\chi(G) \leq \triangle(G) + 1$.

**Proof** Let $G$ be any graph with $n$ vertices. To prove the result, we induct on $n$. For $n = 1, G = K_1$ and $\chi(G) = 1$ and $\triangle(G) = 0$. Therefore, the result is true for $n = 1$.

Assume that the result is true for all graphs with $n-1$ vertices and therefore by induction hypothesis, $\chi(G) \leq \triangle(G-v) + 1$. This shows that $G - v$ can be coloured by using $\triangle(G-v) + 1$ colours. Since $\triangle(G)$ is the maximum degree of a vertex in $G$, vertex $v$ has at most $\triangle(G)$ neighbours in $G$. Thus these neighbours use up at most $\triangle(G)$ colours in the colouring of $G - v$.

Now, if $\triangle(G) = \triangle(G-v)$, then there is at least one colour not used by $v$'s neighbours and that can be used to colour $v$ giving a $\triangle(G) + 1$ colouring for $G$.

In case $\triangle(G) \neq \triangle(G-v)$, then $\triangle(G-v) < \triangle(G)$. Therefore, using a new colour for $v$, we have a $\triangle(G-v)+2$ colouring of $G$ and clearly, $\triangle(G-v)+2 \leq \triangle(G)+1$. Hence in both cases, it follows that $\chi(G) \leq \triangle(G)+1$.                                                                  ❑

**Remarks**

1.  Clearly, Theorem 7.10 is a simple consequence of Theorem 7.7. This is because if $G$ is $k-$ chromatic, then Theorem 7.4 gives $\triangle \geq k-1$, that is, $\chi \leq \triangle+1$.

2.  The equality in Theorem 7.10 holds if $G = C_{2n+1}$, $n \geq 1$ and if $G = K_m$.

## 7.3  Brook's Theorem

**Greedy colouring algorithm:**  The greedy colouring with respect to a vertex ordering $v_1, v_2, \ldots, v_n$ of $V(G)$ is obtained by colouring vertices in the order $v_1, v_2, \ldots, v_n$ assigning to $v_i$ the smallest $-$ indexed colour not already used on its lower $-$ indexed neighbours. This is reported in West [263].

The following recolouring technique as noted in Clark and Holton [60] is due to Kempe [128].

**Kempe Chain argument:**  Let $G$ be a graph with a colouring using at least two different colours represented by $i$ and $j$. Let $H(i, j)$ denote the subgraph of $G$ induced by all the vertices of $G$ coloured either $i$ or $j$ and let $K$ be a connected component of the subgraph $H(i, j)$. If we interchange the colours $i$ and $j$ on the vertices of $K$ and keep the colours of all other vertices of $G$ unchanged, then we get a new colouring of $G$, which uses the same colours with which we started. This subgraph $K$ is called a Kempe chain and the recolouring technique is called the Kempe chain argument (Fig. 7.5).

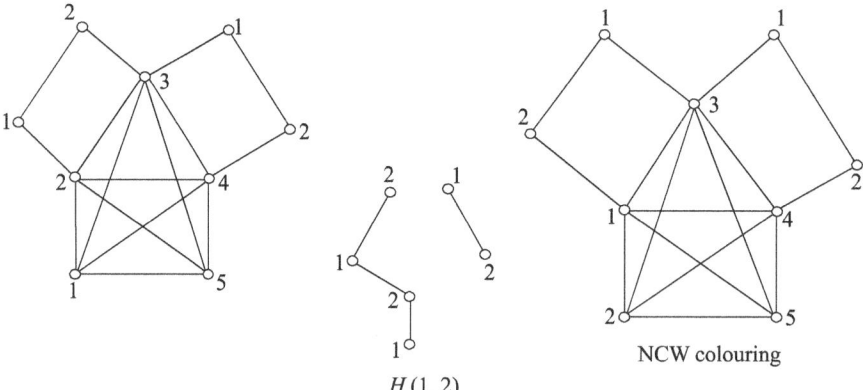

$H(1, 2)$

NCW colouring

**Fig. 7.5**

The following result due to Brooks [39] is an improvement of the bounds obtained in Theorem 7.10. We give two proofs of Brooks theorem, the first given by Lovasz [150] uses greedy colouring, and the second proof uses Kempe chain argument.

**Theorem 7.11 (Brooks [39])** If $G$ is a connected graph which is neither complete nor an add cycle, then $\chi(G) \le \triangle(G)$.

**Proof** Let $G$ be a connected graph with vertex set $V = v_1, v_2, \ldots, v_n$ which is neither a complete graph, nor an odd cycle and let $\triangle = k$. Since $G$ is a complete graph for $k = 1$ and $G$ is an odd cycle or a bipartite graph for $k = 2$, let $k \ge 3$.

Assume $G$ is not $k$-regular. Then there exists a vertex say $v = v_n$ such that $d(v) < k$. Since $G$ is connected, we form a spanning tree of $G$ starting from $v_n$ and whose vertices are arranged in the order $v_n, v_{n-1}, \ldots, v_1$ (Fig. 7.6).

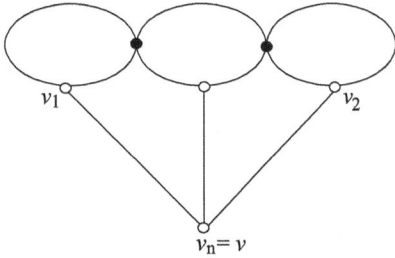

**Fig. 7.6**

Clearly, each vertex $v_i$ other than $v_n$ in the resulting order $v_n, v_{n-1}, \ldots, v_1$ has a higher indexed neighbour along the path to $v_n$ in the tree. Therefore, each vertex $v_i$ has atmost $k - 1$ lower indexed neighbours and the greedy colouring needs at most $k$ colours (Fig. 7.7).

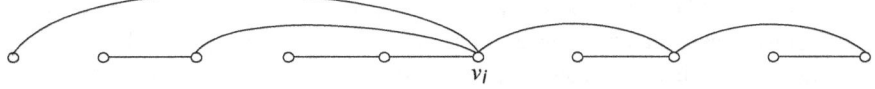

**Fig. 7.7**

Now, let $G$ be $k$-regular. Assume $G$ has a cut vertex say $x$ and let $G'$ be a subgraph containing a component of $G - x$ together with the edges of $G - x$ to $x$. Clearly, $d(x|G') < k$. Therefore, by using the above argument, we have a $k$-colouring of $G'$. By making use of the permutations of the colours, it can be seen that this is true for all such subgraphs. Thus $G$ is $k$-colourable.

Now, let $G$ be 2-connected. We claim that $G$ has an induced 3-vertex path, with vertices say $v_1, v_2, v_n$ in order, such that $G - v_1, v_2$ is connected.

To prove the claim, let $x$ be any vertex of $G$. If $k(G - x) \ge 2$, let $v_1$ be $x$ and let $v_2$ be a vertex with distance two from $x$, which clearly exists, as $G$ is regular and not a complete graph. If $k(G - x) = 1$, then $x$ has a neighbour in every end block of $G - x$, since $G$ has no cut vertex. Let $v_1$ and $v_2$ be the neighbours of $x$ in two such blocks. Clearly $v_1$ and $v_2$ are non adjacent. Also, since blocks have no cut vertices, $G - \{x, v_1, v_2\}$ is connected. As $k \geq 3$, so $G - \{v_1, v_2\}$ is connected and we let $x = v_n$, proving the claim.

Now arrange the vertices of a spanning tree of $G - \{v_1, v_2\}$ as $v_3, v_4, \ldots, v_n$. As before, each vertex before $n$ has atmost $k - 1$ lower indexed neighbours. The greedy colouring uses at most $k - 1$ colours on neighbours of $v_n$, since $v_1$ and $v_2$ get the same colour.

**Second Proof (Using Kempe chain argument)**   Let $G$ be a connected graph with $n$ vertices which is neither complete nor an odd cycle. Let $\triangle(G) = k$. For $k = 1$, $G$ is complete and for $k = 2$, $G$ is an odd cycle or a bipartite graph. Therefore, assume $k \geq 3$.

We induct on n. Since $k \geq 3$, the induction starts from $n = 4$. As $G$ is not complete, then for $n = 4$, $G$ is one of the graphs given in Figure 7.8.

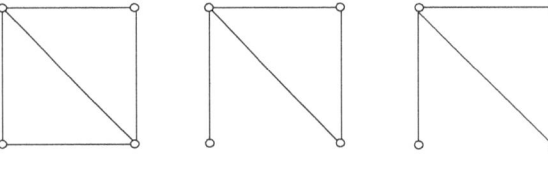

**Fig. 7.8**

Clearly, for each such graph, the chromatic index is at most three.

Now, let $n \geq 5$, and assume the result to be true for all graphs with fewer than $n$ vertices.

If $G$ has a vertex $v$ of degree less than $k$, then it follows from Theorem 7.10 that $G$ can be coloured by $k$ colours, since the neighbours of $v$ use up atmost $k - 1$ colours. Therefore the result is true in this case.

Now assume that degree of every vertex of $G$ is $k$, that is, $G$ is $k$-regular. We show that $G$ has a $k$-colouring.

Let $v$ be any vertex of $G$. Then by induction hypothesis, the subgraph $G - v$ has a $k$-colouring. If the neighbours of $v$ in $G$ do not use all the $k$ colours in the $k$-colouring of $G - v$, then any unused colour is assigned to $v$ giving a $k$-colouring of $G$. Assume that the $k$ neighbours of $v$ are assigned all the $k$ colours in the $k$-colouring of $G - v$. Let the neighbours of $v$ be $v_1, v_2, \ldots, v_k$ which are coloured by the colours $1, 2, \ldots, k$ respectively.

Let the Kempe chains $H_{v_i}(i, j)$ and $H_{v_j}(i, j)$ containing the neighbours $v_i$ and $v_j$ be different. That is, $v_i$ and $v_j$ are in different components of the subgraph $H(i, j)$ induced by the colours $i$ and $j$. Therefore, using Kempe chain argument, the colours in $H_{v_i}(i, j)$ are interchanged to give a $k$-colouring of $G - v$, where now $v_i$ has been assigned the colour $j$. This implies that the neighbours of $v$ use less than $k$ colours and the unused colour assigned to $v$ gives a $k$-colouring of $G$.

Now assume that for each $i$ and $j$, the neighbours $v_i$ and $v_j$ are in the same Kempe chain, which is briefly denoted by $H$. If the degree of $v_i$ in $H$ is greater than one, then $v_i$ is adjacent to at least two vertices coloured $j$. Therefore there is a third colour, say $\ell$, not used in colouring the neighbours of $v_i$. Recolour $v_i$ by $\ell$ and colour $v$ by $i$, giving the $k$-colouring of $G$. Assume that $v_i$ and $v_j$ both are of degree one in $H$. Let $P$ be a path from $v_i$ to $v_j$ in $H$ and let there be a vertex in $P$ with degree at least three in $H$ (Fig. 7.9). Let $u$ be the first such vertex and coloured $i$. (If $u$ is coloured $j$, the same argument is used as in case of $i$). Then at least three neighbours of $u$ are coloured $j$ and therefore there is a colour, say $\ell$, not used by these neighbours. Recolour $u$ by $\ell$ and interchange colours $i$ and $j$ on the vertices of $P$ from $v_i$ upto $u$, excluding $u$. So we get a colouring of $G-v$, where $v_i$ and $v_j$ are now both coloured $j$. This allows $v$ to be coloured by $i$.

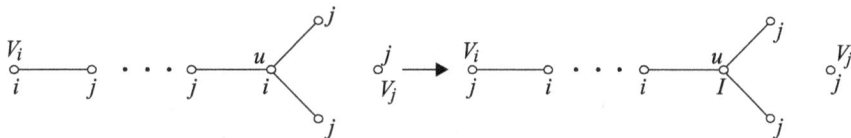

**Fig. 7.9**

Let all the vertices on a path $v_i$ to $v_j$, excluding the end vertices $v_i$ and $v_j$, be of degree two in $H$. Clearly $H$ contains a single path from $v_i$ to $v_j$.

Let all the Kempe chains be paths. Let $H$ and $K$ be such chains corresponding to $v_i$, $v_j$ and $v_i$, $v_\ell$ respectively, with $j \neq \ell$. Let $w \neq v_i$ be a vertex present in both the chains (Fig. 7.10). Then $w$ is coloured $i$, has two neighbours coloured $j$ and two neighbours coloured $\ell$. Therefore there is a fourth colour, say $s$, not used by the neighbours of $w$. Now colour $w$ by $s$, and interchange colours $\ell$ and $i$ on the vertices of $K$ beyond $w$ upto and including $v_\ell$, we get a colouring of $G-v$, where $v_i$ and $v_\ell$ are now both coloured $i$. This allows $v$ to be coloured by $\ell$.

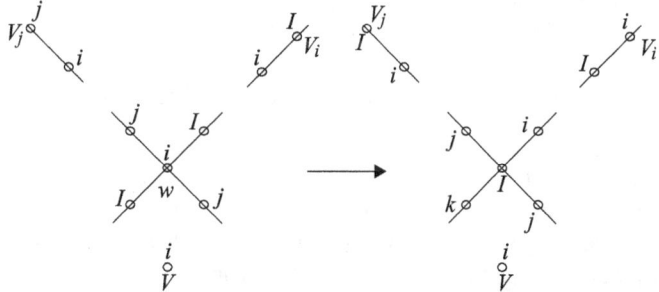

**Fig. 7.10**

Thus assume that two such Kempe chains meet only at their common end vertex $v_i$.

Let $v_i$ and $v_j$ be two neighbours of $v$ which are nonadjacent and let $x$ be the vertex coloured $j$, adjacent to $v_i$ on the Kempe chain $H$ from $v_i$ to $v_j$. With $\ell \neq j$, let $K$ denote the Kempe chain from $v_i$ to $v_j$. Then by the Kempe chain argument, we interchange the colours in $K$, without changing the colours of the other vertices. This results in $v_i$ coloured $\ell$, and $v_\ell$ coloured $i$. Since $x$ is adjacent to $v_i$, it is in the Kempe chain for colours $\ell$ and $j$. However, it is also the Kempe chain for colours $i$ and $j$. This contradicts the assumption that Kempe chains have at most one vertex in common, the end vertex. This contradiction implies that any two $v_i$ and $v_j$ are adjacent. In other words, all neighbours of $v$ are also neighbours of each other. This shows that $G$ is the complete graph $K_k$, a contradiction to the hypothesis of $G$. ❑

**Definition:** In the depth first search tree (DFS), a search tree $T$ is used to represent the edge examination process. In DFS a new adjacent vertex is selected, which is incident with the first edge incident with $v$. In other words, in DFS we leave $v$ as quickly as possible, examining only one of its incident edges and replacing $v$ by a new vertex, which is adjacent to $v$.

The following result is due to Chartrand and Kronk [51].

**Theorem 7.12 (Chartrand and Kronk [51])** Let $G$ be a connected graph every depth-first search tree of which is a Hamiltonian path. Then $G$ is a cycle, a complete graph, or a complete bipartite graph $K_{n,n}$.

**Proof** Let $P$ be a Hamiltonian path of $G$, with origin $u$. Because the path $P - u$ extends to a Hamiltonian path of $G$, the path $P$ extends to a Hamiltonian cycle $C$ of $G$.

When $C$ has no chord, $G = C$ is a cycle. So let $uv$ be a chord of $C$. Then $u^-v^-$ is one too, because $u^-CvuC^{-1}v^-$ is a Hamiltonian path of $G$, likewise, $u^-v^-$ is a chord of $C$ (where $u^-$ denotes the successor of $u$ on $C$ and $u^{--}$ is the successor of $u^-$). And if the length of $uCv$ is at least four, $uv$ and $u^-v^-$ are also chords of $C$, in view of the Hamiltonian path $u^{--}Cv^-u^-C^{-1}v^-u^-uv$ and the fact that $u^-v^- = (u^-)^-v^-$.

When $C$ has a chord $uw$ of length two, let $v = u^-(= w^-)$. Then $vw^- \in E$. Moreover, if $vw^- \in E$, then $vw^{-(-1)} \in E$ in view of the Hamiltonian path $w^{-(-1)}CuwCw^-v$. It follows that $v$ is adjacent to every vertex of $G$. But then $G$ is complete, because $u^-w^-$ is a chord of length two for all $i$. If $C$ has no chord of length two, every chord of $C$ is odd, moreover, every odd chord must be present. Thus, $G = K_{n,n}$, where $|V(G)| = 2n$. $\square$

The following is the third proof of Brook's theorem which is due to Bondy [35].

**Bondy's Proof** Suppose first that $G$ is not regular. Let $u$ be a vertex of degree $\delta$ and let $T$ be a search tree of $G$ rooted at $u$. Colour the vertices with the colours $1, 2, \ldots, \triangle$ according to the greedy heuristic, selecting at each step a pendent vertex of the subtree of $T$ induced by the vertices not yet coloured, assigning to it the smallest available colour and ending with the root $u$ of $T$. When each vertex $v$ different from $x$ is coloured, it is adjacent (in $T$) to at least one uncoloured vertex and so is adjacent to at most $d(v) - 1 < \triangle - 1$ coloured vertices. It is therefore assigning one of the colours $1, 2, \ldots, \triangle$, because $d(u) = \delta \leq \triangle - 1$. The greedy heuristic therefore produces a $\triangle$-colouring of $G$.

Now, let $G$ be regular. If $G$ has a cut vertex $u$, then $G = G_1 \cup G_2$, where $G_1$ and $G_2$ are connected and $G_1 \cap G_2 = \{u\}$. Because the degree of $u$ in $G$ is less than $\triangle(G)$, neither subgraph of $G$ is regular, so $\chi(G) \leq \triangle(G_i) = \triangle(G)$, $i = 1, 2$ and $\chi(G) = \max\{\chi(G_1), \chi(G_2)\} \leq \triangle(G)$. We may assume, therefore, that $G$ is 2-connected.

If every depth-first search tree of $G$ is a Hamiltonian path, then $G$ is a cycle, a complete graph, or a complete bipartite graph $K_{n,n}$, by Theorem 7.12. Since by hypothesis, $G$ is neither an odd cycle nor a complete graph, $\chi(G) = 2 \leq \triangle(G)$. Suppose then, that $T$ is a depth-first search tree of $G$, but not a path. Let $u$ be a vertex of $T$ with at least two children, $v$ and $w$. Because $G$ is 2-connected, both $G - v$ and $G - w$ are connected. Thus there are proper descendants of $v$ and $w$, each of which is joined to an ancestor of $u$, and it follows that $G - \{v, w\}$ is connected. Consider a search tree $T$ with root $u$ in $G$. By colouring $v$ and $w$ with colour 1 and then the vertices of $T$ by the greedy heuristic as above, ending with the root $u$, we obtain a $\triangle$-colouring of $G$. ❏

Brooks theorem and the observation that in a graph $G$ containing $K_n$ as a subgraph, $\chi(G) \geq n$, provide estimates for the chromatic number. For instance, in Figure 7.11(a) for the graph $G_1$, $\triangle(G_1) = 8$ and $G_1$ has $K_4$ as a subgraph. Therefore, $4 \leq \chi(G_1) \leq 8$. It can be easily seen that $\chi(G_1) = 4$. Similarly for $G_2$ in Fig. 7.11(b) known as the Birkhoff diamond, $\triangle(G_2) = 5$ and $G_2$ has $K_3$ as a subgraph. So, $3 \leq \chi(G_2) = 5$. In fact, $\chi(G_2) = 4$.

**Independent set:** A set of vertices in a graph $G$ is independent if no two of them are adjacent. The largest number of vertices in such a set is called the vertex independence number of $G$ and is denoted by $\alpha_0(G)$ or $\alpha_0$. Analogously, an independent set of edges of $G$ has no two of its edges adjacent and the maximum cardinality of such a set is the edge independence number $\alpha_1(G)$ or $\alpha_1$. For the complete graph $K_n$, $\alpha_0 = 1$, $\alpha_1 = \left[\dfrac{n}{2}\right]$. In the graph of Figure 7.12, $\alpha_0(G) = 2$ and $\alpha_1(G) = 3$.

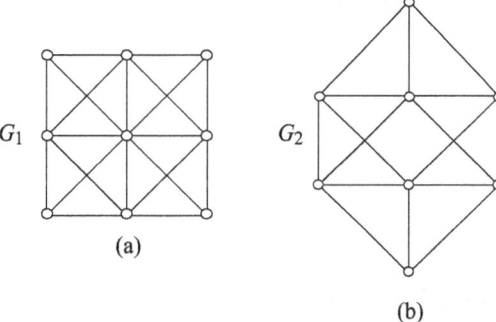

$G_1$    $G_2$

(a)

(b)

**Fig. 7.11**

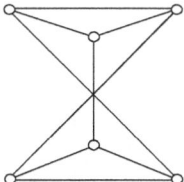

**Fig. 7.12**

A lower bound, noted in Berge [22] and Ore [178] and an upper bound by Harary and Hedetniemi [106] involve the vertex independent number $\alpha_0$ of a graph.

**Theorem 7.13**    For any graph $G$, $\dfrac{n}{\chi(\overline{G})} \leq \dfrac{n}{\alpha_o} \leq \chi(G) \leq n - \alpha_o + 1$.

**Proof**    If $\chi(G) = k$, then $V$ can be partitioned into $k$ colour classes $V_1, V_2, \ldots, V_k$, each of which is an independent set of vertices.

If $|V_i| = n_i$, then every $n_i \leq \alpha_o$, so that $n = \sum n_i \leq k \alpha_o$. This proves the middle inequality.

Now, let $S$ be a maximal independent set containing $\alpha_o$ vertices. Clearly, $\chi(G - S) \geq \chi(G) - 1$.

As $G - S$ has $n - \alpha_o$ vertices, $\chi(G - S) \leq n - \alpha_o$, and $\chi(G) \leq \chi(G - S) + 1 \leq n - \alpha_o + 1$, proving the last inequality.

As $\overline{G}$ has a complete subgraph of order $\alpha_o(G)$,

$$\chi(\overline{G}) \geq \alpha_o(G), \text{ or } \frac{n}{\chi(\overline{G})} \leq \frac{n}{\alpha_o(G)}, \text{ proving the first inequality.} \qquad \square$$

We now present the result due to Nordhaus and Gaddum [172] giving the bounds on the sum and product of the chromatic numbers of a graph and its complement.

**Theorem 7.14 (Nordhaus and Gaddum [172])**    For any graph of order $n$,

$$2\sqrt{n} \leq \chi + \overline{\chi} \leq n + 1, \text{ and} \tag{7.14.1}$$

$$n \leq \chi\overline{\chi} \leq \left(\frac{n+1}{2}\right)^2, \tag{7.14.2}$$

where $\chi = \chi(G)$ and $\overline{\chi} = \chi(\overline{G})$.

**Proof**    Evidently from the Theorem 7.13, we have

$$\chi\overline{\chi} \geq n. \tag{7.14.3}$$

Since the arithmetic mean is greater than or equal to the geometric mean,

$$\frac{\chi + \overline{\chi}}{2} \geq \sqrt{\chi\overline{\chi}}. \tag{7.14.4}$$

Combining (7.14.3) and (7.14.4), we get

$$\frac{\chi + \overline{\chi}}{2} \geq n.$$

Therefore, the left inequalities of (7.14.1) and (7.14.2) are proved.

Now, let $d_1 \geq d_2 \geq \ldots \geq d_n$ be the degree sequence of $G$. Then, $\bar{d}_1 \geq \bar{d}_2 \geq \ldots \geq \bar{d}_n$, where $\bar{d}_i = n - 1 - d_{n+1-i}$, is the degree sequence of $\overline{G}$. Then by using Theorem 7.9, we have

$$\chi(G) + \chi(\overline{G}) \leq \max_i \min\{d_i + 1, i\} + \max_i \min\{n - d_{n+1-i}, i\}$$

$$= \max_i \min\{d_i + 1, i\} + (n+1) - \min_i \max\{d_{n+1-i} + 1, n+1-i\}$$

$$= \max_i \min\{d_i + 1, i\} + (n+1) - \min_j \max\{d_j + 1, j\}.$$

Thus, $\chi(G) + \chi(\overline{G}) \leq n + 1$.

Also, $\dfrac{\chi(G) + \chi(\overline{G})}{2} \geq \sqrt{\chi\overline{\chi}}$. Therefore, $\sqrt{\chi\overline{\chi}} \leq \dfrac{\chi + \overline{\chi}}{2} \leq \dfrac{n+1}{2}$. Thus, $\chi\overline{\chi} \leq \left(\dfrac{n+1}{2}\right)^2$.

**Second Proof**  Let $G$ be $k$-chromatic, and let $v_1, v_2, \ldots, v_k$ be the colour classes of $G$, where $|V_i| = n_i$ Then, $\Sigma n_i = n$, and $\max n_i \geq n/k$. Since each $V_i$ induces a complete subgraph of $\overline{G}, \overline{\chi} \geq \max n_i \geq n/k$, so that $\chi\overline{\chi} \geq n$. As the geometric mean of two positive numbers is always less or equal to their arithmetic mean, it follows that $\chi + \overline{\chi} \geq 2\sqrt{n}$.

To prove $\chi + \overline{\chi} \leq n + 1$, we induct on $n$. Clearly, the equality holds for $n = 1$. We assume that $\chi(G) + \overline{\chi}(G) \leq n$ for all graphs $G$ having fewer than $n$ vertices. Let $H$ and $\overline{H}$ be complementary graphs with $n$ vertices and let $v$ be a vertex of $H$. Then, $G = H - v$ and $\overline{G} = \overline{H} - v$ are complementary graphs with $n - 1$ vertices. Let the degree of $v$ in $H$ be $d$, so that degree of $v$ in $\overline{H}$ is $n - d - 1$. Clearly, $\chi(H) \leq \chi(G) + 1$ and $\overline{\chi}(H) \leq \overline{\chi}(G) + 1$. If either $\chi(H) < \chi(G) + 1$ or $\chi(H) < \chi(G) + 1$, then $\chi(H) + \overline{\chi}(H) \leq n + 1$. If $\overline{\chi}(H) = \overline{\chi}(G) + 1$ and $\overline{\chi}(H) = \overline{\chi}(G) + 1$, then the removal of $v$ from $H$, producing $G$, decreases the chromatic number, so that $d \geq \chi(G)$. Similarly, $n - d - 1 \geq \overline{\chi}(G)$. Thus, $\chi(G) + \overline{\chi}(G) \leq n - 1$. Therefore we always have $\chi(H) + \overline{\chi}(H) \leq n + 1$. Applying now the inequality $4\chi\overline{\chi} \leq (\chi + \overline{\chi})^2$, we get

$$\chi\overline{\chi} \leq \left(\dfrac{n+1}{2}\right)^2. \qquad \qquad \Box$$

We now have the following result.

**Theorem 7.15**  If a connected $k$-chromatic graph has exactly one vertex of degree exceeding $k - 1$, then it is minimal.

**Proof**  Let $G$ be a connected $k$-chromatic graph having exactly one vertex of degree exceeding $k - 1$. Let $e$ be any edge of $G$. Then, $\delta(G - e) \leq k - 2$ (otherwise, $G$ will have at least two vertices of degree exceeding $k - 1$).

For every induced subgraph $H$ of $G - e$, we have $\delta(H) \leq k - 2$. Thus, by Theorem 7.7, $\chi(G - e) \leq k - 1$ and hence $\chi(G - e) \leq k - 1$. Since $e$ is arbitrary, therefore $G$ is minimal.  $\Box$

We observe from Theorem 7.4(b) that the number of edges $m$ of a $k$-critical graph is at least $n(k-1)/2$. Dirac extended this to the inequality $2m \geq n(k+1) - 2$ for a $(k+1)$-critical graph, the proof of which can be found in Bollobas [29].

## 7.4   Edge colouring

An edge colouring of a nonempty graph $G$ is a labelling $f : E(G) \rightarrow \{1, 2, \ldots\}$; the labels are called *colours*, such that adjacent edges are assigned different colours. A $k$-edge colouring of $G$ is a colouring of $G$ which consists of $k$ different colours and in this case $G$ is said to be *k-edge colourable*.

The definition implies that the $k$-edge colouring of a graph $G = (V, E)$ partitions the edge set $E$ into $k$ independent sets $E_1, E_2, \ldots, E_k$ such that $E = E_1 \cup E_2 \cup \ldots \cup E_k$. The independent sets $E_i, 1 \le i \le k$ are called the *colour classes* and the function $f : E(G) \rightarrow \{1, 2, \ldots, k\}$ such that $f(e) = i$, for each $e \in E_i$, $1 \le i \le k$, is called the *colour function*. The minimum number $k$ for which there is a $k$-colouring of $G$ is called the *edge chromatic number* (or edge chromatic index) and is denoted by $\chi'(G)$.

We have the following observations.

1. If $H$ is a subgraph of a graph $G$, then $\chi'(H) \le \chi'(G)$.

2. For any graph $G, \chi'(G) \ge \triangle(G)$.

   If $v$ is any vertex of $G$ with $d(v) = \triangle(G)$, then the $\triangle(G)$ edges incident with $v$ have a different colour in any edge colouring of $G$.

3. $\chi'(C_n) = \begin{cases} 2, & \text{if n is even}, \\ 3, & \text{if n is odd}. \end{cases}$

The following is a recolouring technique for edge colouring, called *Kempe chain argument*.

Let $G$ be a graph with an edge colouring using at least two different colours say $i$ and $j$. Let $H(i, j)$ represent the subgraph of $G$ induced by all the edges coloured either $i$ or $j$. Let $K$ be a connected component of the subgraph $H(i, j)$. It can be easily verified that $K$ is a path whose edges are alternately coloured by $i$ and $j$. Now, if the colours on these edges are interchanged and the colours on all other edges of $G$ are kept unchanged, the result is a new colouring of $G$, using the same initial colours. The component $K$ is called Kempe chain and this recolouring method is called the Kempe chain argument.

**Definition:**   Let $i$ be a colour used in the edge colouring of a graph $G$. If there is an edge coloured $i$ incident at the vertex $v$ of $G$, we say i is *present* at $v$, and if there is no edge coloured $i$ at $v$, we say $i$ is *absent* from $v$.

The following result is due to Konig [136].

**Theorem 7.16 (Konig [136])**   For a nonempty bipartite graph $G, \chi'(G) = \triangle(G)$.

**Proof**   The proof is by induction on the number of edges of $G$. If $G$ has only one edge, the result is trivial.

Let $G$ have more than one edge and assume that the result is true for all nonempty bipartite graphs having fewer edges than $G$. Since $\triangle(G) \le \chi'(G)$, it is enough to prove that $G$ has a $\triangle(G)$-edge colouring. We let $\triangle(G) \le k$. Let $e = uv$ be an edge of $G$. Then $G - e$ is bipartite with less edges than $G$. Therefore, by induction hypothesis, $G - e$ has a $\triangle(G-e)$-edge colouring. Since $\triangle(G-e) \leq \triangle(G) = k$, $G - e$ has a $k$-colouring. We show that the same $k$ colours are used to colour $G$.

Now, $d(u) \leq k$ in $G$ and the edge $e$ is uncoloured, therefore there is at least one of the $k$ colours absent from $u$. Similarly, at least one of these colours is absent from $v$.

If one of the colours absent at $u$ and $v$ is same, then we use this to colour $e$ and we get a $k$-edge colouring of $G$.

Now take the case of a colour $i$ present at $u$, but absent from $v$ and a colour $j$ present at $v$, but absent from $u$.

Let $K$ be the Kempe chain containing $u$ in the subgraph $H(i, j)$ induced by the edges coloured $i$ or $j$. We claim that $v$ does not belong to the Kempe chain $K$.

For if $v$ belongs to $K$, then there is a path $P$ in $K$ from $u$ to $v$. Since $u$ and $v$ are adjacent, they do not belong to the same bipartition subset of the bipartite graph $G$ and therefore the length of the path $P$ is odd. As the colour $i$ is present at $u$, the first edge of $P$ is coloured $i$. Since the edges of $P$ are alternately coloured $i$ and $j$, and $P$ is of odd length, therefore the last edge of $P$, which is incident at $v$, is also coloured $i$. This is a contradiction, as $i$ is absent from $v$, proving our claim.

Using Kempe chain argument on $K$, the interchanging of colours now makes $i$ absent from $u$ and does not affect the colours of the edges incident at $v$. Therefore, $i$ is absent from both $u$ and $v$ in this new edge colouring and colouring edge $e$ by $i$ gives a $k$-edge colouring of $G$. $\qquad \square$

The next result gives edge chromatic number of complete graphs.

**Theorem 7.17** If $G = K_n$ is a complete graph with $n$ vertices, $n \geq 2$, then

$$\chi'(G) = \begin{cases} n-1, & \text{if } n \text{ is even}, \\ n, & \text{if } n \text{ is odd}. \end{cases}$$

**Proof** Let $G = K_n$ be a complete graph with $n$ vertices.

Assume that $n$ is odd. Draw $G$ so that its vertices form a regular polygon. Clearly, there are $n$ edges of equal length on the boundary of the polygon. Colour the edges along the boundary using a different colour for each edge. Now, each of the remaining internal edges of $G$ is parallel to exactly one edge on the boundary. Each such edge is coloured with the same colour as the boundary edge. So two edges have the same colour if they are parallel and therefore we have the edge colouring of $G$. Since it uses $n$ colours, we have shown that $\chi'(G) \leq n$.

Now, let $G$ have an $(n-1)$-colouring. From the definition of an edge colouring, the edges of one particular colour form a matching in G (set of independent edges). Since $n$ is odd, therefore the maximum possible number of these is $(n-1)/2$. This implies that there are atmost $(n-1)(n-1)/2$ edges in $G$. This is a contradiction, as $K_n$ has $n(n-1)/2$ edges. Thus $G$ does not have an $(n-1)$ colouring. Hence, $\chi'(G) = n$.

Now, let $n$ be even, and let $v$ be any vertex of $G$. Clearly $G - v$ is complete with $n - 1$ vertices. Since $n - 1$ is odd, $G - v$ has an $(n-1)$-colouring. With this colouring, there is a colour absent from each vertex and different vertices having different absentees. Reform $G$

from $G - v$ by joining each vertex $w$ of $G - v$ to $v$ by an edge and colour each such edge by the colour absent from $w$. This gives an $(n-1)$-colouring of $G$ and therefore $\chi'(G) = n - 1$.

❑

The above result is illustrated by taking $K_5$ and $K_6$ in Figure 7.13.

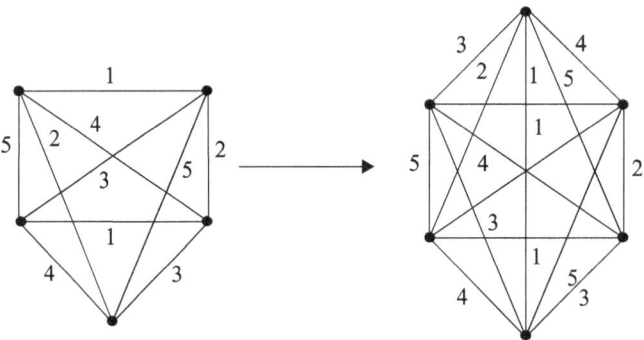

**Fig. 7.13**

Since $\triangle(G) = n - 1$ in a complete graph, Theorem 7.17 shows that $\chi'(K_n)$ is either $\triangle(G)$ or $\triangle(G) + 1$. The next result obtained by Vizing [258] and independently by Gupta [94] gives the tight bounds for edge chromatic number of a simple graph.

**Theorem 7.18 (Vizing [258])**    For any graph $G, \triangle(G) \le \chi'(G) \le \triangle(G) + 1$.

**Proof**    Let $G$ be a simple graph, we always have $\triangle(G) = \chi'(G)$.

To prove $\chi'(G) \le \triangle(G) + 1$, we use induction on the number of edges of $G$. Let $\triangle(G) = k$. If $G$ has only one edge, then $k = 1 = \chi'(G)$. Therefore, assume that $G$ has more than one edge and that the result is true for all graphs having fewer edges than $G$.

Let $e = v_1 v_2$ be an edge of $G$. Then by induction hypothesis the subgraph $G - e$ has $(k+1)$-edge colouring and let the colours used be $1, 2, \ldots, k+1$.

Since $d(v_1) \le k$ and $d(v_2) \le k$, out of these $k+1$ colours at least one colour is absent from $v_1$ and at least one colour is absent from $v_2$. If there is a common colour absent from both $v_1$ and $v_2$, then we use this to colour $e$ and get a $(k+1)$-colouring of $G$. Therefore in this case, $\chi'(G) \le k + 1$.

We now assume that there is a colour, say 1, absent from $v_1$ but present at $v_2$ and there is a colour, say 2, absent from $v_2$ but present at $v_1$. We start from $v_1$ and $v_2$ and construct a sequence of distinct vertices $v_1, v_2, \ldots, v_j$, where each $v_i$ for $i \ge 2$ is adjacent to $v_1$. Let $v_1 v_3$ be coloured 2. This $v_3$ exists, because 2 is present at $v_1$. We observe that not all the $k+1$ colours are present at $v_3$ and assume that the colour 3 is absent from $v_3$. But the colour 3 is present at $v_1$ and choose the vertex $v_4$ so that $v_1 v_4$ is coloured 3. Continuing in this way, we choose a new colour $i$ absent from $v_i$ but present at $v_1$, so that $v_1 v_{i+1}$ is the edge coloured $i$. In this way, we get a sequence of vertices $v_1, v_2, v_3, \ldots, v_{j-1}, v_j$ such that

a. $v_i$ is adjacent to $v_1$ for each $i > 1$,

b. the colour $i$ is absent from each $i = 1, 2, \ldots, j-1$ and

c. the edge $v_1 v_{i+1}$ is coloured i for each $i = 1, 2, \ldots, j-1$.

This is illustrated in Figure 7.14.

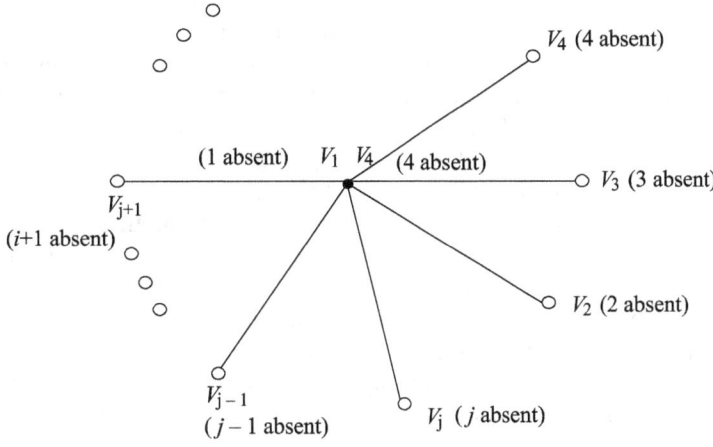

**Fig. 7.14**

As $d(v_1) \le k$, (a) implies that such a sequence has at most $k+1$ terms, that is, $j \le k+1$. Assume that $v_1, v_2, \ldots, v_j$ is a longest such sequence, that is, the sequence for which it is not possible to find a new colour $j$, absent from $v_j$, together with a new neighbour $v_{j+1}$ of $v_1$ such that $v_1 v_{j+1}$ is coloured $j$.

We first assume that for some colour $j$ absent from $v_j$ there is no edge of that colour present at $v_1$. We colour the edge $e = v_1 v_2$ by colour 2 and then recolour the edges $v_1 v_j$ by colour $i$, for $i = 3, \ldots, j-1$. Since $i$ was absent from $v_i$, for each $i = 2, \ldots, j-1$, this gives a $(k+1)$-colouring of the subgraph $G - v_1 v_j$. Now as the colour $j$ is absent from both $v_j$ and $v_1$, recolour $v_1 v_j$ by the colour $j$. This gives a $(k+1)$-colouring of $G$ (Fig. 7.15).

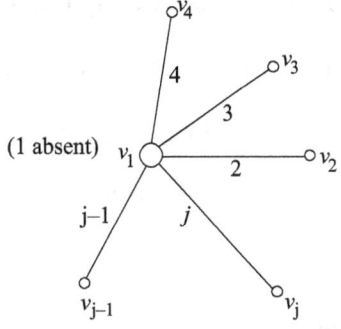

**Fig. 7.15**

Now assume that whenever $j$ is absent from $v_j$, $j$ is present at $v_1$. If $v_{j+1}$ is a new neighbour of $v_1$ so that $v_1 v_{j+1}$ is coloured by $j$, then we have extended our sequence to $v_1, v_2, \ldots, v_j, v_{j+1}$ which is a contradiction to the assumption that $v_1, v_2, \ldots, v_j$ is the longest sequence. Thus one of the edges $v_1 v_3, \ldots, v_1 v_{j-1}$ is to be coloured by $j$, say $v_1 v_\ell$, with $3 \le \ell \le j - 1$. Now colour $e = v_1 v_2$ by 2, and for $i = 3, \ldots, \ell - 1$, recolour each of the edges $v_1 v_i$ by $i$ while unaltering the colours of the edges $v_1 v_i$, for $i = \ell + 1, \ldots, j$. Removing the colour $j$ from $v_1 v_\ell$, we have a $(k+1)$-colouring of the edge deleted subgraph $G - v_1 v_\ell$.

Let $H(l, j)$ represent the subgraph of $G$ induced by the edges coloured 1 or $j$ in this partial colouring of $G$. Since the degree of every vertex in $H(1, j)$ is either 1 or 2, each component of $H(1, j)$ is either a path or a cycle. As 1 is absent from $v_1$ and $j$ is absent from both $v_j$ and $v_\ell$, it follows that all these three vertices do not belong to the same connected component of $H(l, j)$. Therefore, if $K$ and $L$ represent the corresponding Kempe chains containing $v_j$ and $v_\ell$ respectively, then either $v_1 \notin K$ or $v_1 \notin L$. Let $v_1 \notin L$. Then interchanging the colours of $L$, the Kempe chain argument gives a $(k+1)$-colouring of $G - v_1 v_\ell$ in which 1 is missing from both $v_1$ and $v_\ell$. colouring $v_1 v_\ell$ by 1 gives a $(k+1)$-colouring of $G$.

Now, let $v_1 \notin K$. Colour the edge $v_1 v_\ell$ by $\ell$, recolour the edges $v_1 v_i$ by $i$, for $i = \ell, \ldots, j - 1$, and remove the colour $j - 1$ from $v_1 v_j$. Then, from the definition of the sequence $v_1, v_2, \ldots, v_j$, we get a $(k+1)$-colouring of $G - v_1 v_j$ without affecting two coloured subgraph $H(l, j)$. Using Kempe chain argument to interchange the colours of $K$, we obtain a $(k+1)$-colouring of $G - v_1 v_j$ in which 1 is absent from both $v_1$ and $v_j$. Therefore, again colouring $v_1 v_j$ by 1 gives $(k+1)$-colouring of $G$.                                                                          ❑

The following result is due to Vizing [259] and Alavi and Behzad [2].

**Theorem 7.19**  Let $G$ be a graph of order $n$ and let $\overline{G}$ be the complement of $G$. Then,

a. $n - 1 \le \chi'(G) + \chi'(\overline{G}) \le 2(n - 1)$,

$0 \le \chi'(G)\chi'(\overline{G}) \le (n - 1)^2$, for even $n$,

b. $n \le \chi'(G) + \chi'(\overline{G}) \le 2n - 3$,

$0 \le \chi'(G)\chi'(\overline{G}) \le (n - 1)(n - 2)$, for odd $n$.

Further, the bounds are the best possible for every positive integer $n(n \ne 2)$.

**Proof**  Let $G$ be a graph of order $n$ and let $\overline{G}$ be the complement of $G$. Then, clearly,

$\triangle(\overline{G}) \ge n - 1 - \triangle(G)$, so that $\triangle(G) + \triangle(\overline{G}) \ge n - 1$.

Therefore, combining with $\chi'(G) \ge \triangle(G)$, we get

$\chi'(G) + \chi'(\overline{G}) \ge n - 1$.

b. If $n$ is odd, we have $\chi'(G) + \chi'(\overline{G}) \ge n$, since $\chi'(G) + \chi'(\overline{G}) < n$ implies $\chi'(K_n) < n$, which is a contradiction.

Obviously, $\chi'(G)\chi'(\overline{G}) \geq 0$. It can be seen that the lower bounds are attained in complete graphs $K_n$.

We now prove that $\chi'(G) + \chi'(\overline{G}) \leq 2n - 3$ and

$$\chi'(G) + \chi'(\overline{G}) \leq (n-1)(n-2).$$

Clearly, for $n = 1, 3$, the inequalities $\chi'(G) + \chi'(\overline{G}) \leq 2n - 3$ are true. So, let $n \geq 5$. If $\triangle(G) + \triangle(\overline{G}) \leq 2n - 5$, then by Vizing's theorem, we get $\chi'(G) + \chi'(\overline{G}) \leq 2n - 3$.

Otherwise, we have the following cases.

i. $\triangle(G) = n - 1$ and $\triangle(\overline{G}) = n - 2$. So $G$ has a pendant vertex $v$. Then, $\triangle(G-v) = n - 2$ and $\chi'(G-v) = n - 2$. But $\chi'(G-v) \leq \chi'(K_{n-1}) = n - 2$. Therefore, $\chi'(G-v) = n - 2$. Thus, $\chi'(G) = n - 1$.

   As $\overline{G}$ is the disjoint union of an isolated vertex and a subgraph of $K_{n-1}$, $\chi'(\overline{G}) = n - 2$. Hence, $\chi'(G) + \chi'(\overline{G}) = 2n - 3$.

ii. $\triangle(G) = n - 2$ and $\triangle(\overline{G}) = n - 2$. Again, $G$ has a pendant vertex $v$ and as before, $\chi(G-v) \leq n - 2$ and $\chi(G) = n - 2$. Similarly, $\chi(\overline{G}) = n - 2$. Thus, $\chi'(G) + \chi'(\overline{G}) = 2n - 4 < 2n - 3$.

iii. $\triangle(G) = n - 1$ and $\triangle(\overline{G}) = n - 3$. In this case, $G$ has a vertex $v$ of degree two and so $\chi'(G-v) = n - 2$.

   Let $vu$ and $vw$ be the edges with $v$ in $G$, with $d(u) = n - 1$. In $(n-2)$-edge colouring of $G - v$, change the colour $i$ of an edge $uu'(u' \neq w)$ to a new colour $n - 1$, and now colour $vu$ by $i$ and $vw$ by $n - 1$. This gives an $(n-1)$-colouring of $G$. Now, by Vizing's theorem, $\chi'(\overline{G}) \leq n - 2$.

   Together, we get $\chi'(G)\chi'(\overline{G}) \leq 2n - 3$. Since $\chi/(G) + \chi'(\overline{G}) \leq 2n - 3$, clearly we have $\chi'(G)\chi'(\overline{G}) \leq (n-1)(n-2)$.

   We observe that in graph $K_{1,n-1}$, the upper bounds are attained.

a. Let $n$ be even. Then, $\chi'(G) + \chi'(\overline{G}) \geq n - 1$. Also, $\chi'(G)\chi'(\overline{G}) \geq 0$.
   The lower bounds are attained for complete graphs.
   To get the upper bounds, since $G$ and $\overline{G}$ are subgraphs of $K_n$ and $\chi'(K_n) = n - 1$ for all even $n$, $\chi'(G) + \chi(\overline{G}) \leq 2(n-1)$ and $\chi(G)\chi(\overline{G}) \leq (n-1)^2$. These upper bounds are attained in the complete bipartite graphs $K_{1,n-1}, \neq 2$. $\qquad\square$

## 7.5  Region Colouring (Map Colouring)

A region colouring of a planar graph is a labeling of its regions $f : R(G) \rightarrow \{1, 2, \ldots\}$; the labels called *colours*, such that no two adjacent regions get the same colour. A $k$-region colouring of a planar graph $G$ consists of $k$ different colours and $G$ is then called *k-region colourable*. From the definition, it follows that the $k$-region colouring of a planar graph $G$ partitions the region set $R$ into $k$ independent sets $R_1, R_2, \ldots, R_k$, so that $R = R_1 \cup R_2 \cup \ldots \cup R_k$. The independent sets are called the *colour classes*, and the function $f : R(G) \rightarrow$

$\{1, 2, \ldots, k\}$ such that $f(r) = i$, for each $r \in R_i$, $1 \le i \le k$, is called the *colour function*. The minimum number $k$ for which there is a k-region colouring of the planar graph $G$ is called the *region-chromatic number* of $G$, and is denoted by $\chi''(G)$. The colouring of regions is also called *map colouring*, because of the fact that in an atlas different countries are coloured such that countries with common boundaries are shown in different colours.

**The four colour problem:** Any map on a plane surface (or a sphere) can be coloured with at most four colours so that no two adjacent regions have the same colour.

Now coming to the origin of the four colour problem, there have been reports that Mobius was familiar with the problem in 1840. But the problem was introduced in 1852 by Francis Guthrie, student of Augustus DeMorgan and the problem first appeared in a letter (October 23, 1852) from DeMorgan to Sir William Hamilton. DeMorgan continued the discussion of the problem with other mathematicians and in the years that followed attempts were made to prove or disprove the problem by top mathematical minds of the world. In 1878, Cayley announced the problem to the London Mathematical Society, and in 1879, Alfred Kempe announced that he had found a proof. An error in Kempe's proof was discovered by P. J. Heawood in 1890. Kempe's idea was based on the alternating paths and Heawood used this idea to prove that five colours are sufficient. Kempe's argument did not prove the four colour problem, but did contain several ideas which formed the foundation for many later attempts at the proof, including the successful attempts by Appel and Haken. In 1976, K. Appel and W. Haken [5, 6, 7] with the help of J. Koch established what is now called four colour theorem. Their proof made use of large scale computers (using over 1000 hours of computer time) and this is the first time in the history of mathematics that a mathematical proof depended upon the external factor of the availability of a large scale computing facility. Though the Appel-Haken proof is accepted as valid, mathematicians are still in search of alternative proof. Robertson, Sanders, Seymour and Thomas [225] have given a short and clever proof, but their proof still requires a number of computer calculations. Saaty [230] presents thirteen colourful variations of four colour problem.

In the year 2000, Ashay Dharwadkar [64] has given a new proof of four the colour theorem, which will be discussed in details in Chapter 14.

The following observations are immediate from the definitions introduced above.

1. A planar graph is $k$-vertex colourable or $k$-region colourable if and only if its components have this property.

2. A planar graph is $k$-vertex colourable or $k$-region colourable if and only if its blocks have this property.

These observations imply that for studying vertex colourings or region colourings, it suffices to consider the graph to be a block.

**Theorem 7.20**

    a. A planar graph $G$ is $k$-region colourable if and only if its dual $G$ is $k$-vertex colourable.

    b. If $G$ is a plane connected graph without loops, then $G$ has a $k$-vertex colouring if and only if its dual $G^*$ has a $k$-region colouring.

**Proof**

    a. Let the regions and edges of $G$ be respectively denoted by $r_1, \ldots, r_t$ and $e_1, \ldots, e_m$. Let the vertices of $G^*$ be $r_1^*, \ldots, r_t^*$ and edges be $e_1^*, \ldots, e_m^*$. Then the vertices and edges of $G^*$ are in one-to-one correspondence with the regions and edges of $G$, and two vertices $r^*$ and $s^*$ in $G^*$ are joined by an edge $e^*$ if and only if the corresponding regions $r$ and $s$ in $G$ have the corresponding edge $e$ as a common edge on their boundary.

       Now, let $G$ be $k$-region colourable. We colour the vertices in $G^*$ such that each vertex in $G^*$ gets the same colour as assigned to the region $r$ in $G$. Since the vertices $r^*$ and $s^*$ are only adjacent in $G^*$ if the corresponding regions $r$ and $s$ are adjacent in $G$, $G^*$ is $k$-vertex colourable.

       Conversely, let $G^*$ be $k$-vertex colourable. Now colour the regions of $G$ such that the region $r$ in $G$ gets the same colour as the vertex $r^*$ in $G^*$. This gives a $k$-region colouring of $G$, since the regions $r$ and $s$ are adjacent in $G$ only if the corresponding vertices $r^*$ and $s^*$ are adjacent in $G^*$.

    b. Since $G$ has no loops its dual $G^*$ has no bridges, and therefore $G^*$ is planar. Thus by (a), $G^*$ is $k$-region colourable if and only if the double dual $G^{**}$ is $k$-vertex colourable. Since $G$ is connected, $G$ is isomorphic to $G^{**}$ and hence the result follows.    ❑

**Remarks**    A graph has a dual if and only if it is planar and this implies that colouring the regions of a planar graph $G$ is equivalent to colouring the vertices of its dual $G^*$ and vice versa. It also follows from Theorem 7.17 that if $G^*$ is the dual of the planar graph $G$, then $\chi(G) = \chi'(G^*)$ and $\chi(G^*) = \chi''(G)$. These observations give the dual form of the four colour problem which states that, every planar graph is 4-vertex colourable.

Since loops and multiple edges are not allowed in vertex colourings, it may be assumed that no two regions have more than one boundary edge in common, for region colouring of a planar graph

As every triangulation is a planar graph (in fact, a maximal planar graph) and every planar graph is a subgraph of a triangulation, the four colour problem is true if and only if every triangulation is 4-colourable.

We now have the following result.

**Theorem 7.21**    Every planar graph is 6-colourable.

**Proof**    Let $G$ be a planar graph and $H$ be the dual of $G$. Then it is sufficient to prove that $H$ has a vertex colouring of at most 6 colours. More generally, we prove that any graph $H$ is 6-colourable.

To prove the result, we use induction on $n$, the order of $H$. The result is trivial if $H$ has at most six vertices. So assume $n \geq 7$.

Let all planar graphs with fewer than $n$ vertices be 6-colourable. Obviously, $H$ has a vertex, say $v$, so that $d(v) \leq 5$. Therefore $v$ has at most five neighbours in $H$ and these neighbours evidently need at most five colours for colouring. The vertex deleted subgraph $H - v$ is planar with $n - 1$ vertices and therefore by induction hypothesis is 6-colourable. Since at most five colours are used for colouring the neighbours of $v$, therefore assigning v the sixth colour not used by the its neighbours gives the 6-colouring of $H$.                ❏

The following result is a consequence of Theorem 7.20.

**Theorem 7.22**    A planar graph $G$ is 2-colourable if and only if it is an Euler graph.

**Proof**    Let $G$ be a planar graph which is 2-colourable. Then, if $G^*$ is the geometric dual of $G$, we have $\chi(G^*) = 2$. Therefore $G^*$ is bipartite and thus $G^{**}$ (the dual of $G^*$) is an Euler graph. Since $G$ and $G^{**}$ are isomorphic, therefore $G$ is an Euler graph.

Conversely, let $G$ be an Euler graph. Then its double dual $G^{**}$ is an Euler graph and thus $G^*$ is bipartite. Therefore, $\chi(G^*) = 2$ and hence the planar graph $G$ is 2-colourable.                ❏

The next result due to Heawood [113] is called *five colour theorem* and Heawood used the Kempe chain argument in proving it.

**Theorem 7.23 (Heawood [113])**    Every planar graph is 5-colourable.

**Proof**    Let $G$ be a planar graph with $n$ vertices. We use induction on $n$, the order of $G$. The result is obvious for $n \leq 5$. So, let $n \geq 6$. Assume the result to be true for all planar graphs with fewer than $n$ vertices.

Let $G'$ be the graph obtained from $G$ by deleting the vertex $v$ and removing all the edges incident with $v$. The graph $G'$ with order $n - 1$ is clearly planar and by induction hypothesis is 5-colourable. Let the colours used to colour $G'$ be $c_1, c_2, c_3, c_4, c_5$.

We know for a planar graph with $n \geq 6$ vertices, there exists a vertex, say $v$, such that $d(v) \leq 5$. Thus, $v$ has atmost five neighbours in $G$ and all of these neighbours are the already coloured vertices in $G'$.

If in $G'$ less than five colours are used to colour these neighbours, then the 5-colouring of $G$ is obtained by using the colouring for $G'$ on all vertices, and by colouring $v$ with the colour not used to colour the neighbours of $v$ (Fig. 7.16).

Now, let all the five colours be used in $G'$, to colour the neighbours of $v$. This implies that there are exactly five neighbours of $v$, say $u_1, u_2, u_3, u_4$ and $u_5$. Assume without loss of generality that $u_i$ is coloured with $c_i$, for each $i$, $1 \leq i \leq 5$.

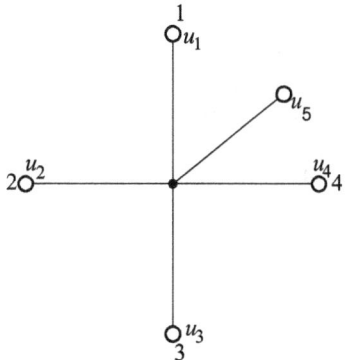

**Fig. 7.16**

Consider all the vertices of $G'$ that are coloured with colour $c_1$ and $c_3$. If $u_1$ and $u_3$ are in different components of the Kempe subgraph $H(c_1, c_3)$ induced by those vertices coloured $c_1$ and $c_3$, then using the Kempe chain argument to interchange the colours $c_1$ and $c_3$ in the component containing $u_1$ leaves $c_1$ unused on the set $u_1$, $u_2$, $u_3$, $u_4$, $u_5$. We colour $v$ by $c_1$ to get a 5-colouring of $G$.

Finally, if $u_1$ and $u_3$ are in the same component of $H(c_1, c_3)$, and $u_2$ and $u_4$ are in the same component of $H(c_2, c_4)$, then the $c_1 - c_3$ Kempe chain from $u_1$ to $u_3$ crosses the $c_2 - c_4$ Kempe chain from $u_2$ to $u_4$. This is impossible as the triangulation is a plane graph (Fig. 7.17). ❑

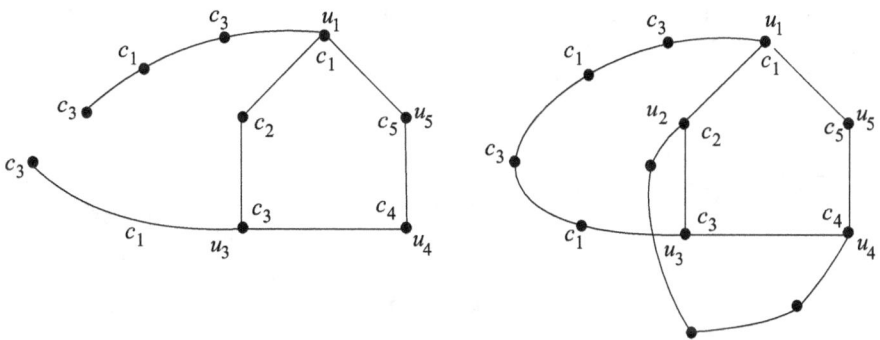

**Fig. 7.17**

The following result is due to Grunbaum [91].

**Theorem 7.24 (Grunbaum [91])** Every planar graph with fewer than four triangles is 3-colourable.

The next result due to Grotzsch [90] immediately follows from Theorem 7.24.

**Theorem 7.25 (Grotzsch [90])** Every planar graph without triangles is 3-colourable.

While making various attempts to solve the four colour problem, the problem got translated into several equivalent conjectures and sometimes conjectures were made which implied the four colour problem. The details of several such conjectures can be found in Saaty [229] and Saaty and Kainen [230]. Finch and Sachs [78] proved that every planar graph with at most 21 triangles is 4-colourable. Ore and Stemple [179] showed that all planar graphs with upto 39 regions are 4-colourable.

The following result is one such equivalence.

**Theorem 7.26**    Every planar graph is four colourable if and only if every cubic bridgeless planar graph is 4-colourable.

**Proof**    We observe that every planar graph is four colourable if and only if every bridgeless planar graph (without cut edges) is four colourable. This is because if $G$ has a bridge $e$ and $G'$ is obtained from $G$ by contracting $e$, then $\chi'(G) = \chi''(G)$ (as the elementary contraction of identifying the end vertices of a bridge affects neither the number of regions in the planar graph nor the adjacency of any of the regions).

We now prove that every bridgeless planar graph is 4-colourable if and only if every cubic bridgeless planar graph is 4-colourable. If every bridgeless planar graph is 4-colourable, then evidently every cubic bridgeless planar graph is 4-colourable.

Conversely, let all cubic bridgeless planar graphs be 4-colourable. Let $G$ be a bridgeless planar graph. We obtain $G'$ from $G$ by performing the following operations. In case $G$ has a vertex $v$ of degree two, let $xv$ and $yv$ be the edges incident with $v$. Subdivide $xv$ at $u$ and $yv$ at $w$. Take two new vertices $a$ and $b$ and add the edges $au, aw, bu, bw$ and $ab$. Now remove $v$ (and $uv$ and $uw$). In doing so, we have replaced $v$ by a $K_4 - e$ and we see that each new vertex has degree three (Fig. 7.18).

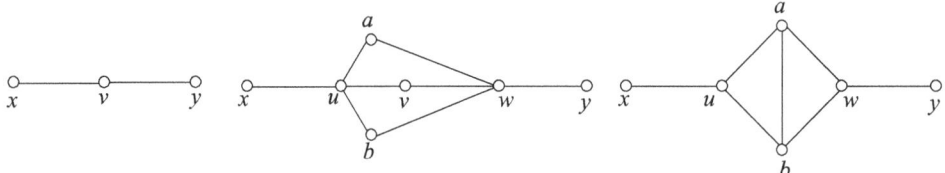

**Fig. 7.18**

If $G$ has a vertex $v$ so that $d(v) \geq 4$, then let $vx_1, vx_2, \ldots, vx_d$ be the edges incident with $v$. Subdivide each $vx_i$ producing a new vertex $vi$, for $1 \geq i \geq d$. We then remove $v$ and add the new edges $v_1 v_2, v_2 v_3, \ldots, v_{d-1} v_d, v_d v_1$. Here we have replaced $v$ by $C_d$ and again the degree of each new vertex is three (Fig. 7.19).

In both cases the graph $G'$ formed is a bridgeless cubic graph. If these $K_4 - e$ and $C_d$ introduced are contracted, we get the original graph $G$. Hence $G$ is 4-colourable, since we have assumed that $G'$ is 4-colourable.                                                                    ❑

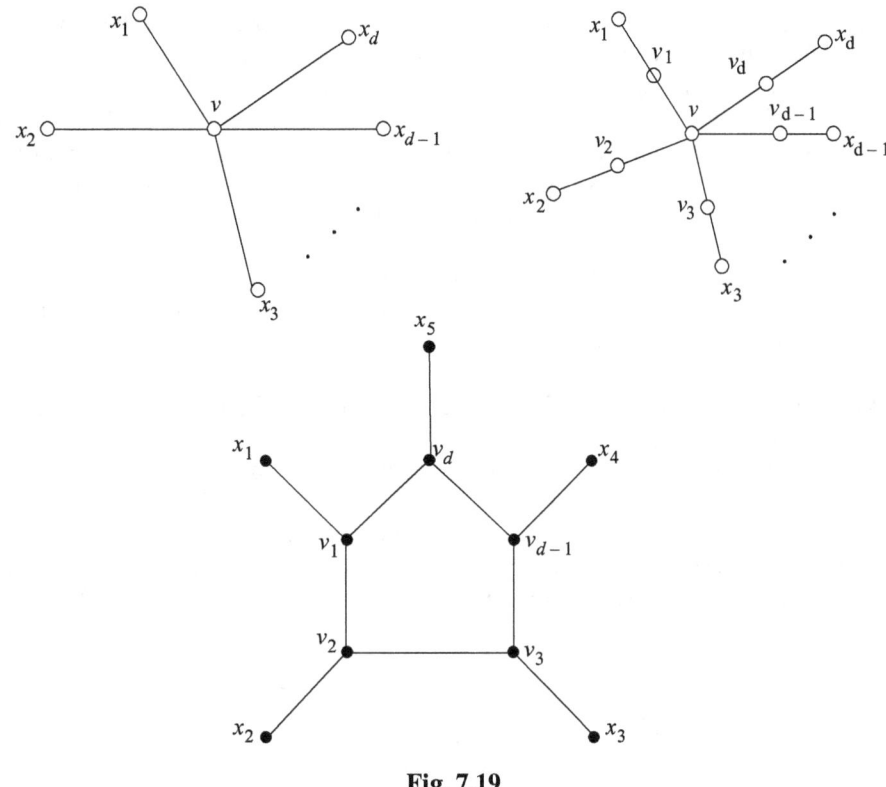

**Fig. 7.19**

The next result is due to Tait [238] and the details can also be found in Bondy and Chvatal [36] and Bollobas [29].

**Theorem 7.27 (Tait [238])**    Every planar graph $G$ is 4-colourable if and only if $\chi'(G) = 3$, for every bridgeless cubic planar graph $G$.

**Proof**    As seen in Theorem 7.26, the statement that every planar graph is 4-colourable is equivalent to the statement that every cubic bridgeless planar graph is 4-colourable. Therefore, to prove the result, we prove that a cubic bridgeless planar graph $G$ is 4-colourable if and only if $\chi'(G) = 3$.

Now assume that $G$ is a bridgeless cubic planar graph which is 4-colourable. For the set of colours we choose the elements of the Klein four group $Q = \{c_0, c_1, c_2, c_3\}$, where addition in $Q$ is defined by $c_i + c_i = c_0$ and $c_1 + c_2 = c_3$, with $c_0$ being the identity element. Now, define the colour of an edge to be the sum of the colours of two distinct regions which are incident with that edge. We see that the edges are coloured with elements of the set $\{c_1, c_2, c_3\}$ and that no two adjacent edges get the same colour. Hence, $\chi'(G) = 3$.

Conversely, let $G$ be a bridgeless cubic planar graph with $\chi'(G) = 3$ and colour its edges with the three non-zero elements of $Q$. Consider a region, say $R_0$, and give the colour $c_0$ to it. Let $R$ be any other region of $G$ and let $C$ be any curve in the plane joining the interior of

$R_0$ with the interior of $R$, so that $C$ does not pass through a vertex of $G$. Then the colour of $R$ is defined to be the sum of the colours of those edges which intersect $C$.

That the colours of the regions are well defined follows from the fact that the sum of the colours of the edges which intersect any simple closed curve not passing through a vertex of $G$ is $c_0$. Suppose $S$ is such a curve and assume $q_1, q_2, \ldots, q_n$ to be the colours of the edges which intersect $S$. Also assume $r_1, r_2, \ldots, r_m$ be the colours of those edges interior to $S$. If $c(v)$ denotes the sum of the colours of the three edges incident with $v$, then we see that $\sum c(v) = c_0$. Thus, for all vertices $v$ interior to $S$, we have $\sum c(v) = c_0$. While on the other hand, we have $\sum c(v) = q_1 + q_2 + \ldots + q_n + 2(r_1 + r_2 + \ldots + r_m) = q_1 + q_2 + \ldots + q_n$ as every element of $Q$ is self-inverse. Thus, $q_1 + q_2 + \ldots + q_n = c_o$. Hence we get the 4-colouring of the regions of $G$ and the colours used are $c_0, c_1, c_2, c_3$.                    ❏

**Remarks**    Because of Theorem 7.27, a three colouring of the edges of a cubic graph is called a *Tait colouring*.

In an attempt to solve the four colour problem, Tait considered edge colourings of bridgeless cubic planar graphs and proved that every such graph is 3-edge colourable. In 1880, Tait tried to give a proof of the four colour problem by using Theorem 7.27 and based on the wrong assumption that any bridgeless cubic planar graph is Hamiltonian. A counter example called the Tuttle graph to Tait's assumption was given by Tutte in 1946 and is shown in Figure 7.20.

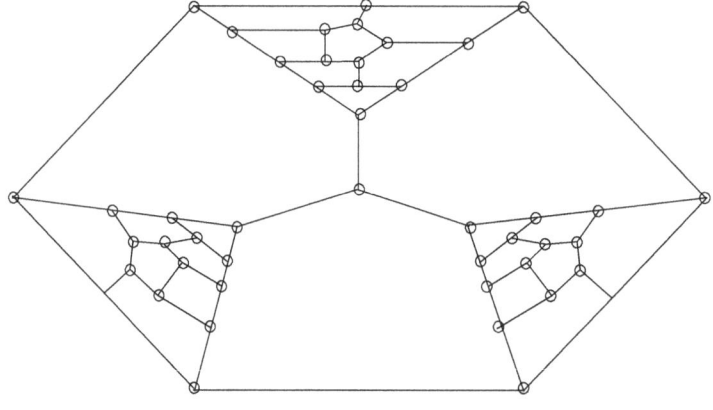

**Fig. 7.20**

# 7.6  Heawood Map-Colouring Theorem

Let $S_p$ be the orientable surface of genus $p$, so that $S_p$ is topologically equivalent to a sphere with p handles. The chromatic number of $S_p$, denoted by $\chi(S_p)$, is the maximum chromatic number among all graphs which can be embedded on $S_p$. The surface $S_o$ is clearly the sphere. The four colour problem states that $\chi(So) = 4$.

**Definition:** The genus $\gamma(G)$ of a graph is the minimum number of handles which must be added to a sphere so that $G$ can be imbedded on the resulting surface. Clearly, $\gamma(G) = 0$ if and only if $G$ is planar. Further, for a polyhedron, $n - m + f = 2 - 2\gamma$.

For $n \geq 3$, the genus of the complete graph $K_n$ is

$$\gamma(K_n) = \frac{(n-3)(n-4)}{12}.$$

We now give the inequality which is due to Heawood [113].

**Theorem 7.28 (Heawood [113])** The chromatic number of the orientable surface of positive genus $p$ has the upper bound

$$\chi(S_p) \leq \frac{7 + \sqrt{1 + 48p}}{2}, \ p > 0.$$

**Proof** Let $G$ be an $(n, m)$ graph embedded on $S_p$. Since any graph can be augmented to a triangulation of the same genus by adding edges without reducing $\chi$, we assume $G$ to be a triangulation. Let $d'$ be the average degree of the vertices of $G$. Then $n$, $m$ and $f$ (the number of regions) are related by the equations

$$d'n = 2m = 3f.$$

Solving for $m$ and $f$ in terms of $n$ and using Euler's equation $n - m + f = 2 - 2\gamma$, we get

$$d' = \frac{12(p-1)}{n} + 6. \tag{7.28.1}$$

As $d' \leq n - 1$, this gives $n - 1 \geq \dfrac{12(p-1)}{n} + 6$.

Solving for $n$ and taking the positive square root, we obtain $n \geq \dfrac{7 + \sqrt{1 + 48p}}{2}$.

Let $H(p) = \dfrac{7 + \sqrt{1 + 48p}}{2}$. Then we show that $H(p)$ colours are sufficient to colour the vertices of $G$. If $n = H(p)$, obviously we have sufficient colours. In case $n > H(p)$, we substitute $H(p)$ for n in (7.28.1) and obtain

$$d' < \frac{12(p-1)}{H(p)} + 6 = H(p) - 1.$$

Therefore, when $n > H(p)$, there is a vertex $v$ of degree at most $H(p) - 2$. Identify $v$ and any adjacent vertex by an elementary contraction to obtain a new graph $G'$. If $n' = n - 1 = H(p)$, then $G'$ can be coloured in $H(p)$ colours. If $n' > H(p)$, repeat the argument and evidently we get an $H(p)$ colourable graph. It is then easy to see that the colouring of this graph induces a colouring of the preceding one in $H(p)$ colours, and so forth, so that $G$ itself is $H(p)$–colourable. Hence the result follows.                                                                    ❑

The next result is called Heawood map colouring theorem, the proof of which is due to Ringel and Youngs [223].

**Theorem 7.29 (Heawood map-colouring theorem)**    For every positive integer $p$, the chromatic number of the orientable surface of genus $p$ is given by

$$\chi(S_p) = \frac{7 + \sqrt{1 + 48p}}{2}, p > 0.$$

**Proof**    Let $G$ be an $(n, m)$ graph embedded on $S_p$. Then,

$$\chi(S_p) = \frac{7 + \sqrt{1 + 48p}}{2}, p > 0.$$

Now, if the complete graph $K_n$ is embedded in $S_p$, then

$$p = \chi(K_n) = \frac{(n-3)(n-4)}{12}. \tag{7.29.1}$$

Setting $n$ to be the largest integer satisfying (7.29.1), we have

$$\frac{(n-3)(n-4)}{12} \leq \left[\frac{(n-3)(n-4)}{12}\right] \leq p \leq \left[\frac{(n-2)(n-3)}{12}\right] - 1 < \frac{(n-2)(n-3)}{12}.$$

Solving for $n$, we get

$$\frac{5 + \sqrt{1 + 48p}}{2} < n \leq \frac{7 + \sqrt{1 + 48p}}{2}$$

Thus, $n = \dfrac{7 + \sqrt{1 + 48p}}{2}$.

Since $\chi(K_n) = n$, we have found a graph with genus $p$ and chromatic number equal to $H(p)$. This shows that $H(p)$ is a lower bound for $\chi(Sp)$, completing the proof.                                    ❑

## 7.7 Uniquely Colourable Graphs

A graph $G = (V, E)$ is said to be *uniquely k-vertex-colourable* (or uniquely *k*-colourable) if there is a unique *k*-part partition of the vertex set $V$ into independent subsets. That is, in the uniquely *k*-colouring of a graph $G$, every *k*-colouring of $G$ induces the same partition of $V$. The graph of Figure 7.21(a) is uniquely 3-colourable, since every 3-colouring of $G$ has the partition $\{v_1\}$, $\{v_2, v_4\}$, $\{v_3, v_5\}$. The graph of the Figure 7.21(b) is not uniquely 3-colourable. A pentagon is not uniquely 3-colourable, as five different partitions of its vertex set are possible.

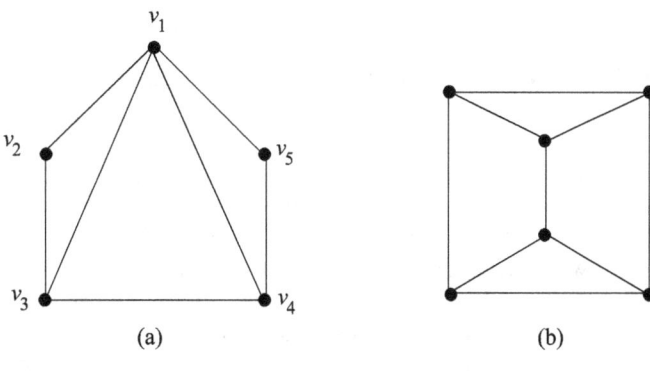

**Fig. 7.21**

We observe that the empty graphs $\overline{K_n}$ are the only uniquely 1-colourable graphs and the connected bipartite graphs are the only uniquely 2-colourable graphs. We note that unique $k$-colourability is not defined for $k > n$. Further, we see that $K_n$ is uniquely $n$-colourable. These observations imply to assume $3 \leq k \leq n$, for studying uniquely $k$-colourable graphs.

First, we have the following observation.

**Theorem 7.30**    If $G$ is uniquely $k$-colourable, then $\delta(G) \geq k - 1$.

**Proof**    Let $G$ be uniquely $k$-colourable graph. Then every vertex $v$ of $G$ is adjacent to at least one vertex of every colour different from that assigned to $v$. For otherwise, a different $k$-colouring of $G$ is obtained by recolouring $v$. This implies that $d(v) \geq k - 1$, for every $v$. □

**Corollary 7.2**    If $G$ is a uniquely $k$-colourable graph with $n$ vertices and $m$ edges, then $2m \geq n(k - 1)$.

The next result, due to Cartwright and Harary [45], gives a necessary condition for a graph to be uniquely colourable.

**Theorem 7.31 (Cartwright and Harary [45])**    The subgraph induced by the union of any two colour classes in a $k$-colouring of a uniquely $k$-colourable graph is connected.

**Proof**    Let $G$ be uniquely $k$-colourable graph. Let $C_1$ and $C_2$ be two classes in the $k$-colouring of the graph $G$. Assume the subgraph $S$ of $G$ induced by $C_1 \cup C_2$ be disconnected, and let $S_1$ and $S_2$ be components of $S$. Then each of $S_1$ and $S_2$ contain vertices of both $C_1$ and $C_2$. Now interchanging the colour of the vertices in $C_1 \cap S_1$ by the colour of the vertices in $C_2 \cap S_2$ gives a different $k$-colouring of $G$. This contradicts the hypothesis that $G$ is uniquely $k$-colourable. Hence $S$ is connected.                                                                        □

The converse of Theorem 7.31 is not true. To see this, consider the 3-chromatic graph G of Fig. 7.21(b). The graph $G$ has the property that in any 3-colouring, the subgraph induced by the union of any 2 colour classes is connected. But $G$ is not uniquely colourable. It follows from Theorem 7.30 that every uniquely $k$-colourable graph, $k \geq 2$, is connected.

The following stronger result is due to Chartrand and Geller [49].

**Theorem 7.32 (Chartrand and Geller [49])**    Any uniquely $k$-colourable graph is $(k-1)$-connected.

**Proof**    Let $G$ be a uniquely k-colourable graph. In case $G$ is $K_k$, then it is $(k-1)$-connected. Now assume $G$ to be an incomplete graph which is not $(k-1)$-connected. Therefore there exists a set $U$ of $k-2$ vertices whose removal disconnects $G$. Then in any $k$-colouring of $G$, there are at least two distinct colours, say $c_1$ and $c_2$, not assigned to any vertex of $U$. By Theorem, 7.30, a vertex coloured $c_1$ is connected to any vertex coloured $c_2$ by a path all of whose vertices are coloured $c_1$ or $c_2$. Therefore the set of vertices of $G$ coloured $c_1$ or $c_2$ lies within the same component of $G-U$, say $G_1$. Another k-colouring of $G$ can thus be obtained by taking any vertex of $G-U$ which is not in $G_1$ and recolouring it with either $c_1$ or $c_2$. This contradicts the hypothesis that $G$ is uniquely $k$-colourable. Hence $G$ is $(k-1)$-connected.                                                                                           ❑

**Corollary 7.3**    In any $k$-colouring of uniquely $k$-colouring graph, the subgraph induced by the union of any $h$ colour classes, $2 \leq h \leq k$, is $(h-1)$-connected.

The following result due to Bollobas [30] gives a sufficient lower bound for uniquely colourable graphs.

**Theorem 7.33 (Bollobas [30])**    If $G$ is a $k$-colourable graph of order $n(k \geq 2)$ with $\delta(G) > n(3k-5)/(3k-2)$, then $G$ is uniquely $k$-colourable.

The following result is due to Harary, Hedetniemi and Robinson [107].

**Theorem 7.34 (Harary, Hedetniemi and Robinson [107])**    For all $k \geq 3$, there is a uniquely $k$-colourable graph which contains no subgraph isomorphic to $K_k$.

Clearly, a graph is uniquely 1-colourable if and only if it is 1-colourable, that is, totally disconnected. Also, a graph is uniquely 2-colourable if and only if $G$ is 2-chromatic and connected.

The converse of Theorem 7.36 is not true. This is because a uniquely 3-colourable planar graph may have more than one region which is not a triangle, as shown in Figure 7.22.

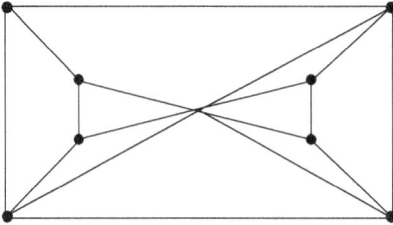

**Fig. 7.22**

**Theorem 7.35**   If $G$ is a uniquely 3-colourable planar graph with at least four vertices, then $G$ contains at least two triangles.

**Theorem 7.36**   Every uniquely 4-colourable planar graph is maximal planar.

**Proof**   Let a 4-colouring be given to uniquely 4-colourable planar graph $G$ with the colour classes denoted by $V_i, 1 \leq i \leq 4$, where $|V_i| = n_i$. Since the subgraph induced by $V_i \cup V_j, i \neq j$, is connected, $G$ has at least $\sum (n_i + n_j - 1)$ edges, $1 \leq i < j < 4$. Clearly, $\sum (n_i + n_j - 1) = n_1 + n_2 - 1 + n_1 + n_3 - 1 + n_1 + n_4 - 1 + n_2 + n_3 - 1 + n_2 + n_4 - 1 + n_3 + n_4 - 1 = 3(n_1 + n_2 + n_3 + n_4) - 6 = 3n - 6$.
Therefore, $m \geq 3n - 6$.
Hence $G$ is maximal planar.   $\square$

The next result for uniquely 5-colourable graphs is due to Hedetniemi [114].

**Theorem 7.37 (Hedetniemi [114])**   No planar graph is uniquely 5-colourable.

**Theorem 7.38**   A necessary and sufficient condition that a connected planar graph is 4-colourable is that $G$ be the sum of three subgraphs $G_1, G_2$ and $G_3$ such that for each vertex $v$, the number of edges of each $G_i$ incident with $v$ are all even or odd.

The following equivalence is due to Whitney [264].

**Theorem 7.39 (Whitney [264])**   The four colour problem holds if and only if every Hamiltonian planar graph is 4-colourable.

## 7.8  Hajos Conjecture

Hajos [97] made the following conjecture.
   If a graph is $k$-chromatic, then it contains a subdivision of $K_k$.

When $k = 1$, or 2, the conjecture trivially holds. As for $k = 3$, every chromatic graph contains an odd cycle which is a subdivision of $K_3$, therefore proving the validity of the conjecture. The validity of the conjecture for $k = 4$, as noted in Parthasarthy [180] is due to Dirac [66].

**Theorem 7.40 (Dirac [66])**   Hajos conjecture is true for $k = 4$.

**Proof**   Assume without loss of generality that $G$ is a 4-minimal graph. So $G$ is a block and $\delta = 3$. In case $n = 4$, $G$ is $K_4$ and the result is obvious. Therefore assume $n = 5$. We induct on $n$.
   Let $G$ have a 2-vertex cut $S = \{u, v\}$. Then by Theorem 7.5, $G = G_1 \cup G_2$, where $G_1$ is of type 1, and $G_2$ is of type 2 and $G_1 + uv$ is 4-minimal. By induction hypothesis, $G_1 + uv$ contains a subdivision of $K_4$. Here we replace $uv$ by a $u - v$ path $P$ in $G_2$ and so $G_1 \cup P$ contains a subdivision of $K_4$. Thus $G$ also contains a subdivision of $K_4$. Now, let $G$ be 3-

connected. Since $\delta(G) \geq 3$, therefore it has a cycle $C$ of length at least 4. If $u$ and $v$ are any two non-consecutive vertices on $C$, then $G - \{u, v\}$ is still connected and therefore there exist vertices $x$, $y$ on $C$, and an $x - y$ path $P$ in $G - \{u, v\}$. Similarly, there exists a $u - v$ path $Q$ in $G - \{x, y\}$. If $P$ and $Q$ have no vertices in common, then $C \cup P \cup Q$ is a subdivision of $K_4$ in $G$. Otherwise, let $w$ be the first vertex of $P$ on $Q$, then $C \cup P_{xw} \cup Q$ is a subdivision of $K_4$ in $G$.

Hence in all cases $G$ contains a subdivision of $K_4$.                                                                □

For $k \geq 5$, Hajos conjecture implies four colour problem. This is because if $G$ is a planar graph which is not colourable by 4 colours, its chromatic number is at least 5 and thus contains a subdivision of $K_5$, and so cannot be planar, a contradiction. The four colour theorem implies that a 5-chromatic graph contains a homeomorph of $K_5$ or $K_{3,3}$. For $k = 5$, Hajos conjecture makes the stronger assertion that it contains a homeomorph of $K_5$. For $k = 5$, or 6, the conjecture has not been settled, but for $k = 7$, it is disproved by Catlin. The counter example of Catlin is the graph $H = L(3C_5) - \{v_1, v_2\}$, where $3C_5$ is the multigraph obtained from $C_5$ by replacing each edge by three edges, $L$ represents the edge graph, and $v_1$ and $v_2$ are any two non-adjacent vertices of $L(3C_5)$ (Fig. 7.23).

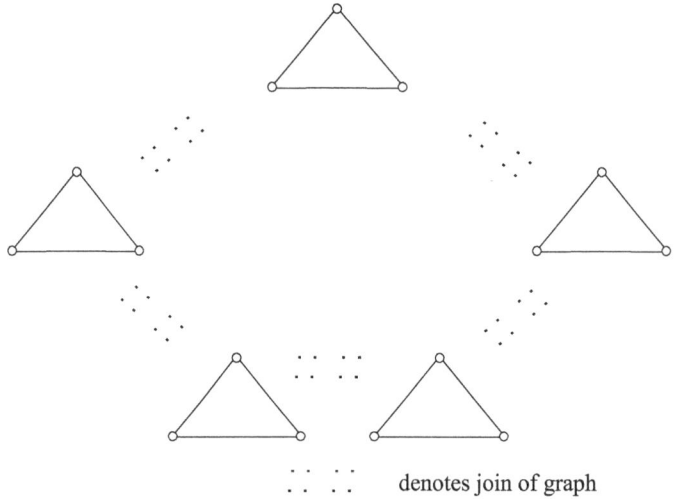

  $\vdots \ \ \vdots$    denotes join of graph

**Fig. 7.23**

The largest integer for which a given graph $G$ contains a $TK_n$ is called the subdivision number (or topological clique number or Hajos number) of $G$ and is denoted by $tw(G)$. With this, Hajos conjecture is equivalent to $tw(G) \geq \chi(G)$. In the above example, we see that $H$ contains a $K_6$ and from any vertex outside this $K_6$ there are no six internally disjoint paths to the vertices of the $K_6$. Therefore, $tw(H) = 6$. A maximum independent set of $H$ has cardinality two, so that $\chi(H) \geq \left\lceil \dfrac{13}{2} \right\rceil = 7$ and a 7-colouring of $H$ is shown in the Fig. 7.23. Thus, $\chi(H) = 7$ and hence $H$ is a counter example for Hajos conjecture. We now observe that if $G$ is a counter example for Hajos conjecture for $k$, then $G+v$ is a counter example for Hajos conjecture for $k+1$. This can be seen from the fact that $tw(G+v) = tw(G)+1$ and $\chi(G+v) = \chi(G)+1$. Hence Hajos conjecture is false for all $k = 7$. Erdos and Fajtlowicz [72] proved that almost every graph is a counter example to Hajos conjecture. Bollobas and Catlin [31] proved that $tw(G)$ is approximately $2\sqrt{n}$ for n-vertex graphs.

The following conjecture involving contradictions is due to Hadwiger [95].

**Hadwiger's conjecture [95]**    Every $k$-chromatic graph contains $K_k$ as a subcontraction.
  Hadwiger's conjecture is trivially true for $k = 1$. Since 2-chromatic graphs are the bipartite graphs and 3-chromatic graphs contain an odd cycle, contractible to $K_3$, the conjecture is true for $k = 2$ and 3. Dirac [66] proved the conjecture for $k = 4$. For $k = 5$, this conjecture states that every 5-chromatic graph is contractible to $K_5$ and therefore every such graph is non-planar. Thus, Hadwiger's conjecture for $k = 5$ implies the four colour problem. The converse of this is given by Wagner.

## 7.9  Exercises

1. Prove that $\chi(G) = \triangle(G)+1$ if and only if $G$ is either a complete graph or a cycle of odd length.

2. Show that if $G$ contains exactly one odd cycle, then $\chi(G) = 3$.

3. If $G$ is a graph in which any pair of odd cycles have a common vertex, then prove that $\chi(G) \leq 5$.

4. Find the chromatic number of the Peterson graph and the Birkhoff Diamond.

5. Determine the chromatic number of the graphs in Figure 7.24.

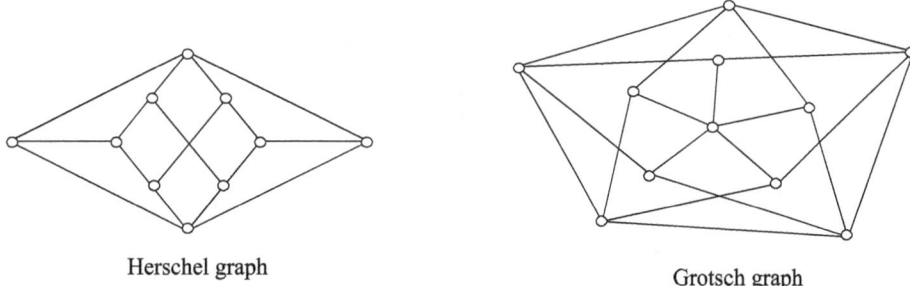

Herschel graph

Grotsch graph

**Fig. 7.24**

6. If $G$ is k-regular, prove that $\chi(G) \geq \dfrac{n}{n-k}$.

7. If $G$ is connected and $m \leq n$, show that $\chi(G) \leq 3$.

8. Prove that the 3-critical graphs are the odd cycles $C_{2n+1}$.

9. Prove that the wheel $W_{2n-1}$ is a 4-critical graph for each $n \geq 2$.

10. Prove that the wheel $W_{2n}$ is a 4-critical graph for each $n \geq 2$.

11. Prove that the Peterson graph has edge chromatic number 4.

12. If $m(G)$ is the number of edges in a longest path of $G$, prove that $\chi(G) \leq 1 + m(G)$.

13. Show that if $G$ is 3-regular Hamiltonian graph, then $\chi'(G) = 3$.

14. Show that a triangulation with a vertex of degree 2 or 3 can be coloured with five colours.

15. Prove that for every $k \geq 1$ there is a $k$-chromatic graph $M_k$ with no triangle subgraphs. (Mycielski, 1955).

16. If $G$ is a graph in which no set of four vertices induces $P_4$ as a subgraph, then prove that $\chi(G) = cl(G)$ (Seinsche, 1974).

17. Obtain proofs of Theorems 2.24 and 2.25.

18. Show that a uniquely 3-colourable graph contains three Hamiltonian cycles.

# 8. Matchings and Factors

Consider the formation of an executive council by the parliament committee. Each committee needs to designate one of its members as an official representative to sit on the council, and council policy states that no senator can be the official representative for more than one committee. For example, let there be five committees 1, 2, 3, 4 and 5. Let $A$, $C$, $D$, $E$ be members of committee 1; $B$, $C$, $D$ be members of committee 2; $A$, $B$, $E$, $F$ be members of committee 3; and $A$, $F$ be members of committee 5. Here, the executive council can be formed by members $A$, $C$, $B$, $F$ and $E$, representing committees 1, 2, 3, 4 and 5 respectively. The problem of formation of the executive council and many related problems can be solved using the concept of matchings in graphs.

## 8.1 Matchings

**Definition:** A *matching* in a graph is a set of independent edges. That is, a subset M of the edge set $E$ of a graph $G = (V, E)$ is a matching if no two edges of $M$ have a common vertex. A matching $M$ is said to be *maximal* if there is no matching $M'$ strictly containing $M$, that is, $M$ is *maximal* if it cannot be enlarged. A matching $M$ is said to be *maximum* if it has the largest possible cardinality. That is, $M$ is maximum if there is no matching $M'$ such that $|M'| > |M|$.

Consider the graph $G$ shown in Figure 8.1. The examples of matchings in $G$ are $M_1 = \{v_1v_5, v_2v_6, v_3v_4, v_7v_8\}$, $M_2 = \{v_1v_5, v_2v_6, v_3v_4, v_7v_9\}$, $M_3 = \{v_1v_5, v_2v_6, v_3v_4, v_7v_{10}\}$, $M_4 = \{v_1v_5, v_2v_6, v_3v_4\}$ and $M_5 = \{v_4v_7, v_1v_6, v_2v_3\}$. Clearly, $M_1$ is a maximum matching, whereas $M_5$ is maximal but not maximum.

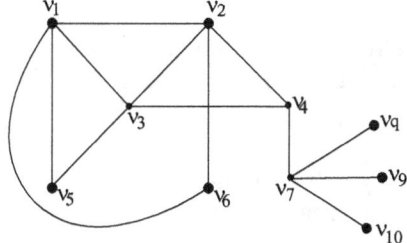

**Fig. 8.1**

The cardinality of a maximum matching is denoted by $\alpha_1(G)$ and is called the *matching number* of $G$ (or the edge-independence number of $G$).

**Definition:** Let $M$ be a matching in a graph $G$. A vertex $v$ in $G$ is said to be *M-saturated* (or saturated by $M$) if there is an edge $e \in M$ incident with $v$. A vertex which is not incident with any edge of $M$ is said to be *M-unsaturated*. In other words, given a matching $M$ in a graph $G$, the vertices belonging to the edges of $M$ are $M$-saturated and the vertices not belonging to the edges of $M$ are $M$-unsaturated. Consider the graph shown in Figure 8.2. Clearly, $M = \{v_1 v_2,\ v_3 v_7,\ v_4 v_5\}$ is a matching and the vertices $v_1$, $v_2$, $v_3$, $v_4$, $v_5$, $v_7$ are $M$-saturated but $v_6$ is $M$-unsaturated.

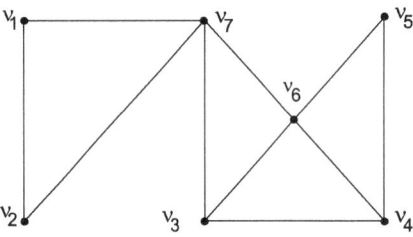

**Fig. 8.2**

**Definition:** A matching $M$ in a graph $G$ is said to be a *perfect matching* if $M$ saturates every vertex of $G$. In Figure 8.3(a), $G_1$ has a perfect matching $M = \{v_1 v_2,\ v_3 v_6,\ v_4 v_5\}$, but $G_2$ has no perfect matching.

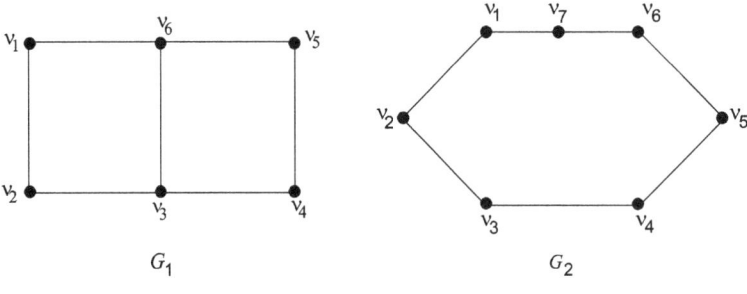

**Fig. 8.3(a)**

**Definition:** If $G$ is a bipartite graph with bipartition $V = V_1 \cup V_2$ and $|V_1| = n_1$, $|V_2| = n_2$, then a matching $M$ of $G$ saturating all the vertices in $V_1$ is called a *complete matching* (or $V_1$ matched into $V_2$).

**Definition:** Let $H$ be a subgraph of a graph $G$. An *H-alternating path (cycle)* in $G$ is a path (cycle) whose edges are alternately in $E(G) - E(H)$ and $E(H)$.

If $M$ is matching in a graph $G$, then an $M$-alternating path (cycle) in $G$ is a path (cycle) whose edges are alternately in $E(G) - M$ and $M$. That is, in an M-alternating path, the edges alternate between $M$-edges and non-$M$-edges. An $M$-alternating path whose end vertices are $M$-unsaturated is said to be an *$M$-augmenting path*.

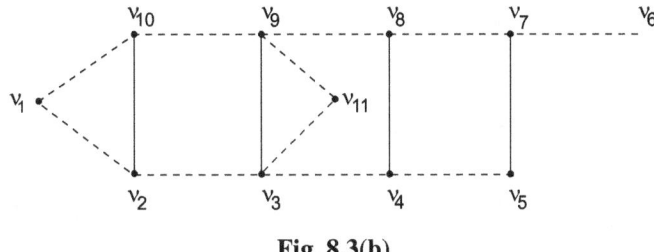

**Fig. 8.3(b)**

Consider the graph $G$ shown in Figure 8.3(b). An example of a matching in $G$ is $M = \{v_2v_{10}, v_3v_9, v_4v_8, v_5v_7\}$. Clearly $v_{10}, v_2, v_3, v_9, v_8, v_4$ is an $M$-alternating path and $v_1, v_{10}, v_2, v_9, v_3, v_{11}$ is an $M$-augmenting path.

The following result due to Berge [21] characterises a maximum matching in a graph. This forms the foundation of an efficient algorithm for obtaining a maximum matching.

**Theorem 8.1 (Berge [21])** A matching $M$ of a graph $G$ is maximum if and only if $G$ contains no $M$-augmenting paths.

**Proof** Let $M$ be a maximum matching in a graph $G$. Assume $P = v_1v_2\ldots v_k$ is an $M$-augmenting path in $G$. Due to the alternating nature of $M$-augmenting path, we observe that $k$ is even and the edges $v_2v_3, v_4v_5, \ldots, v_{k-2}v_{k-1}$ belong to $M$. Also the edges $v_1v_2, v_3v_4, \ldots, v_{k-1}v_k$ do not belong to $M$ (Figure 8.4).

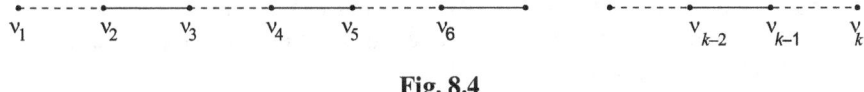

**Fig. 8.4**

Now, let $M_1$ be the set of edges given by

$$M_1 = [M - \{v_2v_3, \ldots, v_{k-2}v_{k-1}\}] \cup \{v_1v_2, \ldots, v_{k-1}v_k\}.$$

Then $M_1$ is a matching and clearly, $M_1$ contains one more edge than $M$. This contradicts the assumption that $M$ is maximum. Thus $G$ contains no M-augmenting paths.

Conversely, assume that $G$ has no $M$-augmenting paths. Let $M'$ be a matching such that $|M'| > |M|$. Let $H$ be a subgraph of $G$ with $V(H) = V(G)$ and $E(H)$ be the set of edges of $G$ that appear in exactly one of $M$ and $M'$. Since every vertex of $G$ lies on at most one edge from $M$ and at most one edge from $M'$, therefore degree (in $H$) of each vertex of $H$ is at most 2. This implies that each connected component of $H$ is either a single vertex, or a path, or a cycle (Fig. 8.5). If a component is a cycle, then it is an even cycle, because the edges alternate between $M$-edges and $M'$-edges. Since $|M'| > |M|$, there is at least one component in $H$ that is a path which begins and ends with edges from $M'$. Clearly, this path is an $M$-augmenting path, contradicting the assumption. Thus, no such matching $M'$ can exist and hence $M$ is maximum.                                                                                         □

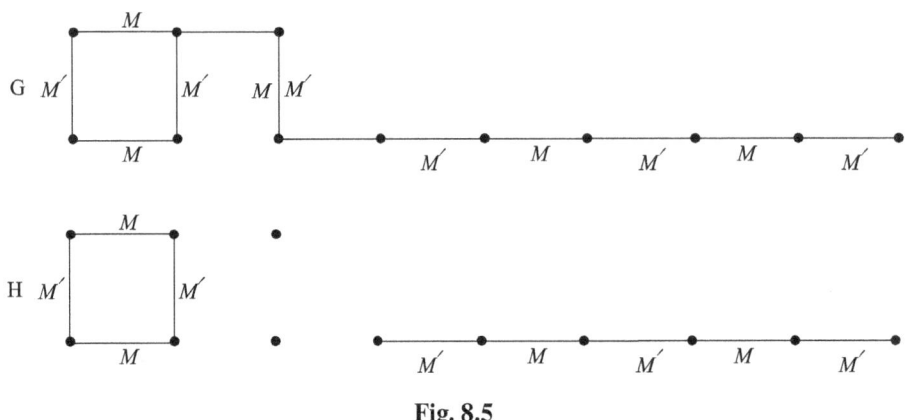

**Fig. 8.5**

**Definition:**    The *neighbourhood* of a set of vertices $S$, denoted by $N(S)$, is the union of the neighbourhood of the vertices of $S$.

The following classic result due to P. Hall [101] characterises a complete matching in a bipartite graph.

**Theorem 8.2 (Hall [101])**    If $G = (V_1, V_2, E)$ is a bipartite graph with $|V_1| \leq |V_2|$, then $G$ has a matching saturating every vertex of $V_1$, if and only if $|N(S)| \geq |S|$, for every subset $S \subseteq V_1$.

**Proof**    Let $G$ have a complete matching $M$ that saturates all the vertices of $V_1$ and let $S$ be any subset of $V_1$. Then every vertex in $S$ is matched by $M$ into a different vertex in $N(S)$, so that $|S| \leq |N(S)|$ (Fig. 8.6 (a)).

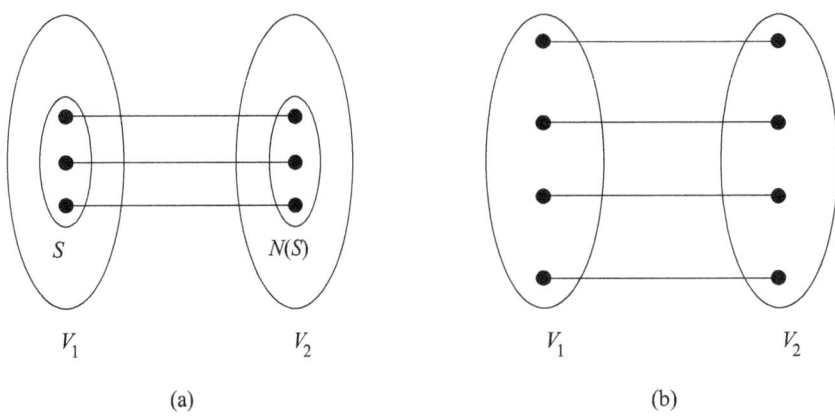

(a)                                                                (b)

**Fig. 8.6**

Conversely, let $|N(S)| \geq |S|$ for all subsets $S$ of $V_1$ and let $M$ be maximum matching. Assume $G$ has no complete matching. Then there exists a vertex, say $v \in V_1$, which is $M$-unsaturated (Fig. 8.6(b)). Let $Z$ be the set of vertices of $G$ that can be joined to $v$ by $M$-alternating paths.

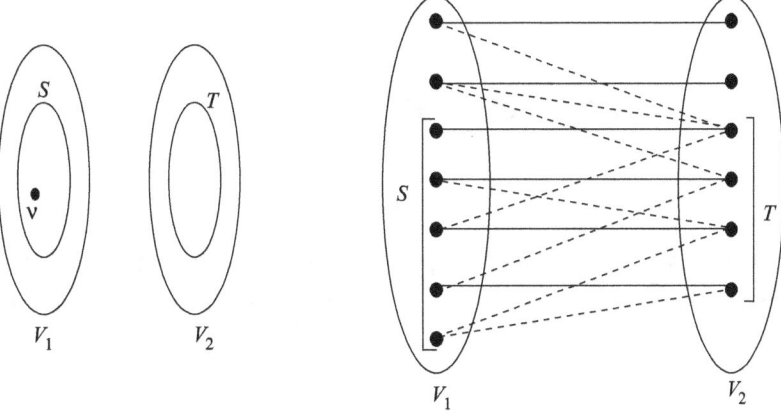

**Fig. 8.7**

Since $M$ is maximum matching, by Theorem 8.1, $G$ has no $M$-augmenting path. This implies that $v$ is the only vertex of $Z$ that is $M$-unsaturated. Let $S = Z \cap V_1$ and $T = Z \cap V_2$ (Fig. 8.7). Then every vertex of $T$ is matched under $M$ to some vertex of $S - v$ and conversely and $N(S) = T$. Since $|T| = |S| - 1$, therefore $|N(S)| = |T| = |S| - 1 < |S|$, contradicting the given assumption. Hence G has a complete matching. $\qquad \square$

The aim of the above theorem was basically to obtain the conditions for the existence of a system of distinct representatives (SDR) for a collection of subsets of a given set. Before giving such conditions, we have the following definition.

**Definition:** If $A = \{A_i : i \in N\}$ is a family of sets, then a system of distinct representatives (SDR) of $A$ is a set of elements $\{a_i : i \in N\}$ such that $a_i \in A_i$ for every $i \in N$ and $a_i \neq a_j$ whenever $i \neq j$. For example, if $A_1 = \{2, 8\}$, $A_2 = \{8\}$, $A_3 = \{5, 7\}$, $A_4 = \{2, 4, 8\}$ and $A_5 = \{2, 4\}$, then the family $\{A_1, A_2, A_3, A_4\}$ has an *SDR* $\{2, 8, 7, 4\}$, but the family $\{A_1, A_2, A_4, A_5\}$ has no SDR.

**Theorem 8.3 (Hall's SDR Theorem)** A family of finite nonempty sets $A = \{A_i : 1 \leq i \leq r\}$ has an SDR if and only if for every $k$, $1 \leq k \leq r$, the union of any $k$ of these sets contains at least $k$ elements.

**Proof** Clearly $\bigcup\limits_{i=1}^{r} A_i$ is finite, since each of the sets $A_1, A_2, \ldots, A_r$ is finite. Let $\bigcup\limits_{i=1}^{r} A_i = \{a_i, a_2, \ldots, a_n\}$.

Construct a bipartite graph $G$ with partite sets $X = \{A_1, \ldots, A_r\}$ and $Y = \{a_1, \ldots, a_n\}$, as shown in Figure 8.8. We take an edge between $A_i$ and $a_j$ if and only if $a_j \in A_i$.

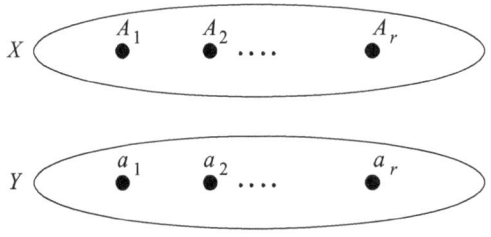

**Fig. 8.8**

Clearly, $A$ has an *SDR* if and only if $G$ has a matching that saturates all the vertices of $X$. Now, Theorem 8.2 implies that $G$ has such a matching if and only if $|S| \leq |N(S)|$ for all subsets $S$ of $X$, that is, if and only if $|S| \leq |\underset{a_i \in S}{\cup} A_i|$, proving the result.                                ❑

**Remarks**

1.  Hall's theorem is often referred to as Hall's marriage theorem.

2.  Since Hall's theorem, there has been remarkable progress in the theory of SDR, and besides other references the reader can refer to the book of Mirsky [161].

The following result due to M. Hall [100] counts the number of complete matchings.

**Theorem 8.4 (M. Hall [100])**    If $G = (V_1, V_2, E)$ is a bipartite graph with $|V_1| = n_1 \leq n_2 = |V_2|$ and satisfying Hall's condition $|S| \leq |N(S)|$ for all $S \subseteq V_1$, then $G$ has at least $r(\delta, n_1) = \prod_{1 \leq i \leq \min(\delta, n_1)} (\delta + 1 - i)$ complete matchings, where $\delta = \min\{d(u) : u \in V_1\}$.

**Proof**    Induct on $n_1$. The result is trivial for $n_1 = 1$. Assume the result to be true for all values of $n_1 \leq m - 1$ and let $G$ be a bipartite graph satisfying the given conditions and $|V_1| = m$.

**Case 1**    For each nonempty proper subset $S$ of $V_1$, $|S| < |N(S)|$. Now, for a given $u \in V_1$, choose $v \in N(u)$. This $v$ can be chosen in at least $\delta$ ways, since $\delta = \min\{d(u) : u \in V_1\}$. Obviously, $G - \{u, v\}$ is a bipartite graph with $|V_1| = m - 1$, and $\min\{d(x) : x \in V_1\} \geq \delta - 1$. Therefore by induction hypothesis, $G - \{u, v\}$ has at least $r(\delta - 1, m - 1) = \prod_{1 \leq i \leq \min(\delta - 1, m - 1)} (\delta - i)$ complete matchings. These together with $uv$, give at least $\delta[r(\delta - 1, m - 1)] = \overset{\delta}{\Lambda} \prod_{1 \leq i \leq \min(\delta - 1, m - 1)}$ $(\delta - i) = \prod_{1 \leq i \leq \min(\delta, m)} (\delta + 1 - i) = r(\delta, m)$ complete matchings in $G$.

**Case 2**    There is a non empty proper subset $S$ of $V_1$ with $|S| = |N(S)|$. Let $|S| = s$, and $G_1 = <S \cup N(S)>$ and $G_2 = G - <S \cup N(S)>$. Then the values of $n_1$ and $\delta$ in $G_1$ are $s$ and $\delta$, and in $G_2$ are $m-s$ and $\delta_2 = \max\{\delta - s, 1\}$. Therefore $G_1$ and $G_2$ are disjoint graphs with at least $r(\delta, s)$ and $r(\delta_2, m-s)$ complete matchings, by induction hypothesis. Thus $G$ has at least $r(\delta, s)\, r(\delta_2, m-s) = r(\delta, m)$ complete matchings.                                     ❏

The following result is due to Ore [175].

**Theorem 8.5 (Ore [175])**   If a bipartite graph $G = (V_1, V_2, E)$ satisfies $|S| \le d + |N(S)|$ for all $S \subseteq V_1$, where $d$ is a given positive integer, then $G$ contains a matching $M$ with at most $d$ vertices of $V_1$ being $M$-unsaturated.

**Proof**   Let $H = (V_1, U_2, E')$ be the bipartite graph with $U_2 = V_2 \cup W_2$, $|W_2| = d$ and $E' = E \cup [V_1, W_2]$. Then the given condition of $G$ implies Hall's condition for $H$ and so $H$ has a complete matching $M$. Obviously, at most $d$ of the edges of $M$ can be in $E' - E$, proving the result.                                     ❏

The following observation is an immediate consequence of Theorem 8.5.

**Corollary 8.1**   For a bipartite graph $G = (V_1, V_2, E)$, with $|V_1| = n_1$, $\alpha_1(G) = n_1 - \max\{|S| - |N(S)| : S \subseteq V_1\} = \min\{|V_1 - S| + |N(S)| : S \subseteq V_1\}$.

The following result is a consequence of Hall's theorem.

**Theorem 8.6**   A $k(\ge 1)$ regular bipartite graph has a perfect matching.

**Proof**   Let $G$ be a $k$ regular bipartite graph with partite sets $V_1$ and $V_2$. Then $E(G)$ is equal to the set of edges incident to the vertices of $V_1$ and also $E(G)$ is equal to the set of edges incident to the vertices of $V_2$. Therefore, $k|V_1| = k|V_2| = E(G)$ and thus $|V_1| = |V_2|$. If $S \subseteq V_1$, then $N(S) \subseteq V_2$ and thus $N(N(S))$ contains $S$. Let $E_1$ and $E_2$ be the sets of edges of $G$ incident respectively to $S$ and $N(S)$. Then, $E_1 \subseteq E_2$, $|E_1| = k|S|$ and $|E_2| \ge k|N(S)|$. Since $|E_2| \ge |E_1|$, $|N(S)| \ge |S|$. Thus by Theorem 8.2, $G$ has a matching that saturates all the vertices of $V_1$, that is $G$ has a complete matching $M$. Now deleting the edges of $M$ from $G$ gives a $(k-1)$-regular bipartite graph $G'$, with $V(G) = V(G')$. By the same argument, $G'$ has a complete matching $M'$. Deletion of the edges of $M'$ from $G'$ results in a $(k-2)$-regular bipartite graph. Repeating this process $k-1$ times, we arrive at a 1-regular bipartite graph $G^*$ such that $|V(G^*)| = |V(G)|$. Clearly $G^*$ has a perfect matching.                                     ❏

**Definition:**   A set $C$ of vertices in a graph $G$ is said to *cover* the edges of $G$ if every edge of $G$ is incident with at least one vertex of $C$. Such a set $C$ is called a *covering* of $G$. Consider the graphs $G_1$ and $G_2$ shown in Figure 8.9. Clearly, $\{v_1, v_2, v_4, v_5\}$ and $\{v_1, v_5, v_6\}$ are coverings of $G_1$. Also, we observe that there is no covering of $G_1$ with fewer than three vertices. In $G_2$, each of $\{v_1, v_2, v_6\}$ and $\{v_1, v_2, v_3, v_4, v_5, v_6\}$ is a covering.

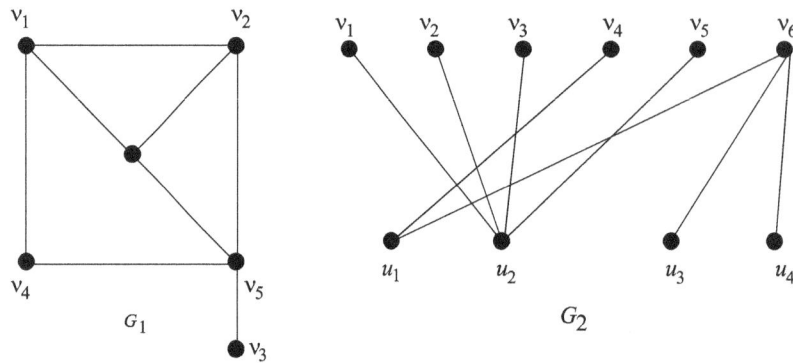

**Fig. 8.9**

The number of vertices in a minimum covering of $G$ is called the *covering number* of $G$ and is denoted by $\beta(G)$.

**Definition:**   An *edge covering* of a graph $G$ is a subset $L$ of $E$ such that every vertex of $G$ is incident to some edge of $L$. Clearly, an edge covering of $G$ exists if and only if $\delta > 0$. The cardinality of a minimum edge covering of $G$ is denoted by $\beta_1(G)$. For example, in the wheel $W_5$ of Figure 8.10, the set $\{v_1 v_5, v_2 v_3, v_4 v_6\}$ is a minimum edge covering.

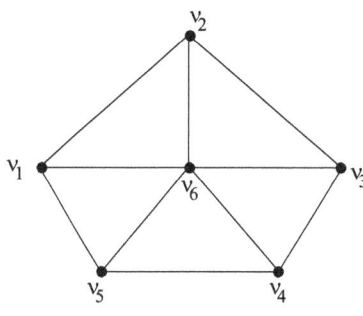

**Fig. 8.10**

The following results are immediate.

**Lemma 8.1**   A subset S of $V$ is independent if and only if $V - S$ is a covering of $G$.

**Proof**   Clearly, S is independent if and only if no two vertices in S are adjacent in $G$. Thus every edge of $G$ is incident to a vertex of $V - S$. This is possible if and only if $V - S$ is a covering of $G$.                                                                                 ❏

**Lemma 8.2**   For any graph $G$, $\alpha + \beta = n$.

**Proof**    If S is a maximum independent set of $G$, then by Lemma 8.1, $V - S$ is a covering of $G$. Therefore, $|V - S| = n - \alpha \geq \beta$. Similarly, if $C$ is a minimum covering of $G$, then $V - C$ is independent and therefore, $|V - S| = n - \beta \leq \alpha$. Combining the two inequalities, we obtain $\alpha + \beta = n$.                                                                                                              ❏

**Remark**    Consider the graph of Figure. 8.11. Clearly the set $E_1 = \{e_3, e_4\}$ is independent and $E - E_1 = \{e_1, e_2, e_5\}$ is not an edge covering of $G$. Also, $E_2 = \{e_1, e_3, e_4\}$ is an edge covering of $G$ and $E - E_2$ is not independent in $G$. These observations imply that the edge analogue of Lemma 8.1 is not true in general.

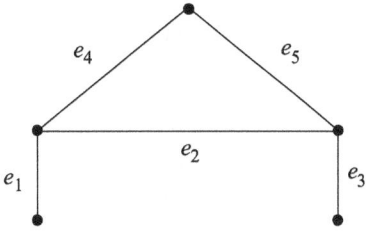

**Fig. 8.11**

The following result gives the relation between $\alpha_1$ and $\beta_1$ of a graph $G$.

**Theorem 8.7(a)**    In a graph $G$ with $n$ vertices and $\delta > 0$, $\alpha_1 + \beta_1 = n$.

**Proof**    Let $M$ be a maximum matching in $G$, so that $|M| = \alpha_1$. Let $X$ be the set of $M$-unsaturated vertices in $G$. Since $M$ is maximum, $X$ is an independent set of vertices with clearly $|X| = n - 2\alpha_1$. As $\delta > 0$, choose one edge for each vertex in $X$ incident with it and let $F$ be the set of these edges chosen. Then $M \cup F$ is an edge covering of $G$. Thus, $|M \cup F| = |M| + |F| = \alpha_1 + n - 2\alpha_1 \geq \beta_1$, and so

$$n \geq \alpha_1 + \beta_1. \tag{8.7.1}$$

Now assume that $L$ is a minimum edge covering of $G$, so that $|L| = \beta_1$. Let $H = G(L)$, the edge subgraph of $G$ defined by $L$ and let $M_H$ be a maximum matching in $H$. Let the set of $M_H$-unsaturated vertices in $H$ be denoted by $X$. Since $L$ is an edge covering of $G$, therefore, $H$ is a spanning subgraph of $G$. Thus, $|L| - |M_H| = |L - M_H| \geq |X| = n - 2|M_H|$, and therefore $|L| + |M_H| \geq n$. Since $M_H$ is a matching in $G$, so $|M_H| \leq \alpha_1$, and hence

$$n \leq \beta_1 + \alpha_1. \tag{8.7.2}$$

Combining (8.7.1) and (8.7.2), we obtain $\alpha_1 + \beta_1 = n$.                                                     ❏

Let $M$ be any matching of a graph $G$ and $C$ be any vertex covering of $G$. Then in order to cover each edge of $M$, at least one vertex of $C$ is to be chosen. Thus, $|M| \leq |C|$. In case, $M'$ is a maximum matching and $C'$ is a minimum covering of G, then $|M'| \leq |C'|$.

We have the following obvious result.

**Lemma 8.3**   If $C$ is any covering and $M$ any matching of a graph $G$ with $M| = |C$, then $C$ is a minimum covering and $M$ is a maximum matching.

**Proof**   Let $M'$ be a maximum matching and $C'$ be a minimum covering of $G$. Then, $|M| \leq |M'|$ and $|C| \geq |C'|$. Therefore, $|M| \leq |M'| \leq |C'| \leq |C|$. As $|M| = |C|$ we have $|M| = |M'| = |C'| = \|C|$, and the proof is complete.                                          ❏

The next result due to Konig [135] is of great significance.

**Theorem 8.7(b) (Konig [135])**   In a loopless bipartite graph $G$, the maximum number of edges in a matching of $G$ is equal to the minimum number of vertices in an edge cover of $G$, that is, $\alpha_1 = \beta$.

**Proof**   Let $G$ be a bipartite graph with partite sets $V_1$ and $V_2$ and let $M$ be a matching of $G$. Let $X$ be the set of all $M$-unsaturated vertices of $V_1$, so that $|M| = |V_1| - |X|$ (Fig. 8.12).

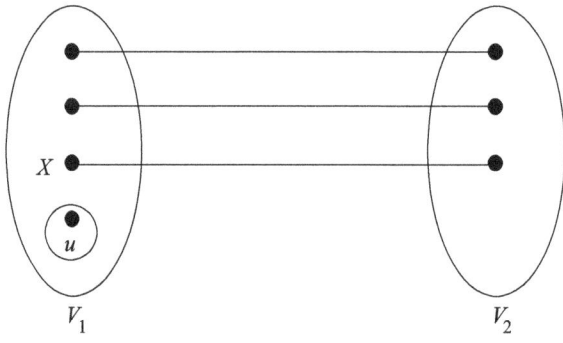

**Fig. 8.12**

Let $A$ be the set of those vertices of $G$ which are connected to some vertex of $X$ by an $M$-alternating path. Further, let $S = A \cap V_1$ and $T = A \cap V_2$. Obviously, $N(S) = T$ and $S - X$ is matched to $T$ so that $|X| = |S| - |T|$. Clearly, $C = (V_1 - S) \cup T$ is a covering of $G$, because if there is an edge $e$ not incident to any vertex in $C$, then one of the end vertices of $e$ is in $S$ and the other in $V_2 - T$, which contradicts the fact that $N(S) = T$. Now, $|C| = |V_1| - |S| + |T| = |V_1| - |X| = |M|$. Then by Lemma 8.3, $M$ is a maximum matching and $C$ is a minimum covering of $G$.                                          ❏

We now give a new proof of Konig's theorem due to Rizzi [224].

**Second proof of Theorem 8.7 (Rizzi [224])**   Let $G$ be a minimal counter-example. Then $G$ is connected but is not a cycle, nor is a path. Therefore, $G$ has a vertex of degree at least three. Let $u$ be such a vertex and $v$ one of its neighbours. If $\alpha_1(G - v) < \alpha_1(G)$, then by minimality, $G - v$ has a cover $W'$ with $|W'| < \alpha_1(G)$. Thus $W' \cup \{v\}$ is a cover of $G$ with cardinality at most $\alpha_1(G)$. Therefore, assume there exists a maximum matching $M$ of $G$ having no edge incident at $v$. Let $f$ be an edge of $G - M$ incident at $u$ but not at $v$. Let $W'$ be a cover of $G - f$ with $|W'| = \alpha_1(G)$. Since no edge of $M$ is incident at $v$, it follows that $W'$ does not contain $v$. Therefore $W'$ contains $u$ and is a cover of $G$. ❏

## 8.2 Factors

We start this section with the following definitions.

**Definition:** A *factor F* of a graph $G$ is a spanning subgraph of $G$. A factor $F$ is said to be an *F–Factor* if the components of $F$ are some graphs in a given collection $F = \{H_1, H_2, \ldots, H_k\}$ of subgraphs of $G$. If $F$ contains a single graph $H$, then $F$ is called an *H-factor*. Consider the graph $G$ shown in Figure 8.13. Clearly $G_1$ is a spanning subgraph whose components belong to the set $F = \{H_1, H_2, H_3\}$ and therefore is an $F$-factor. Each of the components of the spanning subgraph $G_2$ is a graph $K_3$ and so $G_2$ is a $K_3$-factor.

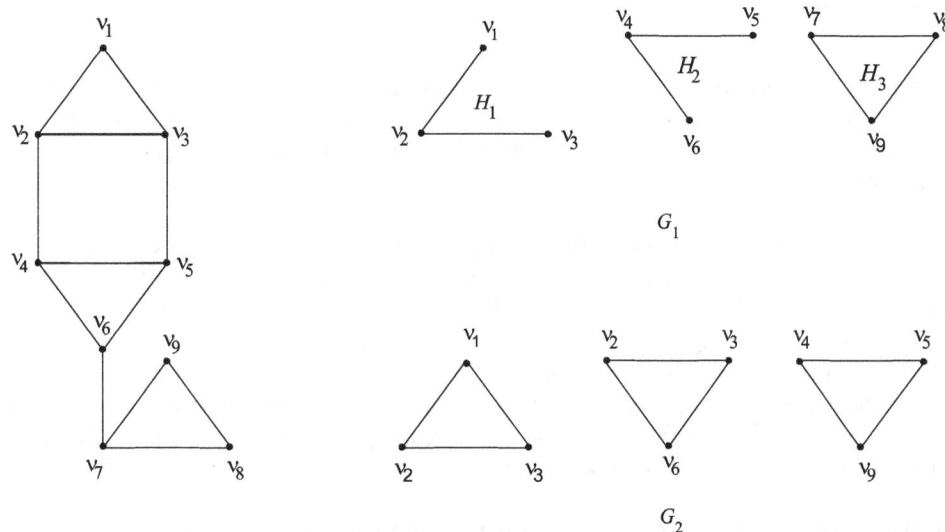

**Fig. 8.13**

We observe that a perfect matching is a $K_2$-factor, since each of the components in a perfect matching is a graph $K_2$.

A *vertex function* of a graph $G = (V, E)$ is a function $f : V \rightarrow N_o$ ($N_o$ being the set of non-negative integers).

**Definition:** A spanning subgraph $F$ of a graph $G = (V, E)$ is called an *f-factor* of $G$ if for a vertex function $f$ on $V$, $d(v|F) = f(v)$ for all $v \in V$. For example, consider the graph

$G$ shown in Figure 8.14. Define $f : V \rightarrow N_0$ by $f(v_i) = 1$ or 2, for all $v \in V$ in $G$. Then the $f$-factor of $G$ is given by graph $H$.

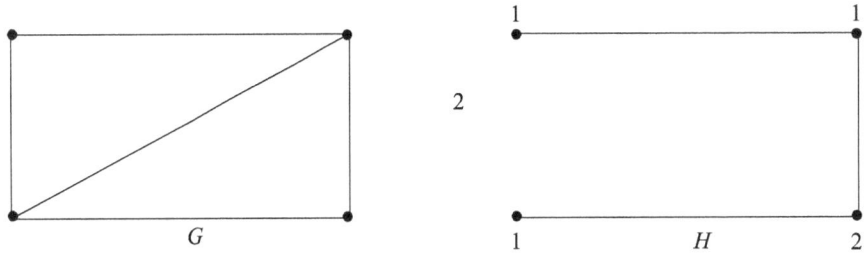

**Figure 8.14**

Further, $F$ is said to be a *partial f–factor* (or $f$-matching) if $d(v|F) \leq f(v)$, for all $v \in V$ and $F$ is said to be a covering $f$-factor of $G$ if $d(v|F) \geq f(v)$, for all $v \in V$. Some of the details of this can be seen in Graver and Jurkat [88]. We note that an $f$-factor exists only if $\sum_{v \in V} f(v)$ is even.

**Definition:** A *k-factor* of a graph $G$ is a factor of G that is $k$-regular. Clearly, a 1-factor is a perfect matching and exists only for graphs with an even number of vertices. A 2-factor of $G$ is a factor of $G$ that is a disjoint union of cycles of $G$ and connected 2-factor is a Hamiltonian cycle.

**Definition:** If the edge set of a graph $G$ is partitioned by the edge sets of a set of factors $F_1, F_2, \ldots, F_q$ of $G$, then these factors are said to constitute a *factorisation* or decomposition of $G$ and we write $G = F_1 \cup F_2 \cup \ldots \cup F_q$. If each factor $F_i$ in the factorisation is isomorphic to the same spanning subgraph $F$ of $G$, then the factorisation is called an *isomorphic factorisation* or an *F-decomposition* of $G$.

The factorisation is called an *f-factorisation* or *k-factorisation* according as each factor $f_i$ of a factorisation is an $f$-factor or $k$-factor. A graph $G$ is $f$-factorable or $k$-factorable if $G$ has respectively $f$-factorisation or $k$-factorisation.

**Example** In Figure 8.15, the three distinct 1-factors of $G_1$ are $< e_1, e_6 >$, $< e_2, e_4 >$ and $< e_3, e_5 >$. The two distinct 2-factors of $G_2$ are $< e_1, e_2, e_3, e_4, e_5 >$ and $< e_6, e_7, e_8, e_9, e_{10} >$.

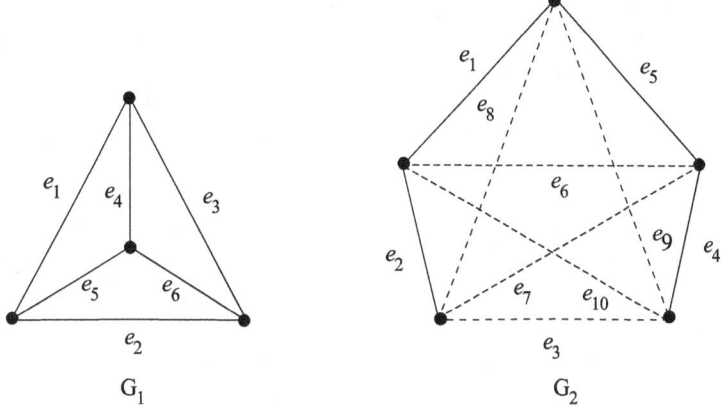

**Fig. 8.15**

**Definition:** A component of a graph is *odd or even* according to whether it has an odd or even number of vertices. The number of odd components of a graph $G$ is be denoted by $k_0(G)$.

The following result of Tutte [249] characterises the graphs with a 1-factor. Several proofs of this result exist in the literature and the proof given here is due to Lovasz [150].

**Theorem 8.8 (Tutte [249])** A graph $G$ has a 1-factor if and only if

$$k_0(G - S) \leq |S|, \text{ for } S \subseteq V. \tag{8.8.1}$$

**Proof** Let $G$ be a graph having 1-factor $M$. Let S be an arbitrary subset of $V$ and let $O_1$, $O_2, \ldots, O_k$ be the odd components of $G - S$. For each $i$, the vertices in $O_i$ can be adjacent only to other vertices in $O_i$ and to vertices in $S$. Since $G$ has a 1-factor (perfect matching), at least one vertex from each $O_i$ is to be matched with a different vertex in $S$. Thus, the number of vertices in $S$ is at least $k$, the number of odd components. Hence $|S| \geq k$, so that $k_0(G - S) \leq |S|$ (Fig. 8.16).

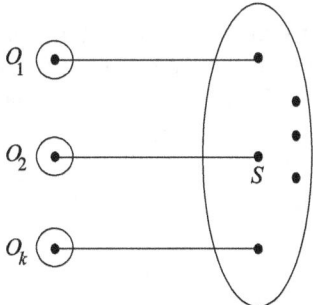

**Fig. 8.16**

Conversely, assume that the condition (8.8.1) holds. Let $G$ have no 1-factor. Join pairs of non-adjacent vertices of $G$ and continue this process till we get a maximal super graph $G^*$ of $G$ having no 1-factor. Since joining two odd components by an edge results in an even component, therefore

$$k_0(G^* - S) \leq k_0(G - S). \tag{8.8.2}$$

Thus for $G^*$ also, we have $k_0(G^* - S) \leq |S|$.

When $S = \phi$ in (8.8.1), we observe that $k_0(G) = 0$ so that $k_0(G^*) = 0$. Therefore, $|V(G^*)| = |V(G)| = n$ is even. Also, for every pair of non-adjacent vertices $u$ and $v$ of $G^*$, $G^* + uv$ has a 1-factor and any such 1-factor definitely contains the edge $uv$.

Now, let $K$ be the set of those vertices of $G^*$ which are having degree $n - 1$. Clearly, $K \neq V$, since otherwise $G^* = K_n$ has a perfect matching.

Claim, that each component of $G^* - K$ is complete. Assume, on the contrary, that in $G^* - K$, there exists a component $G_1$ which is not complete. Then in $G_1$, there are vertices $x$, $y$ and $z$ such that $xy, yz \in E(G^*)$ and $xy \notin E(G^*)$. Also, $y \in V(G_1)$ so that $d(y/G^*) < n - 1$ and thus there exists a vertex $w$ of $G^*$ with $yw \notin E(G^*)$. Evidently, $w \notin K$ (Fig. 8.17).

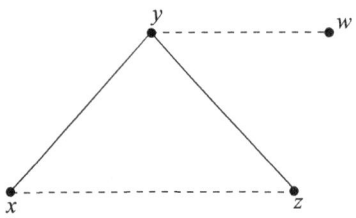

**Fig. 8.17**

Now, by the choice of $G^*$, both $G^* + xz$ and $G^* + yw$ have 1-factors, say $M_1$ and $M_2$ respectively. Clearly, $xz \in M_1$ and $yw \in M_2$. Let $H$ be the subgraph of $G^* + \{xz, yw\}$ induced by the edges that are in $M_1$ and $M_2$ but not in both (that is, in the symmetric difference of $M_1$ and $M_2$). Since $M_1$ and $M_2$ are 1-factors, each vertex of $G^*$ is saturated by both $M_1$ and $M_2$, and $H$ is a disjoint union of even cycles in which the edges alternate between $M_1$ and $M_2$. We have two cases to consider.

**Case 1** Let $xz$ and $yw$ belong to different components of $H$, as shown in Figure 8.18 (a). If $yw$ belongs to the cycle $C$, then the edge of $M_1$ in $C$ together with the edges of $M_2$ not belonging to $C$ form a 1-factor of $G^*$, contradicting the choice of $G^*$.

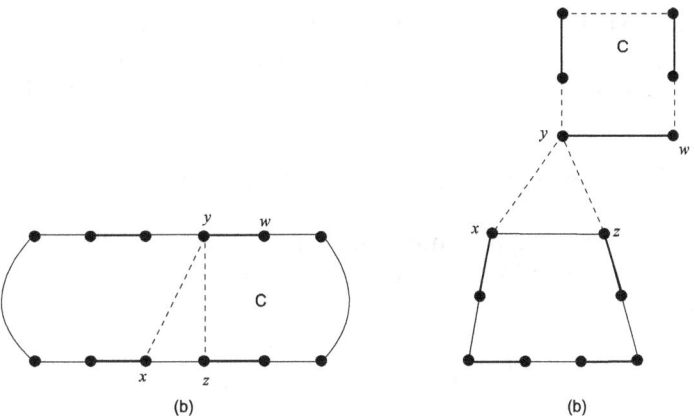

Ordinary lines correspond to the edges of $M_1$, and bold lines to the edges of $M_2$

**Fig. 8.18**

**Case 2** Let $xz$ and $yw$ belong to the same component $C$ of $H$. Since each component of $H$ is a cycle, $C$ is a cycle (Fig. 8.18 (b)). By symmetry of $x$ and $z$, it can be assumed that the vertices $x$, $y$, $w$ and $z$ appear in that order on $C$. Then the edges of $M_1$ belonging to the $yw \ldots z$ section of $C$, together with the edge $yz$, and the edges of $M_2$ not in the $yw \ldots z$ section of $C$ form a 1-factor of $G^*$, again contradicting the choice of $G^*$.

Thus, each component of $G^* - K$ is complete.

Now, by conditions (8.8.2), $k_0(G^* - K) \leq |K|$. Therefore one vertex of each of the odd components of $G^* - K$ is matched to a vertex of $K$. This is because each vertex of $K$ is adjacent to every other vertex of $G^*$. Also the remaining vertices in each of the odd and even components of $G^* - K$ can be matched among themselves (Fig. 8.19). The total number of vertices thus matched is even. As $|V(G^*)|$ is even, the remaining vertices of $K$ can be matched among themselves, thus giving a 1-factor of $G^*$, which is a contradiction. ❏

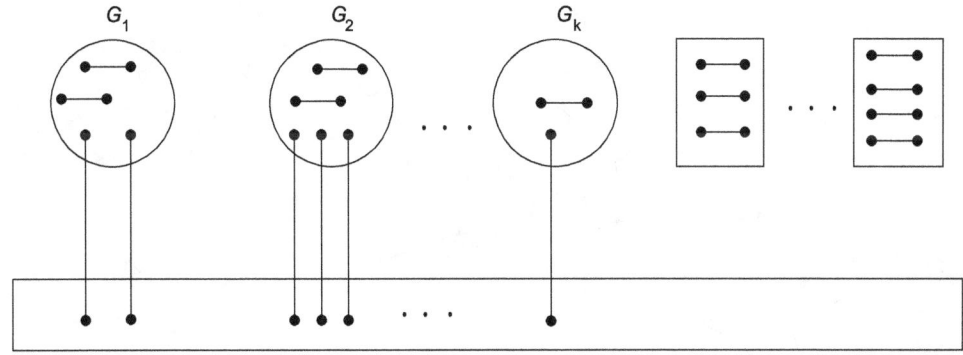

**Fig. 8.19**

**Remark** Among the various proofs of Theorem 8.8, one proof due to Anderson [4] uses Hall's theorem.

Tutte's characterisation of graphs admitting 1-factors has been used to obtain several results including those for regular graphs and many of these are reported in the review article by Akiyama and Kano [1]. We present some of these results. The first is due to Peterson [181].

**Theorem 8.9 (Peterson [181])**    Every cubic graph without cut edges has a 1-factor.

**Proof**    Let $G$ be a cubic graph without cut edges and let $S \subseteq V$. Let $O_1, O_2, \ldots, O_k$ be the odd components of $G - S$ and let $m_i$ be the number of edges of $G$ having one end in $V(O_i)$ and the other end in $S$. As $G$ is cubic,

therefore,         $\sum\limits_{v \in v(O_i)} d(v) = 3\,|V(O_i)|$                                       (8.9.1)

and             $\sum\limits_{v \in S} d(v) = 3\,|S|.$                                                (8.9.2)

Now, $E(O_i) = [V(O_i), V(O_i) \cup S] - [V(O_i), S]$, where [A, B] denotes the set of edges having one end in $A$ and the other end in $B$, $A \subseteq V$, $B \subseteq V$. Thus, $m_i = |[V(O_i), S]| = \sum\limits_{v \in v(O_i)} d(v) - 2|E(O_i)|$. Since $d(v) = 3$ for each $v$ and $V(O_i)$ is odd component, $m_i$ is odd for each $i$. Also, since $G$ has no cut edges, $m_i \geq 3$. Therefore, $k_0(G - S) = k \leq \frac{1}{3} \sum\limits_{v \in S} d(v) < \frac{1}{3} 3\,|S| = |S|.$ Hence by Theorem 8.8, $G$ has a 1-factor.                                                      ❑

**Remark**    A cubic graph with cut edges may not have a 1-factor. To see this, consider the graph $G_1$ of Figure 8.20 (a). If we take $S = \{v\}$, then $k_0(G - S) = 3 > 1 = |S|$ and hence $G_1$ has no 1-factor.

Also, a cubic graph with a 1-factor may have cut edges. Consider the graph $G_2$ of Figure 8.20 (b). Clearly, $< e_1, e_2, e_3, e_4, e_5 >$ is a 1-factor and $e_3$ is a cut edge of $G_2$.

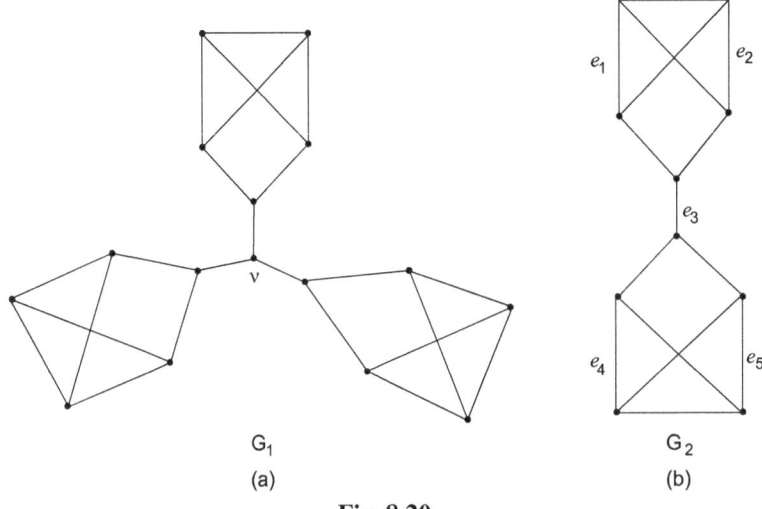

$G_1$                                          $G_2$

(a)                                          (b)

**Fig. 8.20**

The next result is due to Cunningham [62].

**Corollary 8.2**    The edge set of a simple 2-edge-connected cubic graph $G$ can be partitioned into paths of length three.

**Proof**    By Theorem 8.9, $G$ is a union of a 1-factor and 2-factor. Orient the edges of each cycle of the 2-factor in any manner so that each cycle becomes a directed cycle. Then, if $e$ is any edge of the 1-factor and $f_1$, $f_2$ are the two arcs of $G$ having their tails at the end vertices of $e$, then $\{e, f_1, f_2\}$ form typical 3-path of the edge partition of $G$ (Fig. 8.21).    ❑

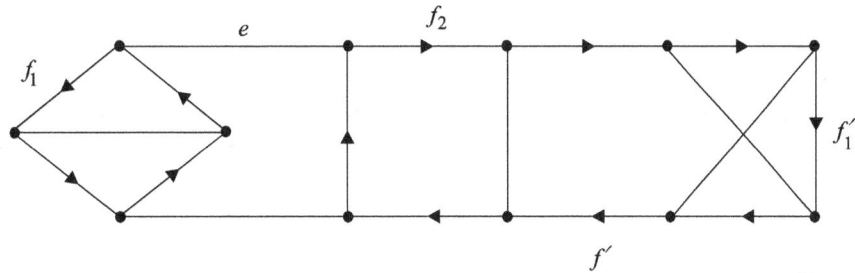

**Fig. 8.21**

Berge [23] obtained the following estimate for $\alpha_1(G)$ of a graph $G$, the proof of which is given by Bollobas [29].

**Theorem 8.10** For any simple graph $G$, $\alpha_1(G) = (n-d)/2$, where $d = \max\{k_0(G-S) - |S| : S \subseteq V\}$.

**Proof**    Clearly at most $n-d$ vertices can be saturated in any matching of $G$. In order to show that this bound is attained, we use the following construction. Let $H = GVK_d$. Then $H$ satisfies Tutte's condition. For if $S'$ is any subset of $V(H)$, then $k_0(H - S') = 0$ or 1 if $S' \not\supseteq V(K_d)$. Also, if $S' \supseteq V(K_d)$, then $k_0(H - S') = k_0(G - S)$, where $S = S' \cap V(G)$, so that $k_0(H - S') - |S'| = k_0(G - S) - |S| - d = 0$, by definition of $d$. Therefore, by Tutte's theorem, $H$ has a 1-factor $M$. The restriction of $M$ of $G$ is a matching in $G$ which does not cover at most $d$ vertices of $G$.    ❑

**Remark**    Berge actually used the theory of alternating paths to prove Theorem 8.10 and then deduced the Tutte's theorem.

The restatement of Theorem 8.10 is as follows.

**Corollary 8.3**    If $G$ is a graph with $n$ vertices and $\alpha_1(G) = a$, then there is a set $S$ of $s$ vertices such that $k_0(G - S) = n + s - 2a$.

The following result is due to Babler [10].

**Theorem 8.11 (Babler [10])**   Every $k$-regular, $(k-1)$-edge-connected graph with even order has a 1-factor.

**Proof**   Let $G$ be a $k$-regular, $(k-1)$-edge-connected graph with even order and let $S$ be any subset of $V$. Let $O_1, O_2, \ldots, O_r$ be the odd components of $G - S$ with order $n_i$ and size $m_i$, and let $k_i$ edges join $O_i$ to $S$, for $1 \leq i \leq r$. Then, $k_i \geq k - 1$,                   (8.11.1)

$$\sum_{v \in v(O_i)} d(v) = kn_i = 2m_i + k_i \tag{8.11.2}$$

and   $\displaystyle\sum_{v \in S} d(v) = k|S| \geq \sum_{1}^{r} k_i + 2|E(S)|.$                   (8.11.3)

Now, (8.11.2) gives $k_i = kn_i - 2m_i$, so that if $k$ is even, so is $k_i$, and thus $k_i > k - 1$ implies $k_i \geq k$. Similarly, if $k$ is odd, so is $k_i$, and again, $k_i > k - 1$ implies $k_i \geq k$. Summing this over $r$ odd components, we obtain by using (8.11.3)

$$rk \leq \sum_{i=1}^{r} k_i \leq \sum_{i=1}^{r} k_i + 2|E(S)| \leq k|S|.$$

Therefore, $k_0(G - S) = r \leq |S|$ and by Tutte's theorem, $G$ has a 1-factor.   ❏

## 8.3   Antifactor Sets

**Definition:**   A subset $S \subseteq V$ of a graph $G = (V, E)$ such that $k_0(G - S) > |S|$ is called an *antifactor* set. An antifactor set $S$ such that no proper subset of $S$ is an antifactor set is called a *minimal antifactor set*. For example, in the graph of Figure 8.20(a), $\{v\}$ is an antifactor set.

**Definition:**   A vertex $v \in V$ is a *claw centre* if and only if at least three vertices of $G$ adjacent to $v$ form an independent set. Clearly, vertex $v$ is a claw centre in the graph of Figure 8.20(a).

   Now assume that $G$ is a graph of even order $n$ and let $S$ be an antifactor set of $G$. Let $k_0(G - S) = k|$ and $O_1, O_2, \ldots, O_k$ be the odd components of $G - S$. Since $n$ is even, $|S|$ and $k$ have the same parity. Choose a vertex $u_i \in V(O_i)$, $1 \leq i \leq k$. Then, $|S \cup \{u_1, u_2, \ldots, u_k\}|$ is even, that is, $k + |S| \equiv 0 \pmod 2$. This implies that $k \equiv |S| \pmod 2$ and therefore $k_0(g - S) \equiv |S| \pmod 2$. Thus, we have the following result.

**Lemma 8.4**   If $S$ is an antifactor set of a graph $G$ of even order, then $k_0(G - S) \geq |S| + 2$.

   Well, Tutte's theorem implies that a graph $G$ has either a 1-factor or an antifactor set. It is necessary to mention here that the computational complexity is high in the verification of Tutte's condition for every subset $S$ of $V$. In order to reduce the computations, the antifactor sets are of great significance. The following result on antifactors is due to Sumner [236].

**Lemma 8.5 (Sumner [236])** If $S$ is a minimal antifactor set of a connected graph $G$ of even order and having no 1-factor, and if $k_0(G - S) = r$ and $|S| = s$, then

   i. $r \geq s + 2$,

   ii. each vertex of $S$ is adjacent to vertices in at least $r - s + 1$ distinct odd components of $G - S$,

   iii. every vertex of $S$ is a claw centre.

**Proof**

   i. Already established (Lemma 8.4).

   ii. If $v \in S$ is adjacent to $h$ distinct odd components say $O_1$, $O_2$, ..., $O_h$ with $h \leq r - s$, then $G - \{S - \{v\}\}$ has the odd components $O_{h+1}$, $O_{h+2}$, ..., $O_r$. Then, if $h$ is odd, $k_0(G - S') = r - h \geq s > s - 1 = |S'|$, and if $h$ is even, then $\{O1 \cup O_2 \cup ... \cup O_h\} \cup \{v\}$ is also an odd component of $G - S'$ so that $k_0(G - S') = r - h + 1 \geq s + 1 > s - 1 = |S'|$, where $S' = S - \{v\}$. That is, $S'$ is an antifactor set, contradicting the minimality of $S$. Thus, $h \geq r - s + 1$.

   iii. Since any $v \in S$ is adjacent to at least $r - s + 1 \geq 3$ (by (i)) distinct odd components, $v$ is a claw centre, by definition. □

**Remark** The above observations imply that Tutte's theorem is equivalent to the following.

A graph $G$ has a 1-factor if and only if $G$ does not have a set $S$ of claw centres such that $k_0(G - S) > |S|$. Obviously, this form of Tutte's theorem decreases the number of subsets $S$ of $V$ to be verified in Tutte's condition.

An improvement of Theorem 8.11 is due to Plesnik [209].

**Theorem 8.12 (Plesnik [209])** If $G$ is a $k$-regular, $(k-1)$-edge-connected graph of even order and $G_1$ is obtained from $G$ by removing any $k - 1$ edges, then $G_1$ has a 1-factor.

**Proof** Assume that $G_1$ has no 1-factor. Then $G_1$ has an antifactor set $S$ with $k_0(G_1 - S) > |S|$. Therefore by Lemma 8.4,

$$k_0(G_1 - S) \geq |S| + 2. \tag{8.12.1}$$

Let $O_1$, $O_2$, ..., $O_r$ be the odd components and $O_{r+1}$, $O_{r+2}$, ..., $O_{r+s}$ be the even components of $G_1 - S$. For a component $O_i$, let $a_i$ and $m_i$ be the number of edges respectively of $E(G) - E(G_1)$ and $E(G_1)$ joining it to $S$ and $b_i$ be the number of edges of $E(G) - E(G_1)$ joining it to the other $O_j$'s. Then the total number of edges going out of $O_i$ is $a_i + b_i + m_i$. Since $G$ is $(k-1)$-edge-connected, $a_i + b_i + m_i \geq k - 1$. By the argument used in Theorem 8.11, this implies that $a_i + b_i + m_i \geq k$, for an odd component $O_i$. Summing over the $r$ odd components, we obtain

$$\sum_{i=1}^{r} a_i + \sum_{i=1}^{r} b_i + \sum_{i=1}^{r} m_i \geq rk. \tag{8.12.2}$$

Now, the number of edges between the $O_i$'s, and between $O_i$'s and $S$, removed from $G$ is $\sum_{i=1}^{t} a_i + \left(\frac{1}{2}\right) \sum_{i=1}^{t} b_i$, where $t = r + s$. As this is at most $k - 1$, we have

$$2 \sum_{i=1}^{t} a_i + \sum_{i=1}^{t} b_i \leq 2(k-1). \tag{8.12.3}$$

Also, $\sum_{i=1}^{t} a_i + \sum_{i=1}^{t} m_i \leq \sum_{v \in S} d(v) = k|S|.$ \hfill (8.12.4)

Adding (8.12.3) and (8.12.4), we get

$$3 \sum_{i=1}^{t} a_i + \sum_{i=1}^{t} b_i + \sum_{i=1}^{t} m_i \leq k(|S|+2) - 2. \tag{8.12.5}$$

As the sum on the left of (8.12.2) is less than the sum on the left of (8.12.5), we have

$$rk \leq k(|S|+2) - 2 < k(|S|+2), \text{ giving } r < |S| + 2, \text{ which contradicts (8.12.1).} \qquad \square$$

We now have the following observation.

**Corollary 8.4**    A $(k-1)$-regular simple graph on $2k$ vertices has a 1-factor.

**Proof**    Assume that $G$ is a $(k-1)$-regular simple graph with $2k$ vertices having no 1-factor. Then $G$ has an antifactor set $S$ and by Lemma 8.4, $k_0(G-S) \geq |S| + 2$. Therefore, $|S| + (|S|+2) \leq 2k$ and so $|S| \leq k-1$. Let $|S| = k-r$. Then $r \neq 1$. For if $r = 1$, then $|S| = k-1$ and therefore $k_0(G-S) = k+1$. Thus each odd component of $G-S$ is a singleton and therefore each such vertex is adjacent to all the $k-1$ vertices of $S$, as $G$ is $(k-1)$ regular. This implies that every vertex of $S$ is of degree at least $k+1$, a contradiction. Thus, $|S| = k-r$, $2 \leq r \leq k-1$. If $G_1$ is any component of $G-S$ and $v \in V(G_1)$, then $v$ can be adjacent to at most $|S|$ vertices of $S$. Since $G$ is $(k-1)$-regular, $v$ is adjacent to at least $(k-1) - (k-r) = r-1$ vertices of $G_1$. So, $|V(G_1)| \geq r$. Counting the vertices of $S$, we obtain $(|S|+2)r + |S| = 2k$, or $(k-r+2)r + (k-r) \leq 2k$. This gives $(r-1)(r-k) \geq 0$, violating the condition on $r$. $\qquad \square$

The next result due to Sumner [236] is of great importance.

**Theorem 8.13 (Sumner [236])**    If $G$ is a connected graph of even order $n$ and claw-free (that is, contains no $K_{1,3}$ as an induced subgraph), then $G$ has a 1-factor.

**Proof**    If $G$ has no 1-factor, then $G$ contains a minimal antifactor set $S$ of $G$. Also, there is an edge between $S$ and each odd component of $G-S$. If $v \in S$ and $vx$, $vy$ and $vz$ are edges of $G$ with $x$, $y$ and $z$ belonging to distinct odd components of $G-S$, then a $K_{1,3}$ is induced in $G$. This is not possible, by hypothesis.

Since $k_0(G-S) > |S|$, there exists a vertex $v$ of $S$ and edges $vu$ and $vw$ of $G$, with $u$ and $w$ in distinct odd components of $G-S$. Assume $O_u$ and $O_w$ to be the odd components containing $u$ and $w$ respectively. Then $< O_u \cup O_w \cup \{v\} >$ is an odd component of $G-S_1$, where $S_1 = S - \{v\}$. Also, $k_0(G-S_1) = k_0(G-S) - 1 > |S| - 1 = |S_1|$ and therefore $S_1$ is an antifactor set of $G$ with $|S_1| = |S| - 1$, a contradiction to the choice of $S$. So $G$ has a 1-factor. (Clearly, by Lemma 8.4, the case $|S| = 1$ and $k_0(G-S) = 2$ does not exist). ❏

The following results are immediate, as each type of graph in these cases does not contain $K_{1,3}$ as an induced subgraph.

**Corollary 8.5** Every connected even order edge graph has a 1-factor.

**Corollary 8.6** If $G$ is a connected graph of even order, then $G_2$ has a 1-factor.

**Corollary 8.7** Every connected total graph of even order has a 1-factor.

**Corollary 8.8** Every connected cubic graph in which every vertex lies on a triangle has a 1-factor.

**Corollary 8.9** If a connected even order graph $G$ has less than $\kappa(G)$ claw centres, then $G$ has a 1-factor.

The following result is the generalisation of Theorem 8.13.

**Theorem 8.14** If $G$ is $n$-connected even order graph with no induced $K_{1,n+1}$, then $G$ has a 1-factor.

**Proof** Assume $G$ has no 1-factor. Then $G$ has a minimal antifactor set $S$. Let $|S| = s$ and let the odd components of $G-S$ be $O_1, O_2, \ldots, O_k$. If $S_i' = \{v \in S : v$ is adjacent to some vertex to $O_i\}$, then $S_i'$ is a vertex cut of $G$ and therefore $|S_i'| \geq n$. Thus there are at least $n$ edges from $O_i$ incident with $n$ distinct vertices of $S$. The number of such edges as $i$ ranges from 1 to $k$ is at least $nk$ and this is greater or equal to $n(s+2)$, by Lemma 8.4. Thus at least one vertex $v$ of $S$ is incident with at least $n+1$ of these edges. But then, $v$ is the centre of an induced $K_{1,n+1}$, contradicting the assumption. ❏

It can be observed that a graph can have more than one 1-factors and different 1-factors can either have no edge in common or have some edges in common. Depending on the nature of edges, we have the following definition.

**Definition:** In a given graph, two 1-factors are said to be *disjoint* if they have no edge in common and two 1-factors are said to be *distinct* if they have at least one different edge. For example, in the graph of Figure 8.22, the 1-factors $\{e_1, e_2, e_3\}$ and $\{e_4, e_5, e_6\}$ are disjoint, while as the 1-factors $\{e_1, e_2, e_3\}$ and $\{e_1, e_7, e_5\}$ are distinct.

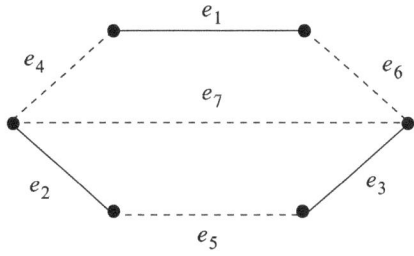

**Fig. 8.22**

Consider maximal sets $S$ which satisfy Tutte's condition in graphs admitting a 1-factor. If such a graph is connected, then definitely there exist subsets $S$ such that $k_0(G-S) = |S|$. For such a graph will have a vertex $v$ which is not a cut vertex, and $G-v$ being connected and of odd order, $k_0(G-S) = 1 = |v|$. The maximal subsets $S$ for which $k_0(G-S) = |S|$ holds is called a maximal factor set.

A result on counting distinct 1-factors for bipartite graphs is given by Theorem 8.4. The next result as reported in Bollobas [29] gives the bounds for distinct 1-factors in a graph..

**Theorem 8.15**   Let $G$ be a graph admitting a 1-factor and let $S_0$ be a maximal factor set of $G$ with $|S_0| = m$ and $O_1, O_2, \ldots, O_m$ be the odd components of $G-S_0$. Further, let $d = \min \{|N(O_i) \cap S_o| : 1 \leq i \leq m\}$ and $d'$ be the minimum number of edges joining $O_i$ to $S_0$, $1 \leq i \leq m$. Then the number of $F(G)$ of 1-factors of $G$ satisfies

i. $F(G) \geq r(d,m) = \displaystyle\prod_{i=1}^{\min(d,m)} (d+1-i)$ and

ii. $F(G) \geq \tilde{r}(d',m)$, where $\tilde{r}(d',m)$ is the minimum number of 1-factors in a bipartite multigraph with $m$ vertices in each class, in which $d'$ is the minimum degree in the first vertex class.

**Proof**   Let $S_0$ be a maximal factor set of $G$ so that $k_0(G-S_0) = |S_0|$.

Using Hall's theorem for a complete matching in a bipartite graph, we can show that $|S_0|$ odd components of $G-S_0$ can be paired with the vertices of $S_0$. The sufficient condition is that any $k$ such odd components should be joined in $G$ to at least $k$ vertices in $S_0$. Clearly, this is satisfied here. For if some set of $k$ odd components are joined in $G$ only to a subset $T$ of $S_0$ with h vertices, $h < k$, then $k_0(G-T) \geq k > h = |T|$, contradicting the hypothesis. Therefore one vertex each of $|S_0|$ odd components can be matched to one vertex of $S_0$. Let $O$ be an odd component of which a vertex has been matched with a corresponding vertex $s$ of $S_0$. Clearly, $O' = O - a$ is a graph on an even number of vertices. Now we verify Tutte's condition on $O'$. If there is a $T \subseteq V' = V(O')$ such that $k_0(O'-T) > |T|$, then $k_0(O'-T) > |T|+2$. Let $S = S_0 \cup T \cup \{a\}$. Then, $k_0(G-S) = k_0(O'-T) + k_0(G-S_0) - 1 = k_0(O'-T) + |S_0| - 1 > |T| + |S_0| + 1 = |S|$, contradicting the choice of $S_0$. Thus Tutte's condition is satisfied for $O'$ and $O'$ has a 1-factor.

If $G-S_0$ has an even component $C$ and $a \in V(C)$, then $G-(S_0 \cup \{a\})$ has at least one more odd component than $G-S_0$, contradicting the choice of $S_0$ or the Tutte condition. Thus $G-S_0$ has no even components.

Therefore, we have seen that for each $o_i \in O_i$ such that there is an $s_i \in S_0$ with $o_i s_i \in E$, each $O_i - o_i$ has a 1-factor forming part of a 1-factor of $G$. The number of such choices of $m$ independent edges $o_i s_i$ equals the number of 1-factors in a bipartite graph $H = (V_1 \cup V_2, E)$, where $V_1$ has $m$ vertices, one corresponding to each odd component $O_i$ of $G - S_0$ and $V_2$ has the $m$ vertices of $S_0$. If we take the edge set $E$ to be such that whenever $s_j \in N(O_i) \cap S_0$, we take exactly one $o_i s_i \in E$, then we get an ordinary bipartite graph for $H$ in which the minimum degree in $V_1$ is $d = \min\{|N(O_i) \cap S_0|\}$ and then (i) follows from Theorem 8.4. However, if we take $E$ to be such that for each edge joining $O_i$ to $s_j$ we take an edge $o_i s_j$, then we get a bipartite multigraph for $H$ in which the least degree of a vertex in $V_i$ is $d'$ as defined and then (ii) follows. ❏

The following observations can be seen immediately.

**Corollary 8.10**   For a $k$-edge-connected graph with a unique 1-factor, $F(G) \geq k$.

**Corollary 8.11**   If $G$ is a connected graph with a unique 1-factor, then $G$ has a cut edge which belongs to a 1-factor.

The details of the following results can be found in Bollobas [29].

**Theorem 8.16**   If $\delta(n)$ is the minimum degree of a graph of order $2n$ with exactly one 1-factor, then $\delta(n) \leq [\log 2(n+1)]$.

**Theorem 8.17**   If $e(n)$ is the maximum size of a graph of order $2n$ with exactly one 1-factor, then $e(n) = n^2$.

## 8.4   The $f$-factor Theorem

The $f$-factor Theorem gives a set of necessary and sufficient conditions for the existence of an f-factor in a general graph. This result was proved by Tutte [250] for loopless graphs, using the theory of alternating paths, and later again by Tutte [251], using the 1-factor Theorem.

First of all, we have the following definitions.

**Definition:**   A *vertex function* for a general graph $G = (V, E)$ is a function $f : V \to N_0$ such that $0 \leq f(v) \leq d(v)$, for every $v \in V$.

For any function $f : V \to N_0$ and for any $U \subseteq V$, we set $f(U) = \sum_{v \in U} f(v)$.

**Definition:**   Three pair-wise disjoint (possibly empty) subsets $S$, $T$ and $U$ of the vertex set $V$ of a general graph $G$, such that $V = S \cup T \cup U$, is called a *decomposition* $D = (S, T, U)$ of $G$. The components, say $C_1, C_2, \ldots, C_k$, of the vertex-induced subgraph $< U >$ of $G$ are called the components of $U$. For each $C_i$, let $h_i = f(C_i) + q_G[T, C_i]$, where $f$ is a given vertex function of $G$, and $q_G[T, C_i]$ denotes the number of edges of $G$, whose one end is in $T$ and the other end is in $C_i$. Then $C_i$ is said to be an odd or even component of $U$, depending upon $h_i$ being odd or even, and the number of odd components of $U$ is denoted by $k_0'(D, f)$.

The *f-deficiency* of $D$ for the given $G$ is defined as

$$\delta(D, f) = k_0'(D, f) - f(S) + f(T) - d(T) + q[S, T].$$

The decomposition $D$ is defined to be an *f-barrier* of $G$ if $\delta(D, f) > 0$.

The term decomposition is used by Graver and Jurkat [88], and Tutte calls it a *G-triple*. These decompositions play the same role as the antifactor sets in the theory of 1-factors.

Before going into the details, we first state the $f$-factor Theorem.

**Theorem 8.18**  If $f$ is a vertex function for a graph $G$, then either $G$ has an $f$-factor or an $f$-barrier but not both.

It can be seen that both the statement and the proof of the $f$-factor Theorem as in Tutte [253] definitely have similarities to those of the 1-factor Theorem. Since there are few complications in the constructions involved which make the proof given by Tutte [253] some what lengthy, we follow here the earlier proof by Tutte [251].

Tutte [250] defines an ordinary graph $G'$ corresponding to a given graph $G$ such that $G$ has an $f$-factor if and only if $G'$ has a 1-factor and shows that the 1-factor condition corresponds to the $f$-factor condition, and derives the $f$-factor Theorem as a consequence of the 1-factor Theorem. Berge [23] uses the same graph construction to derive the maximum $f$-matching Theorem. Since we have general graphs under consideration, we construct $G'$ with suitable changes as reported by Parthasarthy [180], and then find that the proof still holds.

Let $G$ be a general graph with the vertex function $f$. This defines for each vertex $v_i$, its degree $d_i = d(v_i|G)$, $f_i = f(v_i)$ and $k_i = d_i - f_i$. For each vertex $v_i$ of $G$, take two subsets $X_i$ and $Y_i$ of vertices of $G'$, where $|X_i| = d_i$ and $|Y_i| = k_i$. Let the vertices of $Y_i$ be labelled $j_j^i$, $1 \leq j \leq k_i$, arbitrarily. Now, label the vertices of $X_i$ in the following manner.

Label the edges of $G$ arbitrarily as $1, 2, \ldots, m$. Since $|X_i| =$ number of link edges $l_i$ incident with $v_i$ plus twice the number of loops $\ell_i^0$ incident with $v_i$, a vertex $x_j^i \in X_i$ is taken corresponding to each link edge $j$ incident with $v_i$ and two vertices $x_h^i$, $x_{h'}^i$ are taken corresponding to each loop $h$ incident at $v_i$. As for edges, if link $j$ joins $v_i$ and $v_k$, then $x_j^i x_j^k$ is made an edge. Corresponding to loop $h$ incident with $x_h^i$ the edge $x_h^i x_{h'}^i$ is taken. In addition, all the edges of the complete bipartite subgraph with vertex partition $\{X_i, Y_i\}$ are taken for each $i$. These constitute the graph $G'$. The induced subgraph $< X_i \cup Y_i >$ of $G$ is called the *star graph* at $i$ and is denoted by $St(i)$. For multi graph $G$, it is a complete bipartite graph, but for general graphs it will contain edges in $< X_i >$ which we call *loop edges*. The other edges of $St(i)$ are called *star edges* and the remaining edges of $G'$ are called as *link edges*.

This construction is illustrated in Figure 8.23. In the general graph $G$, $d(v_1) = 5$, $d(v_2) = 3$ and $d(v_3) = 2$. Let $f = (3, 2, 1)$ be the vertex function, so that $f(v_1) = 3$, $f(v_2) = 2$ and

$f(v_3) = 1$. Therefore, $k_1 = 2$, $k_2 = 1$ and $k_3 = 1$. Here $|X_1| = 5$, $|X_2| = 3$ and $|X_3| = 2$, and $|Y_1| = 2$, $|Y_2| = 1$ and $|Y_3| = 1$.

Label the edges of $G$ as 1, 2, 3, 4 and 5 as shown. Then, $Y_1 = \{y_1^1, y_2^1\}$, $Y_2 = \{y_1^2\}$ and $Y_3 = \{y_1^3\}$, and we have $X_1 = \{x_1^1, x_2^1, x_3^1, x_4^1\}$, $X_2 = \{x_2^2, x_3^2, x_5^2\}$ and $X_3 = \{x_4^3, x_5^3\}$. The vertex set of $G'$ is $X_i \cup Y_i$, $1 \le i \le 3$. The edges of $G'$ are $x_1^1 x_{1'}^1$, $x_2^1 x_2^2$, $x_3^1 x_3^2$, $x_4^1 x_4^3$, $x_5^2 x_5^3$ (shown by dotted lines) and the star edges (shown by bold lines). The dotted lines in $G$ form an $f$-factor with $f = (3, 2, 1)$.

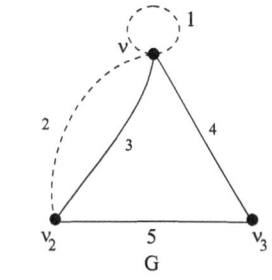

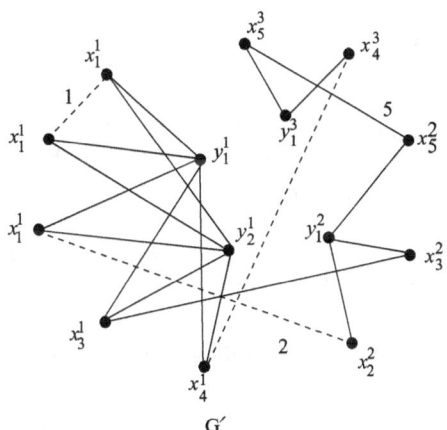

**Fig. 8.23**

The following result gives a relation between $f$-factors of $G$ and the 1-factors of $G'$.

**Theorem 8.18**  A general graph $G$ has an $f$-factor if and only if $G'$ has a 1-factor.

**Proof**  First, assume that $G$ has an $f$-factor $F$ with link edges $L$ and loop edges $L^0$. Let $F$ have $g_i$ link edges and $f_i^o$ loop edges at the vertex $v_i$. Then, $f_i = g_i + 2f_i^0$ $|X_i| = d_i = l_i + 2\lambda_i^0$ and $|Y_i| = k_i = d_i - f_i = (l_i + 2\lambda_i^0) - (g_i + 2f_i^0) = (l_i - g_i) + 2(2\lambda_i^0 - f_i^0)$, where $l_i$ is the number of link edges incident with $v_i$ and $\lambda_i^0$ is the number of loops incident with $v_i$. Let $L'$ and $(L^0)'$ be the set of corresponding set of link and loop edges in $G'$. The matching consisting of these edges, keeps unsaturated all the vertices of $Y_i$ and a set of $l_i - g_i + 2(l_i^o - f_i^o) = k_i$

vertices of $X_i$, for $1 \le i \le n$. Now match the vertices of $Y_i$ with these unsaturated vertices of $X_i$ using the star edges $St(i)$ and let $S'$ be this set of edges. Clearly $L' U (L^0)' \cup S'$ is a 1-factor in $G'$.

Conversely, assume $G'$ has a 1-factor $H$. As the vertices in $Y_i$ are matched only to vertices in $X_i$, there are exactly $d_i - (d_i - f_i) = f_i = g_i + 2f_i^0$ vertices of $X_i$ which are not matched by $H$ to vertices in $Y_i$. Clearly, $2f_i^0$ vertices from these unmatched vertices of $X_i$ correspond to loop edges of $< X_i >$ and thus are matched in pairs amongst themselves. The remaining $g_i$ vertices are matched by the link edges of $H$ to $X_j$ vertices of other star graphs $St(i)$ of $G'$. If $F$ is the set of edges of $G$ corresponding to the loop edges and link edges, then $d$ $(v_i / < F >) = g_i + 2f_i^0$. Thus $f$ is an $f$-factor of $G$.                                                      □

Now, before moving ahead, the following definition will be required for the graph $G'$ constructed as above.

**Definition:**   A subset $W$ of the vertex set $V'$ of the graph $G'$ is said to be *simple* if the following conditions get satisfied.

   i.  $(X_i \cap W) \ne \varphi$ implies $X_i \subseteq W$,

   ii.  $(Y_i \cap W) \ne \varphi$ implies $X_i \subseteq W$ and

   iii.  At most one of $X_i$ and $Y_i$ is a subset of $W$, implies that if $Y_i = \varphi$, then $Y_i \subseteq W$, that is $X_i \cap W = \varphi$.

Now assume $W$ be a simple set of $V'$, $S$ be the set of vertices $v_i$ of $G$, for each $X_i \subseteq W$ and $T$ be the set of vertices $v_j$ of $G$ for each $Y_j \subseteq W$. We observe that $S \cap T = \varphi$, and $S \cup T \cup (V - S - T)$ is a decomposition $D$ of $V$.

Denoting by $d(G, S) = k_0(G - S) - |S|$, for any subset $S$ of the vertex set $V$ of a graph, Theorem 8.8 can be restated in the following form.

**Theorem 8.19**   A simple graph $G$ has a 1-factor if and only if it has no subset $S \subseteq V$ with $d(G, S) > 0$.

**Theorem 8.20**   If $W$ is a simple subset of $V'$ and $D$, the corresponding decomposition of $V$, then $\mathrm{d}(G', W) = \delta(D, f)$.

**Proof**   Since $W$ is a simple subset of $V'$ and $D$ is the corresponding decomposition of $V$,

$$|W| = \sum_{v_j \in S} |X_j| + \sum_{v_j \in T} |Y_j| = \sum_{v_j \in S} d_j + \sum_{v_j \in T} (d_j - f_j) = d(S) + d(T) - f(T). \qquad (8.20.1)$$

Thus, $d(G', W) = k_0(G' - W) - d(S) - d(T) + f(T)$.

We proceed to find $k_0(G' - W)$ and in doing so we consider the components $K$ of $G' - W$. We have the following possibilities.

   i.  The component $K$ contains a single vertex of the form $x_r^i$. This is possible when $Y_i \subseteq W$ and the edge $r$ is a link edge whose other end is in $Y_j \subseteq W$. Thus, in this case, $K$

is an odd component, and the number of such odd components is equal to the number of such link edges and equals $q[S, T]$.

ii. The component $K$ contains a single vertex of the form $y_x^i$, and this is possible when $X_i \subseteq W$. This again is an odd component and the number of such odd components is equal to

$$\sum_{i \in S} |Y_i| = \sum_{i \in S} (d_i - f_i) = d(S) - f(S).$$

iii. $K$ contains a single edge of the form $x_x^i \, x_x^i$ corresponding to a loop of $G$ at $x$. This is possible when $Y_i \subseteq W$. Clearly, this is an even component of $G' - W$.

iv. $K$ contains a single edge of the form $x_r^i x_r^j$, where $r$ is the link edge of $G$ joining vertices $v_i$ and $v_j$. Clearly, this happens when $Y_i \subseteq W$ and $Y_j \subseteq W$. This also is an even component.

v. $K$ contains more than one edge and we call such components large components. Clearly, they have one star graph $St(i)$ as a subgraph. If $K$ contains more than one such star graph as a proper subgraph, it will also contain all edges of $G'$ between such star graphs. The other edges of $K$ are of the form $x_r^i x_r^j$, where $r$ is an edge of $G$ joining $v_i$ and $v_j$, and $St(i) \subseteq K$ and $Y_j \subseteq T$. Let $v_{i_1}, v_{i_2}, \ldots, v_{i_k}$ be the vertices of $G$ corresponding to the star graphs $St(i_j)$, $1 \le j \le k$, which are subgraphs of $K$. Let $C$ be the induced subgraph $< v_{i_1}, v_{i_2}, \ldots, v_{i_k} >$ of $G$. Then, $C$ is a component of $G - S - T$ and therefore the number of vertices of the large component $K$ is given by

$$n(K) = |V(K)| = \sum_{v_i \in C} |V(St(i))| + q[C, T]$$

$$= \sum_{v_i \in C} (d_i + d_i - f_i) + q[C, T]$$

$$= \sum_{v_i \in C} f_i + q[C, T] \pmod 2$$

$$= f(C) + q[C, T] \pmod 2.$$

Therefore, $K$ is an odd component of $G' - W$ if and only if $C$ is an odd component of $G - S - T$. So the number of large odd components of $G' - W$ is $k_0'(D, f)$.

Combining the above results, we observe that the number of odd components of $G' - W$ is given by

$$k_0(G' - W) = q[S, T] + d(S) - f(S) + k_0'(D, f). \tag{8.20.2}$$

Now, $\delta(D, f) = k_0'(D, f) - f(S) + f(T) - d(T) + q[S, T]$

$$= k_0(G' - W) - d(S) + f(T) - d(T), \qquad \text{(by using (8.20.2))}$$

$$= d(G', W). \qquad \text{(by using (8.20.1))} \qquad \square$$

Now, we have the following main result.

**Theorem 8.21**   A general graph $G$ has an $f$-factor if and only if it has no decomposition $D$ with $\delta(D, f) > 0$.

## Proof

*Necessity*   Assume there are decompositions $D$ for which $\delta(D, f) > 0$ and choose one such decomposition for which $|S|$ is least. Claim that for each $v_i \in S$, $f(v_i) < d(v_i)$. If not, there is a $v \in S$ such that $f(v) = d(v)$. Consider the decomposition $D' = (S', T', U')$, where $S' = S - \{v\}$, $T' = T \cup \{v\}$ and $U' = U$. Then the components of $< V - S' - T' >$ are the same as those of $< V - S - T >$. Also, for such a component $C$, $f(C)$ is unchanged, but $q[C, T'] = q[C, T] + q[C, v]$. Therefore at most $q[v, U]$ components can change from odd to even. That is, $k_0'(G - S' - T') \geq k_0'(G - S - T) - q[v, U]$.

Thus, $\delta(D', f) = k_0'(G - S' - T') + f(T') - f(S') - d(T') + q[S', T']$

$$= k_0'(G - S' - T') + f(T) + f(v) - f(S) + f(v) - d(T) - d(v) +$$

$$q[S, T] - q[v, T] + q[v, S]$$

$$= k_0'(G - S' - T') + f(T) - f(S) - d(T) + q[S, T] + 2f(v) - d(v)$$

$$-q[v, T] + q[v, S]$$

$$\geq k_0'(G - S - T) - q[v, U] + f(T) - f(S) - d(T) + q[S, T] + d(v)$$

$$-q[v, T] + q[v, S] \qquad (\text{as } d(v) = f(v))$$

$$= \delta(D, f) + d(v) - q[v, U] - q[v, T] + q[v, S]$$

$$= \delta(D, f) + 2q[v, S] > \delta(D, f) > 0.$$

This contradicts the choice of $D$ and thus the claim is established.

Now, suppose $W = \{X_i : v_i \in S\} \cup \{Y_j : v_j \in T\}$.

Then $W$ is simple, as $S$ and $T$ are disjoint. Therefore by Theorem 8.20, $d(G', W) = \delta(D, f) > 0$ and $G'$ has no 1-factor. Thus by Theorem 8.18, $G$ has no $f$-factor.

*Sufficiency*   Assume that $G$ has no $f$-factor, so that $G'$ has no 1-factor. Therefore, by Theorem 8.19, there is a subset $W$ of $V'$ with $d(G', W) > 0$. We choose a $W$ with least cardinality and prove that

**Claim 1** $Y_j \cap W \neq \varphi$, and $Y_j \subseteq W$.

If this is not true, then there is a $Y_j$ such that $Y_j \cap W \neq \varphi$, and $Y_j \cap (V' - W) \neq \varphi$. Define $Q = W - (Y_j \cap W)$ (Fig. 8.24).

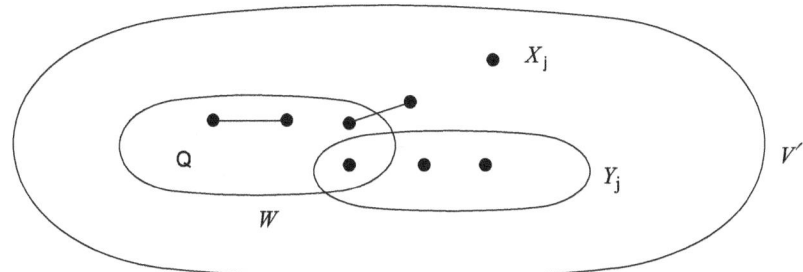

**Fig. 8.24**

If $X_j \not\subseteq W$, then $G' - W$ and $G' - Q$ differ in one component only, the component star edges.

If $X_j \subseteq W$, then each component of $G' - W$ is also a component of $G' - Q$.

In either case, $k_0(G' - Q) \geq k_0(G' - W) - 1$ and $|Q| \leq |W| - 1$. Therefore, $d(G', Q) > 0$, contradicting the choice of $W$.

**Claim 2** $X_i \cap W \neq \varphi$ implies $Y_i \subseteq W$.

If this is not true, then there is an $i$ such that $X_i \cap W \neq \varphi$ and $Y_i \subseteq W$. Let $a \in X_i \cap W$ and define $Q = W - \{a\}$. Then $G' - W$ can have at most one component with a vertex belonging to $Y_i$ which is adjacent to $a$ in $G$. Thus,

$$k_0(G' - W) \leq k_0(G' - Q) + 1 \text{ and } |Q| = |W| - 1.$$

Therefore, $d(G', Q) \geq d(G', W)$, contradicting the choice of $W$.

Now claim (1) and claim (2) together imply that both $X_i$ and $Y_i$ are not subsets of $W$.

**Claim 3** $X_i \cap W \neq \varphi$ implies $X_i \subseteq W$.

If not, there is an $i$ such that $X_i \cap W \neq \varphi$ and $X_i \cap (V' - W) \neq \varphi$. Let $a \in X_i \cap W$ and define $Q = W - \{a\}$. By claims (2) and (1), $Y_i \subseteq G' - W$, so that all vertices of $Y_i$ belong to a single component of $G' - W$. Also there is at most one component of $G' - Q$ which has a vertex not belonging to $Y_i$ which is adjacent to an $a$ in $G$ (Fig. 8.25). Therefore $G' - W$ and $G' - Q$ differ in at most two components.

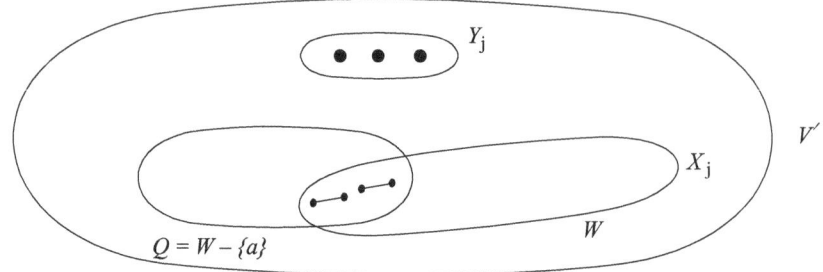

**Fig. 8.25**

Thus, $k_0(G' - Q) \geq k_0(G' - W) - 2$

and $d(G', Q) \geq d(G', W) - 1$.                                                  (8.21.1)

But $k_0(G' - W) + |W| \equiv |V'|$ and

$k_0(G' - Q) + |Q| \equiv |V'|$.

So, $k_0(G' - W) - |W| \equiv |V'| \equiv k_0(G' - Q) - |Q|$

and $d(G', W) \equiv d(G', Q)$.

Therefore, (8.21.1) becomes $d(G', Q) \geq d(G', W) > 0$, contradicting the choice of $W$.

Hence we have proved that $W$ is a simple set and by Theorem 8.20, for corresponding decomposition $D$ of $G$, $\delta(D, f) > 0$.                                          □

**Remark** Theorem 8.21 shows that the 1-factor Theorem implies the $f$-factor Theorem. The converse is also true as can be seen in Theorem 8.22. Before that we have the following definition.

**Definition:** Suppose $D = (S, T, U)$ is a decomposition of $V$ and $x \in S$, $y \in T$. The decomposition $D_1 = (S - \{x\}, T, U \cup \{x\})$ is said to be obtained by an $x$-transfer from $S$ and the decomposition $D_2 = (S, T - \{y\}, U \cup \{y\}$ is said to be obtained by $y$-transfer from $T$.

The following result is an immediate consequence of the above definition and can be easily established.

**Lemma 8.6** An $x$-transfer from $S$ does not reduce the deficiency of $D$ if $f(x) = d(x)$ or $d(x) - 1$ and a $y$-transfer does not reduce the deficiency of $D$ if $f(y) = 0$ or 1.

**Theorem 8.22** The $f$-factor Theorem implies the 1-factor Theorem.

**Proof** Clearly, by the $f$-factor Theorem with $f(v) = 1$, for all $v \in V$, $G$ has a 1-factor if and only if it has a 1-barrier $D = (S, T, U)$. But by Lemma 8.6, $D_1 = (S, \varphi, V - S)$ has no less deficiency than $D$ (by $y$-transfer from $T$). Therefore $G$ has no 1-factor if and only if there is a 1-barrier of the form $D_1$. But $\delta(D_1, f) = k_0(G - S) - f(S) - d(T) + f(T) + q[S, T] = k_0(G - S) - |S|$ and $D_1$ is a 1-barrier if and only if $\delta(D_1, f) > 0$, that is $k_0(G - S) > |S|$. Hence the 1-factor Theorem. ❑

**Corollary 8.12** The Erdos-Gallai Theorem 2.4 on degree sequences follows from the $f$-factor Theorem.

**Proof** Theorem 2.4 gives a necessary and sufficient condition for a non-increasing sequence $d = [d_i]_1^n$ of non-negative integers to be the degree sequence of a simple graph. In order to get this condition, we observe that $d$ is graphic if and only if the complete graph $K_n$ has a $d$-factor, where $d(v_i) = d_i$, given $\sum_{i=1}^{n} d_i = 2m$.

By the $f$-factor Theorem, $d$ fails to be graphic if and only if $K_n$ has a $d$-barrier $D = (S, T, U)$. Since for $K_n$, $U$ can have at most one component, $k_0(D, d) = 0$ or 1, further $\delta(D, d) > 0$ is even, as $d(K_n) = \sum d_i = 2m$ is even. Therefore the condition $\delta(D, d) > 0$ becomes

$$q_{k_n} = [S, T] - d(S) - (n-1)|T| + d(T) > 0. \tag{8.12.1}$$

Now transfer of elements $t$ from $T$ to $U$, or elements $s$ from $S$ to $U$ does not decrease the deficiency $\delta(D, d)$. So elements $t \in T$ with $d(t) < |T| + |U|$ and elements $s \in S$ with $d(s) > |T|$ in $S$ can be transferred to $U$ without decreasing the deficiency. Also it can be seen that elements $u \in U$ with $d(u) \geq |T| + |U|$ can be transferred to $S$ without decreasing the deficiency. Therefore the $f$-barrier can be made such that $d(t) > |T| + |U|$ for all $t \in T$, $|T| < d(u) < |T| + |U|$ for all $u \in U$ and $d(s) \leq |T|$ for all $s \in S$, and $d$ is non-graphical if and only if $K_n$ has such a $d$-barrier $D$. But for such a $D$, $q[S, T] = r(n - r - |U|)$, where $|T| = r$, $d(T) = \sum_1^r d_i$, and thus condition (8.12.1)

$$r(n - r - |U|) - d(S) - r(n-1) + d(T) > 0.$$

That is, $\sum_1^r d_i > r(n-1) + d(S) - r(n - r - |U|)$

$$= r(r-1) + r(n-r) + d(S) - r(n - r - |U|)$$

$$= r(r-1) + r|U| + d(S)$$

$$> r(r-1) + \sum_{i=r+1}^{n} \min\{r, d_i\},$$

which is the Erdos-Gallai condition. ❑

## 8.5  Degree Factors

Let $d = [d_i]_1^n$ and $r = [d_i - k_i]_1^n$ be graphical sequences. The problem we consider here is the existence of a graph $G$ realising $d$ with a spanning subgraph (factor) realising $k = [k_i]_1^n$. This problem was asked by Rao and Rao [215] as $k$-factor conjecture, when $k \le k_i \le k+1$ for each $i$, and solved by Kundu [143]. An improvement of this was due to Kleitman and Wang [129] and further improvement by Kundu [141]. For $k_i = 1$, $1 \le i \le n$, the problem was earlier posed by Grunbaum [93] and proved by Lovasz [149], and this proof also applies when $k_i = k$, $1 \le i \le n$.

**Theorem 8.23 (K-factor Theorem)**  If $[d_i]_1^n$ is a graphical sequence such that $[d_i - k]_1^n$ is also graphical, then there is a realisation of $[d_i]_1^n$ with a $k$-factor.

**Proof**  Let $G_1$ and $G_2$ be two graphs on $V = \{v_1, v_2, \ldots, v_n\}$ such that $d(v_i|G_1) = d_i$ and $d(v_i|G_2) = d_i - k$. Assume that $G_1$ and $G_2$ are chosen such that $G_2$ has maximum edges common with $G_1$. We show that all edges of $G_2$ are in $G_1$. Let $E_1 = E(G_1)$ and $E_2 = E(G_2)$.

**Claim**

If $v_i v_j \in E_2 - E_1$ and $v_i v_k, v_j v_k \in E_1 - E_2$ ($i$, $j$, $k$, $r$ being distinct), then $v_k v_r \in E_1 - E_2$.

If $v_k v_r \in E_2$, then an EDT in $G_2$ switching the pair $v_i v_j$, $v_k v_r$ to the positions $v_i v_k$, $v_j v_r$ gives a graph degree-equivalent to $G_2$ with more edges common with $G_1$, contradicting the choice of $G_1$ and $G_2$. If $v_k v_r \notin E_1$, then an EDT in $G_1$ switching the pair $v_i v_k$, $v_j v_r$ to the positions $v_i v_j$, $v_k v_r$ gives a graph degree-equivalent to $G_1$, such that $G_2$ has more common edges with this graph, again a contradiction. Thus, the claim is established.

Let $G_2$ has some edges which are not in $G_1$ and let $v_1$ be a vertex with a maximum number $t$ of edges of $E_2 - E_1$ incident with it. By the degree stipulation on $G_1$ and $G_2$, there are $t + k$ vertices $v_2, v_3, \ldots, v_{t+k+1}$ which are adjacent to $v_1$ in $G_1$, but not in $G_2$. Assume $v_1 v_k \in E_2 - E_1$ and $v_k v \in E_1 - E_2$, and consider the following two cases.

**Case 1**  Let $v \ne v_i$ for $i \le 2 \le i \le t+k+1$. Then by the above claim, $v v_i \in E_1 - E_2$ for $2 \le i \le t+k+1$, so that there are $t + k + 1$ edges in $E_1 - E_2$ adjacent with $v$. By the degree-stipulation on $G_1$ and $G_2$, there are $t + 1$ edges in $E_2 - E_1$ incident with $v$ and contradicting the choice of $v$.

**Case 2**  Let $v = v_i$ for some $i$, $2 \le i \le t+k+1$, and without loss of generality, take $v = v_2$. Then again, by the above claim, $v v_i \in E_1 - E_2$ for $3 \le i \le t+k+1$. Also, $v v_1, v v_k \in E_1 - E_2$. Therefore, there are $t + k + 1$ edges in $E_1 - E_2$ incident with $v$, again a contradiction.

Hence $G_2$ has no edges which are not in $G_1$. But then $< E(G_1) - E(G_2) >$ is a $k$-factor of $G_1$.                                                                                      ❑

**Lemma 8.7**   If $[k_i]_1^n$ is a sequence of non-negative integers such that $k \leq k_i \leq k+1$ for $1 \leq i \leq n$ and for some $k$, $0 \leq k \leq n-2$, and $\sum_1^n k_i$, is even, then $[k_i]_1^n$ is a graphical sequence.

**Proof**   Induct on $S = \sum_1^n k_i$. Clearly, the result is trivial for $S = 0$. Assume it is true for all values of $\sum_1^n k_i < S-1$ and let $[k_i]_1^n$ be a sequence with $\sum_1^n k_i = S$ (even). Then by Theorem 2.1, $[k_i]_1^n$ is graphical if and only if $[k_2 - 1, k_3 - 1, \ldots, k_{k_1+1} - 1, k_{k_1+2}, \ldots, k_n]$ is graphical, where $k_1 \geq k_2 \geq \ldots \geq k_n$. But the sum of the terms in this sequence is $S - 2k_1$, which is even, and by induction hypothesis is graphical. Hence by Theorem 2.1, $[k_i]_1^n$ is graphical.   ❑

The proof given here of the following $k$-factor Theorem of Kundu, is due to Chen [57].

**Theorem 8.24**   If $[d_i]_1^n$ and $[d_i - k]_1^n$ are two graphical sequences satisfying $k \leq k_i \leq k+1$, for $1 \leq i \leq n$ and some $k > 0$, then there is a graph realising $[d_i]_1^n$ and having a $k$-factor $k = [k_i]_1^n$.

**Proof**   Since $[d_i]_1^n$ and $[d_i - k]_1^n$ are both graphical, $[k_i]_1^n$ is also graphical by Lemma 8.7.

Clearly, there exists a graph with degree sequence $[d_i]_1^n$ containing a $k$-factor $k = [k_i]_1^n$ if and only if there exists a graph with degree sequence $[n - 1 - k_i]_1^n$ containing a $k'$-factor $k' = [n - 1 - d_i]_1^n$. This can be easily established by taking the complementary graphs. Thus, the conditions can be shifted from $k_i$'s to $d_i$'s, so that it is enough to prove that if $[d_i]_1^n$ and $[d_i - k]_1^n$ are graphical with $k \leq d_i \leq k+1$ for $1 \leq i \leq n$ and the $k_i$'s unrestricted, then there is a graph with degree sequence $[d_i]_1^n$ and a $k$-factor $k = [k_i]_1^n$.

Let $G_1$ and $G_2$ be graphs with degree sequences $[k_i]_1^n$ and $[d_i - k]_1^n$ respectively, such that the multigraph $G$, obtained by superimposing $G_1$ and $G_2$ has a minimum number of multiple edges. Obviously, in the superimposed graph, there are at most two edges between any pair of vertices, one of $G_1$ and one of $G_2$. If there are no such multiple edges, then $G_1$ is a factor of $G$.

Assume there are multiple edges in $G$ and let $uv$ be one such multiple edge. Since $d(u|G) = d(u|G_1) + d(u|G_2) \leq n - 1$ and $uv$ is a multiple edge, there is a vertex $w$ not adjacent to $u$ in $G$. Let $V$ be the common vertex set of $G_1$, $G_2$ and $G$. If for every $x \in V - u$, $q[v, x] \geq q[w, x]$, then $d(v|G) - 2 \geq d(w|G)$, contradicting the choice that each $d_i$ is $k$ or $k+1$. Therefore, there is a vertex $x \in V - u$ such that $q[v, x] < q[w, x]$ (Fig. 8.26).

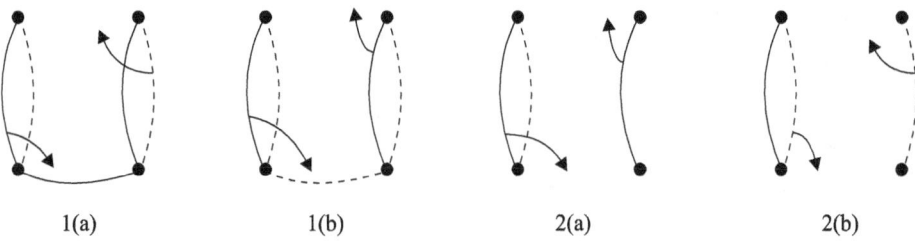

1(a)          1(b)          2(a)          2(b)

**Fig. 8.26**

The following two cases arise:

1. $q[v, x] = 1$ and $q[w, x] = 2$ ,

2. $q[v, x] = 0$ and $q[w, x] = 0$.

In case 1, we have either 1(*a*) $vx \in G_1$, or 1(*b*) $vx \in G_2$, and in case 2, either 2(a) $wx \in G_1$, or 2(*b*) $wx \in G_2$. For all these cases, make an EDT switching the pair of edges $vu$, $wx$ to the positions $vx$, $uw$. In 1(a) and 2(b), this is done in $G_2$ to obtain a graph $G_2'$ degree-equivalent to $G_2$ and with less number of multiple edges in $G_1 \cup G_2'$. In 1(b) and 2(a), this is done in $G_1$ to get a graph $G_1'$ degree-equivalent to $G_1$ and with less multiple edges in $G_1' \cup G_2$. This contradicts the choice of $G_1$ and $G_2$.                                                          ❏

**Remark**    If $d_i$'s and $k_i$'s, being allowed to be arbitrary with $[d_i]_1^n$, $[d_i - k_i]_1^n$, are graphical, and so $[k_i]_1^n$ is graphical, then $[d_i]_1^n$ need not be realised with a $k$-factor. To see this, consider the following example of Lovasz [149].

Let $d = [8, 8, 3, 2, 2, 2, 2, 2, 3]$ and $k = [4, 4, 0, 0, 0, 1, 2, 2, 3]$ so that $d - k = [4, 4, 3, 2, 2, 1, 0, 0, 0]$. Then $d$ and $d - k$ are graphical, realised by graphs $G_1$ and $G_2$ of Figure 8.27, and evidently $k$ is same as $d - k$ with a different labeling. Now, it is easy to see that any realisation of $k = d - k$ needs an edge joining the two vertices of degree 4 in it. But if this $uv$ is an edge in $G_2$, it will not be an edge in $G_1 - G_2$, so that there is no realisation of $d$ with a $k$-factor.

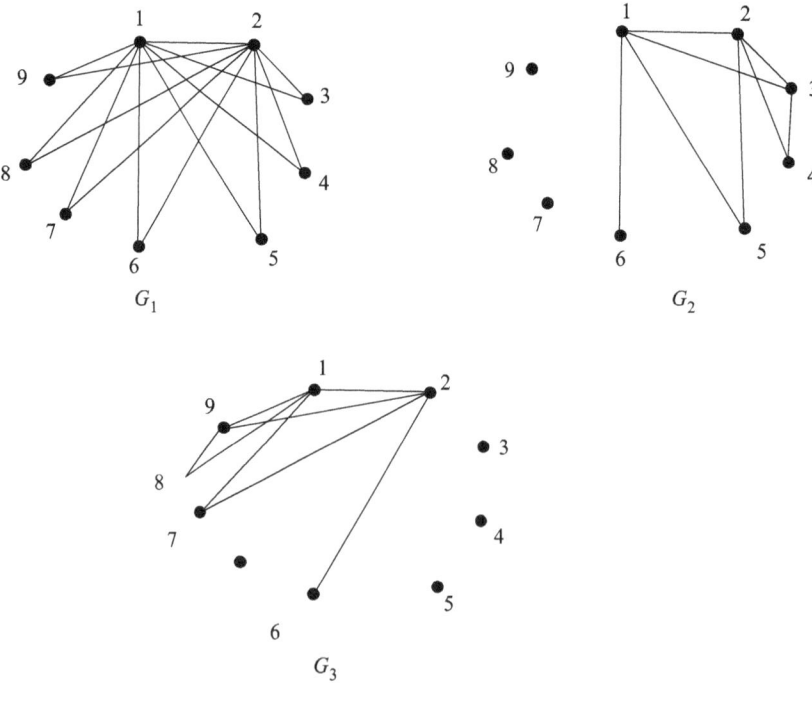

**Fig. 8.27**

# 8.6 $(g, f)$ and $[a, b]$-factors

In this section, we give a brief description of $(g, f)$ and $[a, b]$ factors in a graph.

**Definition:** If $f$ and $g$ are two vertex functions of a graph $G$ such that $g(v) \leq f(v)$ for all $v \in V$ and $F$ is a factor of $G$ such that $g(v) \leq d(v|F) \leq f(v)$ for all $v \in V$, then $F$ is called a $(g, f)$-*factor* of $G$.

If $H$ is a $(g, f)$-factor of $G$, its $f$-*deficiency* is defined as $\Delta_H(G, f) = f(V) - d(H)$, where $d(H) = \sum_{v \in V} d(v|H)$. Clearly, $\Delta_H(G, f) \equiv f(V)(\mathrm{mod}\, 2)$.

If $D = (S, T, U)$ is a decomposition of $G$, then $k''(D, f, g)$ denotes the number of odd components of $U$ with respect to $D$ and $f$ for which $f(c) = g(c)$, for every vertex $c$ of the component.

We now state the $(g, f)$-factor Theorem of Lovasz [148].

**Theorem 8.25 (Lovasz [148])** If $f$ and $g$ are vertex functions of a graph $G = (V, E)$ such that $0 \leq g(v) \leq f(v) \leq d(v/G)$ for all $v \in V$, then either $G$ has a $(g, f)$-factor or it has a decomposition $D = (S, T, U)$ satisfying $k''(D, f, g) > f(S) + d(T) - g(T) - q[S, T]$, but not both, where $k''(D, f, g)$ is the number of odd components of $< U >_G$ with $f(v) = g(v)$ for every vertex $v$ of the component.

Next, we state the $(g, f)$-factor Theorem due to Las Vergnas [255].

**Theorem 8.26 (Las Vergnas [255])** Let $g, f$ be vertex functions for a graph $G = (V, E)$ such that $0 \leq g(v) \leq 1 \leq f(v) \leq d(v|G)$ for all $v \in V$. For $S \subseteq V$, let $k'(G - S)$ be the number of odd components $C$ of $G - S$ which are such that either $C = \{v\}$ and $g(v) = 1$ or $< V(C) >$ is odd and at least 3, with $g(v) = f(v) = 1$ for all $v \in V(C)$. Then $G$ has a $(g, f)$-factor if and only if $f(S) \geq k'(G - S)$ for all $S \subseteq V$.

The following variation of the $(g, f)$-factor Theorem is due to Kano and Saito [124].

**Theorem 8.27 (Kano and Saito [124])** If $g$ and $f$ are vertex functions of a graph $G = (V, E)$ and $\theta$ a real number such that $0 \leq \theta \leq 1$, and if $g(v) < f(v)$, $g(v) \leq \theta d(v/G) \leq f(v)$ for all $v \in V$, then $G$ has a $(g, f)$-factor.

**Definition:** If $a$ and $b$ are non-negative integers with $a \leq b$ and $F$ is a factor of $G$ such that $a \leq d(v|F) \leq b$ for all $v \in V$, then $F$ is called an $[a, b]$-*factor* of $G$. We note that an $[a, b]$-factor is a $(g, f)$-factor for constant vertex functions $g(v) = a$ for all $v \in V$ and $f(v) = b$ for all $v \in V$.

The following $[a, b]$-factor Theorem can easily be derived from the $(g, f)$-factor Theorem.

**Theorem 8.28** If $G = (V, E)$ is a graph, and $a$ and $b$ are distinct integers with $a < b$, then either $G$ has an $[a, b]$-factor $H$ or a decomposition $D = (S, T, U)$ such that $q[S, T] > d(T) + b|S| - a|T|$, but not both.

The following result is due to Las Vergnas [255].

**Theorem 8.29 (Las Vergnas [255])**   If $G$ is a graph and $k(\geq 2)$ an integer, then $G$ has a $[1, k]$-factor if and oly if $k|N(S)| \geq |S|$, for all independent sets $S$ of $G$.

The next result can be found in Kano and Saito [125].

**Theorem 8.30 (Kano and Saito [125])**   If $k$, $r$, $s$ and $t$ are integers such that $0 \leq k \leq r$, $0 \leq s$, $1 \leq t$ and $ks \leq rt$, then every $[r, r+s]$-graph has a $[k, k+1]$-factor.

The following result on the existence of $[a, b]$-factors is given by Kouider and Long [139].

**Theorem 8.31 (Kouider and Long [139])**   Let $b \geq a+1$ and let $G$ be a graph with minimum degree 8.

$$\text{If} \quad a(G) \leq \begin{cases} \dfrac{4b(\delta - a + 1)}{(a+1)^2}, & \text{for } a \text{ odd} \\ \dfrac{4b(\delta - a + 1)}{a(a+2)}, & \text{for } a \text{ odd} \end{cases}$$

then $G$ has an $[a, b]$-factor.

## 8.7   Exercises

1. Prove that a tree can have at most one perfect matching.

2. Give an example of a cubic graph having no 1-factor.

3. Show that $K_{n,n}$ and $K_{2n}$ are 1-factorable, and further show that the number of 1-factors of $K_{n,n}$ and $K_{2n}$ are respectively $n!$ and $\dfrac{(2n)!}{2^n n!}$.

4. Prove that the Peterson graph is not 1-factorable.

5. Let $G = (V_1, V_2, E)$ be a bipartite graph with $|V_1| = n_1$ and $|V_2| = n_2$, and the vertices $x_i \in V_1$ and $y_j \in V_2$ indexed such that $d(x_1) \leq d(x_2) \leq \ldots \leq d(x_n)$ and $d(y_1) \leq d(y_2) \leq \ldots \leq d(y_n)$. Then show that a sufficient condition for $V_1$ to be matched into $V_2$ is that $n_1 \leq n_2$, $d(x_1) > 0$ and $\sum\limits_{i=1}^{k} d(x_i) > \sum\limits_{i=1}^{k-1} d(y_j)$, $2 \leq k \leq n_1$.

6. Prove that a bipartite graph $G = (V_1, V_2, E)$ has a matching saturating all the vertices of maximum degree.

7. Show that the following statements are equivalent for a bipartite graph $G = (V_1, V_2, E)$.

   a. $G$ is connected and each edge of $G$ is contained in a 1-factor.

b. $|V_1| = |V_2|$ and for each non-empty subset $S$ of $V_1$, $|S| < |N(S)|$.

c. For each $u \in V_1$ and $v \in V_2$, $G - u - v$ has a 1-factor.

8. Show that a tree $T$ has a perfect matching if and only if $k_0(T - v) = 1$ for every vertex $v$ of $T$.

9. If $G$ is a cubic graph without cut edges or with all cut edges on a single suspended path, then prove that $G$ has a 1-factor.

10. Prove directly that if $G$ is a connected graph with no induced subgraph isomorphic to $K_{1, 3}$, then $G$ has an edge $uv$ such that $G - \{u, v\}$ is connected.

11. If $M$ and $N$ are matchings in a graph $G$ and $|M| > |N|$, then prove that there exist matchings $M'$ and $N'$ in $G$ such that $|M'| = |M| - 1$, $|N'| = |N| + 1$ and $M'$ and $N'$ have the same union and intersection (as edge sets) as $M$ and $N$.

12. Prove that deleting any perfect matching from the Peterson graph leaves the subgraph $C_5 + C_5$.

13. Determine the minimum size of a maximal matching in the cycle $C_n$.

14. If $G = (V_1, V_2, E)$ is a bipartite graph that has a matching of $V_1$ into $V_2$, then prove that $G$ has at most $\binom{|V_1|}{2}$ edges belonging to no matching of $V_1$ and $V_2$.

15. Describe complete tripartite graphs of the form $K_{n, n, n}$ that have perfect matching.

16. For what values of $n \geq 4$ does the wheel $W_n$ have a perfect matching?

17. Prove that a graph has an $f$-factor if and only if it has an $f'$-factor, where $f'(v) = d(v) - f(v)$ for every $v \in V$.

18. Show that a $k$-regular, $(k-1)$-edge connected graph of even order has a 1-factor.

19. If $G$ is a connected graph of even order, then show that $G^2$ has a 1-factor.

20. If $G$ is a connected cubic graph without a 1-factor and $S$ is a minimal antifactor set, then prove that $S$ is an independent set of claw centers and $G - S$ has exactly $|S| + 2$ odd components and no even components.

# 9. Edge Graphs and Eccentricity Sequences

Many authors discovered edge graphs independently and gave it a different name, for example, interchange graph by Ore [177], derivative by Sabidussi [231], derived graphs by Beineke [18], edge-to-vertex dual by Seshu and Reed [233], covering graph by Kasteleyn [127] and adjoint by Menon [159].

## 9.1 Edge Graphs

**Definition:** Let $G = (V, E)$ be a graph with $V = \{v_1, v_2, \ldots, v_n\}$ and $E = \{e_1, e_2, \ldots, e_m\}$. The *edge graph* $L(G)$ of $G$ has the vertex set $E$ and two vertices $e_i$ and $e_j$ are adjacent in $L(G)$ if and only if the corresponding edges $e_i$ and $e_j$ of $G$ are adjacent in $G$. For example, in Figure 9.1, $(L(G)$ is the edge graph of G. A graph $G$ is an edge graph if it is isomorphic to the edge graph $L(H)$ of some graph $H$.

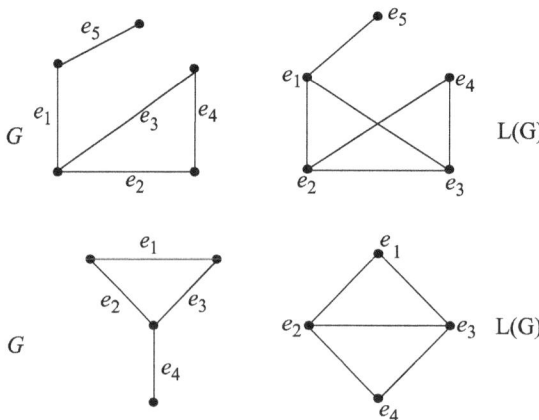

**Fig. 9.1**

Since isolated vertices do not contribute anything to the study of edge graphs, we assume that the graphs contain no isolated vertices. Besides this, the graphs under consideration will be without loops. We have the following observations about edge graphs.

1. A graph $G$ is connected if and only if $L(G)$ is connected.

2. If $H$ is a subgraph of $G$, then $L(H)$ is a subgraph of $L(G)$.

3. The edges incident at a vertex of $G$ form a maximal complete subgraph of $L(G)$.

4. In $G$, if $e = uv$ is an edge, then the degree of $e$ in $L(G)$ is the number of edges of $G$ adjacent to $e$ in $G$. Clearly, $d_{L(G)}(e) = d_G(u) + d_G(v) - 2$.

5. For $n > 1$, $L^n(G) = L(L^{n-1}(G))$ and $L^0(G) = G$.

The following result determines the number of edges in an edge graph.

**Theorem 9.1** The number of edges $m'$ in $L(G)$ when $G$ has degree sequence $[d_i]_1^n$ is given by

$$m' = \frac{1}{2} \left( \sum_{i=1}^{n} d_i^2 \right) - m.$$

**Proof** Let $[d_i]_1^n$ be the degree sequence of the graph $G$ and let $L(G)$, the edge graph of $G$, have $m'$ edges.

As the degree of the vertex $v_i$ in $G$ is $d_i$, there are $d_i$ edges incident on $v_i$. From these $d_i$ edges, any two are adjacent at $v_i$ in $G$. Hence the number of edges contributed by $v_i$ to $L(G)$ is $\binom{d_i}{2}$.

Thus,
$$m' = \sum_{i=1}^{n} \binom{d_i}{2} = \sum_{i=1}^{n} \frac{d_i(d_i-1)}{2} = \frac{1}{2} \sum_{i=1}^{n} (d_i^2 - d_i)$$

$$= \frac{1}{2} \sum_{i=1}^{n} d_i^2 - \frac{1}{2} \sum_{i=1}^{n} d_i$$

$$= \frac{1}{2} \left( \sum_{i=1}^{n} d_i^2 \right) - m. \qquad \square$$

The following observation is immediate.

**Theorem 9.2** The edge graph of a graph $G$ is a path if and only if $G$ is a path.

**Proof** Let $G$ be a graph with $n$ vertices. Assume that $G$ is a path $P_n$. Then $L(G)$ is the path $P_{n-1}$ with $n-1$ vertices.

Conversely, let $L(G)$ be a path. Then no vertex of $G$ has degree greater than two. For, if $G$ has a vertex $v$ of degree greater than two, the edges incident to $v$ form a complete subgraph of $L(G)$ with at least three vertices. Therefore, $G$ is either a cycle or a path. But $G$ cannot be a cycle, since the edge graph of a cycle is a cycle.                                                              ❏

We now have the following stronger result.

**Theorem 9.3**    A connected graph is isomorphic to its edge graph if and only if it is a cycle.

**Proof**    Let $G$ be a connected graph with $n$ vertices, $m$ edges and with degree sequence $[d_i]_1^n$. Let $L(G)$ be the edge graph of $G$. The number of vertices in $L(G)$ is $m$. The number of edges $m'$ in $L(G)$ is given by

$$m' = \frac{1}{2}\left(\sum_{i=1}^{n} d_i^2\right) - m.$$

Clearly, $L(G)$ is connected and $L(C_n) = C_n$.

Conversely, let $G \cong L(G)$.

Then $G$ and $L(G)$ have the same number of vertices and edges.

So, $n = m$ and $m = \frac{1}{2}\left(\sum_{i=1}^{n} d_i^2\right) - m.$

Therefore, $n = m$, and $\sum_{i=1}^{n} d_i^2 = 4m.$

Thus, variance

$$\{[d_i]\} = \frac{1}{n}\sum_{i=1}^{n} d_i^2 - \left(\frac{1}{n}\sum_{i=1}^{n} d_i\right)^2$$

$$\left[\text{Because } Var = \frac{1}{N}\sum_i f_i x_i^2 - \left(\frac{1}{N}\sum_i f_i x_i\right)^2 \text{ and we have } f_i = 1\right]$$

$$= \frac{1}{n}4m - \frac{1}{n^2}(2m)^2 = \frac{4m}{m} - \frac{4m^2}{m^2} = 4 - 4 = 0.$$

Therefore, the $d_i$'s are equal and $G$ is regular of degree, $d$, say.

So $nd = 2m$ implies that $d = \dfrac{2m}{n} = \dfrac{2m}{m} = 2$.

Thus $G$ is a 2-regular connected graph, that is, $C_n$.                                         ❏

The next result is about the isomorphism of edge graphs.

**Theorem 9.4**   If the graphs $G_1$ and $G_2$ are isomorphic, then $L(G_1)$ and $L(G_2)$ are also isomorphic.

**Proof**   Assume $(\phi, \theta)$ to be an isomorphism of $G_1$ onto $G_2$. Then $\theta$ is a bijection of $E(G_1)$ onto $E(G_2)$. We show that $\theta$ is an isomorphism of $L(G_1)$ to $L(G_2)$ by showing that $\theta$ preserves adjacency. Let $e_i$ and $e_j$ be two adjacent vertices of $L(G_1)$. So there exists a vertex $v$ of $G_1$ incident to both $e_i$ and $e_j$, and therefore $\phi(v)$ is a vertex incident to both $\theta(e_i)$ and $\theta(e_j)$. Thus, $\theta(e_i)$ and $\theta(e_j)$ are adjacent vertices in $L(G_2)$.

Let $\theta(e_i)$ and $\theta(e_j)$ be adjacent vertices in $L(G_2)$. Then they are adjacent edges in $G_2$ and therefore there exists a vertex $v'$ of $G_2$ incident to both $\theta(e_i)$ and $\theta(e_j)$. Then $\phi^{-1}(v')$ is a vertex of $G_1$ incident to both $e_i$ and $e_j$, and thus $e_i$ and $e_j$ adjacent vertices of $L(G_1)$.

Therefore $e_i$ and $e_j$ are adjacent vertices of $L(G_1)$ if and only if $\theta(e_i)$ and $\theta(e_j)$ are adjacent vertices of $L(G_2)$. Hence $\theta$ is an isomorphism of $L(G_1)$ onto $L(G_2)$.                 ❏

The converse of Theorem 9.4 is not true. To see this, consider the graphs $K_{1,3}$ and $K_3$ whose edge graphs are $K_3$. But $K_{1,3}$ is not isomorphic to $K_3$, since there is a vertex of degree three in $K_{1,3}$ while there is no such vertex in $K_3$. However, it was shown by Whitney [265] that the converse holds unless one is $K_{1,3}$ and the other is $K_3$. The proof of this result is due to Jung [123].

**Theorem 9.5**   Let $G$ and $G'$ be connected graphs with isomorphic edge graphs. Then $G$ and $G'$ are isomorphic unless one is $K_3$ and the other is $K_{1,3}$.

**Proof**   First suppose that $n(G)$ and $n(G')$ are less than or equal to 4. A necessary condition for $L(G)$ and $L(G')$ to be isomorphic is that $m(G) = m(G')$. The only nonisomorphic connected graphs on at most four vertices are those shown in Figure 9.2.

In Figure 9.2, graphs $G_4$, $G_5$ and $G_6$ are the three graphs having three edges each. We have already seen that $G_4$ and $G_6$ have isomorphic edge graphs, namely $K_3$. The edge graph of $G_5$ is a path of length 2 and hence $L(G_5)$ cannot be isomorphic to $L(G_4)$ or $L(G_6)$. Further, $G_7$ and $G_8$ are the only two graphs in the list having four edges each.

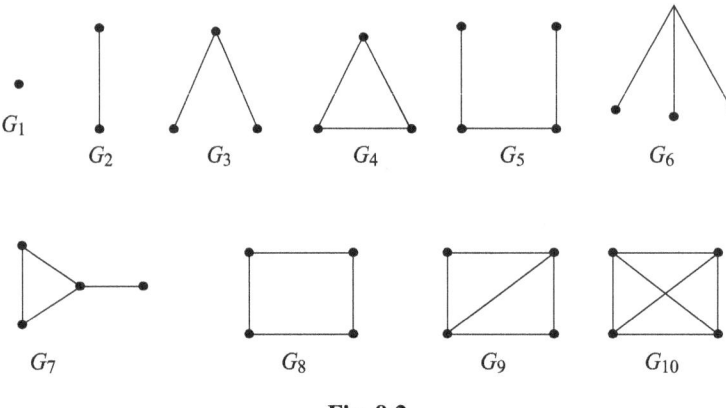

**Fig. 9.2**

Now, $L(G_8) \cong G_8$ and $L(G_7)$ is isomorphic to $G_9$. Thus, the edge graphs of $G_7$ and $G_8$ are not isomorphic. No two of the remaining graphs have the same number of edges. Hence the only non-isomorphic graphs with at most four vertices having isomorphic edge graphs are $G_4$ and $G_6$.

We now suppose that either $G$ or $G'$, say $G$, has at least five vertices and that $L(G)$ and $L(G')$ are isomorphic under an isomorphism $\phi_1$. So $\phi_1$ is a bijection from the edge set of $G$ onto the edge set of $G'$.

We now prove that $\phi_1$ transforms an induced $K_{1,3}$ subgraph of $G$ onto a $K_{1,3}$ subgraph of $G'$. Let $e_1 = uv_1$, $e_2 = uv_2$ and $e_3 = uv_3$ be the edges of an induced $K_{1,3}$ subgraph of $G$. As $G$ has at least five vertices and is connected, there exists an edge $e$ adjacent to only one or all three of edges $e_1$, $e_2$ and $e_3$, as illustrated in Figure 9.3.

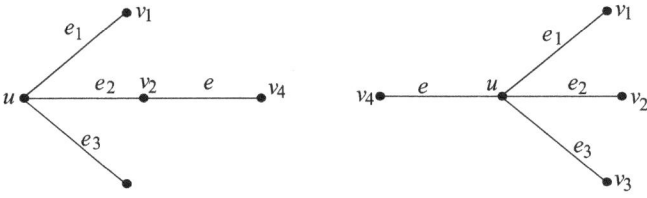

**Fig. 9.3**

Now, $\phi_1(e_1)$, $\phi_1(e_2)$ and $\phi_1(e_3)$ form either a $K_{1,3}$ subgraph or a triangle in $G'$. If $\phi_1(e_1)$, $\phi_1(e_2)$ and $\phi_1(e_3)$ form a triangle in $G'$, $\phi_1(e)$ can be adjacent to precisely two of $\phi_1(e_1)$, $\phi_1(e_2)$ and $\phi_1(e_3)$ (since $L(G')$ is simple), whereas $\phi_1(e)$ must be adjacent to only one or all the three. This contradiction shows that $\{\phi_1(e_1), \phi_1(e_2), \phi_1(e_3)\}$ is not a triangle in $G'$ and therefore forms a $K_{1,3}$ in $G'$.

It is clear that a similar result holds as well for $\phi_1^{-1}$, since it is an isomorphism on $L(G')$ onto $L(G)$.

Let $S(u)$ denote the star subgraph of $G$ formed by the edges of $G$ incident at a vertex $u$ of $G$. We shall prove that $\phi_1$ maps $S(u)$ onto the unique star subgraph $S(u')$ of $G'$.

i. First suppose that the degree of $u$ is at least 2. Let $f_1$ and $f_2$ be any two edges incident at $u$. The edges $\phi_1(f_1)$ and $\phi_1(f_2)$ of $G'$ have an end vertex $u'$ in common. If $f$ is any other edge of $G$ incident with $u$, then $\phi_1(f)$ is incident with $u'$, and conversely, for every edge $f'$ of $G'$ incident with $u'$, $\phi_1^{-1}(f')$ is incident with $u$. Thus, $S(u)$ in $G$ is mapped to $S(u')$ in $G'$.

ii. Let the degree of $u$ be 1 and $e = uv$ be the unique edge incident with $u$. As $G$ is connected and $n(G) \geq 5$, degree of $v$ must be at least 2 in $G$, and therefore, by (i), $S(v)$ is mapped to a star $S(v')$ in $G'$. Also $\phi_1(uv) = u'v'$ for some $u' \in V(G')$. Now, if the degree of $u'$ is greater than 1, by (i), the star at $u'$ in $G'$ is transformed by $\phi_1^{-1}$ either to the star at $u$ in $G$ or to the star at $v$ in $G$. But as the star at $v$ in $G$ is mapped to the star at $v'$ in $G'$ by $\phi_1$, $\phi_1^{-1}$ should map the star at $u'$ to the star at $u$ only. As $\phi_1^{-1}$ is $1-1$, this means that $\deg u \geq 2$, a contradiction. Therefore, $\deg u' = 1$ and so $S(u)$ in $G$ is mapped to $S(u')$ in $G'$.

We now define $\phi : V(G) \to V(G')$ by setting $\phi(u) = u'$ if $\phi_1(S(u)) = S(u')$. Since $S(u) = S(v)$ only when $u = v$ ($G \neq K_2$, $G' \neq K_2$), $\phi$ is $1-1$. $\phi$ is also onto since, for $v'$ in $G'$, $\phi_1^{-1}(S(v')) = S(v)$ for some $v \in V(G)$, and by the definition of $\phi$, $\phi(v) = v'$. Finally, if $uv$ is an edge of $G$, then $\phi_1(uv)$ belongs to both $S(u')$ and $S(v')$, where $\phi_1(S(u)) = S(u')$ and $\phi_1(S(v)) = S(v')$. This means that $u'v'$ is an edge of $G'$. But $u' = \phi(u)$ and $v' = \phi(v)$. Consequently, $\phi(u)\phi(v)$ is an edge of $G'$. If $u$ and $v$ are nonadjacent in $G$, $\phi(u)\phi(v)$ must be nonadjacent in $G'$. Otherwise, $\phi(u)\phi(v)$ belongs to both $S(\phi(u))$ and $S(\phi(v))$ and hence $\phi_1^{-1}(\phi(u)\phi(v)) = uv \in E(G)$, a contradiction. Thus $G$ and $G'$ are isomorphic under $\phi$. ❏

The following result shows that $K_{1,3}$ is not an edge graph and thus $K_{1,3}$ is of great significance in studying edge graphs as will be seen in further discussions.

**Lemma 9.1**  The star $K_{1,3}$ is not an edge graph.

**Proof**  Assume that $K_{1,3}$ is an edge graph. Then, $K_{1,3} = L(H)$ for some graph $H$. Since $K_{1,3}$ has four vertices, therefore $H$ has four edges. Also $H$ is connected. All the connected graphs with four edges are given in Figure 9.4.

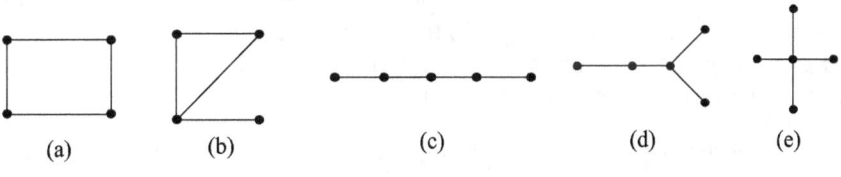

(a)        (b)        (c)        (d)        (e)

**Fig. 9.4**

$H$ is neither graph (a) nor (b), because $L(C_4) = C_4$ and $L(K_{1,3} + x) = K_4 - x$. These are shown in Figure 9.5.

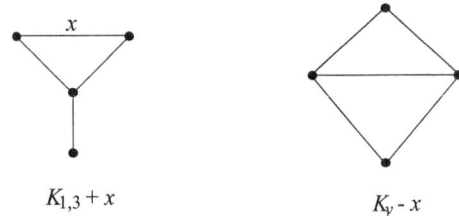

$K_{1,3} + x$       $K_y - x$

**Fig. 9.5**

Thus, $H$ is one of the three trees as given in (c), (d) and (e). But the edge graphs of these trees are the path $P_4$, the graph $K_3$, $K_2$ and $K_4$, given in Figure 9.6.

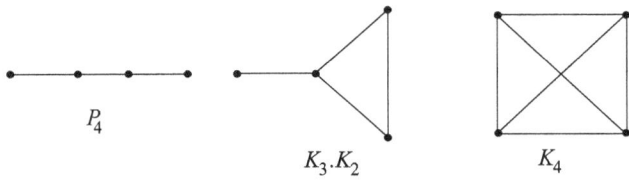

$P_4$       $K_3.K_2$       $K_4$

**Fig. 9.6**

This shows that $H$ is none of the trees (c), (d) or (e). Hence it follows that $K_{1,3}$ is not an edge graph.       ❑

Now, we proceed to give a characterisation of edge graphs which is due to Krausz [140].

**Theorem 9.6 (Krausz [140])** A graph $G$ is the edge graph of some graph if and only if the edges of $G$ can be partitioned into cliques such that no vertex appears in more than two cliques.

### Proof

*Necessity* Let $G$ be an edge graph of a graph $H$. Assume without loss of generality that $H$ has no isolated vertices. Then the edges in the star $K_{1,3}$ at each vertex of $H$ induce a clique of $G$ and every edge of $H$ belongs to the stars of exactly two vertices of $H$, therefore no vertex of $G$ is in more than two of these cliques.

*Sufficiency* Let the edges of $G$ be partitioned into the cliques $S_1$, $S_2$, ..., $S_k$ such that no vertex of $G$ belongs to more than two of these cliques. We construct $H$ such that $L(H) = G$. As isolated vertices of $G$ become isolated edges of $H$, therefore, assume that $\delta(G) \geq 1$. Let $v_1$, $v_2$, ..., $v_\ell$ be the vertices of $G$ (if any) that appear in exactly one of $S_i$. The vertices of $H$ correspond to the set $S = \{S_1, S_2, ..., S_k, \{v_1\}, \{v_2\}, ..., \{v_\ell\}\}$, with one vertex for each member of $S$. Any two of these vertices are adjacent whenever their corresponding sets intersect. Each vertex of $G$ appears in exactly two sets in $S$ and no two vertices appear in the same pair of sets. Thus, $H$ is a simple graph with one edge for each vertex of $G$.

If vertices are adjacent in $G$, then they appear together in some $S_i$ and the corresponding edges of $H$ share the vertex corresponding to $S_i$. Hence, $G = L(H)$.                                       ❑

Krausz characterisation of edge graphs is close to the definition. Since it characterises edge graphs by the existence of a special edge partition, it does not directly give an efficient test. This has been improved by Van Rooij and Wilf [227] by describing the structural criterion for a graph to be an edge graph. Before we take the Van Rooij and Wilf characterisation, we have the following definition.

**Definition:**   An *induced subgraph* is a subgraph which is maximal on its vertex set. A triangle $T$ of a graph $G$ is said to be *odd*, if there is a vertex of $G$ adjacent to an odd number of vertices of $T$, otherwise $T$ is said to be *even*. That is, $T$ is odd, if $|V(T) \cap N(v)|$ is odd, for some $v \in V(G)$, and $T$ is even if $|V(T) \cap N(v)|$ is even, for every $v \in V(G)$. An induced copy of $K_4 - e$ is a double triangle and clearly has two triangles with a common edge.

We now present Van Rooij and Wilf characterisation.

**Theorem 9.7**   A graph $G$ is the edge graph of some graph if and only if $G$ does not contain an induced subgraph $K_{1,3}$ and no double triangle of $G$ has two odd triangles.

### Proof

*Necessity*   Let $G = L(H)$. Clearly, $G$ does not contain an induced subgraph $K_{1,3}$, since $K_{1,3}$ itself is not an edge graph. Now we observe that the vertices of a double triangle in $G$ correspond to the edges of a $K_{1,3} + e$ in $H$. In particular, one of these double triangles in $G$ is generated by a triangle in $H$. Obviously a triangle in $G$ generated by a triangle in $H$ is even, since an edge incident to a triangle in $H$ intersects exactly two edges of the triangle in $H$ (Fig. 9.7).

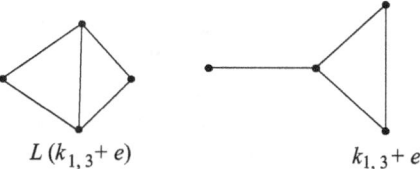

$L(k_{1,3} + e)$                              $k_{1,3} + e$

**Fig. 9.7**

*Sufficiency*   Let $G$ not contain an induced subgraph $K_{1,3}$ and let no double triangle of $G$ have two odd triangles. Assume that $G$ is connected, for otherwise, we apply the construction to each component. In case $G$ is $K_{1,3}$−free and has a double triangle with both triangles even, then $G$ is one of the graphs given in Figure 9.8.

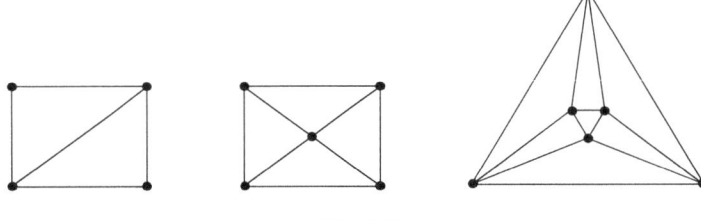

**Fig. 9.8**

Thus, we consider the case when every double triangle of $G$ has exactly one odd triangle. To prove the result, it suffices by Theorem 9.6, to partition $E(G)$ into cliques that cover each vertex at most twice. Now, let $S_1, S_2, \ldots, S_k$ be the maximal cliques of $G$ that are not even triangles and let $T_1, T_2, \ldots, T_\ell$ be the edges that belong to one even triangle and no odd triangle. We claim that $B = \{S_i\} \cup \{T_j\}$ partitions $E(G)$ into cliques using each vertex at most twice.

Now, every edge appears in a maximal clique, but every triangle in a clique with more than three vertices is odd. Therefore $T_j$ is not in any clique $S_i$. Also $S_i$ and $S_{i'}$ have no common edge, because $G$ has no double triangles with both triangles odd. Thus the cliques in $B$ are edge-disjoint. If $e \in E(G)$, then $e$ belongs to some $S_i$ unless the only maximal clique containing $e$ is an even triangle. In this case $e$ is a $T_j$, since the double triangles do not have both triangles even (Fig. 9.9).

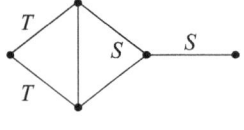

**Fig. 9.9**

We now show that each $v \in G$ appears at most twice in $B$. Assume that $v$ belongs to $B_1$, $B_2$, $B_3 \in B$. Edge-disjointness implies that $v$ has neighbours $x$, $y$, $z$ with each belonging to only one of $B_1$, $B_2$, $B_3$. Since $G$ has no induced $K_{1,3}$, assume that $xy$ is an edge. Now by edge-disjointness, the triangle $vxy$ does not belong to a member of $B$. Therefore $vxy$ is an even triangle. Thus $z$ has exactly one other edge to $vxy$, say $zx$, while $zy$ is not an edge. But now the same argument shows that $zvx$ is an even triangle and we have a double triangle with both triangles even. This contradicts our supposition and hence each $v \in G$ appears at most twice in $B$.                                                                                          $\square$

The next characterisation due to Beineke [149] displays those subgraphs which are not present in edge graphs. These subgraphs other than $K_{1,3}$ are vertex-minimal $K_{1,3}$-free graphs containing a double triangle with both triangles odd. Each such graph has a double triangle and one or two additional vertices that make the triangles odd by having one or three neighbours in the triangle.

**Theorem 9.8**    A graph $G$ is an edge graph of some graph if and only if $G$ does not contain an induced subgraph of any one of the graphs in Figure 9.10.

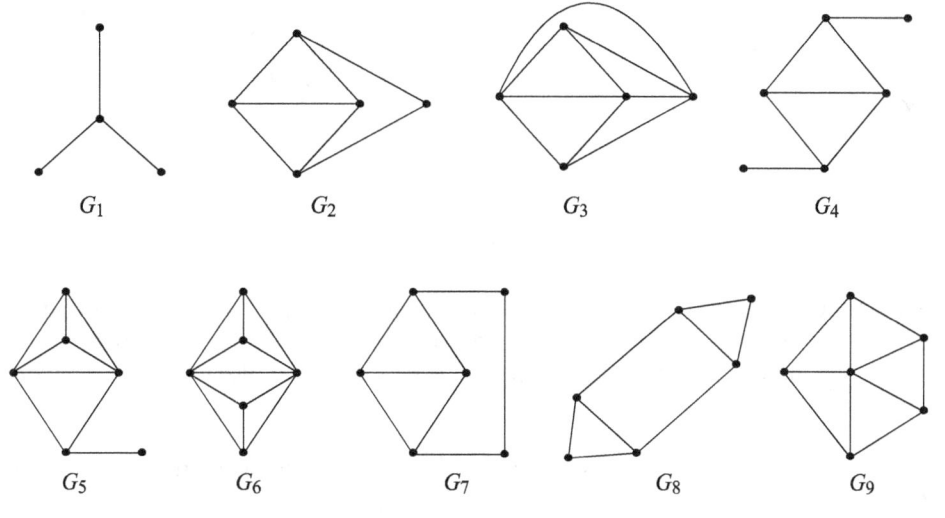

**Fig. 9.10**

## Proof

*Necessity*   Let $G$ be the edge graph of some graph $H$ so that $G = L(H)$. Then by Theorem 9.6, the edges of $G$ can be partitioned into cliques such that every vertex appears in at most two cliques. We observe that none of these nine graphs have such a partition. Since every induced subgraph of an edge graph is itself an edge graph, $G$ does not contain an induced subgraph of any one of the nine graphs, in Figure 9.10.

*Sufficiency*   Let $G$ not contain an induced subgraph of any one of these nine graphs. We prove that no double triangle of $G$ has two odd triangles. Assume to the contrary that $G$ has a double triangle both of which are odd. Let these triangles be *abc* and *abd* with *c* and *d* non adjacent. We discuss two cases, one in which there is a vertex *v* adjacent to an odd number of vertices of both odd triangles and second when there is no such vertex.

**Case 1**   Assume there is a vertex *v* which is adjacent to an odd number of vertices in each of the triangles *abc* and *abd*. Now two possibilities arise; either *v* is adjacent to exactly one vertex of each of these triangles, or it is adjacent to more than one vertex of one of them. If *v* is adjacent to exactly one vertex of each of these triangles, then either *v* is adjacent to *a* or *b* giving $G_1$, or to both *c* and *d* giving $G_2$. If *v* is adjacent to more than one vertex of one of the triangles, then *v* is adjacent to all four vertices of the two triangles, giving $G_3$ as an induced subgraph of $G$ (Fig. 9.11).

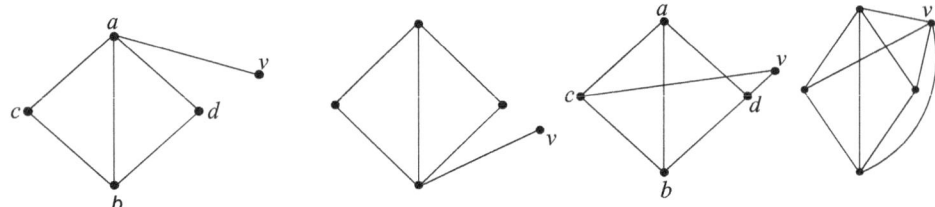

**Fig. 9.11**

**Case 2**    Now, let there be no vertex adjacent to an odd number of vertices of both triangles. Assume that the vertex $u$ is adjacent to an odd number of vertices of triangle $abc$ and the vertex $v$ is adjacent to an odd number of vertices of triangle $abd$. We consider three subcases.

*Case 2.1*    $u$ is adjacent to exactly one vertex of $abc$ and $v$ is adjacent to exactly one vertex of $abd$.

*Case 2.2*    One of $u$ or $v$ is adjacent to all three vertices of its triangle and the other to only one vertex of its triangle.

*Case 2.3*    $u$ is adjacent to all three vertices of $abc$ and $v$ is adjacent to all three vertices of $abd$.

We observe that if $u$ or $v$ is adjacent to $a$ or $b$, then it is also adjacent to $c$ or to $d$, since otherwise $G_1$ is an induced subgraph. Also, neither $u$ nor $v$ is adjacent to both $c$ and $d$, since otherwise $G_2$ or $G_3$ is induced.

*Case 2.1*    Let $uc, vd \in G$. Then for $uv \in G$, $G_7$ is induced, and for $uv \notin G$, $G_4$ is induced. Now, let $ub, vd \in G$. Then it follows from the above observations that $ud \in G$ while $vc \notin G$. Therefore for $uv \notin G$, the vertices $a, d, u, v$ induce $G_1$ and for $uv \in G$, the vertices $a, b, c, d$, $u, v$ induce $G_8$. Next, let $ub, va \in G$, then clearly $ud, vc \in G$. So, when $uv \notin G$, $G_8$ is induced, and when $uv \in G$, $G_2$ is induced. Finally, let $ub, vb \in G$, then again $ud, vc \in G$. Therefore, when $uv \in G$, $G_9$ is induced, and when $uv \notin G$, $G_1$ is induced (Fig. 9.12, Case 2.1).

*Case 2.2*    Let $ua, ub, uc \in G$. If $ud \in G$, then $G_3$ is induced. Take $ud \notin G$. Then either $vd \in G$ or $vb \in G$. If $vd \in G$, then for $uv \in G$, $G_2$ is induced, and for $uv \notin G$, $G_5$ is induced. If $vb \in G$, then $G_3$ or $G_1$ is induced depending on whether or not $v$ is adjacent to both $c$ and $u$ (Fig. 9.12, Case 2.2).

*Case 2.3*    If $ud, vc$ or $uv \in G$, then $G_3$ is induced. The only other possibility gives $G_6$ (Fig. 9.12, Case 2.3).    ❏

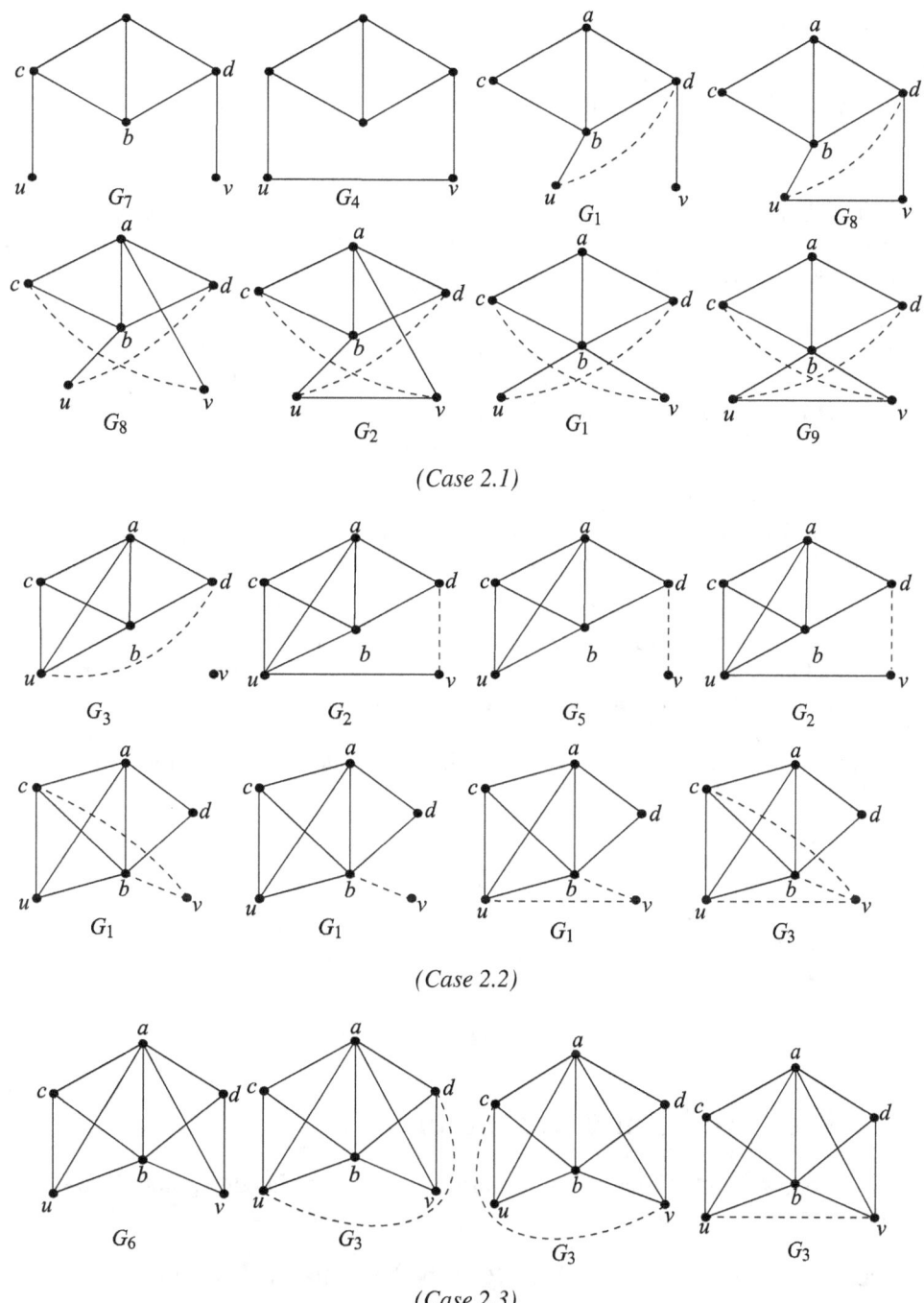

*(Case 2.1)*

*(Case 2.2)*

*(Case 2.3)*

**Fig. 9.12**

The following result due to Chartrand [53] characterises the edge graphs of a tree.

**Theorem 9.9 (Chartrand [53])**    A graph is the edge graph of a tree if and only if it is a connected block graph in which each cut vertex is on exactly two blocks.

### Proof

*Necessity*    Let $T$ be any tree and let $G = L(T)$. Then $G$ is also $B(T)$ since the edges and blocks of a tree coincide. Each cut vertex $w$ of $G$ corresponds to a bridge $uv$ of $T$ and is on exactly those two blocks of $G$ which correspond to the stars at $u$ and $v$.

*Sufficiency*    Let $G$ be a block graph in which each cut vertex is on exactly two blocks. Since each block of a block graph is complete, there exists a graph $H$ such that $L(H) = G$, by Theorem 9.6. If $G = K_3$, we can take $H = K_{1,3}$. If $G$ is any other block graph, then we show that $H$ is a tree. Assume that $H$ is not a tree, so that it contains a cycle. If $H$ is itself a cycle, then by Theorem 9.3, $L(H) = H$, but the only cycle which is a block graph is $K_3$, a case not under consideration. Thus $H$ properly contains a cycle, implying that $H$ has a cycle $Z$ and an edge $e$ adjacent to two edges of $Z$, but not adjacent to some edge $f$ in $Z$. The vertices $e$ and $f$ of $L(H)$ lie on a cycle of $L(H)$ and they are not adjacent. This contradicts the fact that $L(H)$ is a block graph. Hence $H$ is a tree.    ❑

Consider the block graph $G$ of Figure 9.13(a) in which each cut vertex lies on just two blocks. Figure 9.13(b) shows the tree $T$ of which $G$ is the edge graph, is constructed by first forming the block graph $B(G)$ and then adding new vertices for the non-cut vertices of $G$, and the edges joining each block with its non-cut vertices.

The edge graphs of complete graphs were independently characterised by Chang [47] and Hoffman [116, 117], while the edge graphs of complete bipartite graphs were characterised by Moon [163] and Hoffman [118].

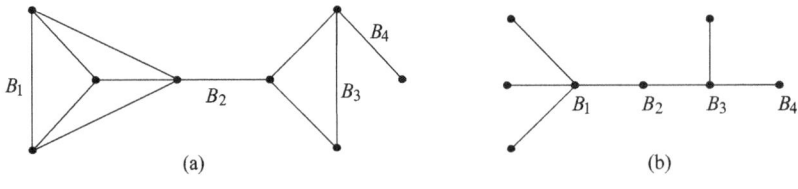

**Fig. 9.13**

## 9.2   Edge Graphs and Traversability

In this section, we study Eulerian and Hamiltonian property in edge graphs. We start with the following result.

**Theorem 9.10**    If $G$ is Eulerian, then $L(G)$ is both Eulerian and Hamiltonian.

**Proof**    Let $G$ be Eulerian and let $\{e_1, e_2, \ldots, e_m\}$ be the edge sequence of an Euler line in $G$. Let the edge $e_i$ in $G$ be represented by the vertex $v_i$ in $L(G)$, $1 \le i \le m$. Then $v_1 v_2 \ldots v_m v_1$

is a Hamiltonian cycle of $L(G)$. Now, if $e = u_i u_j \in E(G)$ and the vertex $v$ in $L(G)$ represents the edge $e$, then $d_{L(G)}(v) = d_G(u_i) + d_G(u_j) - 2$, which is obviously even and greater than or equal to two, since both $d_G(u_i)$ and $d_G(u_j)$ are even (and $\geq 2$). Thus in $L(G)$ every vertex is of even degree ($\geq 2$). Hence $L(G)$ is Eulerian. $\qquad\qquad\qquad \square$

The converse of Theorem 9.11 is not true. To see this, consider the graph $G$ shown in Figure 9.14. Clearly $L(G)$ is both Eulerian and Hamiltonian, but $G$ is not Eulerian.

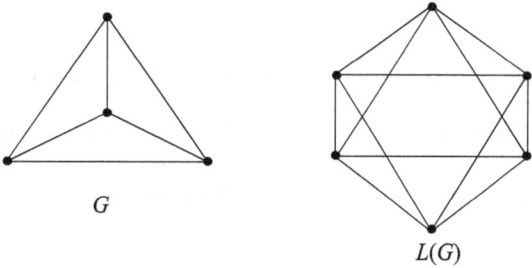

$G$

$L(G)$

**Fig. 9.14**

**Definition:** A *dominating walk* of a graph $G$ is a closed walk $W$ in $G$ (which can be just a single vertex) such that every edge of $G$ not in $W$ is incident with $W$. For example, the walk $v_1\, v_2\, v_3\, v_4$ in the graph of Figure 9.15 is a dominating walk.

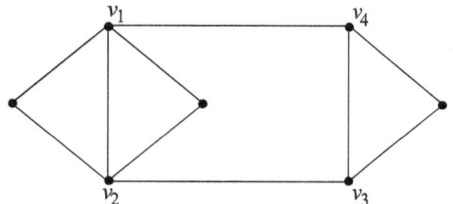

**Fig. 9.15**

The following characterisation of graphs that contain Hamiltonian edge graphs is due to Harary and Nash-Williams [108].

**Theorem 9.11 (Harary and Nash-Williams [108])** The edge graph of a graph $G$ with at least three edges is Hamiltonian if and only if $G$ has a dominating walk.

**Proof** Let $W$ be a dominating walk of $G$ which is represented by the edge sequence $\{e_1, e_2, \ldots, e_k\}$. Let $e_1$ and $e_2$ be incident at $v_1$. Replace the subsequence $\{e_1, e_2\}$ by the sequence $\{e_1, e_{11}, e_{12}, \ldots, e_{1r_1}, e_2\}$, where $e_{11}, e_{12}, \ldots, e_{1r_1}$ are the edges other than $e_1$ and $e_2$ incident at $v_1$. Continuing this process for all subsequences $\{e_i, e_{i+1}\}$, $1 \leq i \leq k$ with $e_{k+1} = e_1$, we obtain a sequence of edges $e_1 e_{11} e_{12} \ldots e_{1r_1}\, e_2 e_{21} e_{22} \ldots e_{2r_2}\, e_3 \ldots e_k e_{k1} e_{k2} \ldots e_{kr_k} e_1$. This clearly gives the Hamiltonian cycle $u_1 u_{11} u_{12} \ldots u_{1r_1}\, u_2 u_{21} u_{22} \ldots u_{2r_2}\, u_3 \ldots u_k u_{k1} u_{k2} \ldots u_{kr_k}\, u_1$ in $L(G)$, with $u_1$ being the vertex of $L(G)$ that corresponds to the edge $e_1$ of $G$, and so on.

Conversely, let $L(G)$ contain a Hamiltonian cycle $C = u_1 u_2 \ldots u_m u_1$ and let $e_i$ be the edge of $G$ corresponding to the vertex $u_i$ of $L(G)$. Let $W_0$ be the edge sequence $e_1 e_2 \ldots e_m e_1$. We delete edges from $W_0$ in succession in the following way. If $e_i \, e_j \, e_k$ are the first three distinct consecutive edges of $W_0$ that have a common vertex, then delete $e_j$, and let $W'_0 = W_0 - e_j = e_1 e_2 \ldots e_i e_k \ldots e_m e_1$. Now starting with $W'_0$, apply the same process as is applied in $W$, to get $W_0$. Continue in this way, till no such three consecutive edges exist. Clearly, the resulting subsequence of $W_0$ is a dominating walk or a pair of adjacent edges incident at a vertex, say $v_0$. In the later case, all the edges of $G$ are incident at $v_0$ and hence $v_0$ is the dominating walk of $G$.                                                                                          ❑

The following results are simple consequences of Theorem 9.11.

**Corollary 9.1**    The edge graph of a Hamiltonian graph is Hamiltonian.

**Proof**    Let $G$ be a Hamiltonian graph with Hamiltonian cycle $C$. Then $C$ is a dominating walk of $G$, and hence, $L(G)$ is Hamiltonian.

We note that the converse of Corollary 9.1 is not true in general. To see this, consider the graph $G$ as shown in Figure 9.16. Clearly $L(G)$ is Hamiltonian but $G$ is not.

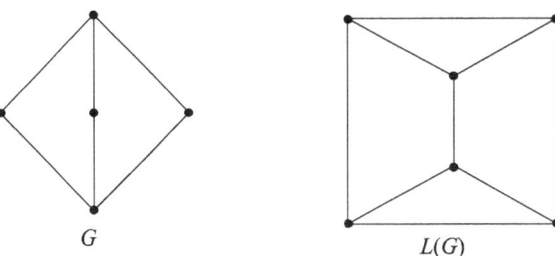

$G$ $\qquad\qquad$ $L(G)$

**Fig. 9.16**

**Corollary 9.2**    If $G$ is a connected graph and each edge of $G$ belongs to a triangle, then $L(G)$ is Hamiltonian.

**Proof**    This follows from Theorem 9.11.                                                   ❑

The following result is due to Chartrand and Wall [55].

**Theorem 9.12 (Chartrand and Wall [55])**    If $G$ is connected and $\delta(G) \geq 3$, then $L^2(G)$ is Hamiltonian.

**Proof**   Since $\delta(G) \geq 3$, each vertex of $L(G)$ belongs to a clique of size at least three and hence each edge of $L(G)$ belongs to a triangle. Then the result follows by applying Corollary 9.2.                                                                          ❏

The next result is due to Nebesky [170].

**Theorem 9.13 (Nebesky [170])**   If $G$ is a connected graph with at least three vertices, then $L(G^2)$ is Hamiltonian.

**Proof**   Since $G$ is a connected graph with at least three vertices, every edge of $G^2$ belongs to a triangle. Hence by Corollary 9.2, $L(G^2)$ is Hamiltonian.                           ❏

**Theorem 9.14**   Let $G$ be a connected graph in which every edge belongs to a triangle. If $e_1$ and $e_2$ are edges of $G$ such that $G - \{e_1, e_2\}$ is connected, then there exists a spanning walk of $G$ with $e_1$ and $e_2$ as its initial and terminal edges.

**Proof**   Consider the longest walk $W$ of $G$ with $e_1$ and $e_2$ as its initial and terminal edges. Then proceed as in Theorem 9.11.                                                               ❏

The following result is due to Jaeger [121].

**Theorem 9.15 (Jaeger [121])**   The edge graph of a 4-connected graph is Hamiltonian.

**Proof**   Let $G$ be a 4-edge connected graph. By Theorem 9.11, it suffices to show that $G$ contains a spanning Eulerian subgraph.

Now, $G$ contains two edge-disjoint spanning trees $T_1$ and $T_2$. Let $S$ be the set of vertices of odd degree in $T_1$. Then $|S|$ is even. Let $|S| = 2k$, $k \geq 1$. By Theorem 9.12, there exists a set of $k$ pairwise edge-disjoint paths $\{P_1, P_2, \ldots, P_k\}$ in $T_2$ with the property stated in Theorem 9.12. Then $G_0 = T_1 U (P_1 \cup P_2 \cup \ldots \cup P_k)$ is a connected spanning subgraph of $G$ in which each vertex is of even degree. Hence $G_0$ is a spanning Eulerian subgraph of $G$.                               ❏

Let every edge in a graph $G$ be subdivided and let $S(G)$ be the subdivision graph. If the graph obtained from $G$ by inserting $n$ new vertices of degree two into every edge of $G$ be denoted by $S_n(G)$ and taking $S(G) = S_1(G)$, we define $L_n(G) = L(S_{n-1}(G))$. We see that in general $L_n(G) \cong L^n(G)$ (Fig. 9.17).

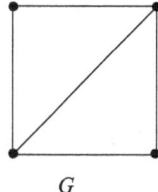
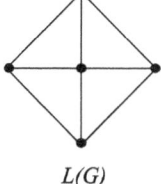
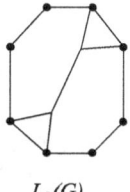
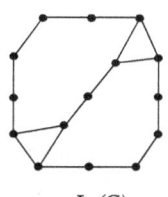

$G$        $L(G)$        $L_2(G)$        $L_3(G)$

**Fig. 9.17**

An improvement of Theorem 9.13 is seen in the following result of Harary and Nash-Williams [108].

**Theorem 9.16 (Harary and Nash-Williams [108])**   A sufficient condition for $L_2(G)$ to be Hamiltonian is that $G$ be Hamiltonian and a necessary condition is that $L(G)$ be Hamiltonian.

Now, we have the following consequence.

**Corollary 9.3**   A graph $G$ is Eulerian if and only if $L_3(G)$ is Hamiltonian.

The following result is given by Chartrand [48].

**Theorem 9.17 (Chartrand [48])**   If $G$ is a non-trivial connected graph with $n$ vertices which is not a path, then $L^k(G)$ is Hamiltonian for all $k \geq n - 3$.


## 9.3   Total Graphs

Let $G = (V, E)$ be a graph. The total graph $T(G)$ of $G$ has vertex set $V \cup E$ and two vertices of $T(G)$ are adjacent if and only if one of the following is true.

   i. the vertices are $v_i$, $v_j \in V$ and $v_i v_j$ is an edge in $E$.

  ii. one vertex is $v \in V$ and the other $e \in E$ and the edge $e$ of $G$ is incident with the vertex v of $G$.

 iii. the edges are $e_i$, $e_j \in E$ and the edges $e_i$ and $e_j$ have a vertex in common in $G$.

**Example**   The total graph of a graph $G$ is shown in Figure 9.18.

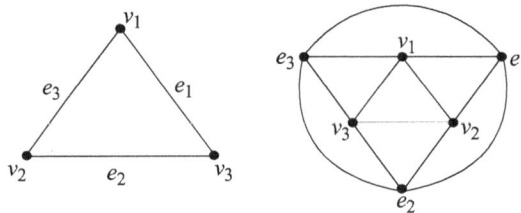

**Fig. 9.18**

It can be seen that $G$ and $L(G)$ are induced subgraphs of $T(G)$ and that the remaining edges of $T(G)$ form a graph homeomorphic to $G$.

## 9.4 Eccentricity Sequences and Sets

The concept of eccentricity has been introduced in chapter 1 and has been further discussed in chapter 4. Now, we study eccentricity sequences in graphs.

**Definition:**  A positive sequence $[e_i]_1^n$ is called an *eccentricity sequence* if it is an eccentricity sequence of some graph. The graph is said to realise the sequence. A set of positive integers is called an *eccentricity set* if it is an eccentricity set of some graph. The graph is said to realise the set. (The set of distinct eccentricities in a graph is called the eccentricity set of that graph.)

When eccentricity set is written in the increasing order $\{e_1, e_2, ..., e_k\}$ with $e_1 < e_2 < ... < e_k$ and the eccentricity sequence is then expressed as $[e_1^{n_1}, e_2^{n_2}, ..., e_k^{n_k}]$, where $n_1, n_2, ..., n_k$ are respectively the number of occurrences of $e_1, e_2, ..., e_k$, the sequence $n_1, n_2, ..., n_k$ is called the *eccentricity frequency sequence* of the graph.

Here it can be noted that $e_1 = r$ (radius) and $e_k = d$ (diameter) of the graph.

Therefore, $r \leq d \leq 2r$ gives

$$e_1 \leq e_k \leq 2e_1,$$

which is a necessary condition for a positive sequence to be an eccentricity sequence.

Now, we have the following observation.

**Theorem 9.18**    If $uv$ is an edge of a connected graph $G$, then $|e(u) - e(v)| \leq 1$.

**Proof**    Let $w$ be an eccentric vertex of $u$ (i.e., $w$ is the farthest vertex from $u$). Then by the triangle inequality for the metric $d$ (distance), we have

$$d(u, w) \leq d(u, v) + d(v, w)$$

so that $e(u) \leq d(u, v) + d(v, w)$.                                          (9.18.1)

But $u$ and $v$ are adjacent, therefore $d(u,v) = 1$.

Also, $e(v) \geq d(v,w)$ so that $d(v,w) \leq e(v)$.

Thus, from (9.18.1) we have

$$e(u) \leq 1 + d(v, w) \text{ so that } e(u) \leq 1 + e(v).$$

Thus,    $e(u) - e(v) \leq 1$.                                          (9.18.2)

Similarly, by considering an eccentric vertex of $v$, we have

$$e(v) - e(u) \leq 1.$$                                          (9.18.3)

From (9.18.2) and (9.18.3) it follows that

$$|e(u) - e(v)| \leq 1. \qquad \qquad \qquad \qquad \qquad \qquad \qquad \qquad \qquad \qquad \qquad \qquad \qquad \Box$$

**Note**   The above result shows that the eccentricities of two adjacent vertices are either equal or differ by 1 as $|e(u) - e(v)| \leq 1$ gives $|e(u) - e(v)| = 0$ or $|e(u) - e(v)| = 1$.

An important consequence of Theorem 9.18 is as follows.

**Corollary 9.4**   If $u_0 \, u_1 \, u_2 \ldots u_m$ is a path in a connected graph and $e(u_0) < e(u_m)$ and $k$ is any integer such that $e(u_0) < k < e(u_m)$, then there exists an integer $j$ $(0 < j < m)$ such that $e(u_j) = k$.

**Proof**   We know that the difference of eccentricities of any two adjacent vertices along the path $u_0 \, u_1 \, u_2 \ldots u_m$ is always less or equal to 1. Therefore, every integer between $e(u_0)$ and $e(u_m)$ occurs as the eccentricity of some vertex in this path. This can also be seen in the following way.

Assume, $e(u_0) < e(u_1) < \ldots < e(u_{j-1}) < k$,

so that $j = 1 + \max\{i : e(u_i) < k\}$, and $j - 1 = \max\{i : e(u_i) < k\}$.

Thus, $e(u_{j-1}) < k$.

Therefore, $|e(u_j) - e(u_{j-1})| \leq 1$ gives

$$e(u_j) \leq e(u_{j-1}) + 1 < k + 1,$$

so that $e(u_j) \leq k$.                                                                                    (9.4.1)

But by the choice of $j$, we have $e(u_j) \geq k$.                                                        (9.4.2)

Hence from (9.4.1) and (9.4.2), we get $e(u_j) = k$.                                              $\Box$

The following necessary condition for a positive sequence to be an eccentricity sequence is due to Lesniak [146].

**Theorem 9.19 (Lesniak [146])**   If a non decreasing sequence $[e_i]_1^n$ of positive integers is an eccentric sequence then

   i. $2e_1 \leq n$,

   ii. $e_n \leq \min\{n - 1, 2e_1\}$ and

iii. for every integer $k$ such that $e_1 < k \leq e_n$, there exists an integer $i$ ($2 \leq i \leq n-1$) such that $e_i = e_{i+1} = k$.

## Proof

i. Let the vertices of $G$ be labelled as $v_1, v_2, \ldots, v_n$ such that $e(v_i) = e_i$. Then $G$ has a spanning tree $T$ which preserves the distance from $v_1$. This gives

$$e_G(v_1) = e_T(v_1)$$

and $e_G(v_i) \leq e_T(v_i)$, for $2 \leq i \leq n$ (since removal of edges cannot reduce distances)

Thus, if $[a_1, a_2, \ldots, a_n]$ is the eccentricity sequence of $T$, we have $a_1 = e_1$. So it is enough to prove that $2a_1 \leq n$. We prove this for any tree $T$.

Now let $T$ be any tree with eccentricity sequence $[a_1, a_2, \ldots, a_n]$.

If $n = 2$, then $a_1 = a_2 = 1$, and the result is true. So assume that $n \geq 3$.

Let $u$ be a central vertex of $T$. Then, $e(u) = a_1$. Also $u$ is a cut vertex of $T$.

Suppose $a_1 = e(u) \geq \dfrac{n+1}{2}$. $\qquad [2a_1 \geq n+1]$

Since an eccentric vertex $\bar{u}$ of $u$ should lie in a component of $T - u$, there is at least one component $C$ of $T - u$ with $|V(C)| \geq \dfrac{n+1}{2}$.

Now, let $v$ be the vertex adjacent to $u$ in $C$. Then for any vertex $w$ in $C$, we have $d(v, w) = d(u, w) - 1$. So, $d(v, w) < e(u)$, because $d(u, w) \leq e(u)$ and so $d(u, w) - 1 < e(u)$.

Now, for every vertex $w$ in $V(T) - V(C)$, we have $d(v, w) - d(u, w) = 1$,

so that $d(v, w) = d(u, w) + 1$.

Total vertices in $T$ is $n$, $|V(C)| \geq \dfrac{n+1}{2}$, therefore number of vertices in

$$V(T) - V(C) \leq n - \left(\frac{n+1}{2}\right) = \frac{n-1}{2}.$$

That is, $|V(T) - V(C)| \leq \dfrac{n-1}{2}$.

Therefore, $d(u, w) \leq \dfrac{n-1}{2} - 1 = \dfrac{n-3}{2}$.

Thus, $d(v, w) \leq \dfrac{n-3}{2} + 1 = \dfrac{n-1}{2}$

So, $d(v, w) < e(u)$.

Hence for all vertices $w$, we have $d(v, w) < e(u)$, and thus $e(v) < e(u)$, so that $e(v) < a_1$, which is a contradiction as $a_1$ is the least eccentricity of a vertex of $T$. Thus, $a_1 \geq \dfrac{n+1}{2}$ is wrong and so, $a_1 \leq \dfrac{n}{2}$.

ii. The maximum distance possible in an $n$–vertex graph is $n - 1$. So, $e_n \leq n - 1$. Also, $e_n \leq 2e_1$, $r \leq d \leq 2r$, and here $d = e_n$, $r = e_1$.

  Hence, $e_n \leq \min\{(n-1), 2e_1\}$.

iii. We have to prove that each integer between $e_1$ and $e_n$ ($e_n$ inclusive) occurs at least twice in the sequence. Let $u_1$ be the central vertex and $u_k$ be the peripheral vertex of $G$. Then $e(u_1) = e_1$ and $e(u_k) = e_n$. Since $G$ is connected, there exists a $u_1 - u_k$ path. Now by Corollary 9.4, if $k$ is any integer between $e_1$ and $e_k$, there exists a vertex $u_j$ in this path with $e(u_j) = k$. This gives the existence of a vertex whose eccentricity is $k$.

  Now, if $e(w) > e_1$, we show there is a vertex $u$ other than $w$ such that $e(u) = e(w)$. Let $\overline{w}$ be an eccentric vertex of $w$, that is, $d(w, \overline{w}) = e(w) = k$, say. As we have assumed that $u_1$ is the central vertex of $G$, let $P = u_1 \ldots u_m (u_m = \overline{w})$ be a $u_1 - \overline{w}$ distance path in $G$. Since $e(u_1) = e_1 < e(w) = d(\overline{w}, w) \leq e(\overline{w})$, applying Corollary 9.4, there is a vertex $u_j$ in this path such that $e(u_j) = k$. But $d(\overline{w}, u_j) \leq m - 1 = d(u_1, \overline{w}) \leq e(u_1) = e_1 < e(w) = d(\overline{w}, w)$. Therefore, $d(\overline{w}, u_j) < d(\overline{w} \, w)$. Thus, $u_j \neq w$ and the result is proved.

The following characterisation of eccentricity sequences of trees is again due to Lesniak [146].

**Theorem 9.20 (Lesniak [146])**   A non-decreasing sequence $[e_i]_1^n$ of positive integers is the eccentric sequence of a tree if and only if

  i. For each integer $k$ with $e_1 < k \leq e_n$, we have

  $$e_i = e_{i+1} = k, \text{ for some } i, \ 2 \leq i \leq n - 1,$$

  ii. Either $e_1 = \dfrac{e_n}{2}$ and $e_1 \neq e_2$, or $e_1 = \dfrac{e_n + 1}{2}$, $e_1 = e_2$ and $e_2 \neq e_3$.

## Proof

*Necessity*   Let the nondecreasing sequence $[e_i]_1^n$ of positive integers be the eccentric sequence of a tree. Then (i) follows from condition (iii) of the previous result.

Let $r$ be the radius and $d$ the diameter of the tree, so that $e_1 = r$ and $e_n = d$. Since a tree is either unicentral or bicentral, we have

  $d = 2r$ for unicentral,

and $d = 2r - 1$ for bicentral.

In case the tree is unicentral, then the eccentricity of the center $v_1$ is $e_1$ and $e_1 \neq e_2$. Thus, $e_n = 2e_1$ which implies that $e_1 = \dfrac{e_n}{2}$ and $e_1 \neq e_2$.

In case the tree is bicentral, then $e_1 = e_2$ and $d = 2r - 1$ gives $e_n = 2e_1 - 1$, so that $e_1 = \dfrac{e_n + 1}{2}$ with $e_2 \neq e_3$.

*Sufficiency*  Let the nondecreasing sequence $[e_i]_1^n$ of positive integers satisfy conditions (i) and (ii). We construct a tree with eccentric sequence $[e_i]_1^n$ in the following way.

Let $P$ be a path of length $e_n = d$. Then eccentric sequence of $P$ is

$$S_1 = \frac{d}{2}, \left(\frac{d}{2} + 1\right)^2, \left(\frac{d}{2} + 2\right)^2, \dots, \left(\frac{d}{2} + \frac{d}{2}\right)^2, \qquad \text{if } d \text{ is even,}$$

or  $$S_2 = \left(\frac{d+1}{2}\right)^2, \left(\frac{d+1}{2} + 1\right)^2, \dots, \left(\frac{d+1}{2} + \frac{d+1}{2} - 1\right)^2, \quad \text{if } d \text{ is odd,}$$

that is, $S_1 = r, (r+1)^2, (r+2)^2, \dots, (r+r)^2$, where $r = \dfrac{d}{2}$,

or $S_2 = r^2, (r+1)^2, (r+2)^2, \dots, (r+r-1)^2$, where $r = \dfrac{d+1}{2}$,

where powers denote repetition of eccentricity.

Let the given sequence $[e_i]_1^n$ be written in power notation $\pi = r^{i_1}(r+1)^{i_2} \dots d^{i_k}$, where $i_1 = 1$ or 2, according as $d$ is even or odd. If $i_j > 2$ for any $j$, $1 < j \leq k$, we attach $i_j - 2$ vertices to any vertex with eccentricity $r + j - 2$ in the path $P$. This does not alter the eccentricities of the vertices of $P$ and the resulting tree $T$ has eccentric sequence $[e_i]_1^n$.  ❑

**Example**  Construct a tree with eccentric sequence $[4^2, 5^4, 6^3, 7^4]$.

First draw a path $P$ say $u_0\, u_1 u_2\, u_3\, u_4\, u_5\, u_6\, u_7$ of length 7. Then the eccentricities of these vertices are 7, 6, 5, 4, 4, 5, 6, 7.

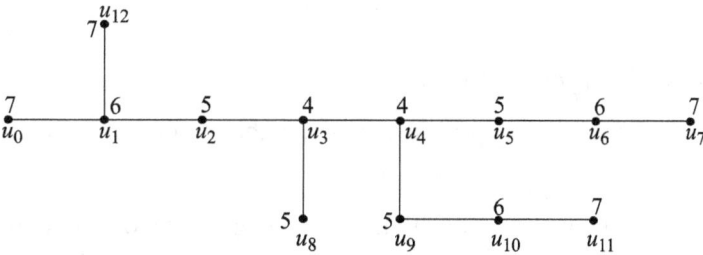

**Fig. 9.19**

To get two more vertices of eccentricities 5, attach a new vertex each to $u_3$ and $u_4$. Let these new vertices be $u_8$ and $u_9$. (Here, $i_2 = 4 > 2$, and $i_2 - 2 = 4 - 2 = 2$). So 2 vertices one each are attached to the vertices of eccentricities $r + j - 2 = r + 2 - 2 = 4 + 2 - 2 = 4$, i.e., $u_3$ and $u_4$). Now $i_3 = 3 > 2$ and $i_3 - 2 = 3 - 2 = 1$. So one vertex is to be attached to the vertex of eccentricity $r + j - 2 = 4 + 3 - 2 = 5$, i.e., the vertex $u_2$ or $u_8$ or $u_9$ or $u_5$. Let this new vertex $u_{10}$ be attached to $u_9$ say. Now $i_4 = 4 > 2$ and $i_4 - 2 = 4 - 2 = 2$. So two new vertices are to be attached, one each among the vertices with eccentricities $r + j - 2 = 4 + 4 - 2 = 6$, i.e., to the vertices $u_1$, $u_6$, $u_{10}$. Let these new vertices be $u_{11}$ attached to $u_{10}$ and $u_{12}$ attached to $u_1$. The resulting tree is shown in Figure 9.19.

**Remark**   Clearly, there are many trees realising this eccentric sequence.

**Neighbourhood**   Let $v$ be any vertex of a connected graph $G$. The $i$th *neighbourhood* of $v$ denoted by $N_i(v)$ is the set of all those vertices in $V$ whose distance from $v$ is $i$.

i.e., $N_i(v) = \{u \in V: d(v, u) = i\}$.

**Example**   Consider the graph in Figure 9.20.

Here, $N_3(v) = \{u_1, u_2, u_3\}$. We set $N_0(v) = \{v\}$.

We denote $N_1(v)$ by $N(v)$ and call it the *neighbourhood* of $v$.

We have $N_1(v) = N(v) = \{u_4, u_5, u_6, u_7\}$.

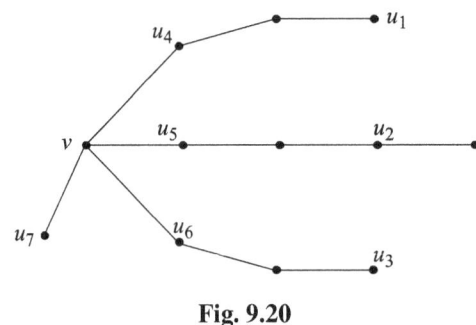

**Fig. 9.20**

We have the following observations.

**Lemma 9.2**   Any two vertices with the same neighbourhood in a graph have the same eccentricity.

**Proof**   Let $u$ and $v$ be two vertices with the same neighbourhood. So $N(u) = N(v)$. Therefore the path lengths from $u$ and $v$ to the other vertices of the graph are equal. Clearly, $u$ and $v$ are not adjacent.                                                                                     ❑

**Lemma 9.3** If $u$ and $v$ are adjacent in $G$ and $N(u) - \{v\} = N(v) - \{u\}$, then $u$ and $v$ have the same eccentricity in $G$.

**Proof** Let $G$ be a graph in which $u_v = e$ and $N(u) - \{v\} = N(v) - \{u\}$.

Let $H = G - e$. Then $u$ and $v$ are not adjacent in $H$ so that $u$ and $v$ have the same neighbourhood in $H$.

Therefore, $e(u|H) = e(v|H)$ $(e(u|H)$ means eccentricity of vertex $u$ in graph $H$). If $e(u|G) = 1$, then $e(v|G) = 1$ also. If not, then $e(u|G) = e(u|H) = e(v|H) = e(v|G)$. ❑

**Definition:** Let $v$ be a vertex of a graph $G$ and let $H$ be a graph obtained from $G - v$ by adding edge to each vertex of a new graph $K_p$ (or $\overline{K}_p$) to every vertex of $G - v$ to which $v$ was adjacent in $G$. Then $H$ is said to be obtained from $G$ by replacing $v$ by $K_p$ (or $\overline{K}_p$). This operation is illustrated in Figure 9.21.

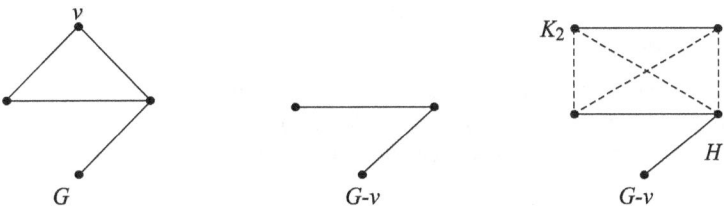

**Fig. 9.21**

The operation of replacing $v$ in $G$ by $\overline{K}_p$ is called *multiplication* of the vertex in $G$. The operation of replacing $v$ in $G$ by $K_p$ is called the *linked multiplication* of $v$ in $G$. Lesniak observed that multiplication or linked multiplication of one or more vertices of a graph is an operation preserving the eccentricity set of the graph.

**Lemma 9.4** If $H$ is the graph obtained by replacing a vertex $v$ of a graph $G$ with diameter greater than one, by a $K_p$ or $\overline{K}_p$ (for any positive integer $p$), then $G$ and $H$ have the same eccentricity sets.

**Proof** Since $v$ and any vertex of $\overline{K}_p$ have the same neighbourhood in $H$,

$$e(u|H) = e(u|G), \text{ for every } u \in \overline{K}_p.$$

For replacement by $K_p$, we have for any two adjacent vertices $u_i$ and $u_j$ in $K_p$,

$$N(u_i) - \{u_j\} = N(u_j) - \{u_i\}.$$

Therefore, $e(u_i|H) = e(u_j|H)$.

So, $e(u|H) = e(u|G)$, for every $u \in K_p$.

Thus, $e(u|H) = e(u|G)$, for every $u \in K_p$ ($\overline{K}_p$) and obviously

$e(w|H) = e(w|G)$, for all other vertices.                                                        ❑

The above result is illustrated in Figure 9.22, where we choose $K_2$ and $\overline{K}_2$

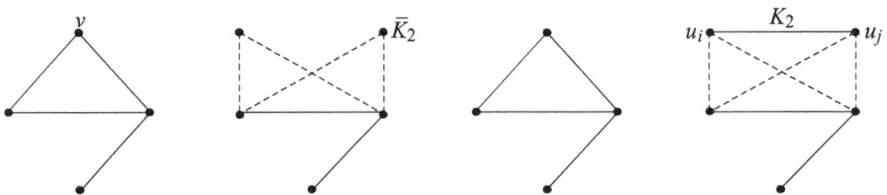

**Fig. 9.22**

Now, we give the necessary and sufficient conditions for an eccentricity sequence of a graph, due to Lesniak [146].

**Theorem 9.21 (Lesniak [146])**    A nondecreasing sequence of positive integers $[e_1^{r_1}, e_2^{r_2}, ...,$
$e_k^{r_k}]$ is an eccentricity sequence if and only if some of its subsequence $[e_1^{s_1}, e_2^{s_2}, ..., e_k^{s_k}]$ with
$s_i \le r_i$ and no $s_i = 0$ for $1 \le i \le k$ is an eccentricity sequence.

## Proof

*Necessity*    Since every sequence is its own subsequence, necessity follows.

*Sufficiency*    Let $[e_1^{s_1}, e_2^{s_2}, ..., e_k^{s_k}]$ be the eccentricity sequence of a graph $G$. Let $v_i$ be a vertex
of $G$ with $e(v_i) = e_i, 1 \le i \le k$.

Let $n_1 = r_1 - s_1 + 1$ so that $r_1 = n_1 + s_1 - 1$.

Now let $G_1$ be obtained from $G$ by replacing $v_1$ by $K_{n_1}$ or $\overline{K}_{n_1}$. Then every one of the
vertices of this $K_{n_1}$ or $\overline{K}_{n_1}$ has eccentricity $e(v_1|G)$ and the eccentricities of the vertices of $G$
are unaltered in $G_1$.

Thus, $G_1$ has eccentricity sequence $[e_1^{r_1}, e_2^{s_2}, ..., e_k^{s_k}]$.

Let $G_2$ be obtained from $G_1$ by replacing $v_2$ by $K_{n_2}$ or $\overline{K}_{n_2}$. Then by similar argument, $G_2$
has eccentricity sequence

$[e_1^{r_1}, e_2^{r_2}, ..., e_k^{s_k}]$.

Proceeding in this way, by successively replacing $v_3, v_4, ..., v_k$ by $K_{n_3}$ ($\overline{K}_{n_3}$), $K_{n_4}$ ($\overline{K}_{n_4}$),
$..., K_{n_k}$ ($\overline{K}_{n_k}$), we get a graph $G_k$ with eccentricity sequence

$[e_1^{r_1}, e_2^{r_2}, ..., e_k^{r_k}]$.                                                          ❑

**Remarks**

1. It is assumed that $e_k > 1$.

2. The construction of $G_k$ is not unique.

3. This result keeps unsolved the problem of characterising minimal eccentricity sequences, that is, those eccentricity sequences which have no proper eccentric subsequences.

The next result characterises eccentricity sets and is due to Behzad and Simpson [17].

**Theorem 9.22 (Behzad and Simpson [17])**  A non-empty set $S = \{e_1, e_2, \dots, e_k\}$ of positive integers arranged in increasing order is an eccentricity set if and only if $k \le e_1 + 1$ and $e_{i+1} = e_i + 1$ for $1 \le i \le k - 1$.

**Proof**

*Necessity*  Let $S$ be an eccentricity set. Then by (iii) of Theorem 4.24, $e_{i+1} = e_i + 1$ for each $i$, $1 \le i \le k - 1$. This gives $e_k = e_1 + k - 1$. Since $e_k \le 2e_1$, we get $k \le e_1 + 1$.

*Sufficiency*  If $e_1 = 1$, then $k = 1$, or 2 and $S = \{1\}$ or $\{1, 2\}$. In this case, $K_2$ and $K_{1,n}$ realise the sets.

For $e_1 > 1$, let $G$ be the graph obtained by identifying a vertex of a cycle $C_{2e_1}$ with an end vertex of a path $P_k$. Let $e_1 - k + 1 = d$. Then, $d \ge 0$, and the eccentricity sequence of $G$ is easily verified to be $[e_1^{2d+1}, (e_1 + 1)^3, (e_1 + 2)^3, \dots, (e_1 + k - 2)^3, (e_1 + k - 1)^3]$. Hence $S$ is the eccentricity set. $\qquad\qquad \square$

## 9.5 Distance Degree Regular and Distance Regular Graphs

Let $G$ be a connected graph and let $v$ be any vertex of $G$. Let $e$ be the eccentricity of vertex $v$ and $N_j(v)$ be the $j$th neighbourhood of $v$. Assume, $n_j(v) = |N_j(v)|$, for $0 \le j \le e$. The sequence $D(G, v) = [n_0(v), n_1(v), \dots, n_e(v)]$ is called the *distance degree sequence* (DDS) of $v$ in $G$. If all the vertices of $G$ have the same distance degree sequence $D(G) = [n_0, n_1, \dots, n_d]$, then $G$ is said to be *distance degree regular*.

If $G$ is not distance degree regular, the $n$ vectors $D(G, v_i)$, $1 \le i \le n$, arranged lexicographically in an array with variable row sizes is called the *distance degree array* of $G$ (DDA(G)).

The *distance* of vertex $v$ is defined by

$$D(v) = \sum_{j=1}^{e(v)} n_j(v).$$

Let $D_i$ be the distance of the vertex $v_i$. The sequence $DS(G) = [D_1, D_2, \dots, D_n]$ in non-decreasing order is called the *distance sequence* of the graph.

A vertex $v$ with minimum distance $D(v)$ is called a *median* of $G$ and the subgraph induced by the set $M$ of median vertices of $G$ is called the *median subgraph* of $G$.

Clearly, if $G$ is distance degree regular, then all vertices in $G$ have the same eccentricity, and $G$ is a *self-centered graph*. Also, $n_0 = 1$ and $n_1 = |N_1(v)|$ for every $v \in G$, so that $G$ is $n_1$-regular.

Randic [214] conjectured that two trees are isomorphic if and only if they have same DDA and Slater [187] disproved this giving the counter example as shown in Figure 9.23.

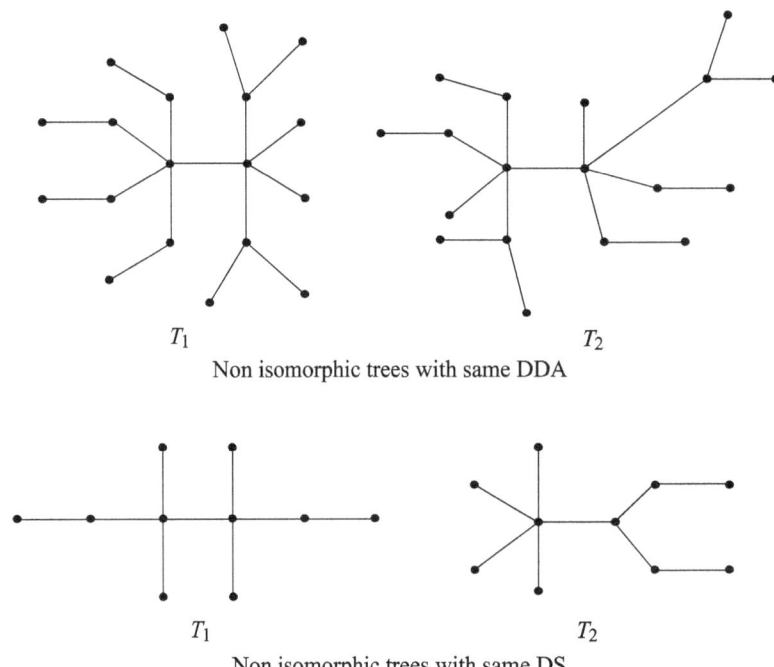

Non isomorphic trees with same DDA

Non isomorphic trees with same DS

**Fig. 9.23**

**Definition:** Let $G$ be a connected graph with diameter $d$ and let $k = b_0, b_1, \ldots, b_{d-1}$; $l = c_1, c_2, \ldots, c_d$ be $2d$ non-negative integers. Then $G$ is said to be *distance regular* (DR) if for every pair of vertices $u, v$ in $G$ with $d(u, v) = j$, we have, (i) the number of vertices in $N_{j-1}(v)$ adjacent to $u$ is $c_j$, $1 \leq j \leq d$ and (ii) the number of vertices in $N_{j+1}(v)$ adjacent to $u$ is $b_j$, $0 \leq j \leq d - 1$.

The sequence $[b_0, b_1, \ldots, b_{d-1}, c_1, c_2, \ldots, c_d]$ is called the *intersection array* of $G$.

Clearly, DR graphs are $k$-regular and self-centered. The examples of distance regular graphs are $K_n$, $K_{n,n}$ and the cubes $Q_n$.

A graph $G$ is said to be *strongly regular* (SR) with parameters $(n, k, \lambda, \mu)$ if it is a $k$-regular graph of order $n$ in which every pair of adjacent vertices are mutually adjacent to $\lambda$ vertices and every pair of non-adjacent vertices are mutually adjacent to $\mu$ vertices.

The Petersen graph is strongly regular with parameters $(10, 3, 0, 1)$, that is $n = 10$, $k = 3$, $\lambda = 0$, $\mu = 1$ (Fig. 9.24).

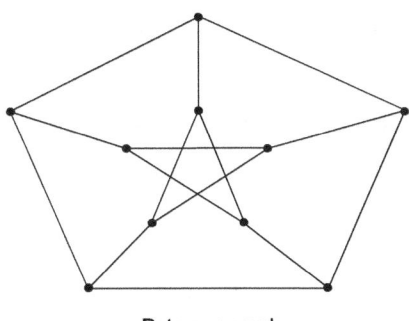

Petersen graph

**Fig. 9.24**

## 9.6 Isometry

The concept of isometry as in Chartrand and Stewart [52] is as follows.

Let $G_1$ and $G_2$ be connected graphs with vertex sets $V_1$ and $V_2$ respectively. Then $G_2$ is said to be *isometric* from $G_1$ if for each $v \in V_1$, there is a one-one map $\phi_v : V_1 \to V_2$ such that $\phi_v$ preserves distances from $v$, that is $d_{G_2}(u, v) = d_{G_1}(\phi_v(V), \phi_v(u))$ for every $u \in V_1$.

Two graphs $G_1$ and $G_2$ are said to be *isometric* if they are isometric from each other.

**Example** Consider the graphs shown in Figure 9.25, we have

$$\phi_1 = \phi_4 = \phi_5 \, (1 \to a, \, 2 \to b, \, 3 \to c, \, 4 \to d, \, 5 \to e),$$

$$\phi_2 = (2 \to e, \, 1 \to a, \, 3 \to d, \, 4 \to c, \, 5 \to b), \, \phi_3 = (3 \to e, \, 4 \to a, \, 2 \to d, \, 1 \to b, \, 5 \to c).$$

Here, $G_2$ is isometric from $G_1$.

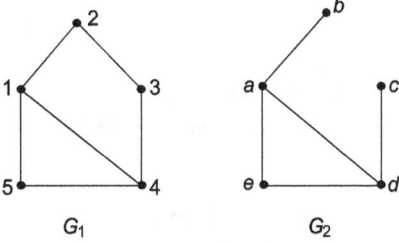

**Fig. 9.25**

**Remarks**    Isometry between graphs as defined above does not imply isomorphism (Fig. 9.26). A pair of isometric graphs may even have same degree sequence and yet be non-isomorphic (Fig. 9.27).

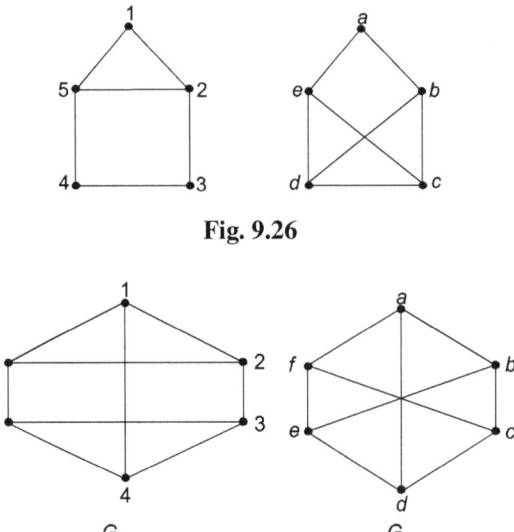

**Fig. 9.26**

**Fig. 9.27**

We now have the following results.

**Theorem 9.23**    If $G_1$ and $G_2$ are $k$-regular graphs of order $n$, where $k \geq n - 1/2$, then $G_1$ and $G_2$ are isometric.

**Proof**    Since $G_1$ is a $k$-regular graph with $k \geq n - 1/2$, $d(G_1) \leq 2$.
   Let $u \in V(G_1)$ and $v \in V(G_2)$ be any two vertices and define

$$\phi_u : V(G_1) \to V(G_2) \text{ by } \phi_u(u) = v.$$

For $i = 1, 2, \ldots, k$, let $u_i \in N_1(u)$ and $v_i \in N_1(v)$ and define $\phi_u(u_i) = v_i$.

For $i = k+1, \ldots, n-1$, let $u_i \in N_2(u)$ and $v_i \in N_2(v)$ and again let $\phi_u(u_i) = v_i$.

The neighbourhoods are in the appropriate graphs. Then $\phi_u$ is an isometry of $G_2$ from $G_1$ at $u$. Since $u$ and $v$ are arbitrary, it is easily seen that $G_2$ is isometric from $G_1$, and $G_1$ is isometric from $G_2$.                                                                                  ❑

**Theorem 9.24**    A necessary condition for two graphs to be isometric is that they have the same degree set and the same eccentricity set.

**Proof** Let $G_1 = (V_1, E_1)$ and $G_2 = (V_2, E_2)$ be isometric graphs. As $G_2$ is isometric from $G_1$, let $\phi_v$ be the one-one mapping from $V(G_1)$ to $V(G_2)$. Therefore, $d(v|G_1) = d(\Phi_v(v)|G_2)$. Also $\phi_v$ has the property of preserving distance, therefore $e(v|G_1) = e(\phi_v(v)|G_2)$. So the eccentricity set of $G_1$ is included in the eccentricity set of $G_2$.

Again, as $G_1$ is isometric from $G_2$, therefore, the degree set and eccentricity set of $G_2$ are included respectively in the degree set and eccentricity set of $G_1$.

Hence, the degree sets are equal in $G_1$ and $G_2$ and the eccentricity sets are equal in $G_1$ and $G_2$. ❑

## 9.7 Exercises

1. Show that the edge graph of $K_{1,n}$ is $K_n$.

2. Show that the edge graph of $K_5$ is the complement of the Petersen graph.

3. Show that if $L(G)$ is connected and regular, then either $G$ is regular or $G$ is a bipartite graph in which vertices of the same partite set have the same degree.

4. If $G$ is $k$-edge-connected, then prove that $L(G)$ is $k$-connected and $(2k-2)$-edge-connected.

5. Show that the graph $L_2(G)$ is Hamiltonian if and only if $G$ has a closed spanning walk.

6. Show that the graph $L_2(G)$ is Hamiltonian if and only if there is a closed walk in $G$ which includes at least one vertex incident with each edge of $G$.

7. Prove that $T(K_n) \cong L(K_{n+1})$.

8. If $G$ is Hamiltonian, then prove that $T(G)$ is Hamiltonian.

9. If $G$ is Eulerian, then prove that $T(G)$ is both Eulerian and Hamiltonian.

10. Prove that $T(G)$ of every nontrivial connected graph $G$ contains a spanning Eulerian subgraph.

11. Show that the edge graph of a graph $G$ has a Hamiltonian path if and only if $G$ has walk $W$ such that every edge of $G$ not in $W$ is incident with $W$.

12. If $G$ is any connected graph with $\delta(G) \geq 4$, then prove that $L^2(G)$ is Hamiltonian-connected.

13. Construct graphs with eccentricity sequences

$$[3^3, 4^6, 2^2 3^2, 3^4, 2^4].$$

14. If $G$ is a connected graph with diameter 3 and $e(u|G) = 3$, then show that $e(u|\overline{G}) = 2$.

15. If $[e_1, e_2, \ldots, e_n]$ with $e_n < 2e_1 - 1$ is an eccentricity sequence, then show that each central vertex lies on a cycle.

16. If $[e_i]_1^n$ is the eccentricity sequence of an $(n, m)$ graph, show that

$$m \leq \frac{1}{2}\left(n^2 - \sum_{i=1}^{n} e_i\right).$$

17. For a distance regular graph, prove the following

    a. If $1 \leq i \leq \frac{1}{2d}$, then $b_i \geq c_i$.

    b. If $1 \leq i \leq \frac{1}{d-1}$, then $b_1 \geq c_i$.

    c. $c_2 \geq k - 2b_1$.

# 10. Graph Matrices

Since a graph is completely determined by specifying either its adjacency structure or its incidence structure, these specifications provide far more efficient ways of representing a large or complicated graph than a pictorial representation. As computers are more adept at manipulating numbers than at recognising pictures, it is standard practice to communicate the specification of a graph to a computer in matrix form. In this chapter, we study various types of matrices associated with a graph, and our study is based on Narsing Deo [63], Foulds [82], Harary [104] and Parthasarathy [180].

## 10.1  Incidence Matrix

Let $G$ be a graph with $n$ vertices, $m$ edges and without self-loops. The incidence matrix $A$ of $G$ is an $n \times m$ matrix $A = [a_{ij}]$ whose $n$ rows correspond to the $n$ vertices and the $m$ columns correspond to $m$ edges such that

$$a_{ij} = \begin{cases} 1, & \text{if jth edge } m_j \text{ is incident on the ith vertex} \\ 0, & \text{otherwise.} \end{cases}$$

It is also called *vertex-edge incidence matrix* and is denoted by $A(G)$.

**Example**  Consider the graphs given in Figure 10.1. The incidence matrix of $G_1$ is

$$
A(G_1) = \begin{array}{c} \\ v_1 \\ v_2 \\ v_3 \\ v_4 \\ v_5 \\ v_6 \end{array}
\begin{array}{c}
e_1 \ e_2 \ e_3 \ e_4 \ e_5 \ e_6 \ e_7 \ e_8 \\
\left[ \begin{array}{cccccccc}
0 & 0 & 0 & 1 & 0 & 1 & 0 & 0 \\
0 & 0 & 0 & 0 & 1 & 1 & 1 & 1 \\
0 & 0 & 0 & 0 & 0 & 0 & 0 & 1 \\
1 & 1 & 1 & 0 & 1 & 0 & 0 & 0 \\
0 & 0 & 1 & 1 & 0 & 0 & 1 & 0 \\
1 & 1 & 0 & 0 & 0 & 0 & 0 & 0
\end{array} \right]
\end{array} .
$$

The incidence matrix of $G_2$ is

$$A(G_2) = \begin{array}{c} \\ v_1 \\ v_2 \\ v_3 \\ v_4 \end{array} \begin{array}{ccccc} e_1 & e_2 & e_3 & e_4 & e_5 \\ \begin{bmatrix} 1 & 1 & 1 & 0 & 0 \\ 1 & 0 & 0 & 1 & 1 \\ 0 & 1 & 1 & 1 & 1 \\ 0 & 0 & 1 & 0 & 0 \end{bmatrix} \end{array}.$$

The incidence matrix of $G_3$ is

$$A(G_3) = \begin{array}{c} \\ v_1 \\ v_2 \\ v_3 \\ v_4 \end{array} \begin{array}{ccccc} e_1 & e_2 & e_3 & e_4 & e_5 \\ \begin{bmatrix} 1 & 1 & 0 & 0 & 1 \\ 1 & 1 & 1 & 0 & 0 \\ 0 & 0 & 0 & 1 & 0 \\ 0 & 0 & 1 & 1 & 1 \end{bmatrix} \end{array}.$$

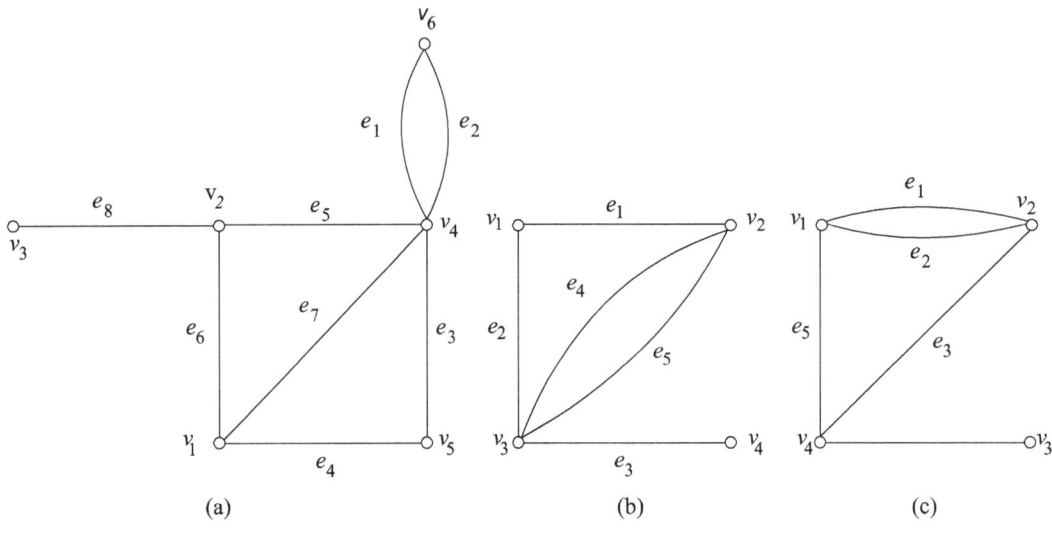

**Fig. 10.1**

The incidence matrix contains only two types of elements, 0 and 1. This clearly is a binary matrix or a (0, 1)-matrix.

We have the following observations about the incidence matrix $A$.

1. Since every edge is incident on exactly two vertices, each column of $A$ has exactly two one's.

2. The number of one's in each row equals the degree of the corresponding vertex.

3. A row with all zeros represents an isolated vertex.

4. Parallel edges in a graph produce identical columns in its incidence matrix.

5. If a graph is disconnected and consists of two components $G_1$ and $G_2$, the incidence matrix $A(G)$ of graph $G$ can be written in a block diagonal form as

$$A(G) = \begin{bmatrix} A(G_1) & 0 \\ 0 & A(G_2) \end{bmatrix},$$

where $A(G_1)$ and $A(G_2)$ are the incidence matrices of components $G_1$ and $G_2$. This observation results from the fact that no edge in $G_1$ is incident on vertices of $G_2$ and vice versa. Obviously, this is also true for a disconnected graph with any number of components.

6. Permutation of any two rows or columns in an incidence matrix simply corresponds to relabeling the vertices and edges of the same graph.

**Note**  The matrix $A$ has been defined over a field, Galois field modulo 2 or $GF(2)$, that is, the set $\{0, 1\}$ with operation addition modulo 2 written as + such that $0 + 0 = 0$, $1 + 0 = 1$, $1 + 1 = 0$ and multiplication modulo 2 written as "." such that $0.0 = 0$, $1.0 = 0 = 0.1$, $1.1 = 1$.

The following result is an immediate consequence of the above observations.

**Theorem 10.1**  Two graphs $G_1$ and $G_2$ are isomorphic if and only if their incidence matrices $A(G_1)$ and $A(G_2)$ differ only by permutation of rows and columns.

**Proof**  Let the graphs $G_1$ and $G_2$ be isomorphic. Then there is a one-one correspondence between the vertices and edges in $G_1$ and $G_2$ such that the incidence relation is preserved. Thus, $A(G_1)$ and $A(G_2)$ are either same or differ only by permutation of rows and columns.

The converse follows, since permutation of any two rows or columns in an incidence matrix simply corresponds to relabeling the vertices and edges of the same graph.  ❑

## Rank of the incidence matrix

Let $G$ be a graph and let $A(G)$ be its incidence matrix. Now each row in $A(G)$ is a vector over $GF(2)$ in the vector space of graph $G$. Let the row vectors be denoted by $A_1$, $A_2$, ..., $A_n$. Then,

$$A(G) = \begin{bmatrix} A_1 \\ A_2 \\ \cdot \\ \cdot \\ \cdot \\ A_n \end{bmatrix}.$$

Since there are exactly two ones in every column of $A$, the sum of all these vectors is 0 (this being a modulo 2 sum of the corresponding entries).

Thus, vectors $A_1, A_2, \ldots, A_n$ are linearly dependent. Therefore, rank $A < n$.

Hence, rank $A \leq n - 1$.

We have the following result due to the above observation.

**Theorem 10.2**  If $A(G)$ is an incidence matrix of a connected graph $G$ with $n$ vertices, then rank of $A(G)$ is $n - 1$.

**Proof**  Let $G$ be a connected graph with $n$ vertices and let the number of edges in $G$ be $m$. Let $A(G)$ be the incidence matrix and let $A_1, A_2, \ldots, A_n$ be the row vector of $A(G)$.

$$\text{Then,} \quad A(G) = \begin{bmatrix} A_1 \\ A_2 \\ \cdot \\ \cdot \\ \cdot \\ A_n \end{bmatrix}. \tag{10.2.1}$$

Clearly, rank $A(G) \leq n - 1$. $\tag{10.2.2}$

Now consider the sum of any $m$ of these row vectors, $m \leq n - 1$. Since $G$ is connected, $A(G)$ cannot be partitioned in the form

$$A(G) = \begin{bmatrix} A(G_1) & 0 \\ 0 & A(G_2) \end{bmatrix}$$

such that $A(G_1)$ has $m$ rows and $A(G_2)$ has $n - m$ rows.

Thus, there exists no $m \times m$ submatrix of $A(G)$ for $m \leq n - 1$, such that the modulo 2 sum of these $m$ rows is equal to zero.

As there are only two elements 0 and 1 in this field, the additions of all vectors taken $m$ at a time for $m = 1, 2, \ldots, n - 1$ gives all possible linear combinations of $n - 1$ row vectors.

Thus, no linear combinations of $m$ row vectors of $A$, for $m \leq n - 1$ is zero.

Therefore, rank $A(G) \leq n - 1$. $\tag{10.2.3}$

Combining (10.2.2) and (10.2.3), it follows that rank $A(G) = n - 1$. $\qquad \square$

**Remark**  If $G$ is a disconnected graph with $k$ components, then it follows from the above theorem that rank of $A(G)$ is $n - k$.

Let $G$ be a connected graph with $n$ vertices and $m$ edges. Then the order of the incidence matrix $A(G)$ is $n \times m$. Now, if we remove any one row from $A(G)$, the remaining $(n - 1)$ by $m$ submatrix is of rank $(n - 1)$. Thus, the remaining $(n - 1)$ row vectors are linearly independent. This shows that only $(n - 1)$ rows of an incidence matrix are required to specify the corresponding graph completely, because $(n-1)$ rows contain the same information as the entire matrix. This follows from the fact that given $(n-1)$ rows, we can construct the $n$th row, as each column in the matrix has exactly two ones. Such an $(n-1) \times m$ matrix of $A$ is called a *reduced incidence matrix* and is denoted by $A_f$. The vertex corresponding to the deleted row in $A_f$ is called the *reference vertex*. Obviously, any vertex of a connected graph can be treated as the reference vertex.

The following result gives the nature of the incidence matrix of a tree.

**Theorem 10.3** The reduced incidence matrix of a tree is non-singular.

**Proof** A tree with $n$ vertices has $n-1$ edges and also a tree is connected. Therefore, the reduced incidence matrix is a square matrix of order $n-1$, with rank $n-1$. Thus, the result follows.

Now a graph $G$ with $n$ vertices and $n-1$ edges which is not a tree is obviously disconnected. Therefore, the rank of the incidence matrix of $G$ is less than $(n-1)$. Hence the $(n-1) \times (n-1)$ reduced incidence matrix of a graph is non-singular if and only if the graph is a tree. ❑

## 10.2 Submatrices of A(G)

Let $H$ be a subgraph of a graph $G$, and let $A(H)$ and $A(G)$ be the incidence matrices of $H$ and $G$ respectively. Clearly, $A(H)$ is a submatrix of $A(G)$, possibly with rows or columns permuted. We observe that there is a one-one correspondence between each $n \times k$ submatrix of $A(G)$ and a subgraph of $G$ with $k$ edges, $k$ being a positive integer, $k < m$ and $n$ being the number of vertices in $G$.

The following is a property of the submatrices of $A(G)$.

**Theorem 10.4** Let $A(G)$ be the incidence matrix of a connected graph $G$ with $n$ vertices. An $(n-1) \times (n-1)$ submatrix of $A(G)$ is non-singular if and only if the $n-1$ edges corresponding to the $n-1$ columns of this matrix constitutes a spanning tree in $G$.

**Proof** Let $G$ be a connected graph with $n$ vertices and $m$ edges. So, $m \geq n-1$.

Let $A(G)$ be the incidence matrix of $G$, so that $A(G)$ has $n$ rows and $m$ columns ($m \geq n-1$).

We know that every square submatrix of order $(n-1) \times (n-1)$ in $A(G)$ is the reduced incidence matrix of some subgraph $H$ in $G$ with $n-1$ edges, and vice versa. We also know that a square submatrix of $A(G)$ is non-singular if and only if the corresponding subgraph is a tree.

Obviously, the tree is a spanning tree because it contains $n-1$ edges of the $n$-vertex graph.

Hence $(n-1) \times (n-1)$ submatrix of $A(G)$ is non-singular if and only if $n-1$ edges corresponding to $n-1$ columns of this matrix forms a spanning tree. ❑

The following form of incidence matrix is reported by Biggs [25].

**Definition:**  The matrix $F = [f_{ij}]$ of the graph $G = (V, E)$ with $V = \{v_1, v_2, \ldots, v_n\}$ and $E = \{e_1, e_2, \ldots, e_m\}$, is the $n \times m$ matrix associated with a chosen orientation of the edges of $G$ in which for each $e = (v_i, v_j)$, one of $v_i$ or $v_j$ is taken as positive end and the other as negative end, and is defined by

$$f_{ij} = \begin{cases} 1, & \textit{if } v_i \textit{ is the positive end of } e_j, \\ -1, & \textit{if } v_i \textit{ is the negative end of } e_j, \\ 0, & \textit{if } v_i \textit{ is not incident with } e_j. \end{cases}$$

This matrix $F$ can also be obtained from the incidence matrix $A$ by changing either of the two 1s to $-1$ in each column.

The above arguments amount to arbitrarily orienting the edges of $G$, and $F$ is then the incidence matrix of the oriented graph.

The matrix $F$ is then the modified definition of the incidence matrix $A$.

**Example**   Consider the graph $G$ shown in Figure 10.2, with $V = \{v_1, v_2, v_3, v_4\}$ and $E = \{e_1, e_2, e_3, e_4, e_5\}$.

The incidence matrix is given by

$$A = \begin{array}{c} \\ v_1 \\ v_2 \\ v_3 \\ v_4 \end{array} \begin{array}{c} \begin{array}{ccccc} e_1 & e_2 & e_3 & e_4 & e_5 \end{array} \\ \left[ \begin{array}{ccccc} 1 & 0 & 0 & 1 & 0 \\ 1 & 1 & 0 & 0 & 1 \\ 0 & 1 & 1 & 0 & 0 \\ 0 & 0 & 1 & 1 & 1 \end{array} \right] \end{array}.$$

Therefore,

$$F = \begin{array}{c} \\ v_1 \\ v_2 \\ v_3 \\ v_4 \end{array} \begin{array}{c} \begin{array}{ccccc} e_1 & e_2 & e_3 & e_4 & e_5 \end{array} \\ \left[ \begin{array}{ccccc} 1 & 0 & 0 & 1 & 0 \\ -1 & 1 & 0 & 0 & 1 \\ 0 & -1 & 1 & 0 & 0 \\ 0 & 0 & -1 & -1 & -1 \end{array} \right] \end{array}$$

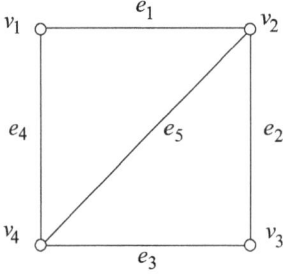

**Fig. 10.2**

**Theorem 10.5**   If $G$ is a connected graph with $n$ vertices, then rank $F = n - 1$.

**Proof**   Let $G$ be a connected graph with $V = \{v_1, v_2, \ldots, v_n\}$ and $E = \{e_1, e_2, \ldots, e_m\}$. Then, the matrix $F = [f_{ij}]_{n \times m}$ is given by

$$f_{ij} = \begin{cases} 1, & \text{if } v_i \text{ is the positive end of } e_j, \\ -1, & \text{if } v_i \text{ is the negative end of } e_j, \\ 0, & \text{if } v_i \text{ is not incident with } e_j. \end{cases}$$

Let $R_j$ be the $j$th row of $F$. Since each column of $F$ has only one $+1$ and one $-1$, as non-zero entries, rank $F < n$. Thus, rank $F \leq n - 1$.

Now, let $\sum_1^n c_j R_j = 0$ be any other linear dependence relation of $R_1, R_2, \ldots, R_n$ with at least one $c_j$ non-zero.

If $c_r \neq 0$, then the row $R_r$ has non-zero entries in those columns which correspond to edges incident with $v_r$. For each such column there is just one row, say $R_s$, at which there is a non-zero entry (with opposite sign to the non-zero entry in $R_r$). The dependence relation thus requires $c_s = c_r$, for all $s$ corresponding to vertices adjacent to $v_r$. Since $G$ is connected, we have $c_j = c$, for all $j = 1, 2, \ldots, n$. Therefore, the dependence relation is $c \left( \sum_1^n R_j \right) = 0$, which is same as the first one. Hence, rank $F = n - 1$.

**Alternative Proof**   Let $G = (V, E)$ be a connected graph with $n$ vertices and $m$ edges. Let $V = \{v_1, v_2, \ldots, v_n\}$ and $E = \{e_1, e_2, \ldots, e_m\}$. Then the matrix $F = [f_{ij}]$ is given by

$$f_{ij} = \begin{cases} 1, & \text{if } v_i \text{ is the positive end of } e_j, \\ -1, & \text{if } v_i \text{ is the negative end of } e_j, \\ 0, & \text{if } v_i \text{ is not incident with } e_j. \end{cases}$$

Let $R_1, R_1, \ldots, R_n$ be the row vectors of $F$.

Then,

$$F = \begin{bmatrix} R_1 \\ R_2 \\ M \\ R_n \end{bmatrix}. \tag{10.5.1}$$

Since each column of $F$ has only one $+1$ and one $-1$ as non-zero entries, $\sum_{j=1}^n R_j = 0$.

Thus,

$$\text{rank } F \leq n - 1. \tag{10.5.2}$$

Consider the sum of any $m$ of these row vectors, $m \leq n-1$. As $G$ is connected, $F$ cannot be partitioned in the form $F = \begin{bmatrix} F_1 & 0 \\ 0 & F_2 \end{bmatrix}$, such that $F_1$ has $m$ rows and $F_2$ as $n-m$ rows. Therefore, there exists no $m \times m$ submatrix of $F$ for $m \leq n-1$, such that the sum of these $m$ rows is equal to zero. Therefore,

$$\sum_{j=1}^{m} R_j \neq 0. \tag{10.5.3}$$

Also, there is no linear combination of $m(m \leq n-1)$ vectors of $F$, which is zero. For, if $\sum_{j=1}^{m} c_j R_j = 0$ is a linear combination, with at least one $c_j$ non-zero, say $c_r \neq 0$, then the row $R_r$ has non-zero entries in those columns which correspond to edges incident with $v_r$. So for each such column there is just one row, say $R_s$ at which there is a non-zero entry with opposite sign to the non-zero entry in $R_r$. The linear combination thus requires $c_s = c_r$ for all $s$ corresponding to vertices adjacent to $v_r$. As $G$ is connected, we have $c_j = c$, for all $j = 1,\ldots,m$. Therefore the linear combination becomes $c\left( \sum_{j=1}^{m} R_j \right) = 0$, or $\sum_{j=1}^{m} R_j = 0$, which contradicts (10.5.3).

Hence, rank $F = n-1$.                                                                                               ❏

**Theorem 10.6**    If $G$ is a disconnected graph with $k$ components, then rank $F = n-k$.

**Proof**    Since $G$ has $k$ components, $F$ can be partitioned as

$$F = \begin{bmatrix} F_1 & 0 & 0 & \cdots & 0 \\ 0 & F_2 & 0 & \cdots & 0 \\ 0 & 0 & F_3 & \cdots & 0 \\ \vdots & & & & \\ 0 & 0 & 0 & \cdots & F_k \end{bmatrix},$$

where $F_1$ is the matrix of the $i$th component $G_i$ of $G$. We have proved that rank $F_i = n_i - 1$, where $n_i$ is the number of vertices in $G_i$.

Thus, rank $F = n_1 - 1 + n_2 - 1 + \ldots + n_k - 1$

$$= n_1 + n_2 + \ldots + n_k - k = n - k,$$

as the number of vertices in $G$ is $n_1 + n_2 + \ldots + n_k = n$.                                    ❏

**Corollary 10.1** A basis for the row space of $F$ is obtained by taking for each $i$, $1 \leq i \leq k$, any $n_i - 1$ rows of $F_i$.

**Theorem 10.7**   The determinant of any square submatrix of the matrix $F$ of a graph $G$ has value 1, $-1$, or zero.

**Proof**   Let $N$ be the square submatrix of $F$ such that $N$ has both non-zero entries $+1$ and $-1$ in each column. Then row sum of $N$ is zero and hence $|N| = 0$. Clearly if $N$ has no non-zero entries, then $|N| = 0$.

   Now let some column of $N$ have only one non-zero entry. Then expanding $|N|$ with the help of this column, we get $|N| = \pm|N'|$, where $N'$ is a matrix obtained by omitting a row and column of $N$. Continuing in this way, we either get a matrix whose determinant is zero, or end up with a single non-zero entry of $N$, in which case $|N| = \pm 1$.   ❏

**Theorem 10.8**   Let $X$ be any set of $n-1$ edges of the connected graph $G = (V, E)$ and $F_x$ the $(n-1) \times (n-1)$ submatrix of the matrix $F$ of $G$, determined by any $n-1$ rows and those columns which correspond to the edges of $X$. Then $F_x$ is non-singular if and only if the edge induced subgraph $< X >$ of $G$ is a spanning tree of $G$.

**Proof**   Let $F'$ be the matrix corresponding to $< X >$. If $< X >$ is a spanning tree of $G$, then $F_x$ consists of $n-1$ rows of $F'$. Since $< X >$ is connected, therefore rank $F_x = n-1$. Hence $F_x$ is non-singular.

   Conversely, let $F_x$ be non-singular. Then $F'$ contains an $(n-1) \times (n-1)$ non singular submatrix. Therefore, rank $F' = n-1$. Since rank + nullity $= m$, for any graph $G$, and $m(< X >) = n-1$ and rank $(< X >) = n-1$, therefore, nullity $(< X >) = 0$. Thus $< X >$ is acyclic and connected and so is a spanning tree of $G$.   ❏

## 10.3   Cycle Matrix

Let the graph $G$ have $m$ edges and let $q$ be the number of different cycles in $G$. The cycle matrix $B = [b_{ij}]_{q \times m}$ of $G$ is a $(0, 1)-$ matrix of order $q \times m$, with $b_{ij} = 1$, if the $i$th cycle includes $j$th edge and $b_{ij} = 0$, otherwise. The cycle matrix $B$ of a graph $G$ is denoted by $B(G)$.

**Example**   Consider the graph $G_1$ given in Figure 10.3.

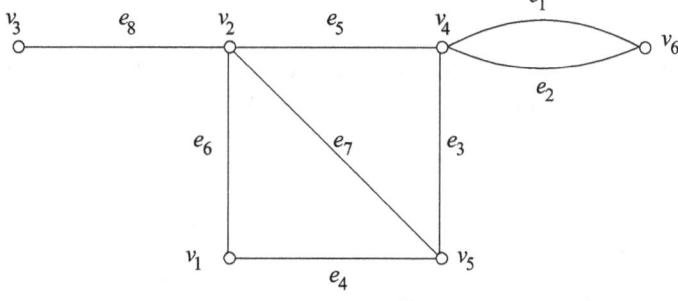

**Fig. 10.3**

The graph $G_1$ has four different cycles $Z_1 = \{e_1, e_2\}$, $Z_2 = \{e_3, e_5, e_7\}$, $Z_3 = \{e_4, e_6, e_7\}$ and $Z_4 = \{e_3, e_4, e_6, e_5\}$.

The cycle matrix is

$$
B(G_1) = \begin{array}{c} \\ z_1 \\ z_2 \\ z_3 \\ z_4 \end{array}
\begin{array}{cccccccc}
e_1 & e_2 & e_3 & e_4 & e_5 & e_6 & e_7 & e_8 \\
\end{array}
\left[
\begin{array}{cccccccc}
1 & 1 & 0 & 0 & 0 & 0 & 0 & 0 \\
0 & 0 & 1 & 0 & 1 & 0 & 1 & 0 \\
0 & 0 & 0 & 1 & 0 & 1 & 1 & 0 \\
0 & 0 & 1 & 1 & 1 & 1 & 0 & 0 \\
\end{array}
\right].
$$

The graph $G_2$ of Figure 10.3 has seven different cycles, namely, $Z_1 = \{e_1, e_2\}$, $Z_2 = \{e_2, e_7, e_8\}$, $Z_3 = \{e_1, e_7, e_8\}$, $Z_4 = \{e_4, e_5, e_6, e_7\}$, $Z_5 = \{e_2, e_4, e_5, e_6, e_8\}$, $Z_6 = \{e_1, e_4, e_5, e_6, e_8\}$ and $Z_7 = \{e_9\}$. The cycle matrix is given by

$$
B(G_2) = \begin{array}{c} \\ z_1 \\ z_2 \\ z_3 \\ z_4 \\ z_5 \\ z_6 \\ z_7 \end{array}
\begin{array}{ccccccccc}
e_1 & e_2 & e_3 & e_4 & e_5 & e_6 & e_7 & e_8 & e_9 \\
\end{array}
\left[
\begin{array}{ccccccccc}
1 & 1 & 0 & 0 & 0 & 0 & 0 & 0 & 0 \\
0 & 1 & 0 & 0 & 0 & 0 & 1 & 1 & 0 \\
1 & 0 & 0 & 0 & 0 & 0 & 1 & 1 & 0 \\
0 & 0 & 0 & 1 & 1 & 1 & 1 & 0 & 0 \\
0 & 1 & 0 & 1 & 1 & 1 & 0 & 1 & 0 \\
1 & 0 & 0 & 1 & 1 & 1 & 0 & 1 & 0 \\
0 & 0 & 0 & 0 & 0 & 0 & 0 & 0 & 1 \\
\end{array}
\right].
$$

We have the following observations regarding the cycle matrix $B(G)$ of a graph $G$.

1. A column of all zeros corresponds to a non cycle edge, that is, an edge which does not belong to any cycle.

2. Each row of $B(G)$ is a cycle vector.

3. A cycle matrix has the property of representing a self-loop and the corresponding row has a single one.

4. The number of ones in a row is equal to the number of edges in the corresponding cycle.

5. If the graph $G$ is separable (or disconnected) and consists of two blocks (or components) $H_1$ and $H_2$, then the cycle matrix $B(G)$ can be written in a block-diagonal form as

$$
B(G) = \begin{bmatrix} B(H_1) & 0 \\ 0 & B(H_2) \end{bmatrix},
$$

where $B(H_1)$ and $B(H_2)$ are the cycle matrices of $H_1$ and $H_2$. This follows from the fact that cycles in $H_1$ have no edges belonging to $H_2$ and vice versa.

6. Permutation of any two rows or columns in a cycle matrix corresponds to relabeling the cycles and the edges.

7. We know two graphs $G_1$ and $G_2$ are 2-isomorphic if and only if they have cycle correspondence. Thus two graphs $G_1$ and $G_2$ have the same cycle matrix if and only if $G_1$ and $G_2$ are 2-isomorphic. This implies that the cycle matrix does not specify a graph completely, but only specifies the graph within 2-isomorphism.

For example, the two graphs given in Figure 10.4 have the same cycle matrix. They are 2-isomorphic, but are not isomorphic.

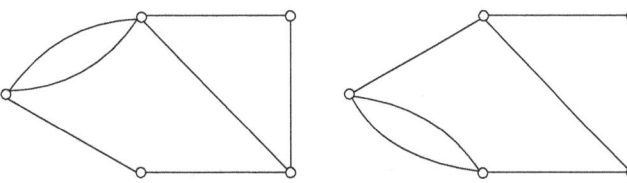

**Fig. 10.4**

The following result relates the incidence and cycle matrix of a graph without self-loops.

**Theorem 10.9** If $G$ is a graph without self-loops, with incidence matrix $A$ and cycle matrix $B$ whose columns are arranged using the same order of edges, then every row of $B$ is orthogonal to every row of $A$, that is $AB^T = BA^T = 0 \pmod{2}$, where $A^T$ and $B^T$ are the transposes of $A$ and $B$ respectively.

**Proof** Let $G$ be a graph without self-loops, and let $A$ and $B$, respectively, be the incidence and cycle matrix of $G$.

We know that in $G$ for any vertex $v_i$ and for any cycle $Z_j$, either $v_i \in Z_j$ or $v_i \notin Z_j$.

In case $v_i \notin Z_j$, then there is no edge of $Z_j$ which is incident on $v_i$ and if $v_i \in Z_j$, then there are exactly two edges of $Z_j$ which are incident on $v_i$.

Now, consider the $i$th row of $A$ and the $j$th row of $B$ (which is the $j$th column of $B^T$).

Since the edges are arranged in the same order, the $r$th entries in these two rows are both non-zero if and only if the edge $e_r$ is incident on the $i$th vertex $v_i$ and is also in the $j$th cycle $Z_j$.

We have $[AB^T]_{ij} = \sum [A]_{ir}[B^T]_{rj} = \sum [A]_{ir}[B]_{jr} = \sum a_{ir}b_{jr}$.

For each $e_r$ of $G$, we have one of the following cases.

  i. $e_r$ is incident on $v_i$ and $e_r \notin Z_j$. Here, $a_{ir} = 1$, $b_{jr} = 0$.

  ii. $e_r$ is not incident on $v_i$ and $e_r \in Z_j$. In this case, $a_{ir} = 0$, $b_{jr} = 1$.

  iii. $e_r$ is not incident on $v_i$ and $e_r \notin Z_j$, so that $a_{ir} = 0$, $b_{jr} = 0$.

    All these cases imply that the $i$th vertex $v_i$ is not in the $j$th cycle $Z_j$ and we have $[AB^T]_{ij} = 0 \equiv 0 \pmod{2}$.

iv. $e_r$ is incident on $v_i$ and $e_r \in Z_j$.

Here we have exactly two edges, say $e_r$ and $e_t$ incident on $v_i$ so that $a_{ir} = 1$, $a_{it} = 1$, $b_{jr} = 1$, $b_{jt} = 1$. Therefore, $[AB^T]_{ij} = \sum a_{ir} b_{jr} = 1 + 1 \equiv 0 \pmod 2$. ❏

We illustrate the above theorem with the following example (Fig. 10.5).

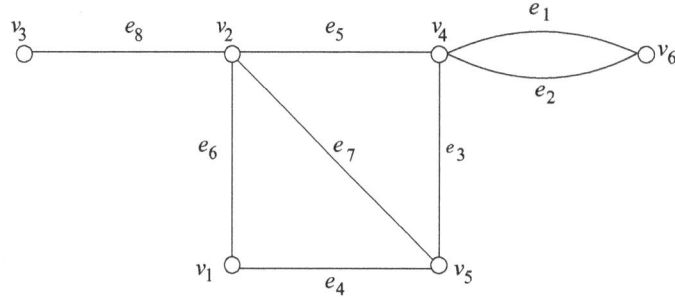

**Fig. 10.5**

Here,

$$
A = \begin{array}{c}
\\
v_1 \\
v_2 \\
v_3 \\
v_4 \\
v_5 \\
v_6
\end{array}
\begin{array}{c}
e_1\ e_2\ e_3\ e_4\ e_5\ e_6\ e_7\ e_8 \\
\left[\begin{array}{cccccccc}
0 & 0 & 0 & 1 & 0 & 1 & 0 & 0 \\
0 & 0 & 0 & 0 & 1 & 1 & 1 & 1 \\
0 & 0 & 0 & 0 & 0 & 0 & 0 & 1 \\
1 & 1 & 1 & 0 & 1 & 0 & 0 & 0 \\
0 & 0 & 1 & 1 & 0 & 0 & 1 & 0 \\
1 & 1 & 0 & 0 & 0 & 0 & 0 & 0
\end{array}\right]
\end{array}.
$$

The cycles of $G$ are $Z_1 = \{e_1, e_2\}$, $Z_2 = \{e_3, e_5, e_7\}$, $Z_3 = \{e_4, e_6, e_7\}$ and $Z_4 = \{e_3, e_4, e_6, e_5\}$. Therefore,

$$
B = \begin{array}{c}
\\
Z_1 \\
Z_2 \\
Z_3 \\
Z_4
\end{array}
\begin{array}{c}
e_1\ e_2\ e_3\ e_4\ e_5\ e_6\ e_7\ e_8 \\
\left[\begin{array}{cccccccc}
1 & 1 & 0 & 0 & 0 & 0 & 0 & 0 \\
0 & 0 & 1 & 0 & 1 & 0 & 1 & 0 \\
0 & 0 & 0 & 1 & 0 & 1 & 1 & 0 \\
0 & 0 & 1 & 1 & 1 & 1 & 0 & 0
\end{array}\right]
\end{array}.
$$

Therefore,

$$AB^T = \begin{bmatrix} 0 & 0 & 0 & 1 & 0 & 1 & 0 & 0 \\ 0 & 0 & 0 & 0 & 1 & 1 & 1 & 1 \\ 0 & 0 & 0 & 0 & 0 & 0 & 0 & 1 \\ 1 & 1 & 1 & 0 & 1 & 0 & 0 & 0 \\ 0 & 0 & 1 & 1 & 0 & 0 & 1 & 0 \\ 1 & 1 & 0 & 0 & 0 & 0 & 0 & 0 \end{bmatrix} \begin{bmatrix} 1 & 0 & 0 & 0 \\ 1 & 0 & 0 & 0 \\ 0 & 1 & 0 & 1 \\ 0 & 0 & 1 & 1 \\ 0 & 1 & 0 & 1 \\ 0 & 0 & 1 & 1 \\ 0 & 1 & 1 & 0 \\ 0 & 0 & 0 & 0 \end{bmatrix}$$

$$= \begin{bmatrix} 0 & 0 & 2 & 2 \\ 0 & 2 & 2 & 2 \\ 0 & 0 & 0 & 0 \\ 2 & 2 & 0 & 2 \\ 0 & 2 & 2 & 2 \\ 2 & 0 & 0 & 0 \end{bmatrix} \equiv 0 (\text{mod} 2).$$

We know that a set of fundamental cycles (or basic cycles) with respect to any spanning tree in a connected graph are the only independent cycles in a graph. The remaining cycles can be obtained as ring sums (i.e., linear combinations) of these cycles. Thus, in a cycle matrix, if we take only those rows that correspond to a set of fundamental cycles and remove all other rows, we do not lose any information. The removed rows can be formed from the rows corresponding to the set of fundamental cycles. For example, in the cycle matrix of the graph given in Figure 10.6, the fourth row is simply the mod 2 sum of the second and the third rows. Fundamental cycles are

$$Z_1 = \{e_1, e_2\}, \quad Z_2 = \{e_3, e_5, e_7\} \text{ and } Z_3 = \{e_4, e_6, e_7\}.$$

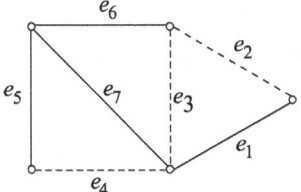

**Fig. 10.6**

$$\begin{array}{c} \\ Z_1 \\ Z_2 \\ Z_3 \end{array} \begin{array}{cccccccc} e_2 & e_3 & e_6 & & e_1 & e_4 & e_5 & e_7 \\ \left[\begin{array}{ccc} 1 & 0 & 0 \end{array}\right. & \vdots & \left. 1 & 1 & 0 & 1 \right] \\ \left[\begin{array}{ccc} 0 & 1 & 0 \end{array}\right. & \vdots & \left. 0 & 1 & 0 & 1 \right] \\ \left[\begin{array}{ccc} 0 & 0 & 1 \end{array}\right. & \vdots & \left. 0 & 0 & 1 & 1 \right] \end{array}$$

A submatrix of a cycle matrix in which all rows correspond to a set of fundamental cycles is called a *fundamental cycle matrix* $B_f$.

The permutation of rows or columns do not affect $B_f$. If $n$ is the number of vertices, $m$ the number of edges in a connected graph $G$, then $B_f$ is an $(m-n+1) \times m$ matrix because the number of fundamental cycles is $m-n+1$, each fundamental cycle being produced by one chord.

Now, arranging the columns in $B_f$ such that all the $m-n+1$ chords correspond to the first $m-n+1$ columns and rearranging the rows such that the first row corresponds to the fundamental cycle made by the chord in the first column, the second row to the fundamental cycle made by the second, and so on. This arrangement is done for the above fundamental cycle matrix.

A matrix $B_f$ thus arranged has the form

$$B_f = [I_\mu : B_t],$$

where $I_\mu$ is an identity matrix of order $\mu = m-n+1$ and $B_t$ is the remaining $\mu \times n - 1$ submatrix, corresponding to the branches of the spanning tree.

From equation $B_f = [I_\mu : B_t]$, we have rank $B_f = \mu = m-n+1$.

Since $B_f$ is a submatrix of the cycle matrix $B$, therefore, rank $B \geq$ rank $B_f$ and thus,

rank $B \geq m-n+1$.

The following result gives the rank of the cycle matrix.

**Theorem 10.10**  If $B$ is a cycle matrix of a connected graph $G$ with $n$ vertices and $m$ edges, then rank $B = m-n+1$.

**Proof**  Let $A$ be the incidence matrix of the connected graph $G$.

Then, $AB^T \equiv 0 \pmod 2$.

Using Sylvester's theorem (Theorem 10.13), we have rank $A+$ rank $B^T \leq m$ so that rank $A+$ rank $B \leq m$.

Therefore, rank $B \leq m-$ rank $A$.

As rank $A = n-1$, we get rank $B \leq m-(n-1) = m-n+1$.

But, rank $B \geq m-n+1$.

Combining, we get rank $B = m-n+1$.                                                        ❑

Theorem 10.10 can be generalised in the following form.

**Theorem 10.11**  If $B$ is a cycle matrix of a disconnected graph $G$ with $n$ vertices, $m$ edges and $k$ components, then rank $B = m-n+k$.

**Proof** Let $B$ be the cycle matrix of the disconnected graph $G$ with $n$ vertices, $m$ edges and $k$ components. Let the $k$ components be $G_1, G_2, ..., G_k$ with $n_1, n_2, ..., n_k$ vertices and $m_1, m_2, ..., m_k$ edges respectively.

Then, $n_1 + n_2 + ... + n_k = n$ and $m_1 + m_2 + ... + m_k = m$.

Let $B_1, B_2, ..., B_k$ be the cycle matrices of $G_1, G_2, ..., G_k$.

$$\text{Then, } B(G) = \begin{bmatrix} B_1(G_1) & 0 & 0 & \cdots & 0 \\ 0 & B_2(G_2) & 0 & \cdots & 0 \\ 0 & 0 & B_3(G_3) & \cdots & 0 \\ \vdots & & & & \\ 0 & 0 & 0 & \cdots & B_k(G_k) \end{bmatrix}.$$

We know rank $B_i = m_i - n_i + 1$, for $1 \le i \le k$.

Therefore, rank $B = \text{rank } B_1 + ... + \text{rank } B_k$

$$= (m_1 - n_1 + 1) + ... + (m_k - n_k + 1)$$

$$= (m_1 + ... + m_k) - (n_1 + ... + nk) + k = m - n + k. \qquad \square$$

**Definition:** Let $A$ be a matrix of order $k \times m$, with $k < m$. The *major determinant* of $A$ is the determinant of the largest square submatrix of $A$, formed by taking any $k$ columns of $A$. That is, the determinant of any $k \times k$ square submatrix is called the major determinant of $A$.

Let $A$ and $B$ be matrices of orders $k \times m$ and $m \times k$ respectively $(k < m)$. If columns $i_1, i_2, ..., i_k$ of $B$ are chosen for a particular major of $B$, then the corresponding major in $A$ consists of the rows $i_1, i_2, ..., i_k$ in $A$.

If $A$ is a square matrix of order $n$, then $AX = 0$ has a non trivial solution $X \ne 0$ if and only if $A$ is singular, that is $|A| = 0$. The set of all vectors $X$ that satisfy $AX = 0$ forms a vector space called the null space of matrix $A$. The rank of the null space is called the nullity of $A$. Further more,

$$\text{rank } A + nullity A = n.$$

These definitions and the above equation also hold when $A$ is a matrix of order $k \times n$, $k < n$.

We now give Binet-Cauchy and Sylvester theorems which will be used in the further discussions.

**Theorem 10.12 (Binet–Cauchy)** If $A$ and $B$ are two matrices of the order $k \times m$ and $m \times k$ respectively $(k < m)$, then $|AB| =$ sum of the products of corresponding major determinants of $A$ and $B$.

**Proof**   We multiply two $(m+k) \times (m+k)$ partitioned matrices to get

$$\begin{bmatrix} I_k & A \\ O & I_m \end{bmatrix} \begin{bmatrix} A & 0 \\ -I_m & B \end{bmatrix} = \begin{bmatrix} 0 & AB \\ -I_m & B \end{bmatrix},$$

where $I_m$ and $I_k$ are identity matrices of order $m$ and $k$ respectively.

Therefore, $\det \begin{bmatrix} A & 0 \\ -I_m & B \end{bmatrix} = \det \begin{bmatrix} 0 & AB \\ -I_m & B \end{bmatrix}.$

Thus, $\det(AB) = \det \begin{bmatrix} A & 0 \\ -I_m & B \end{bmatrix}.$                                                    (10.12.1)

Now apply Cauchy's expansion method to the right side of (10.12.1) and observe that the only non-zero minors of any order in $-I_m$ are its principal minors of that order. Therefore, we see that the Cauchy expansion consists of these minors of order $m - k$ multiplied by their cofactors of order $k$ in $A$ and $B$ together.                                                    ❑

**Theorem 10.13 (Sylvester)**   If $A$ and $B$ are matrices of order $k \times m$ and $n \times p$ respectively, then nullity $AB \leq$ nullity $A+$ nullity $B$.

**Proof**   Since every vector $X$ satisfying $BX = 0$ also satisfies $ABX = 0$, therefore, we have

nullity $AB \geq$ nullity $B \geq 0$.                                                    (10.13.1)

Let nullity $B = s$. So there exists a set of s linearly independent vectors $\{x_1, x_2, \ldots, x_s\}$ forming a basis of the null space of $B$. Therefore,

$BX_i = 0$, for $i = 1, 2, \ldots, s$.                                                    (10.13.2)

Now let nullity $AB = s+t$. Thus, there exists a set of $t$ linearly independent vectors $[X_{s+1}, X_{s+2}, \ldots, X_{s+t}]$ such that the set $\{X_1, X_2, \ldots, X_s, X_{s+1}, \ldots, X_{s+t}\}$ forms a basis for the null space of $AB$. Therefore,

$ABX_i = 0$, for $i = 1, 2, \ldots, s, s+1, \ldots, s+t$.                                                    (10.13.3)

This implies that out of the $s+t$ vectors $x_i$ forming a basis of the null space of $AB$, the first $s$ vectors are made zero by $B$ and the remaining non-zero $BX_i's, i = s+1, \ldots, s+t$ are made zero by $A$.

Clearly, the vectors $Bx_{s+1}, \ldots, Bx_{s+t}$ are linearly independent. For if

$b_1 BX_{s+1} + b_2 BX_{s+2} + \ldots + b_t BX_{s+t} = 0,$

i.e., if $B(b_1 X_{s+1} + b_2 X_{s+2} + \ldots + bt X_{s+t}) = 0,$

then the vector $b_1X_{s+1} + b_2X_{s+2} + \ldots + b_tX_{s+t}$ is the null space of $B$, which is possible only if $b_1 = b_2 = \ldots = b_t = 0$.

Thus, we have seen that there are at least t linearly independent vectors which are made zero by $A$. So, nullity $A \geq t$.

Since $t = (s+t) - s$, therefore $t =$ nullity $AB -$ nullity $B$

Therefore, nullity $AB-$ nullity $B \leq$ nullity $A$, and so

nullity $AB \leq$ nullity $A +$ nullity $B$.　　　　(10.13.4)　　　　❑

**Corollary 10.2**　　We know, rank $A +$ nullity $A = n$, and using this in (10.13.4), we get

$n-$ rank $AB \leq n-$ rank $A + n-$ rank $B$.

Therefore, rank $AB \geq$ rank $A +$ rank $B - n$.

If in above, $AB = 0$, then rank $A +$ rank $B \leq n$.

## 10.4　Cut-Set Matrix

Let $G$ be a graph with m edges and $q$ cutsets. The cut-set matrix $C = [c_{ij}]_{q \times m}$ of $G$ is a $(0, 1)$-matrix with

$$c_{ij} = \begin{cases} 1, & \text{if ith cutset contains jth edge,} \\ 0, & \text{otherwise.} \end{cases}$$

**Example**　　Consider the graphs shown in Figure 10. 7.

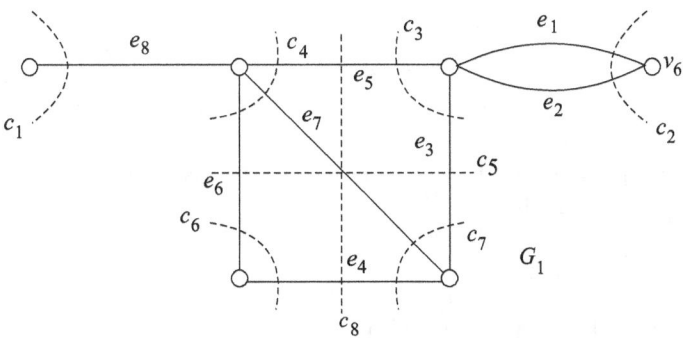

**Fig. 10.7(a)**

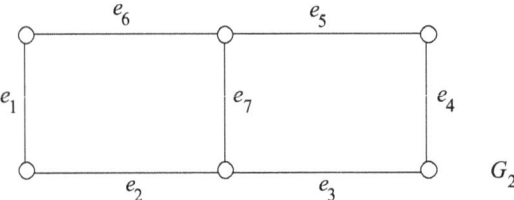

**Fig. 10.7(b)**

In the graph $G_1$, $E = \{e_1, e_2, e_3, e_4, e_5, e_6, e_7, e_8\}$.

The cut-sets are $c_1 = \{e_8\}$, $c_2 = \{e_1, e_2\}$, $c_3 = \{e_3, e_5\}$, $c_4 = \{e_5, e_6, e_7\}$, $c_5 = \{e_3, e_6, e_7\}$, $c_6 = \{e_4, e_6\}$, $c_7 = \{e_3, e_4, e_7\}$ and $c_8 = \{e_4, e_5, e_7\}$.

The cut-sets for the graph $G_2$ are $c_1 = \{e_1, e_2\}$, $c_2 = \{e_3, e_4\}$, $c_3 = \{e_4, e_5\}$, $c_4 = \{e_1, e_6\}$, $c_5 = \{e_2, e_6\}$, $c_6 = \{e_3, e_5\}$, $c_7 = \{e_1, e_4, e_7\}$, $c_8 = \{e_2, e_3, e_7\}$ and $c_9 = \{e_5, e_6, e_7\}$.

Thus, the cut-set matrices are given by

$$
C(G_1) = \begin{array}{c}
\\ c_1 \\ c_2 \\ c_3 \\ c_4 \\ c_5 \\ c_6 \\ c_7 \\ c_8
\end{array}
\begin{array}{c}
\begin{array}{cccccccc} e_1 & e_2 & e_3 & e_4 & e_5 & e_6 & e_7 & e_8 \end{array} \\
\left[\begin{array}{cccccccc}
0 & 0 & 0 & 0 & 0 & 0 & 0 & 1 \\
1 & 1 & 0 & 0 & 0 & 0 & 0 & 0 \\
0 & 0 & 1 & 0 & 1 & 0 & 0 & 0 \\
0 & 0 & 0 & 0 & 1 & 1 & 1 & 0 \\
0 & 0 & 1 & 0 & 0 & 1 & 1 & 0 \\
0 & 0 & 0 & 1 & 0 & 1 & 0 & 0 \\
0 & 0 & 1 & 1 & 0 & 0 & 1 & 0 \\
0 & 0 & 0 & 1 & 1 & 0 & 1 & 0
\end{array}\right]
\end{array}, \text{ and}
$$

$$
C(G_2) = \begin{array}{c}
\\ c_1 \\ c_2 \\ c_3 \\ c_4 \\ c_5 \\ c_6 \\ c_7 \\ c_8 \\ c_9
\end{array}
\begin{array}{c}
\begin{array}{ccccccc} e_1 & e_2 & e_3 & e_4 & e_5 & e_6 & e_7 \end{array} \\
\left[\begin{array}{ccccccc}
1 & 1 & 0 & 0 & 0 & 0 & 0 \\
0 & 0 & 1 & 1 & 0 & 0 & 0 \\
0 & 0 & 0 & 1 & 1 & 0 & 0 \\
1 & 0 & 0 & 0 & 0 & 1 & 0 \\
0 & 1 & 0 & 0 & 0 & 1 & 0 \\
0 & 0 & 1 & 0 & 1 & 0 & 0 \\
1 & 0 & 0 & 1 & 0 & 0 & 1 \\
0 & 1 & 1 & 0 & 0 & 0 & 1 \\
0 & 0 & 0 & 0 & 1 & 1 & 1
\end{array}\right]
\end{array}.
$$

We have the following observations about the cut-set matrix $C(G)$ of a graph $G$.

1. The permutation of rows or columns in a cut-set matrix corresponds simply to re-naming of the cut-sets and edges respectively.

2. Each row in $C(G)$ is a cut-set vector.

3. A column with all zeros corresponds to an edge forming a self-loop.

4. Parallel edges form identical columns in the cut-set matrix.

5. In a non-separable graph, since every set of edges incident on a vertex is a cut-set, therefore every row of incidence matrix $A(G)$ is included as a row in the cut-set matrix $C(G)$. That is, for a non-separable graph $G$, $C(G)$ contains $A(G)$. For a separable graph, the incidence matrix of each block is contained in the cut-set matrix. For example, in the graph $G_1$ of Figure 10.7, the incidence matrix of the block $\{e_3, e_4, e_5, e_6, e_7\}$ is the $4 \times 5$ submatrix of $C$, left after deleting rows $c_1, c_2, c_5, c_8$ and columns $e_1, e_2, e_8$.

6. It follows from observation 5, that rank $C(G) \geq$ rank $A(G)$. Therefore, for a connected graph with $n$ vertices, rank $C(G) \geq n - 1$.

The following is an important result for connected graphs.

**Theorem 10.14**   If $G$ is a connected graph, then the rank of a cut-set matrix $C(G)$ is equal to the rank of incidence matrix $A(G)$, which equals the rank of graph $G$.

**Proof**   Let $A(G)$, $B(G)$ and $C(G)$ be the incidence, cycle and cut-set matrix of the connected graph $G$. Then we have

$$\text{rank } C(G) \geq n - 1. \tag{10.14.1}$$

Since the number of edges common to a cut-set and a cycle is always even, every row in $C$ is orthogonal to every row in $B$, provided the edges in both $B$ and $C$ are arranged in the same order.

$$\text{Thus, } BC^T = CB^T = 0 \text{ (mod 2)}. \tag{10.14.2}$$

Now, applying Sylvester's theorem to equation (10.14.2), we have

rank $B +$ rank $C \leq m$.

For a connected graph, we have rank $B = m - n + 1$.

Therefore, rank $C \leq m -$ rank $B = m - (m - n + 1) = n - 1$.

So, rank $C \leq n - 1$. $\tag{10.14.3}$

It follows from (10.14.1) and (10.14.3), that rank $C = n - 1$.   ❏

## 10.5 Fundamental Cut-Set Matrix

Let $G$ be a connected graph with $n$ vertices and $m$ edges. The fundamental cut-set matrix $C_f$ of $G$ is an $(n-1) \times m$ submatrix of $C$ such that the rows correspond to the set of fundamental cut-sets with respect to some spanning tree. Clearly, a fundamental cut-set matrix $C_f$ can be partitioned into two submatrices, one of which is an identity matrix $I_{n-1}$ of order $n-1$. We have

$$C_f = [C_c : I_{n-1}],$$

where the last $n-1$ columns forming the identity matrix correspond to the $n-1$ branches of the spanning tree and the first $m-n+1$ columns forming $C_c$ correspond to the chords.

**Example**   Consider the connected graphs $G_1$ and $G_2$ given in Figure 10.8. The spanning tree is shown with bold lines. The fundamental cut-sets of $G_1$ are $c_1$, $c_2$, $c_3$, $c_6$ and $c_7$ while the fundamental cut-sets of $G_2$ are $c_1$, $c_2$, $c_3$, $c_4$ and $c_7$.

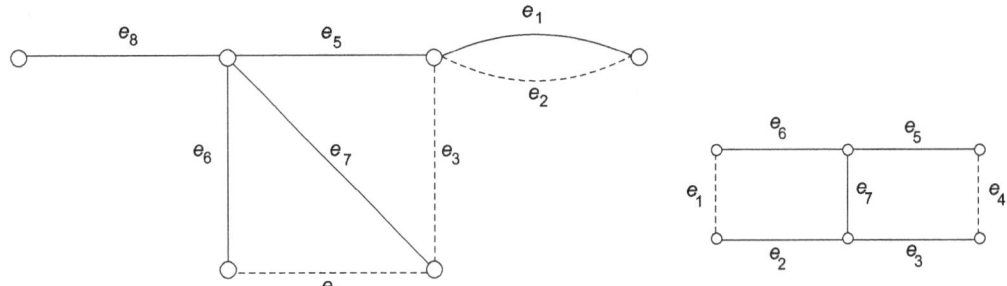

**Fig. 10.8**

The fundamental cut-set matrix of $G_1$ and $G_2$, respectively are given by

$$
C_f =
\begin{array}{c}
\begin{array}{ccccccccc}
e_2 & e_3 & e_4 & & e_1 & e_5 & e_6 & e_7 & e_8
\end{array} \\
\left[
\begin{array}{ccc:ccccc}
1 & 0 & 0 & 1 & 0 & 0 & 0 & 0 \\
0 & 1 & 0 & 0 & 1 & 0 & 0 & 0 \\
0 & 0 & 1 & 0 & 0 & 1 & 0 & 0 \\
0 & 1 & 1 & 0 & 0 & 0 & 1 & 0 \\
0 & 0 & 0 & 0 & 0 & 0 & 0 & 1
\end{array}
\right]
\end{array}
\quad \text{and}
$$

$$
C_f = \begin{array}{c} \\ c_1 \\ c_2 \\ c_3 \\ c_4 \\ c_7 \end{array}
\begin{array}{c}
\begin{array}{ccccccc} e_1 & e_4 & e_2 & e_3 & e_5 & e_6 & e_7 \end{array} \\
\left[ \begin{array}{ccccccc}
1 & 0 & 1 & 0 & 0 & 0 & 0 \\
0 & 1 & 0 & 1 & 0 & 0 & 0 \\
0 & 1 & 0 & 0 & 1 & 0 & 0 \\
1 & 0 & 0 & 0 & 0 & 1 & 0 \\
1 & 1 & 0 & 0 & 0 & 0 & 1
\end{array} \right]
\end{array}
$$

## 10.6 Relations between $A_f$, $B_f$ and $C_f$

Let $G$ be a connected graph and $A_f, B_f$ and $C_f$ be respectively the reduced incidence matrix, the fundamental cycle matrix, and the fundamental cut-set matrix of $G$.

We have shown that

$$
B_f = \begin{bmatrix} I_\mu \ M \ B_t \end{bmatrix} \tag{10.6.i}
$$

and $\quad C_f = [C_c \ M \ I_{n-1}],$ $\tag{10.6.ii}$

where $t$ denotes the submatrix corresponding to the branches of a spanning tree and $C$ denotes the submatrix corresponding to the chords.

Let the spanning tree $T$ in Equations (10.6.i) and (10.6.ii) be the same and let the order of the edges in both equations be same. Also, in the reduced incidence matrix $A_f$ of size $(n-1) \times m$, let the edges (i.e., the columns) be arranged in the same order as in $B_f$ and $C_f$.

Partition $A_f$ into two submatrices given by

$$
A_f = \begin{bmatrix} A_c \vdots A_t \end{bmatrix}, \tag{10.6.iii}
$$

where $A_t$ consists of $n-1$ columns corresponding to the branches of the spanning tree $T$ and $A_c$ is the spanning submatrix corresponding to the $m-n+1$ chords.

Since the columns in $A_f$ and $B_f$ are arranged in the same order, the equation $AB^T = BA^T = 0 \pmod 2$ gives

$$
A_f B_f^T \equiv 0 \pmod 2,
$$

or $\quad [A_c \ M \ A_t] \begin{bmatrix} I_\mu \\ \vdots \\ B_t^T \end{bmatrix} \equiv 0 \pmod 2,$

or $\quad A_c + A_t B_f^T \equiv 0 \pmod 2.$ $\tag{10.6.iv}$

Since $A_t$ is non singular, $A_t^{-1}$ exists. Now, premultiplying both sides of equation (10.6.iv) by $A_t^{-1}$, we have

$$A_t^{-1}A_c + A_t^{-1}A_t B_t^T \equiv 0 (\text{mod } 2),$$

or   $A_t^{-1}A_c + B_t^T \equiv 0 (\text{mod } 2).$

Therefore, $A_t^{-1}A_c = -B_t^T$.

Since in mod 2 arithmetic $-1 = 1$,

$$B_t^T = A_t^{-1}A_c. \tag{10.6.v}$$

Now as the columns in $B_f$ and $C_f$ are arranged in the same order, therefore (in mod 2 arithmetic) $C_f \cdot B_f^T \equiv 0 (\text{mod } 2)$ in mod 2 arithmetic gives $C_f \cdot B_f^T = 0$.

Therefore, $[C_c \; M \; I_{n-1}] \begin{bmatrix} I_\mu \\ \vdots \\ B_t^T \end{bmatrix} = 0$, so that $C_c + B_t^T = 0$, that is, $C_c = -B_t^T$.

Thus, $C_c = B_t^T$ (as $-1 = 1$ in mod 2 arithmetic).

Hence, $C_c = A_t^{-1} A_c$ from (10.6.v).

**Remarks**   We make the following observations from the above relations.

1. If $A$ or $A_f$ is given, we can construct $B_f$ and $C_f$ starting from an arbitrary spanning tree and its submatrix $A_t$ in $A_f$.

2. If either $B_f$ or $C_f$ is given, we can construct the other. Therefore, since $B_f$ determines a graph within 2-isomorphism, so does $C_f$.

3. If either $B_f$ and $C_f$ is given, then $A_f$ in general cannot be determined completely.

**Example**   Consider the graph $G$ of Figure 10.9.

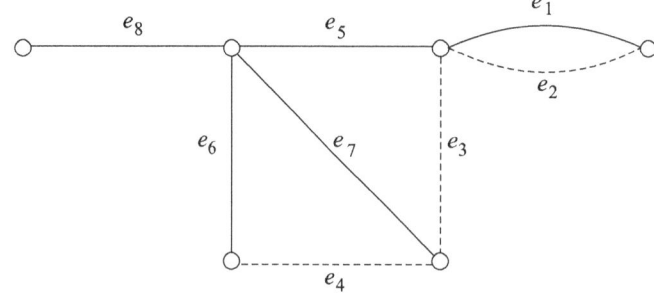

**Fig. 10.9**

Let $\{e_1, e_5, e_6, e_7, e_8\}$ be the spanning tree.

$$
\begin{array}{cccccccc}
e_1 & e_2 & e_3 & e_4 & e_5 & e_6 & e_7 & e_8
\end{array}
$$

We have, $A = \begin{bmatrix} 0 & 0 & 0 & 1 & 0 & 1 & 0 & 0 \\ 0 & 0 & 0 & 0 & 1 & 1 & 1 & 1 \\ 0 & 0 & 0 & 0 & 0 & 0 & 0 & 1 \\ 1 & 1 & 1 & 0 & 1 & 0 & 0 & 0 \\ 0 & 0 & 1 & 1 & 0 & 0 & 1 & 0 \\ 1 & 1 & 0 & 0 & 0 & 0 & 0 & 0 \end{bmatrix}$.

Dropping the sixth row in $A$, we get

$$
\begin{array}{ccccccccc}
 & e_2 & e_3 & e_4 & & e_1 & e_5 & e_6 & e_7 & e_8
\end{array}
$$

$A_f = \begin{bmatrix} 0 & 0 & 1 & : & 0 & 0 & 1 & 0 & 0 \\ 0 & 0 & 0 & : & 0 & 1 & 1 & 1 & 1 \\ 0 & 0 & 0 & : & 0 & 0 & 0 & 0 & 1 \\ 1 & 1 & 0 & : & 1 & 1 & 0 & 0 & 0 \\ 0 & 1 & 1 & : & 0 & 0 & 0 & 1 & 0 \end{bmatrix} = [A_c : A_t]$.

$$
\begin{array}{ccccccccc}
 & e_2 & e_3 & e_4 & & e_1 & e_5 & e_6 & e_7 & e_8
\end{array}
$$

$B_f = \begin{bmatrix} 1 & 0 & 0 & : & 1 & 0 & 0 & 0 & 0 \\ 0 & 1 & 0 & : & 0 & 1 & 0 & 1 & 0 \\ 0 & 0 & 1 & : & 0 & 0 & 1 & 1 & 0 \end{bmatrix} = [I_3 : B_t]$ and

$$
\begin{array}{cccccccc}
e_2 & e_3 & e_4 & e_1 & e_5 & e_6 & e_7 & e_8
\end{array}
$$

$C_f = \begin{bmatrix} 1 & 0 & 0 & 1 & 0 & 0 & 0 & 0 \\ 0 & 1 & 0 & 0 & 1 & 0 & 0 & 0 \\ 0 & 0 & 1 & 0 & 0 & 1 & 0 & 0 \\ 0 & 1 & 1 & 0 & 0 & 0 & 1 & 0 \\ 0 & 0 & 0 & 0 & 0 & 0 & 0 & 1 \end{bmatrix} = [C_c : I_5]$.

Clearly, $B_t^T = C_c$.

We verify $A_t^{-1} A_c = B_t^T$.

Now,

$A_t = \begin{bmatrix} 0 & 0 & 1 & 0 & 0 \\ 0 & 1 & 1 & 1 & 1 \\ 0 & 0 & 0 & 0 & 1 \\ 1 & 1 & 0 & 0 & 0 \\ 0 & 0 & 0 & 1 & 0 \end{bmatrix}, B_t = \begin{bmatrix} 1 & 0 & 0 & 0 & 0 \\ 0 & 1 & 0 & 1 & 0 \\ 0 & 0 & 1 & 1 & 0 \end{bmatrix}$

$$\text{Therefore, } A_t^{-1}Ac = \begin{bmatrix} 1 & 0 & 0 \\ 0 & 1 & 0 \\ 0 & 0 & 1 \\ 0 & 1 & 1 \\ 0 & 0 & 0 \end{bmatrix}. \text{ Hence, } A_t^{-1}A_c = B_t^T.$$

## 10.7  Path Matrix

Let $G$ be a graph with $m$ edges, and $u$ and $v$ be any two vertices in $G$. The path matrix for vertices $u$ and $v$ denoted by $P(u, v) = [p_{ij}]_{q \times m}$, where $q$ is the number of different paths between $u$ and $v$, is defined as

$$p_{ij} = \begin{cases} 1, & \text{if jth edge lies in the ith path,} \\ \\ 0, & \text{otherwise.} \end{cases}$$

Clearly, a path matrix is defined for a particular pair of vertices, the rows in $P(u, v)$ correspond to different paths between $u$ and v, and the columns correspond to different edges in $G$. For example, consider the graph in Figure 10.10.

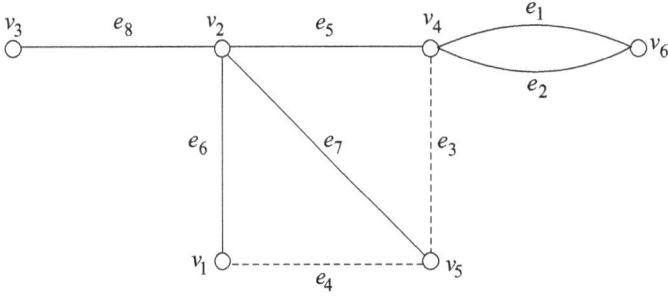

**Fig. 10.10**

The different paths between the vertices $v_3$ and $v_4$ are

$$p_1 = \{e_8, e_5\}, \ p_2 = \{e_8, e_7, e_3\} \text{ and } p_3 = \{e_8, e_6, e_4, e_3\}.$$

The path matrix for $v_3$, $v_4$ is given by

$$P(v_3, v_4) = \begin{array}{c} \begin{array}{cccccccc} e_1 & e_2 & e_3 & e_4 & e_5 & e_6 & e_7 & e_8 \end{array} \\ \begin{bmatrix} 0 & 0 & 0 & 0 & 1 & 0 & 0 & 1 \\ 0 & 0 & 1 & 0 & 0 & 0 & 1 & 1 \\ 0 & 0 & 1 & 1 & 0 & 1 & 0 & 1 \end{bmatrix} \end{array}.$$

We have the following observations about the path matrix.

1. A column of all zeros corresponds to an edge that does not lie in any path between $u$ and $v$.

2. A column of all ones corresponds to an edge that lies in every path between $u$ and $v$.

3. There is no row with all zeros.

4. The ring sum of any two rows in $P(u, v)$ corresponds to a cycle or an edge-disjoint union of cycles.

The next result gives a relation between incidence and path matrix of a graph.

**Theorem 10.15** If the columns of the incidence matrix $A$ and the path matrix $P(u, v)$ of a connected graph are arranged in the same order, then under the product (mod 2).

$$AP^T(u, v) = M,$$

where $M$ is a matrix having ones in two rows $u$ and $v$, and the zeros in the remaining $n - 2$ rows.

**Proof** Let $G$ be a connected graph and let $v_k = u$ and $v_t = v$ be any two vertices of $G$. Let $A$ be the incidence matrix and $P(u, v)$ be the path matrix of $(u, v)$ in $G$.

Now for any vertex $v_i$ in $G$ and for any $u - v$ path $p_j$ in $G$, either $v_i \in p_j$ or $v_i \notin p_j$.

If $v_i \notin p_j$, then there is no edge of $p_j$ which is incident on $v_i$.

If $v_i \in p_j$, then either $v_i$ is an intermediate vertex of $p_j$, or $v_i = v_k$ or $v_t$. In case $v_i$ is an intermediate vertex of $p_j$, then there are exactly two edges of $p_j$ which are incident on $v_i$ and in case $v_i = v_k$ or $v_t$, there is exactly one edge of $p_j$ which is incident on $v_i$.

Now consider the ith row of $A$ and the jth row of $P$ (which is the jth column of $P^T(u, v)$).

As the edges are arranged in the same order, the rth entries in these two rows are both non zero if and only if the edge $e_r$ is incident on the ith vertex $v_i$ and is also on the jth path $p_j$. Let $AP^T(u, v) = M = [m_{ij}]$.

We have, $\left[AP^T\right]_{ij} = \sum_{r=1}^{m} [A]_{ir}[P^T]_{rj}$.

Therefore, $m_{ij} = \sum_{r=1}^{m} a_{ir} p_{jr}$.

For each edge $e_r$ of $G$, we have one of the following cases.

i. $e_r$ is incident on $v_i$ and $e_r \notin p_j$. Here $a_{ir} = 1, b_{jr} = 0$.

ii. $e_r$ is not incident on $v_i$ and $e_r \in p_j$. Here $a_{ir} = 0, b_{jr} = 1$.

iii. $e_r$ is not incident on $v_i$ and $e_r \notin p_j$. Here $a_{ir} = 0, b_{jr} = 0$.

All these cases imply that the $i$th vertex $v_i$ is not in $j$th path $p_j$ and we have $M_{ij} = 0 = 0(\text{mod } 2)$. (Fig. 10.11(a)).

iv. $e_r$ is incident on $v_i$ and $e_r \in p_j$ (Fig. 10.11(b)).

If $v_i$ is an intermediate vertex of $p_j$, then there are exactly two edges say $e_r$ and $e_t$ incident on $v_i$ so that $a_{ir} = 1$, $a_{it} = 1$, $p_{jr} = 1$, $p_{jt} = 1$.

Therefore, $m_{ij} = 1 + 1 = 0(\text{mod } 2)$.

If $v_i = v_k$ or $v_t$ then the edge $e_r$ is incident on either $v_k$ or $v_t$. So, $a_{kr} = 1, p_{jr} = 1$, or $a_{tr} = 1, p_{jr} = 1$.

Thus, $m_{kj} = \Sigma a_{ir} p_{jr} = 1.1 = 1(\text{mod } 2)$, and

$$m_{tj} = \Sigma a_{ir} p_{jr} = 1.1 = 1(\text{mod } 2).$$

Hence $M = [m_{ij}]$ is a matrix, such that under modulo 2,

$$m_{ij} = \begin{cases} 1, & for\ i = k, t, \\ 0, & otherwise. \end{cases}$$

❏

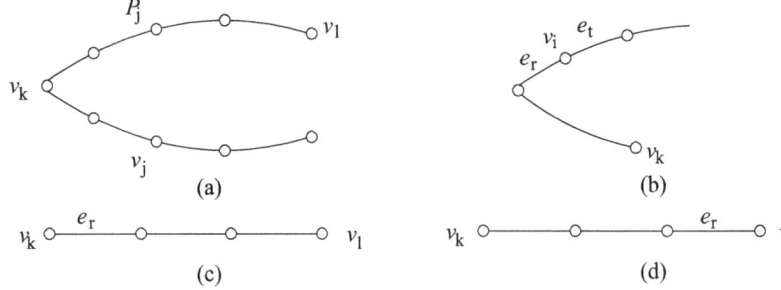

**Fig. 10.11**

**Example** In the graph of Figure 10.10, we have

$$AP^T(v_3, v_4) = \begin{bmatrix} 0 & 0 & 0 & 1 & 0 & 1 & 0 & 0 \\ 0 & 0 & 0 & 0 & 1 & 1 & 1 & 1 \\ 0 & 0 & 0 & 0 & 0 & 0 & 0 & 1 \\ 1 & 1 & 1 & 0 & 1 & 0 & 0 & 0 \\ 0 & 0 & 1 & 1 & 0 & 0 & 1 & 0 \\ 1 & 1 & 0 & 0 & 0 & 0 & 0 & 0 \end{bmatrix} \begin{bmatrix} 0 & 0 & 0 \\ 0 & 0 & 0 \\ 0 & 1 & 1 \\ 0 & 0 & 1 \\ 1 & 0 & 0 \\ 0 & 0 & 1 \\ 0 & 1 & 0 \\ 1 & 1 & 1 \end{bmatrix} = \begin{matrix} v_1 \\ v_2 \\ v_3 \\ v_4 \\ v_5 \\ v_6 \end{matrix} \begin{bmatrix} 0 & 0 & 0 \\ 0 & 0 & 0 \\ 1 & 1 & 1 \\ 1 & 1 & 1 \\ 0 & 0 & 0 \\ 0 & 0 & 0 \end{bmatrix} (\text{mod } 2).$$

## 10.8  Adjacency Matrix

Let $V = (V, E)$ be a graph with $V = \{v_1, v_2, \ldots, v_n\}$, $E = \{e_1, e_2, \ldots, e_m\}$ and without parallel edges. The adjacency matrix of $G$ is an $n \times n$ symmetric binary matrix $X = [x_{ij}]$ defined over the ring of integers such that

$$x_{ij} = \begin{cases} 1, & if\ v_i v_j \in E, \\ 0, & otherwise. \end{cases}$$

**Example**  Consider the graph $G$ given in Figure 10.12.

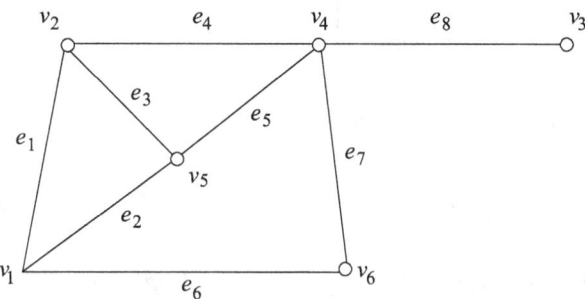

**Fig. 10.12**

The adjacency matrix of $G$ is given by

$$X = \begin{array}{c} \\ v_1 \\ v_2 \\ v_3 \\ v_4 \\ v_5 \\ v_6 \end{array} \begin{array}{c} \begin{array}{cccccc} v_1 & v_2 & v_3 & v_4 & v_5 & v_6 \end{array} \\ \left[ \begin{array}{cccccc} 0 & 1 & 0 & 0 & 1 & 1 \\ 1 & 0 & 0 & 1 & 1 & 0 \\ 0 & 0 & 0 & 1 & 0 & 0 \\ 0 & 1 & 1 & 0 & 1 & 0 \\ 1 & 1 & 0 & 1 & 0 & 0 \\ 1 & 0 & 0 & 1 & 0 & 0 \end{array} \right] \end{array}.$$

We have the following observations about the adjacency matrix $X$ of a graph $G$.

1. The entries along the principal diagonal of $X$ are all zeros if and only if the graph has no self-loops. However, a self-loop at the $i$th vertex corresponds to $x_{ii} = 1$.

2. If the graph has no self-loops, the degree of a vertex equals the number of ones in the corresponding row or column of $X$.

3. Permutation of rows and the corresponding columns imply reordering the vertices. We note that the rows and columns are arranged in the same order. Therefore, when two rows are interchanged in $X$, the corresponding columns are also interchanged.

Thus, two graphs $G_1$ and $G_2$ without parallel edges are isomorphic if and only if their adjacency matrices $X(G_1)$ and $X(G_2)$ are related by

$$X(G_2) = R^{-1}X(G_1)R,$$

where $R$ is a permutation matrix.

4. A graph $G$ is disconnected having components $G_1$ and $G_2$ if and only if the adjacency matrix $X(G)$ is partitioned as

$$X(G) = \begin{bmatrix} X(G_1) & : & 0 \\ .. & : & .. \\ 0 & : & X(G_2) \end{bmatrix},$$

where $X(G_1)$ and $X(G_2)$ are respectively the adjacency matrices of the components $G_1$ and $G_2$.

Obviously, the above partitioning implies that there are no edges between vertices in $G_1$ and vertices in $G_2$.

5. If any square, symmetric and binary matrix $Q$ of order $n$ is given, then there exists a graph $G$ with $n$ vertices and without parallel edges whose adjacency matrix is $Q$.

**Definition:** An *edge sequence* is a sequence of edges in which each edge, except the first and the last, has one vertex in common with the edge preceding it and one vertex in common with the edge following it. A walk and a path are the examples of an edge sequence. An edge can appear more than once in an edge sequence. In the graph of Figure 10.13, $v_1e_1v_2e_2v_3e_3v_4e_4v_2e_2v_3e_5v_5$, or $e_1e_2e_3e_4e_2e_5$ is an edge sequence.

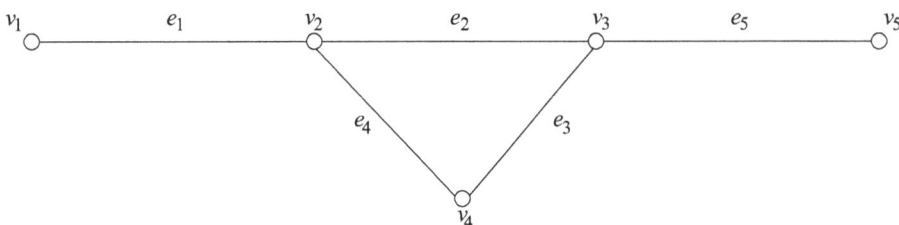

**Fig. 10.13**

We now have the following result.

**Theorem 10.16**    If $X = [x_{ij}]$ is the adjacency matrix of a simple graph $G$, then $[X^k]_{ij}$ is the number of different edge sequences of length $k$ between vertices $v_i$ and $v_j$.

**Proof**    We prove the result by using induction on $k$. The result is trivial for $k = 0$ and 1.

Since $X^2 = X.X$, $X^2$ is a symmetric matrix, as product of symmetric matrices is also symmetric.

For $k = 2, i \neq j$, we have

$[X^2]_{ij}$ = number of ones in the product of $i$th row and $j$th column (or $j$th row) of $X$

= number of positions in which both $i$th and $j$th rows of $X$ have ones

= number of vertices that are adjacent to both $i$th and $j$th vertices

= number of different paths of length two between $i$th and $j$th vertices

Also, $[X^2]_{ii}$ = number of ones in the $i$th row (or column) of $X$

= degree of the corresponding vertex.

This shows that $[X^2]_{ij}$ is the number of different paths and therefore different edge sequences of length 2 between the vertices $v_i$ and $v_j$. Thus, the result is true for $k = 2$. Assume the result to be true for $k$, so that

$[X^k]_{ij}$ = number of different edge sequences of length $k$ between $v_i$ and $v_j$.

We have, $[X^{k+1}]_{ij} = [X^k X]_{ij} = \sum_{r=1}^{n} [X^k]_{ir}[X]_{rj} = \sum_{r=1}^{n} [X^k]_{ir} x_{rj}$.

Now, every $v_i - v_j$ edge sequence of length $k+1$ consists of a $v_i - v_r$ edge sequence of length $k$, followed by an edge $v_t v_j$. Since there are $[X^k]_{ir}$ such edge sequences of length $k$ and $x_{rj}$ such edges for each vertex $v_r$, the total number of all $v_i - v_j$ edge sequences of length $k+1$ is $\sum_{r=1}^{n} [X^k]_{ir} x_{rj}$. This proves the result for $k+1$ also. ❑

We have the following observation about connectedness and adjacency matrix.

**Theorem 10.17** Let $G$ be a graph with $V = \{v_1, v_2, \ldots, v_n\}$ and let $X$ be the adjacency matrix of $G$. Let $Y = [y_{ij}]$ be the matrix

$$Y = X + X^2 + \ldots + X^{n-1}.$$

Then $G$ is connected if and only if for all distinct $i, j$, $y_{ij} \neq 0$. That is, if and only if $Y$ has no zero entries off the main diagonal.

**Proof** We have, $y_{ij} = [Y]_{ij} = [X]_{ij} + [X^2]_{ij} + \ldots + [X^{n-1}]_{ij}$.

Since $[X^k]_{ij}$ denotes the number of distinct edge-sequences of length $k$ from $v_i v_j$,

$y_{ij}$ = number of different $v_i - v_j$ edge sequence of length 1

+ number of different $v_i - v_j$ edge sequences of length 2 + ...

+ number of different $v_i - v_j$ edge sequences of length $n - 1$.

Therefore, $y_{ij}$ = number of different $v_i - v_j$ edge sequence of length less than $n$.

Now let $G$ be connected. Then for every pair of distinct $i$, $j$ there is a path from $v_i$ to $v_j$. Since $G$ has $n$ vertices, this path passes through atmost $n$ vertices and so has length less than $n$. Thus, $y_{ij} \neq 0$ for each $i$, $j$ with $i \neq j$.

Conversely, for each distinct pair $i$, $j$ we have $y_{ij} \neq 0$. Then from above, there is at least 1 edge sequence of length less than $n$ from $v_i$ to $v_j$. This implies that $v_i$ is connected to $v_j$. Since the distinct pair $i$, $j$ is chosen arbitrarily, $G$ is connected.                                       ❑

The next result is useful in determining the distances between different pairs of vertices.

**Theorem 10.18**   In a connected graph, the distance between two vertices $v_i$ and $v_j$ is $k$ if and only if $k$ is the smallest integer for which $[X^k]_{ij} \neq 0$.

**Proof**   Let $G$ be a connected graph and let $X = [x_{ij}]$ be the adjacency matrix of $G$. Let $v_i$ and $v_j$ be vertices in $G$ such that

$$d(v_i, v_j) = k.$$

Then the length of the shortest path between $v_i$ and $v_j$ is $k$.

This implies that there are no paths of length 1, 2, ..., $k - 1$ and so no edge sequences of length 1, 2, ..., $k - 1$ between $v_i$ and $v_j$.

Therefore, $[X]_{ij} = 0$, $[X^2]_{ij} = 0$, ..., $[X^{k-1}]_{ij} = 0$.

Hence $k$ is the smallest integer such that $[X^k]_{ij} \neq 0$.

Conversely, suppose that $k$ is the smallest integer such that $[X^k]_{ij} \neq 0$.

Therefore, there are no edge sequences of length 1, 2, ..., $k - 1$ and in fact no paths of length 1, 2, ..., $k - 1$ between vertices $v_i$ and $v_j$.

Thus, the shortest path between $v_i$ and $v_j$ is of length $k$, so that $d(v_i, v_j) = k$.                                       ❑

**Definition:**   Let $G$ be a graph and let $d_i$ be the degree of the vertex $v_i$ in $G$. The degree matrix $H = [h_{ij}]$ of G is defined by

$$h_{ij} = \begin{cases} 0, & \text{for } i \neq j, \\ d_i, & \text{for } i = j. \end{cases}$$

The following result gives a relation between the matrices $F, X$ and $H$.

**Theorem 10.19**  Let $F$ be the modified incidence matrix, $X$ the adjacency matrix and $H$ the degree matrix of a graph $G$. Then,

$$FF^T = H - X.$$

**Proof**  We have $(i, j)$th element of $FF^T$,

$$[FF^T]_{ij} = \sum_{r=1}^{n} [F]_{ir}[F^T]_{rj} = \sum_{r=1}^{n} [F]_{ir}[F]_{jr}.$$

Now, $[F]_{ir}$ and $[F]_{rj}$ are non-zero if and only if the edge $e_r = v_i v_j$. Then for $i \neq j$,

$$\sum_{r=1}^{m} [F]_{ir}[F]_{jr} = \begin{cases} -1, & \text{if } e_r = v_i v_j \text{ is an edge}, \\ 0, & \text{if } e_r = v_i v_j \text{ is not an edge}. \end{cases}$$

For $i = j$, $[F]_{ir}[F]_{jr} = 1$ whenever $[F]_{ik} = \pm 1$, and this occurs $d_i$ times corresponding to the number of edges incident on $v_i$. Thus,

$$\sum_{r=1}^{m} [F]_{ir}[F]_{jr} = d_i, \text{ for } i = j.$$

Therefore,

$$[FF^T]_{ij} = \begin{cases} -1 \text{ or } 0, & \text{according to whether for } i \neq j, \ v_i v_j \text{ is an edge or not}, \\ d_i, & \text{for } i = j. \end{cases}$$

Also, $[H - X]_{ij} = [H]_{ij} - [X]_{ij}$

$$= \begin{cases} d_i - 0, & \text{for } i = j, \\ 0 - (1 \text{ or } 0), & \text{according as for } i \neq j, \ v_i v_j \text{ is an edge, or } v_i v_j \text{ is not an edge}. \end{cases}$$

$$= \begin{cases} d_i, & \text{for } i = j, \\ -1 \text{ or } 0, & \text{according as for } i \neq j, \ v_i v_j \text{ is an edge, or } v_i v_j \text{ is not an edge}. \end{cases}$$

❑

**Corollary 10.3**  The matrix $Q = FF^T$ is independent of the orientation used for the edges of $G$ in getting $F$.

**Theorem 10.20**  Let $X$, $F$ and $H$ be the adjacency, modified incidence and degree matrices of the graph $G$, and $Q = FF^T = H - X$. Then the matrix of cofactors of $Q$ denoted by adj $Q$ is a multiple of the all ones $n \times n$ matrix $J$.

**Proof**  If $G$ is disconnected, then rank $Q$ =rank $FF^T < n - 1$ and so every cofactors of $Q$ is zero. Therefore, adj $Q = O = O.J$, where $J = [J_{ij}]_{n \times n}$ with $J_{ij} = 1$ for all $i$, $j$.

Now let $G$ be connected. Then rank $Q = n - 1$ and therefore $|Q| = 0$ This implies that every column of adj $Q$ belongs to the kernel (null space) of $Q$.

But nullity $Q = 1$ (as rank $Q+$ nullity $Q = n$). So, if $u$ is the $n$-vector of ones, then

$$(H - X)u = 0.$$

Therefore, $u$ is in the null space of $Q$.

Thus, every other vector in the null space of $Q$ and in particular every column of adj $Q$ is a multiple of $u$.

Since $Q$ and so adj $Q$ are symmetric, the multiplying factor for all columns of adj $Q$ are same.

Hence, adj $Q = cJ$, where $c$ is a constant.  ❏

The next result called Matrix-tree theorem is of great importance as it can be used in finding the complexity of a connected graph.

**Theorem 10.21 (Matrix-tree theorem)**  If $X$, $F$ and $H$ are the adjacency, modified incidence and degree matrices of the connected graph $G$, and $Q = FF^T J$ is the $n \times n$ matrix of ones and $\tau(G)$ is the complexity of $G$, then Adj $Q = \tau(G).J$.

**Proof**  Let $X, F$ and $H$ be the adjacency, modified incidence and degree matrix of a connected graph $G$. We have $Q = FF^T$ and adj $Q =$ matrix of the cofactors of $Q$. Also $\tau(G)J$ is a matrix whose every entry is $\tau(G)$ as $J$ is a matrix whose every entry is unity.

Therefore, to prove adj $Q = \tau(G)J$, it is enough to prove that $\tau(G) =$ any one cofactor of $Q$.

Let $F_0$ be the matrix obtained by dropping the last row from $F$. Then clearly, $|F_o F_o^T|$ is a cofactor of $Q$.

Using Binet-Cauchy theorem of matrix theory, we have

$$\left| F_0 F_0^T \right| = \sum_{x \subseteq E} |F_x| \, |F_x^T|, \tag{10.21.1}$$

where $F_x$ is the square submatrix of $F_o$ whose $n - 1$ columns correspond to $n - 1$ edges in the subset $X$ of $E$, the summation running over all possible such subsets.

We know $|F_x| \neq 0$ if and only if $< X >$ is a spanning tree of $G$ and $|Fx| = \pm 1$.

But, $|F_x^T| = |Fx|$.

Therefore, each $X \subseteq E$ such that $< X >$ is a spanning tree of $G$ contributes one to the sum on the right of (10.21.1) and all other contributions are zero.

Hence, $|F_o F_o^T| = \tau(G)$, proving the theorem.  ❏

**Corollary 10.4**   Prove $\tau(K_n) = n^{n-2}$.

**Proof**   Here, $Q = H - X = (n-1)I - (J-I) = nI - J$. Therefore,

$$Q = \begin{bmatrix} n & 0 & 0 & .. & 0 \\ 0 & n & 0 & .. & 0 \\ 0 & 0 & n & .. & 0 \\ \vdots & & & & \\ 0 & 0 & 0 & .. & n \end{bmatrix} - \begin{bmatrix} 1 & 1 & 1 & .. & 1 \\ 1 & 1 & 1 & .. & 1 \\ 1 & 1 & 1 & .. & 1 \\ \vdots & & & & \\ 1 & 1 & 1 & .. & 1 \end{bmatrix}$$

$$= \begin{bmatrix} n-1 & -1 & -1 & .. & -1 \\ -1 & n-1 & -1 & .. & -1 \\ -1 & -1 & n-1 & .. & -1 \\ \vdots & & & & \\ -1 & -1 & -1 & .. & n-1 \end{bmatrix}.$$

The cofactor of $q_{11}$ is the $(n-1) \times (n-1)$ determinant given by

$$\text{cofactor of } q_{11} = \begin{vmatrix} n-1 & -1 & .. & -1 \\ -1 & n-1 & .. & -1 \\ \vdots & & & \\ -1 & -1 & .. & n-1 \end{vmatrix}.$$

Subtracting the first row from each of the others and then adding the last $n-2$ columns to the first, we get

$$\text{cofactor of } q_{11} = \begin{vmatrix} 1 & -1 & -1 & .. & -1 \\ 0 & n & 0 & .. & 0 \\ 0 & 0 & n & .. & 0 \\ \vdots & & & & \\ 0 & 0 & 0 & .. & n \end{vmatrix}.$$

Expanding with the help of the first column, we have cofactor of $q_{11} = n^{n-2}$. Thus,

$$\tau(k_n) = n^{n-2}. \qquad \qquad \qquad \square$$

## 10.9   Exercises

1. Characterise $A_f$, $B_f$, $C_f$ and $X$ of the complete graph of $n$ vertices.

2. Characterise simple, self-dual graphs in terms of their cycle and cut-set matrices.

3. Show that each diagonal entry in $X^3$ equals twice the number of triangles passing through the corresponding vertex.

4. Characterise the adjacency matrix of a bipartite graph.

5. Prove that a graph is bipartite if and only if for all odd $k$, every diagonal entry of $A^k$ is zero.

6. Similar to the cycle or cut-set matrix, define a spanning tree matrix for a connected graph, and observe some of its properties.

7. If $X$ is the adjacency matrix of a graph $G$ and $L$ is the adjacency matrix of its edge graph $L(G)$ and $A$ and $H$ are the incidence and degree matrices, show that $X = AA^T - H$ and $L = A^T A - 2I$.

8. Use the matrix tree theorem to calculate $\tau(K_4 - e)$.

9. Prove that $\tau(G) = \dfrac{1}{n^2} \det(J + Q)$.

# 11. Digraphs

The concept of digraphs (or directed graphs) is one of the richest theories in graph theory, mainly because of their applications to physical problems. For example, flow networks with valves in the pipes and electrical networks are represented by digraphs. They are applied in abstract representations of computer programs and are an invaluable tools in the study of sequential machines They are also used for systems analysis in control theory. Most of the concepts and terminology of undirected graphs are also applicable to digraphs, and hence in this chapter more emphasis will be given to those properties of digraphs that are not found in undirected graphs.

There are two different treatments of digraphs—one can be found in the book by Harray, Norman and Cartwright [191] and the other in the book by Berge [18]. The former discusses the application of digraphs to sociological problems, and the latter gives a comprehensive mathematical treatment. One can also refer to the books by N. Deo [63], Harary [104], Behzad, Chartrand and Lesniak-Foster [16], and Buckley and Harary [42].

## 11.1 Basic Definitions

**Digraphs (Directed graphs):**  A digraph $D$ is a pair $(V, A)$, where $V$ is a nonempty set whose elements are called the *vertices* and $A$ is the subset of the set of ordered pairs of distinct elements of $V$. The elements of $A$ are called the *arcs* of $D$ (Fig. 11.1(a)).

**Multidigraphs:**  A multidigraph $D$ is a pair $(V, A)$, where $V$ is a nonempty set of vertices and $A$ is a multiset of arcs, which is a multisubset of the set of ordered pairs of distinct elements of $V$. The number of times an arc occurs in $D$ is called its *multiplicity* and arcs with multiplicity greater than one are called *multiple arcs* of $D$ (Fig. 11.1(b)).

**General digraphs:**  A general digraph $D$ is a pair $(V, A)$, where $V$ is a nonempty set of vertices, and $A$ is a multiset of arcs, which is a multisubset of the cartesian product of $V$ with itself. An arc of the form $uu$ is called a *loop* of $D$ and arcs which are not loops are called *proper arcs* of $D$. The number of times an arc occurs is called its multiplicity. A loop with multiplicity greater than one is called a *multiple loop* (Fig. 11.1(c)).

**Oriented graph:**   A digraph containing no symmetric pair of arcs is called an oriented graph (Fig. 11.1(d)).

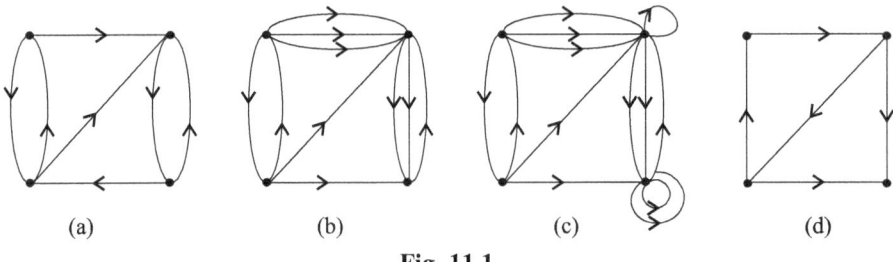

(a)                         (b)                         (c)                         (d)

**Fig. 11.1**

For $u, v \in V$, an arc $a = (u, v) \in A$ is denoted by $uv$ and implies that $a$ is directed from $u$ to $v$. Here, $u$ is the *initial* vertex (tail) and $v$ is the *terminal* vertex (head). Also we say that $a$ joins $u$ to $v$; $a$ is incident with $u$ and $v$; $a$ is incident from $u$ and $a$ is incident to $v$; and $u$ is adjacent to $v$ and $v$ is adjacent from $u$. In case both $uv$ and $vu$ belong to $A$, then $uv$ and $vu$ are called a *symmetric pair of arcs*. For any $v \in V$, the number of arcs incident to $v$ is the *indegree* of $v$ nd is denoted by $d^-(v)$. The number of arcs incident from $v$ is called the *outdegree* of $v$ and is denoted by $d^+(v)$. Berge calls indegree and outdegree as *inner* and *outer-demi degrees*. The *total degree* (or simply degree) of $v$ is $d(v) = d^-(v) + d^+(v)$.

We define $N^+(v)$ and $N^-(v)$ by

$$N^+(v) = \{u \in V : vu \in A\} \text{ and } N^-(v) = \{u \in V : uv \in A\}.$$

If $d(v) = k$ for every $v \in V$, then $D$ is said to be a *k-regular* digraph. If for every $v \in V$, $d^-(v) = d^+(v)$, the digraph is said to be an *isograph* or a *balanced digraph*. We note that an isograph is an even degree digraph, but not necessarily regular. Also, every symmetric digraph is an isograph. Edmonds and Johnson [70] call isographs as asymmetric digraphs and Berge [23] calls them pseudo-symmetric graphs. Kotzig [137, 138] calls an anti-symmetric isograph as oriented-in-equilibrium or a $\rho$-graph.

A vertex $v$ for which $d^+(v) = d^-(v) = 0$ is called an *isolate*. A vertex $v$ is called a *transmitter* or a *receiver* according as $d^+(v) > 0$, $d^-(v) = 0$ or $d^+(v) = 0$, $d^-(v) > 0$. A vertex $v$ is called a *carrier* if $d^+(v) = d^-(v) = 1$.

**Underlying graph of a digraph:**   Let $D = (V, A)$ be a digraph. The graph $G = (V, E)$, where $uv \in E$ if and only if $uv$ or $vu$ or both are in $A$, is called the *underlying graph* of $D$. This is also called the *covering graph* $C(D)$ of $D$. Here we denote $C(D)$ by $G(D)$ or simply by $G$.

In case $G = (V, E)$ is a graph, the digraph with vertex set $V$ and a symmetric $uv$ whenever $uv \in E$, is called the *digraph corresponding* to $G$, and is denoted by $D(G)$, or $D$. Clearly, $D(G)$ is a symmetric digraph. An oriented graph obtained from the graph $G = (V, E)$ by replacing each edge $uv \in E$ by an arc $uv$ or $vu$, but not both is called an *orientation* of $G$ and is denoted by $O(G)$ or $O$.

**Complete symmetric digraph:** A digraph $D = (V, A)$ is said to be *complete* if both $uv$ and $vu \in A$, for all $u, v \in V$. Obviously this corresponds to $K_n$, where $|V| = n$, and is denoted by $K_n^*$. A complete antisymmetric digraph, or a complete oriented graph is called a *tournament*. Clearly, a tournament is an orientation of $K_n$ (Fig. 11.2).

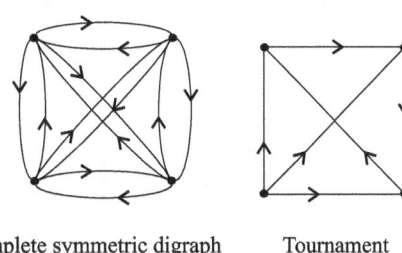

Complete symmetric digraph     Tournament

**Fig. 11.2**

We note that the number of arcs in $K_n^*$ is $n(n-1)$ and the number of arcs in a tournament is $\frac{n(n-1)}{2}$.

## 11.2 Digraphs and Binary Relations

Let $A$ and $B$ be nonempty sets. A *(binary) relation R* from $A$ to $B$ is a subset of $A \times B$. If $R \subseteq A \times B$ and $(a, b) \in R$, where $a \in A$, $b \in B$, we say $a$ "is related to" $b$ by $R$, and we write $aRb$. If $a$ is not related to $b$ by $R$, we write $a\overline{R}b$. A relation $R$ defined on a set $X$ is a subset of $X \times X$. For example, less than, greater than and equality are the relations in the set of real numbers. The property "is congruent to" defines a relation in the set of all triangles in a plane. Also, parallelism defines a relation in the set of all lines in a plane.

Let $R$ define a relation on a nonempty set $X$. If $R$ relates every element of $X$ to itself, the relation $R$ is said to be *reflexive*. A relation $R$ is said to be *symmetric* if for all $x_i \, x_j \in X$, $x_i \, R \, x_j$ implies $x_j \, R \, x_i$. A relation $R$ is said to be *transitive* if for any three elements $x_i$, $x_j$ and $x_k$ in $X$, $x_i R x_j$ and $x_j \, R \, x_k$ imply $x_i \, R \, x_k$. A binary relation is called an *equivalence relation* if it is reflexive, symmetric and transitive.

A binary relation $R$ on a set $X$ can always be represented by a digraph. In such a representation, each $x_i \in X$ is represented by a vertex $x_i$ and whenever there is a relation $R$ from $x_i$ to $x_j$, an arc is drawn from $x_i$ to $x_j$, for every pair $(x_i, x_j)$. The digraph in Figure 11.3 represents the relation is less than, on a set consisting of four numbers 2, 3, 4, 6.

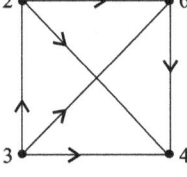

**Fig. 11.3**

We note that every binary relation on a finite set can be represented by a digraph without parallel edges and vice versa.

Clearly, the digraph of a reflexive relation contains a loop at every vertex (Fig. 11.4). A digraph representing a reflexive binary relation is called a reflexive digraph.

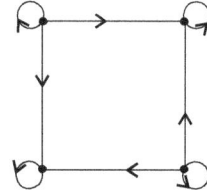

**Fig 11.4**

The digraph of a symmetric relation is a symmetric digraph because for every arc from $x_i$ to $x_j$, there is an arc from $x_j$ to $x_i$. Figure 11.5 shows the digraph of an irreflexive and symmetric relation on a set of three elements.

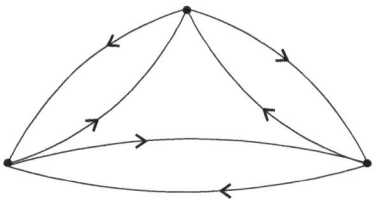

**Fig. 11.5**

A digraph representing a transitive relation on its vertex set is called a *transitive digraph*. Figure 11.6 shows the digraph of a transitive, which is neither reflexive, nor symmetric.

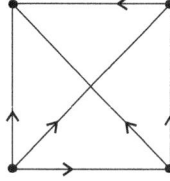

**Fig. 11.6**

A binary relation $R$ on a set $M$ can also be represented by a matrix, called a relation matrix. This is a (0, 1), $n \times n$ matrix $M_R = [m_{ij}]$, where $n$ is the number of elements in $M$, and is defined by

$$m_{ij} = \begin{cases} 1 & \text{if } x_i \, R \, x_j \text{ is true,} \\ 0, & \text{otherwise.} \end{cases}$$

**Isomorphic digraphs:**    Two digraphs are said to be isomorphic if their underlying graphs are isomorphic and the direction of the corresponding arcs are same. Two non-isomorphic digraphs are shown in Figure 11.7.

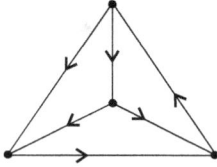 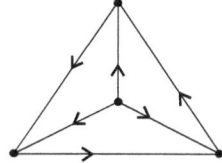

**Fig. 11.7**

**Subdigraph:**    Let $D = (V, A)$ be a digraph. A digraph $H = (U, B)$ is the *subdigraph* of $D$ whenever $U \subseteq V$ and $B \subseteq A$. If $U = V$, the subdigraph is said to be *spanning*.

**Complement of a digraph:**    The *complement* $\overline{D} = (V, \overline{A})$ of the digraph $D = (V, A)$ has vertex set $V$ and $a \in \overline{A}$ if and only if $a \notin A$. That is, $\overline{D}$ is the relative complement of $D$ in $K_n^*$, where $|V| = n$.

**Converse digraph:**    The *converse* $D' = (V, A')$ of the digraph $D = (V, A)$ has vertex set $V$ and $a = uv \in A'$ if and only if $a' = vu \in A$. That is, $A'$ is obtained by reversing the direction of each arc of $D$. Clearly, $(D')' = D'' = D$.

A digraph $D$ is *self-complementary* if $D \cong \overline{D}$ and $D$ is said to be *self-converse* if $D \cong D'$. A digraph $D$ is said to be *self-dual* if $D \cong \overline{D} \cong D'$.

## 11.3  Directed Paths and Connectedness

**Directed walks:**    A *(directed) walk* in a digraph $D = (V \ A)$ is a sequence $v_0 a_1 v_1 a_2 \ldots a_k v_k$, where $v_i \in V$ and $a_i \in A$ are such that $a_i = v_{i-1} v_i$ for $1 \leq i \leq k$, no arc being repeated. As there is only one arc of the form $v_i v_j$, the walk can also be represented by the vertex sequence $v_0 v_1 \ldots v_k$. A vertex may appear more than once in a walk. Clearly, the length of the walk is $k$. If $v_0 \neq v_k$, the walk is *open*, and if $v_0 \neq v_k$, the walk is *closed*. A walk is *spanning* if $V = \{v_0, \ldots, v_k\}$.

A *(directed) path* is an open walk in which no vertex is repeated. A *(directed) cycle* is a closed walk in which no vertex is repeated. A digraph is *acyclic* if it has no cycles.

A *semiwalk* is a sequence $v_0 a_1 v_1 a_2 \ldots a_k v_k$ with $v_i \in V$ and $a_i \in A$ such that either $a_i = v_{i-1} v_i$ or $a_i v_i v_{i-1}$ and no arc is repeated. The length of the semiwalk is $k$. If $v_0 \neq v_k$, the semiwalk is *open*, and if $v_o = v_k$, the semiwalk is *closed*. If no vertex is repeated in an open (closed) semiwalk, it is called a *semi path (semicycle)*.

A spanning path of a digraph is called a  Hamiltonian path and a spanning cycle is called a  Hamiltonian cycle. A digraph with a Hamiltonian cycle is said to be  Hamiltonian.

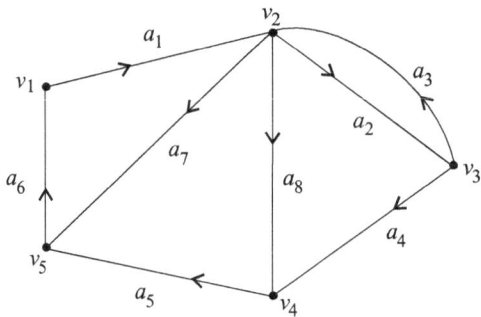

**Fig. 11.8**

In Figure 11.8, $v_1\,a_1\,v_2\,a_2\,v_3\,a_4\,v_4$ is an open walk, $v_1\,a_1\,v_2\,a_3\,v_3\,a_4\,v_4$ is semiwalk, $v_1\,a_1\,v_2\,a_8\,v_4$ $a_5\,v_5$ is a path and $v_1\,a_1\,v_2\,a_8\,v_4\,a_5\,v_5\,a_6\,v_1$ is a cycle.

In a digraph $D = (V, A)$, a vertex $u$ is said to be *joined* to a vertex $v$, if there is a semipath from $u$ to $v$. We note that the relation 'is joined to' is reflexive, symmetric and transitive, and therefore is an equivalence relation on $V$. A vertex $u$ is said to be *reachable* from a vertex $v$, if there is a path from $v$ to $u$. The relation 'is reachable from' is reflexive and transitive, but not symmetric, since there may or may not be a path from $u$ to $v$. A vertex $v$ is called a *source* of $D$ if every vertex of $D$ is reachable from $v$, and $v$ is called a *sink* of $D$, if $v$ is reachable from every other vertex.

## Principle of duality for digraphs

While changing a digraph $D$ to its converse $D'$, we observe that the properties about $D$ get changed to the corresponding properties about $D'$. When $D'$ is changed to $D'' = D$, the original properties of $D$ are obtained. Such type of a pair of properties are called *dual properties*, (transmitter, receiver), (source, sink), (indegree, outdegree), (isolate, isolate) and (carrier, carrier). The dual of a statement $P$ about a digraph is the statement $P'$ obtained from $P$ by changing every concept in $P$ to its dual. For any statements $P$ and $Q$ for digraphs, $P \Rightarrow Q$ in $D$ is true if and only if $P' \Rightarrow Q'$ in $D'$ and for any digraph $D$ there is a converse $D'$. Therefore for every result in digraphs we get a dual result by changing every property to its dual. This is called the *principle of duality* for digraphs.

**Definition:** A digraph is said to be *strongly connected* or strong, if every two of its distinct vertices $u$ and $v$ are such that $u$ is reachable from $v$ and $v$ is reachable from $u$. A digraph is *unilaterally connected* or unilateral, if either $u$ is reachable from $v$ or $v$ is reachable from $u$ and is *weakly connected* or weak, if $u$ and $v$ are joined by a semipath.

In Figure 11.9, (a) shows a strong digraph, (b) a unilateral digraph and (c) a weak digraph.

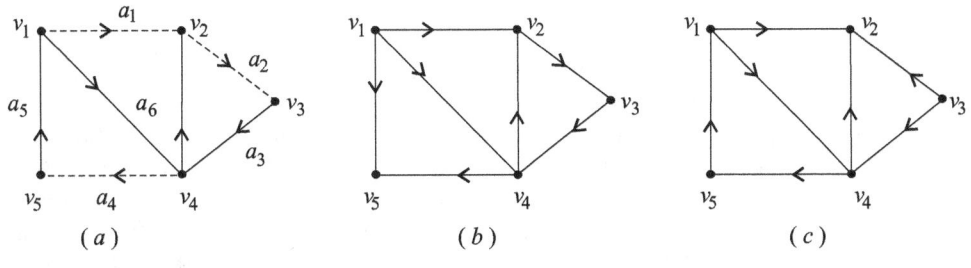

**Fig. 11.9**

A digraph is said to be *disconnected* if it is not even weak. A digraph is said to be *strictly weak* if it is weak, but not unilateral. It is *strictly unilateral*, if it is unilateral but not strong.

Two vertices of a digraph $D$ are said to be

i. 0-connected if there is no semipath joining them,

ii. 1-connected if there is a semipath joining them, but there is no $u-v$ path or $v-u$ path,

iii. 2-connected if there is a $u-v$ or a $v-u$ path, but not both,

iv. 3-connected if there is $u-v$ path and a $v-u$ path.

**Definition:** An *arc sequence* in a digraph $D$ is an alternating sequence of vertices and arcs of $D$.

The following results characterise various types of connectivity in digraphs.

**Theorem 11.1** A digraph is strong if and only if it has a spanning closed arc sequence.

### Proof

*Necessity* Let $D = (V, A)$ be a strong digraph with $V = \{v_1, v_2, \ldots, v_n\}$. Then there is an arc sequence from each vertex in $V$ to every other vertex in $V$. Therefore, there exists in $D$, arc sequences $Q_1, Q_2, \ldots, Q_{n-1}$ such that the first vertex of $Q_i$ is $v_i$ and the last vertex of $Q_i$ is $v_{i+1}$, for $i = 1, 2, \ldots, n-1$. Also, there exists an arc sequence, say $Q_n$, with first vertex $v_n$ and the last vertex $v_1$. Then the arc sequence obtained by traversing the arc sequences $Q_1, Q_2, \ldots, ; Q_n$ in succession is a spanning closed arc sequence of $D$.

*Sufficiency* Let $u$ and $v$ be two distinct vertices of $V$. If $v$ follows $u$ in any spanning closed arc sequence, say $Q$ of $D$, then there exists a sequence of the arcs of $Q$ forming an arc sequence from $u$ to $v$. If $u$ follows $v$ in $Q$, then there is an arc sequence from $u$ to the last vertex of $Q$ and an arc sequence from that vertex to $v$. An arc sequence from $u$ to $v$ is then obtained by traversing these two arc sequences in succession. ❑

**Theorem 11.2** A digraph $D$ is unilateral if and only if it has a spanning arc sequence.

## Proof

*Necessity*   Assume that $D$ is unilateral. Let $Q$ be an arc sequence in $D$ which contains maximum number of vertices and let $Q$ begin at the vertex $v_1$ of $D$ and end at the vertex $v_2$ of $D$. If $Q$ is a spanning arc sequence, there is nothing to prove. Assume that $Q$ is not a spanning arc sequence. Then there exists a vertex, say $u$ of $D$ that is not in $Q$. Also in $D$, there is neither an arc sequence from $u$ to $v_1$, nor from $v_2$ to $u$. Since $D$ is unilateral and does not contain an arc sequence from $u$ to $v_1$, $D$ contains an arc sequence from $v_1$ to $u$.

Let $w(\neq v_2)$ be the last vertex of $Q$ from which an arc sequence from $w$ to $u$ exists in $D$. Let $Q_1$ be an arc sequence from $w$ to $u$ in $D$. Let $z$ be the vertex in $D$ which is the immediate successor of the last appearance of $w$ in $Q$. Clearly, $D$ does not contain an arc sequence from $z$ to $u$. Since $D$ is unilateral, there is an arc sequence, say $Q_2$ from $u$ to $z$ in $D$. Traversing $Q$ from $v_1$ to the last appearance of $w$, then traversing $u$, then traversing $Q_2$ to vertex $z$ and finally traversing $Q$ to $v_2$, we obtain an arc sequence from $v_1$ to $v_2$ which has more distinct vertices than $Q$. This is a contradiction and thus $Q$ is a spanning arc sequence in $D$ (Fig. 11.10).

The sufficiency follows from the definition.                                                      ❑

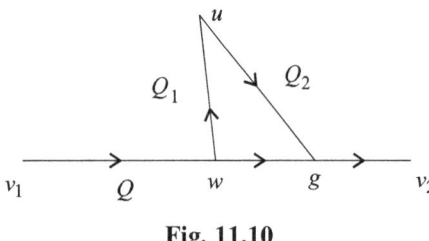

**Fig. 11.10**

**Theorem 11.3**   A digraph is weak if and only if it has a spanning semi arc sequence.

**Proof**   Let $D = (V, A)$ be a weak digraph with $V = \{v_1, v_2, \ldots, v_n\}$. Since $D$ is weak, there is a semi arc sequence, say $Q_i$ from $v_i$ to $v_{i+1}$ in $D$ for $i = 1, 2, \ldots, n-1$. The semi arc sequence obtained by traversing the semi arc sequences $Q_1, Q_2, \ldots, Q_n$ in succession is a spanning semi arc sequence of $D$.

Conversely, let $D$ be a digraph containing a spanning semi arc sequence, say $Q$. Let $v_1$ and $v_2$ be two distinct vertices of $D$. Clearly, $v_1$ and $v_2$ are in $Q$, since $Q$ is spanning. The part of $Q$ which begins at any appearance of $v_1$ $(v_2)$ and ends at any appearance of $v_2$ $(v_1)$ represents a semi arc sequence from $v_1$ to $v_2$ (from $v_2$ to $v_1$) in $D$. Thus, there is either a semi arc sequence from $v_1$ to $v_2$, or from $v_2$ to $v_1$ in $D$. Hence $D$ is weak.

**Definition:**   In a digraph $D$, a *strong component* is a maximal strong subdigraph of $D$. A *unilateral component* is a maximal unilateral subdigraph of $D$ and a *weak component* is a maximal weak subdigraph of $D$. In Figure 11.1(b), the digraph has a strong component induced by the vertex set $\{v_2, v_3, v_4\}$. In Figure 11.1(c), the digraph has a unilateral component induced by the vertex set $\{v_1, v_4, v_2\}$ and a weak component which is the digraph itself.

**Theorem 11.4**

    i. Every vertex and every arc of a digraph $D$ belongs to a unique weak component.

    ii. Every vertex and every arc of a digraph of $D$ belongs to at least one unilateral component.

    iii. Every vertex of a digraph belongs to a unique strong component. Every arc is contained in at most one strong component and it is in a strong component if and only if it is in a cycle.

**Proof**

    i. If a vertex $v$ lies in two weak components $W_1$ and $W_2$, let $v_1$ and $v_2$ be any two vertices of $W_1 - W_2$ and $W_2 - W_1$. Then there is a $vv_1$ and $vv_2$ semi-path. Also there is a $v_1 v_2$ semipath, so that $W_1$ and $W_2$ are in the same weak component. Similar argument holds for an arc.

    ii. Since each vertex and each arc is a unilateral subdigraph, the result follows.

    iii. Let $v$ be any vertex and S be the set of vertices mutually reachable with $v$ including $v$ itself. Then $\langle$ S $\rangle$ is a strong component containing $v$. The uniqueness of this follows as in (i). Also, a similar argument holds for an arc.

If an arc $uv$ is in a cycle, then all vertices on the cycle are pairwise reachable and belong to a strong component containing the arcs of the cycle, in particular $uv$. Conversely, if $uv$ is in a strong component, $u$ and $v$ are mutually reachable and hence there is $vu$ path, which together with $uv$ gives a cycle.    ❑

We note that the vertex sets of the weak components of a digraph give a partition $\pi_w$ of its vertex set $V$ and is called a *weak partition*. Similarly, the vertex sets of the strong components of a digraph give a partition $\pi_s$ of $V$, which is a refinement of the partition $\pi_w$. This $\pi_s$ is called the *strong partition*.

**Definition:** Let $V = S_1 \cup S_2 \cup \ldots \cup S_k$ be a partition $\pi$ of the vertex set $V$ of the digraph $D$ $= (V, A)$. Consider a digraph $D_\pi$ with vertex $V_\pi = \{S_1, S_2, \ldots, S_k\}$ which has an arc $S_i S_j$ if and only if in $D$ there is at least one arc from a vertex of $S_i$ to a vertex of $S_j$. Then $D_\pi$ is called the *contraction* of $D$ with respect to the partition $\pi$.

The contraction of a digraph $D$ with respect to its strong partition $\pi_S$ is called the *condensation $D^*$* of $D$.

**Definition:** The symmetrisation $D^S$ of a digraph $D$ is obtained from $D$ by adding $uv$ to $A$ whenever $uv \notin A$, but $vu \in A$. Equivalently, $D^S = D(G(D))$, which is the digraph corresponding to the underlying graph of $D$.

Figure 11.11 illustrates these operations.

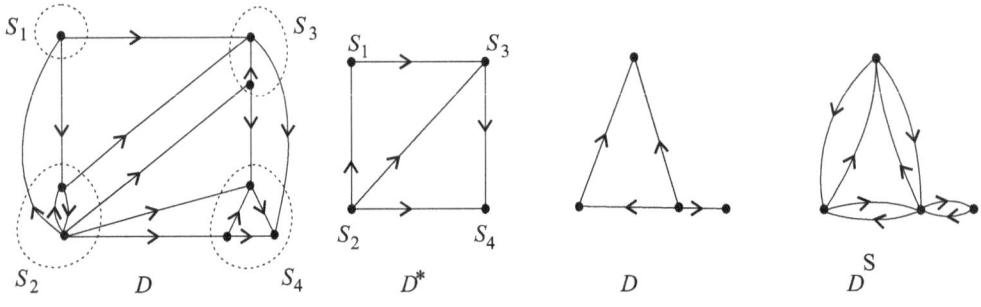

**Fig. 11.11**

We note from the definition of the condensation that a digraph $D$ and its condensation have the same kind of connectedness.

**Theorem 11.5**   If $S_1$ and $S_2$ are two strong components of digraph $D$, and $v_1 \in S_1$ and $v_2 \in S_2$, then there is a $v_1 - v_2$ path in $D$ if and only if there is an $S_1 - S_2$ path in $D^*$.

**Proof**   Let there be a $v_1 - v_2$ path $P$ in $D$. If length of $P$ is one, then there is an $S_1 - S_2$ arc in $D^*$. We induct on the length of $P$. Let the result hold for any path of length $n-1$ in $D$. Assume that $P = v_1 u_1 u_2 \ldots u_{n-1} v_2$ is of length $n$. Let $S(u_{n-1})$ be the strong component containing $u_{n-1}$. Then by induction hypothesis, there is a path $P^*$ from $S_1$ to $S(u_{n-1})$ in $D^*$. If $u_{n-1} \in S_2$, this path serves as an $S_1 - S_2$ path. If not, since $u_{n-1} v_2 \in A$, there is an arc from $S(u_{n-1})$ to $S_2$ in $D^*$, and thus an $S_1 - S_2$ path.

Conversely, let $S_1 S_3 S_4 \ldots S_2$ be an $S_1 - S_2$ path in $D^*$. Then there are arcs $u_1 u_3$, $u_3' u_4$, $u_4' u_5$, ..., $u_n' u_2$ in $D$, $u_i, u_i' \in S_i$. Since the $S_i$'s are strong components, there are $v_1 - u_1$, $u_3 - u_3'$, ..., $u_n - u_n', u_2 - v_2$ paths in $D$. These along with the arcs given above form a $v_1 - v_2$ path in $D$.                                                                                    ❑

The following result can be easily established.

**Theorem 11.6**   If $v_1$, $v_2$ are vertices in different strong components $S_1$, $S_2$ of a digraph $D$, then there is a strict semipath but no path joining $v_1$ to $v_2$ if and only if there is a strict semipath and no path from $S_1$ to $S_2$ in $D^*$.

Now we have the following result.

**Theorem 11.7**   The condensation $D^*$ of any digraph is acyclic.

**Proof**   If $D^*$ contains a cycle, let $S_1 S_2$ be an arc in this cycle. By Theorem 11.1, $S_1 S_2$ lies on a strong component of $D^*$. Therefore $S_1$ and $S_2$ are mutually reachable in $D^*$, and by Theorem 11.5, there are vertices $v_1 \in S_1$ and $v_2 \in S_2$ which are mutually reachable in $D$. Thus, $v_1$ and $v_2$ belong to the same strong component $S$ of $D$. So, $S_1 = S_2 = S$, and $S_1 S_2$ is not an arc in $D^*$, contradicting the assumption.                                                                                    ❑

We note that a digraph $D$ is strong if and only if $D^*$ consists of a single vertex.

**Definition:** A *cut-set* in a digraph $D = (V,A)$ is a set of arcs of $A$, which constitute a cut-set in the multigraph $G = (V,E)$, obtained from $D$ by removing the orientation from each arc of $A$.

## 11.4 Euler Digraphs

A digraph $D$ is said to be Eulerian if it contains a closed walk which traverses every arc of $D$ exactly once. Such a walk is called an *Euler walk*. A digraph $D$ is said to be *unicursal* if it contains an open Euler walk.

The following result characterises Eulerian digraphs.

**Theorem 11.8** A digraph $D = (V,A)$ is Eulerian if and only if $D$ is connected and for each of its vertices $v$, $d^-(v) = d^+(v)$.

### Proof

*Necessity* Let $D$ be an Eulerian digraph. Therefore, it contains an Eulerian walk, say $W$. In traversing $W$, every time a vertex $v$ is encountered we pass along an arc incident towards v and then an arc incident away from $v$. This is true for all the vertices of $W$, including the initial vertex of $W$, say $v$, because we began $W$ by traversing an arc incident away from $v$ and ended $W$ by traversing an arc incident towards $v$.

*Sufficiency* Let for every vertex $v$ in $D$, $d^-(v) = d^+(v)$. For any arbitrary vertex $v$ in $D$, we identify a walk, starting at $v$ and traversing the arcs of $D$ at most once each. This traversing is continued till it is impossible to traverse further. Since every vertex has the same number of arcs incident towards it as away from it, we can leave any vertex that we enter along the walk and the traversal then stops at $v$. Let the walk traversed so far be denoted by $W$. If $W$ includes all arcs of $A$, then the result follows. If not, we remove from $D$ all the arcs of $W$ and consider the remainder of $A$. By assumption, each vertex in the remaining digraph, say $D_1$, is such that the number of arcs directed towards it equals the number of arcs directed away from it. Further, $W$ and $D_1$ have a vertex, say $u$ in common, since $D$ is connected. Starting at $u$, we repeat the process of tracing a walk in $D_1$. If this walk does not contain all the arcs of $D_1$, the process is repeated until a closed walk that traverses each of the arcs of $D$ exactly once is obtained. Hence $D$ is Eulerian. ❑

**Theorem 11.9** A weakly connected digraph $D = (V,A)$ is unicursal if and only if $D$ contains vertices $u$ and $v$ such that $d^+(u) = d^-(u) + 1$, $d^-(v) = d^+(v) + 1$ and $d^+(w) = d^-(w)$, for all $w \in V$, where $w \neq u$, $v$. In this case, the open Euler walk begins at $u$ and ends at $v$.

**Proof** Let $D$ be unicursal. Then $D$ has an Euler walk $W$ that begins at $u$ and ends at $v$. Therefore, as in Theorem 11.8, for every vertex $w$ different from both $u$ and $v$, we have $d^+(w) = d^-(w)$. Also, the first arc of $W$ contributes one to the outdegree of $u$ while every other occurrence of $u$ in $W$ contributes one each to the outdegree of $u$. Thus, $d^+(u) = d^-(u) + 1$.

Similarly, $d^-(v) = d^+(v) + 1$.

Conversely, let $D$ be a weakly connected digraph containing vertices $u$ and $v$ such that $d^+(u) = d^-(u) + 1$, $d^-(v) = d^+(v) + 1$ and for each $w \neq u$, $v$, $d^+(w) = d^-(w)$. In $D$, add a new arc $a$ joining $u$ and $v$. Now, we get a digraph $D_1$ in which the new outdegree of $v$ is one more than its old outdegree, so that $d^-(v) = d^+(v)$. Similarly, $d^-(u) = d^+(u)$, and for every other vertex $w$, $d^-(w) = d^+(w)$. Also, $D_1$ is weakly connected, and since $d - (z) = d + (z)$ for every vertex $z$ in $D_1$, it follows from Theorem 11.8 that $D_1$ is Eulerian. Let $W$ be an Euler walk in $D_1$. Clearly, $D_1$ contains all the arcs of $D$ together with the added arc $a$. Deleting the arc $a$ produces an open Euler walk in $D$. Hence $D$ is unicursal.                    ❏

**Theorem 11.10**   A non-trivial weak digraph is an isograph if and only if it is the union of arc-disjoint cycles.

**Proof**   If the weak digraph $D$ is a union of arc-disjoint cycles, each cycle contributes one to the indegree and one to the outdegree of each vertex on it. Thus, $d^+(v) = d^-(v)$, for all $v \in V$.

Conversely, let $D$ be a non-trivial weak isograph. Then each vertex has positive outdegree and therefore $D$ has a cycle, say $Z$. Removing the edges of $Z$ from $D$, we get a digraph $D_1$ whose weak components are isographs. By using an induction argument, each such non-trivial weak component is a union of arc-disjoint cycles. These cycles together with $Z$ provide a decomposition of the arc set of $D$ into cycles.                    ❏

**Corollary 11.1**   Every weak isograph is strong.

**Proof**   If $u$ and $v$ are any two vertices of the weak isograph $D$, there is a semi path $P$ joining $u$ and $v$, and each arc of this lies on some cycle of $D$. The union of these cycles provides a closed walk containing $u$ and $v$. Thus, $u$ and $v$ are mutually reachable.                    ❏

# 11.5  Hamiltonian digraphs

**Definition:**   A spanning path of a digraph is called a  Hamiltonian path and a spanning cycle a  Hamiltonian cycle. A digraph containing a Hamiltonian cycle is said to be Hamiltonian.

Regarding the results on Hamiltonian digraphs, there are surveys by Bermond and Thomassen [24], and by Jackson. There are a number of results that are analogous to those proved for Hamiltonian graphs in Chapter 3. Two early results on sufficient conditions are due to Ghouila-Houri [87] and Woodall [270]. But now we have a more general result due to Meyniel [160]. The proof of Meyniel's result given here is due to Bondy and Thomassen [38].

First, we have the following observations.

**Lemma 11.1** Let $P = v_1 v_2 \ldots v_k$ be a path in the digraph $D$ and $v$ be any vertex in $V - V(P)$. If there is no $v_1 - v_k$ path with vertex set $V(P) \cup \{v\}$, then $|\{v, V(P)\}| \leq k+1$, where $\{v, V(P)\}$ is the set of all arcs in $D$ with one end in $v$ or $V(P)$ and the other end in $V(P)$ or $v$, respectively.

**Proof** By the assumption on $P$, for any $v \in V - V(P)$, there is no path $v_i v v_{i+1}$ in $D$. Therefore, for each $i$, $1 \leq i \leq k-1$, $|(v_i, v)| + |(v, v_{i+1})| \leq 1$.

Hence, $|\{v, V(P)\}| = \sum_{i=1}^{k-1} (|v_i, v)| + |v, v_{i+1}| + |(v, v_1)| + |(v_k, v)|$

$$\leq (k-1)1 + 2 = k+1. \qquad \square$$

Let in a digraph $D = (V, A)$, $S$ be a proper subset of $V$. A $u - v$ path of length at least two with only $u$ and $v$ in $S$ is called an $S$-path.

**Theorem 11.11** Let $D$ be a strong non-Hamiltonian digraph and $C = v_1 v_2 \ldots v_k v_1$ be a cycle of $D$ such that there is no cycle of $D$ whose vertex set properly contains $V(C) = C$, say. Then there exists a $v \in V - C$, and integers $p$ and $q (1 \leq p, q \leq k)$ such that (i) $v_p v \in A$, (ii) $v v_{p+i} \notin A$ for $i$, $1 \leq i \leq q$ and (iii) $d(v) + d(v_{p+q}) \leq 2n - 1 - q$.

## Proof

**Case 1** $D$ has no $C$-path. Since $D$ is strong and $C$ is a proper subset of $V$, there is a vertex in $C$ joined to some vertex of $V - C$ by a path and joined from some vertex of $V - C$ by a path. By the assumption on $C$, there is a cycle $C'$ having only one vertex, say $v_p$, common with $C$. Let $v$ be the successor of $v_p$ on $C'$ (Fig. 11.12(a)). If $v$ is adjacent to or from any vertex of $C$ other than $v_p$, $D$ has a $C$-path. Therefore, $|\{v, C\}| \leq 2$. Clearly, $|\{v_{p+1}, C\}| \leq 2(k-1)$. For any $u \in V - C$, if $vuv_{p+1}$ or $v_{p+1}uv$ is a 2-path, then these are C-paths in $D$. Thus, such adjacencies are not possible. Therefore, $|\{u, \{v, v_{p+1}\}\}| \leq 2$ for such $u$. Putting these together, we have

$$d(v) + d(v_{p+1}) \leq 2 + 2(k-1) + 2(n-k-1) = 2n-2,$$

and this verifies the statements with $q = 1$.

**Case 2** $D$ has a $C$-path, say $P = v_p u_1 u_2 \ldots u_s v_p$, where $u_i \in V - C$. Choose $P$ such that $r$ is least. By the assumption on $C$, $r > 1$. Let $v = u_1$.

i. By the assumption on $C$ and the minimality of $r$, $v$ is not adjacent to any $v_{p+i}$, $1 \leq i \leq r-1$. Therefore, the path $Q = v_{p+r} v_{p+r+1} \ldots v_1 \ldots v_p$ and $v$ satisfy the hypothesis of Lemma 11.1 and we have

$$|\{v, V(Q)\}| = |\{v, C\}| \leq k - r + 2. \tag{11.11.1}$$

ii. By the minimality of $r$, we observe that for any $u \in V - C$, $vuv_{p+i}$ or $v_{p+i} uv$ are not possible paths in $D$, for $1 \leq i \leq r-1$ (Fig. 11.12(b)). Thus, for any $u \in V - C$,

$$|\{u, \{v, vp + i\}\}| = 2, \text{ for any such } i. \tag{11.11.2}$$

iii. Since $D$ is strong, there are $v_p - v_{p+r}$ and $v_{p+r} - v_p$ paths in $D$. Clearly, $v_{p+r} - v_{p+r+1} \ldots v_k v_1 v_2 \ldots v_p$ is such a $v_{p+r} - v_p$ path. It is possible that there are other $v_{p+r} - v_p$ paths containing all these vertices and some more, say $v_{p+1} v_{p+2} \ldots . v_{p+i-1}$. Let $q$ be the largest integer $i$, $1 \leq i \leq r$ such that there is a $v_{p+r} - v_p$ path with vertex set $S = \{v_{p+r}, v_{p+r+1}, \ldots, v_1, v_2, \ldots, v_p, v_{p+1}, \ldots, v_{p+i-1}\}$ and let $P'$ be such a path. By the assumption on $C$, $S$ cannot contain all the vertices of $C$, since $P \cup P'$ is a cycle. Therefore, $q < r$. It is possible that $q = 1$ (Fig. 11.12 (c)). Now the lemma is applicable for $v_{p+q}$ and $P'$. Therefore,

$$|\{V_{p+q}, V(P')\}| \leq k - r + q + 1. \tag{11.11.3}$$

Using (11.11.2) with $i = q$, we have

$$|\{u, \{v, v_{p+q}\}\}| \leq 2, \text{ for each } u \in V - C. \tag{11.11.4}$$

Also, $v_{p+q}$ can be joined to and from any of the other $r - q - 1$ vertices among

$$\{v_{p+q+1}, \ldots, v_{p+r-1}\}. \tag{11.11.5}$$

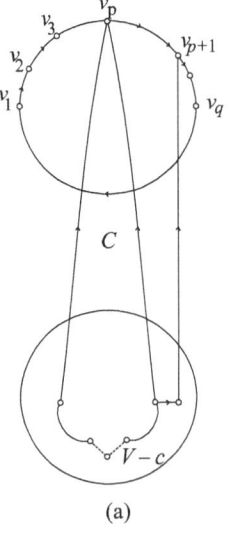

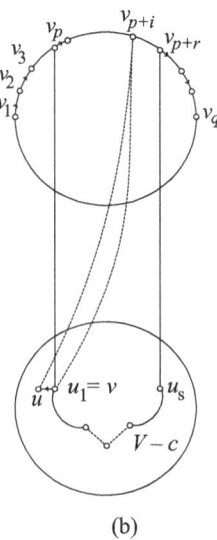

(a)                                                  (b)

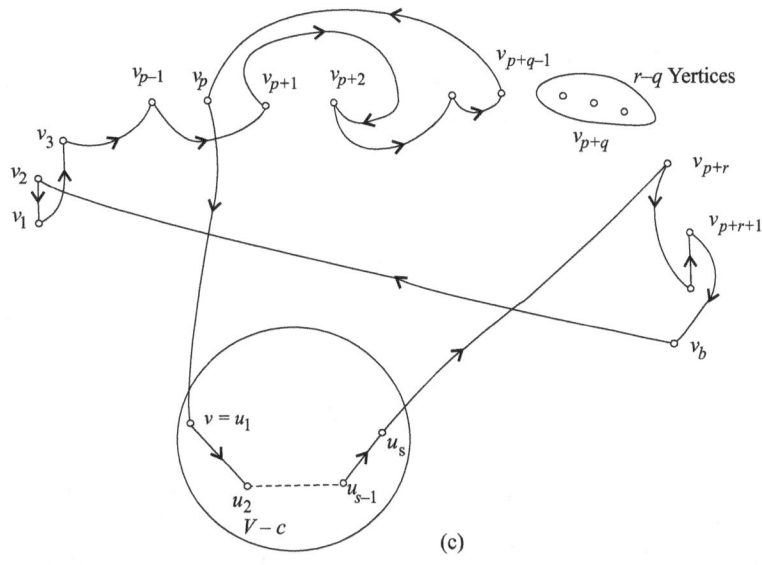

**Fig. 11.12**

Combining all these, we get

$$d(v) + d(v_{p+q}) = |\{v, C\}| + |\{v_{p+q}, V(P')\}| + |\{v_{p+q}, \{v_{p+q+1}, \ldots, v_{p+r-1}\}\}|$$

$$+ \sum_{u \in V-C} |\{u, \{v, v_{p+q}\}\}|$$

$$\leq (k-r+2) + (k-r+q+1) + 2(r-q-1) + 2(n-k-1)$$

$$= 2n - q - 1. \qquad \square$$

**Theorem 11.12 (Meyneil)** If $D$ is a strong digraph of order $n$ such that for any pair of non-adjacent vertices $u$ and $v$, $d(u) + d(v) \geq 2n - 1$, then $D$ is Hamiltonian.

**Proof** If such a $D$ is non-Hamiltonian, by Theorem 11.11, there exists a pair of non-adjacent vertices $v$ and $v_{p+q}$ such that $d(v) + d(v_{p+q}) < 2n - 1$, contradicting the hypothesis.
$\square$

**Corollary 11.2** If $D$ is a strong digraph such that for any vertex $v$, $d(v) = d^+(v) + d^-(v) \geq n$, then $D$ is Hamiltonian.

The direct proof of this result can be found in Berge [18].

## 11.6  Trees with Directed Edges

We know that a tree in undirected graphs is a connected graph without cycles. But in case of digraphs, a structure similar to that of a tree needs absence of cycles as well as absence of semi cycles. We have the following definition.

A *(directed) tree* is a connected digraph without cycles, neither directed cycles nor semi cycles. We observe that a tree with *n* vertices has *n* – 1 directed edges, and has properties analogous to those of trees with undirected edges. A digraph whose weak components are trees is called a *forest*. For example, Figure 11.13 shows a tree.

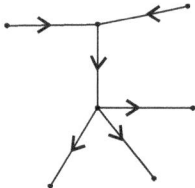

**Fig. 11.13**

**Arborescence:**    A (directed) tree is said to be an *arborescence* if it contains exactly one vertex, called the *root*, with no arcs directed towards it and if all the arcs on any semipath are directed away from the root. For example, the tree in Figure 11.14 is an arborescence. That is, every vertex other than the root has indegree exactly one. Arborescence is also called an *out-tree*. If the direction of every arc in an arborescence is reversed, we get a tree called an *in-tree*.

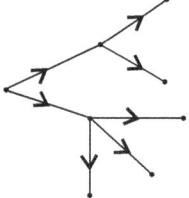

**Fig. 11.14**

**Theorem 11.13**    In an arborescence, there is a directed path from the root *v* to every other vertex. Conversely, a digraph D without cycles is an arborescence if there is a vertex *v* in *D* such that every other vertex is reachable from *v* and *v* is not reachable from any other vertex.

**Proof**    In an arborescence, consider a directed path *P* starting from the root *v* and continuing as far as possible. Clearly, *P* can end only at a pendant vertex, since otherwise, we get a vertex whose indegree is two or more, which is a contradiction. As an arborescence is connected, every vertex lies on some directed path from the root *v* to each of the pendant vertices.

Conversely, since every vertex in *D* is reachable from *v* and *D* has no cycle, *D* is a tree. Further, since *v* is not reachable from any other vertex, $d^-(v) = 0$. Every other vertex is reachable from *v* and therefore indegree of each of these vertices is at least one. The indegree is not greater than one, because there are only $n-1$ arcs in $D$, $n$ being the number of vertices of $D$. ❑

**Ordered trees:** A tree in which the relative order of subtrees meeting at each vertex is preserved is called an *ordered tree* or a planar tree (because the tree can be visualised as rigidly embedded in the plane of the paper). In computer science, the term tree usually means an ordered tree and by convention, a tree is drawn hanging down with the root at the top.

**Spanning trees:** A *spanning tree* is an $n$-vertex connected digraph analogous to a spanning tree in an undirected graph and consists of $n-1$ directed arcs. A *spanning arborescence* in a connected digraph is a spanning tree that is an arborescence. For example, $\{a, b, c, g\}$ is a spanning arborescence in Figure 11.15.

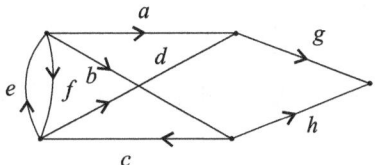

**Fig. 11.15**

**Theorem 11.14** In a connected isograph D of $n$ vertices and $m$ arcs, let $W = (a_1, a_2, \ldots, a_m)$ be an Euler line, which starts and ends at a vertex $v$ (that is, $v$ is the initial vertex of $a_1$ and the terminal vertex of $a_m$). Among the $m$ arcs in $W$ there are $n-1$ arcs that enter each of $n-1$ vertices, other than $v$, for the first time. The subdigraph $D_1$ of these $n-1$ arcs together with the $n$ vertices is a spanning arborescence of $D$, rooted at vertex $v$.

**Proof** In the subdigraph $D_1$, vertex $v$ is of indegree zero, and every other vertex is of indegree one, for $D_1$ includes exactly one arc going to each of the $n-1$ vertices and no arc going to $v$. Further, the way $D_1$ is defined in $W$, implies that $D_1$ is connected and contains $n-1$ arcs. Therefore, $D_1$ is a spanning arborescence in $D$ and is rooted at $v$. ❑

**Illustration:** In Figure 11.16, $W = (b\ d\ c\ e\ f\ g\ h\ a)$ is an Euler line, starting and ending at vertex 2. The subdigraph $\{b, d, f\}$ is a spanning arborescence rooted at vertex 2.

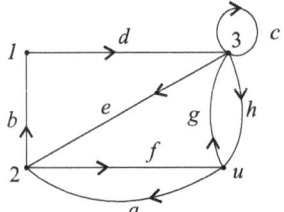

**Fig. 11.16**

The following result is due to Van Aardenne-Ehrenfest and N.G. de Bruijn [257].

**Theorem 11.15**   Let $D$ be an Euler digraph and $T$ be a spanning in-tree in $D$, rooted at a vertex $v$. Let $a_1$ be an arc in $D$ incident out of the vertex $v$. Then a directed walk $W = (a_1, a_2, \ldots, a_m)$ is a directed Euler line, if it is constructed as follows.

i.   No arc is included in $W$ more than once.

ii.  In exiting a vertex the one arc belonging to $T$ is not used until all other outgoing arcs have been traversed.

iii. The walk is terminated only when a vertex is reached from which there is no arc left on which to exit.

**Proof**   The walk $W$ terminates at $v$, since all vertices have been entered as often as they have been left (because $D$ is an isograph). Now assume that there is an arc $a$ in $D$ that has not been included in $W$. Let $u$ be the terminal vertex of $a$. Since $D$ is an isograph, $u$ is also the initial vertex of some arc $b$ not included in $W$. Arc $b$ going out of vertex $u$ is in $T$, according to (i). This omitted arc leads to another omitted arc $c$ in $T$, and so on. Finally we arrive at $v$ and find an outgoing arc not included in $W$. This contradicts (iii).   ❑

Theorem 11.14 provides a method of obtaining a spanning arborescence rooted at any specified vertex, provided the digraph is Eulerian. Conversely, given a spanning arborescence in an Euler digraph, an Euler line can be constructed using Theorem 11.15.

The number of distinct Euler lines formed from a given in-tree $T$ and starting with arc $a_1$ at $v$, can be computed by considering all the choices available at each vertex, after starting with $a_1$. Since there is exactly one outgoing arc in $T$ at each vertex and this arc is to be selected last ((ii), Theorem 11.15), the remaining $d^+(v_i) - 1$ arcs at vertex $v_i$ can be chosen in $(d^+(v_i) - 1)!$ ways. Since these are independent choices, we have altogether $\prod_{i=1}^{n} (d^+(v_i) - 1)!$ different Euler lines that meet (i), (ii) and (iii) of Theorem 11.15.

**Illustration**   Consider Figure 11.17. We apply (i), (ii) and (iii) of Theorem 11.15 to obtain different Euler lines from the in-tree $\{a_2, a_3, a_7, a_{10}, a_{11}\}$, starting with arc $a_1$. The two Euler lines obtained are $(a_1\, a_{12}\, a_5\, a_6\, a_7\, a_8\, a_9\, a_{10}\, a_{11}\, a_2\, a_4\, a_3)$ and $(a_1\, a_{12}\, a_8\, a_9\, a_{10}\, a_{11}\, a_5\, a_6\, a_7\, a_2\, a_4\, a_3)$.

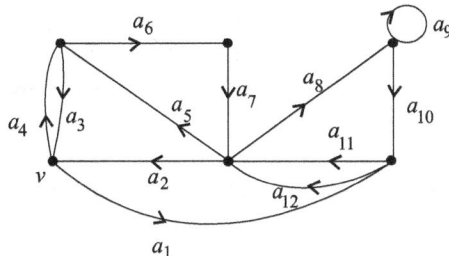

**Fig. 11.17**

Here, $\prod_{i=1}^{n} (d^{+}(v_i) - 1)! = 2.$

It is to be noted that these are not all the Euler lines in the digraph, but only those that are generated by the specific in-tree in accordance with (i), (ii) and (iii) of Theorem 11.15.

**Fundamental cycles in digraphs:**  The arcs of a connected digraph not included in a specified spanning tree $T$ are called the *chords* with respect to $T$. As in undirected graphs, every chord $c_i$ added to the spanning tree $T$ produces a fundamental cycle, which is a directed cycle or a semi cycle.

A *cut-set* in a connected digraph $D$ induces a partitioning of the vertices of $D$ into two disjoint subsets $V_1$ and $V_2$ such that the cut-set consists of all those arcs that have one end vertex in $V_1$ and the other in $V_2$. All arcs in the cut-set can be directed from $V_1$ to $V_2$, or from $V_2$ to $V_1$, or some arcs can be directed from $V_1$ to $V_2$ and others from $V_2$ to $V_1$. A cut-set in which all arcs are oriented in the same direction is called a *directed cut-set*.

Consider the digraph of Figure 11.18. A spanning tree $T = \{a, d, f, h, k\}$ is shown by bold lines. Here, rank = 5, nullity = 4. The chord set with respect to $T$ is $\{b, c, e, g\}$. Fundamental cycles with respect to $T$ are *dfe* (semi cycle), *dkhc* (semi cycle), *khg* (semi cycle) and *adkhb* (directed cycle). The fundamental cut-sets with respect to $T$ are *ab*, *bcde*, *ef*, *bcgk* and *bcgh*.

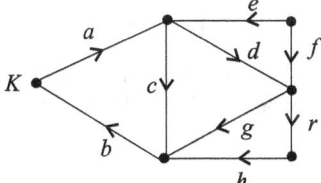

**Fig. 11.18**

# 11.7  Matrices A, B and C of Digraphs

The matrices associated with a digraph are almost similar to those discussed for an undirected graph, with the difference that in matrices of digraphs consist of 1, 0, −1 instead of only 0 and 1 for undirected graphs. The numbers 1, 0, −1 are real numbers and their addition and multiplication are interpreted as in ordinary arithmetic, not modulo 2 arithmetic as in undirected graphs. Thus, the vectors and vector spaces associated with a digraph and its subdigraphs are over the field of all real numbers, but not modulo 2.

**Incidence matrix:**    The incidence matrix of a digraph with $n$ vertices, $m$ arcs and no self-loops is an $n \times m$ matrix $A = [a_{ij}]$, whose rows correspond to vertices and columns correspond to arcs, such that

$$a_{ij} = \begin{cases} 1, & \text{if jth arc is incident out of ith vertex}, \\ -1, & \text{if jth arc is incident into ith vertex}, \\ 0, & \text{if jth arc is not incident on ith vertex}. \end{cases}$$

For example, consider the digraph of Figure 11.19.

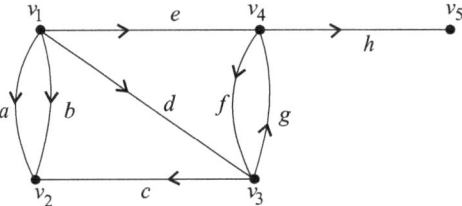

**Fig. 11.19**

The incidence matrix is given by

$$A = \begin{array}{c} \\ v_1 \\ v_2 \\ v_3 \\ v_4 \\ v_5 \end{array} \begin{array}{cccccccc} a & b & c & d & e & f & g & h \\ \left[ \begin{array}{cccccccc} 1 & 1 & 0 & 1 & 1 & 0 & 0 & 0 \\ -1 & -1 & -1 & 0 & 0 & 0 & 0 & 0 \\ 0 & 0 & 1 & -1 & 0 & -1 & 1 & 0 \\ 0 & 0 & 0 & 0 & -1 & 1 & -1 & 1 \\ 0 & 0 & 0 & 0 & 0 & 0 & 0 & -1 \end{array} \right] \end{array}$$

Now, since the sum of each column is zero, the rank of the incidence matrix of a digraph of $n$ vertices is less than $n$. The proof of the following result is almost similar to the result in undirected graphs.

**Theorem 11.16**    If $A(D)$ is the incidence matrix of a connected digraph of $n$ vertices, then rank of $A(D) = n - 1$.

We further note that after deleting any row from $A$, we get $A_f$, the $n - 1 \times m$ reduced incidence matrix. The vertex corresponding to the deleted row is called the reference vertex.

We now have the following result.

**Theorem 11.17**   The determinant of every square submatrix of $A$, which is the incidence matrix of a digraph, is $-1$, or 1, or 0.

**Proof**   Consider a $k \times k$ submatrix $M$ of $A$. If $M$ has any column or row consisting of all zeros, then clearly det $M = 0$. Also, det $M = 0$, if every column of $M$ contains the two non-zero entries, 1 and $-1$.

Now let det $M \neq 0$. Then the sum of entries in each column of $M$ is not zero. Therefore $M$ has a column in which there is a single non-zero element that is either 1 or $-1$. Let this single element be in the $(i, j)$th position in $M$. Thus,

$$\det M = \pm 1 \det M_{ij},$$

where $M_{ij}$ is the submatrix of $M$ with its $i$th row and $j$th column deleted. The $(k-1) \times (k-1)$ submatrix $M_{ij}$ is non-singular (because $M$ is non-singular), therefore $M_{ij}$ also has at least one column with a single non-zero entry, say in the $(p, q)$th position. Expanding det $M_{ij}$ about this element in the $(p, q)$th position, we obtain

$$\det M_{ij} = \pm \, [\text{det of non-singular } (k-2) \times (k-2) \text{ submatrix of } M].$$

Repeated application of this procedure gives

$$\det M = \pm 1. \qquad \qquad \square$$

**Unimodular matrix:** A matrix is said to be unimodular if the determinant of its every square submatrix is 1, $-1$, or 0.

**Cycle matrix of a digraph:**   Let $D$ be a digraph with $m$ arcs and $q$ cycles (directed cycles or semi cycles). An arbitrary orientation (clockwise or counter clockwise) is assigned to each of the $q$ cycles. Then a cycle matrix $B = [b_{ij}]$ of the digraph $G$ is a $q \times m$ matrix defined by

$$b_{ij} = \begin{cases} 1, & \text{if the ith cycle includes the jth arc, and the orientations of the arc} \\ & \text{and cycle coincide,} \\ -1, & \text{if the ith cycle includes the jth arc, and the orientations of the two} \\ & \text{are opposite,} \\ 0, & \text{if the ith cycle does not include the jth arc.} \end{cases}$$

**Example**    Consider the digraph $D$ given in Figure 11.20. The cycle matrix of $D$ is

$$
B = \begin{array}{c} \\ \\ \\ \\ \\ \end{array}
\begin{array}{cccccccc}
a & b & c & d & e & f & g & h \\
\left[\begin{array}{cccccccc}
0 & 0 & 0 & 1 & 0 & 1 & 1 & 0 \\
0 & 0 & 1 & 0 & -1 & 0 & 1 & 0 \\
0 & 0 & 1 & -1 & -1 & -1 & 0 & 0 \\
-1 & 1 & 0 & 0 & 0 & 0 & 0 & 0
\end{array}\right]
\end{array} .
$$

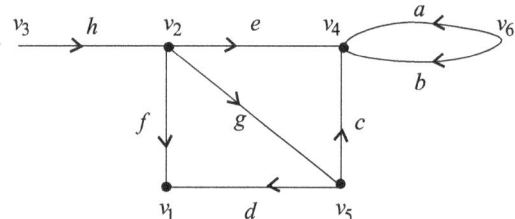

**Fig. 11.20**

We note that the orientation assigned to each of the four cycles is arbitrary. The cycle in the first row is assigned clockwise orientation, in the second row counter-clockwise, in the third counter-clockwise, and in the fourth clockwise. Changing the orientation of any cycle will simply change the sign of every non-zero entry in the corresponding row. Also we observe that if first row is subtracted from second, the third is obtained. Thus the rows are not all linearly independent (in the real field).

The next result gives the relation between incidence matrix and cycle matrix.

**Theorem 11.18**    Let $B$ and $A$ be respectively, the cycle matrix and the incidence matrix of a digraph (without loops) such that the columns are arranged using the same order of arcs. Then,

$$AB^T = BA^T = 0,$$

where $T$ denotes the transposed matrix.

**Proof**    Consider the $p$th row of $A$ and the $r$th row of $B$. The $r$th cycle, say $Z_r$, either (a) does not, or (b) does possess an arc incident with vertex, say $v_p$, represented by the $p$th row of $A$. If (a), the product of the two rows is zero. If (b), there are exactly two arcs, say $a_i$ and $a_j$, of the $r$th cycle incident with $v_p$.

We have the following four possibilities.

i. $a_i$ and $a_j$ are both incident towards $v_p$,

ii. $a_i$ and $a_j$ are both incident away from $v_p$,

iii. the directions of both $a_i$ and $a_j$ coincide with the orientation of $Z_r$ and

iv. the directions of both $a_i$ and $a_j$ do not coincide with the orientation of $Z_r$.

It can be easily verified that in all these four cases, the product of the $p$th row of $A$ and the $r$th row of $B$ is zero. ❏

Figure 11.21 illustrates these four cases.

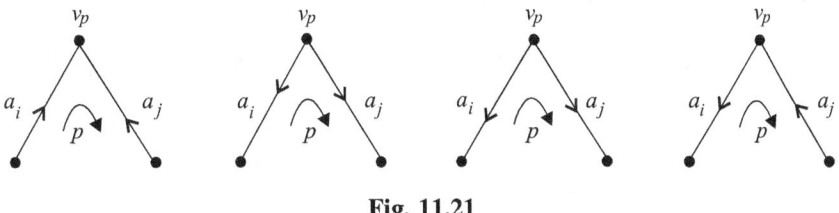

**Fig. 11.21**

Now, using Sylvester's theorem and Theorem 11.18, we can show that

    rank $B+$ rank $A = m$.

If the digraph is connected, then rank $A = n - 1$.

Therefore, rank $B = m - n + 1$.

The following two results can be easily established.

**Theorem 11.19** The non-singular submatrices of order $n - 1$ of $A$ are in one-one correspondence with spanning trees of a connected digraph of n vertices.

**Theorem 11.20** The non-singular submatrices of $B$ of order $\mu = m - n + 1$ are in one-one correspondence with the chord set (complement of the spanning tree) of the connected digraph of $n$ vertices and $m$ edges.

**Sign of a spanning tree:** For a digraph, the determinant of the non-singular submatrix of A corresponding to a spanning tree $T$ has a value either 1 or $-1$. This is referred to as the sign of $T$.

We note that the sign of a spanned tree is defined only for a particular ordering of vertices and arcs in $A$, because interchanging two rows or columns in a matrix changes the sign of its determinant. Thus, the sign of a spanning tree is relative. Once the sign of one spanning tree is arbitrarily chosen, the sign of every other spanning tree is determined as positive or negative with respect to this spanning tree.

**Number of spanning trees:** The following result determines the number of spanning trees in a connected digraph.

**Theorem 11.21**    If $A_f$ is the reduced incidence matrix of a connected digraph, then the number of spanning trees in the graph is equal to

$$\det (A_f . A_f^T).$$

**Proof**    According to Binet-Cauchy theorem,

$$\det (A_f . A_f^T) = \text{sum of the products of all corresponding majors of } A_f \text{ and } A_f^T.$$

Every major of $A_f$ or $A_f^T$ is zero unless it corresponds to a spanning tree, in which case its value is 1 or $-$ 1. Since both majors of $A_f$ and $A_f^T$ have the same value 1 or $-1$, the product is 1 for each spanning tree.                                                                                  ❏

**Fundamental cycle matrix:** The $\mu$ fundamental cycles each formed by a chord with respect to some specified spanning tree, define a fundamental cycle matrix $B_f$ for a digraph. The orientation assigned to each of the fundamental cycles is chosen to coincide with that of the chord. Therefore, $B_f$, a $\mu \times m$ matrix can be expressed exactly in the same form as in the case of an undirected graph,

$$B_f = [I_\mu : B_t],$$

where $I_\mu$ is the identity matrix of order $\mu$ and the columns of $B_t$ correspond to the arcs in a spanning tree. This is illustrated in Figure 11.22.

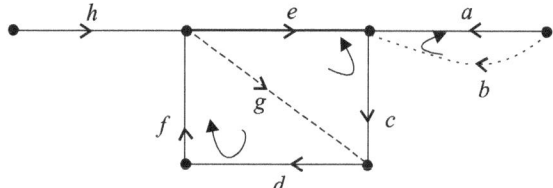

**Fig. 11.22**

$$\text{Here, } B_f = \begin{array}{c} \begin{array}{cccccccc} b & d & g & a & c & e & f & h \end{array} \\ \begin{bmatrix} 1 & 0 & 0 & -1 & 0 & 0 & 0 & 0 \\ 0 & 1 & 0 & 0 & -1 & 1 & 1 & 0 \\ 0 & 0 & 1 & 0 & 1 & -1 & 0 & 0 \end{bmatrix} \end{array} = [I_\mu : B_t].$$

**Cut-set matrix:** Let $D = (V, A)$ be a connected digraph with $q$ cut-sets. The cut-set matrix $C = [c_{ij}]$ of $D$ is a $q \times m$ matrix in which the rows correspond to the cut-sets of $D$ and the columns to the arcs of $D$. Each cut-set is given an arbitrary orientation. Let $R_i$ be the $i$th cut-set of $D$ and let $R_i$ partition $V$ into nonempty vertex sets $V_i'$ and $V_i''$. The orientation can be defined to be either from $V_i'$ to $V_i''$ or from $V_i''$ to $V_i'$. Suppose that the orientation is chosen to be from $V_i'$ to $V_i''$. Then the orientation of an arc $a_j$ of cut-set $R_i$ is said to be the same as that of $R_i$ if $a_j$ is of the form $v_a\, v_b$, where $v_a \in V_i'$ and $v_b \in V_i''$ and opposite, otherwise. Then,

$$c_{ij} = \begin{cases} 1, & \text{if arc } a_j \text{ of } cut-set\ R_i \text{ has the same orientation as } R_i, \\ -1, & \text{if arc } a_j \text{ has the opposite orientation to } R_i, \\ 0, & \text{otherwise}. \end{cases}$$

We have the following observations.

1. A permutation of the rows or columns corresponds to a relabelling of the cut-sets and arcs of $D$ respectively.

2. Rank $C \geq$ rank $A$.

3. Rank $C \geq n-1$, by observation 2.

4. If the arcs of $D$ are arranged in the same column order in $B$ and $C$, then $BC^T = CB^T = 0$.

5. Rank $B+$ rank $C \leq m$.

6. If $D$ is weak, rank $B = m - n + 1$ and rank $C \leq n - 1$.

7. Rank $C = n - 1$, because of (3) and (4).

We observe that the removal of an arc, say $a = v_s v_t$ (also called a branch) of a spanning directed tree of $D$, partitions the vertices of a digraph $D$ into two disjoint sets, say $V_1$ and $V_2$.

The cut-set created by the removal of $a$ is said to be either (i) directed away from $V_1$ and towards $V_2$ if $v_s \in V_1$ and $v_t \in V_2$, or (ii) directed away from $V_1$ and towards $V_2$ if $v_s \in V_2$ and $v_t \in V_1$.

This type of cut-set is called *fundamental cut-set*. Clearly, not all the chords in $R_i$ necessarily have the same orientation as $v_s v_t$. If $v_s v_t$ is directed away from a vertex in $V_1$, there may exist a chord in $R_i$ which is directed towards a vertex in $V_1$. The orientation of a cut-set on the basis of the direction of the branch giving rise to it constitutes a natural way of orienting cut-sets. If all the chords of $R_i$ are oriented as is $v_s v_t$, then $R_i$ is said to be *directed*. Consider the graph shown in Figure 11.23 with $T$ shown by bold lines. The fundamental cut-sets with respect to $T$ are

| branch | cut-set |
|---|---|
| $a_3$ | $\{a_3, a_2\}$ |
| $a_5$ | $\{a_5, a_4\}$ |
| $a_6$ | $\{a_6, a_1, a_4\}$ |
| $a_7$ | $\{a_7, a_1, a_2\}$ |

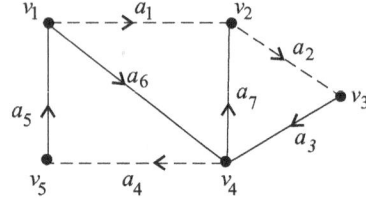

**Fig. 11.23**

A fundamental cut-set is created from C, the cut-set matrix of a connected digraph with the given directed spanning tree $T$, by deleting from C, all rows which do not correspond to fundamental cut-sets with respect to $T$. Therefore $C_f$ is an $(n-1) \times m$ submatrix of C such that each row represents a unique fundamental cut-set with respect to $T$.

The rows of any fundamental cut-set $C_f$ can be permuted to create a matrix of the form $C_f = [C_c : I_{n-1}]$, where $C_c$ is an $(n-1) \times (m-n+1)$ matrix whose columns correspond to the chords of $T$ and $I_{n-1}$ is the identity matrix of order $n-1$ whose columns correspond to the branches of $T$.

**Relation between $\mathbf{B}_f$, $\mathbf{C}_f$ and $\mathbf{A}_r$**    $A_r$ is the reduced incidence matrix in which an arbitrary row has been removed in order to make its rows linearly independent.

We have $B_f = [I_\mu : B_t]$                                                                            (11.7.i)

and $C_f = [C_c : I_{n-1}]$,                                                                              (11.7.ii)

where $t$ corresponds to the branches of $T$ and $c$ to the chords of $T$. Let the arcs be arranged in the same order in (11.7.i) and (11.7.ii) and in $A_r$. Partition $A_r$ as

$$A_r = [A_c : A_t],$$

where $A_c$ is an $(n-1) \times (m-n+1)$ submatrix whose columns correspond to the chords of $T$ and $A_t$ is an $(n-1) \times (n-1)$ submatrix whose columns correspond to the branches of $T$.

Since, $AB^T = 0$, therefore, $A_r B_f^T = 0$.

Thus, $[A_c : A_t] \begin{bmatrix} I_\mu \\ .. \\ B_t^T \end{bmatrix} = 0$, so that $A_c + A_t B_t^T = 0$.

Since $A_t$ is non-singular, we have $A_t^{-1}[A_c + A_t B_t^T] = 0$.

Therefore, $A_t^{-1} A_c + B_t^T = 0$ and so $A_t^{-1} A_c = -B_t^T$. Also, $C_f B_f^T = 0$.

Therefore, $[C_c : I_{n-1}] \begin{bmatrix} I_\mu \\ .. \\ B_t^T \end{bmatrix} = 0$, and so $C_c + B_t^T = 0$.

Thus, $C_c = -B_t^T$ and so $C_c = A_t^{-1} A_c$.

**Example** Consider the digraph given in Figure 11.24, with spanning tree shown in bold lines.

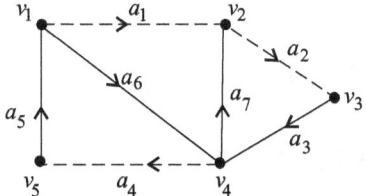

**Fig. 11.24**

We have

$$A_r = [A_c : At] = \begin{array}{cccccccc} a_1 & a_4 & a_2 & & a_5 & a_6 & a_7 & a_3 \end{array} \begin{bmatrix} -1 & 0 & 0 & \vdots & -1 & 1 & 0 & 0 \\ -1 & 0 & 0 & \vdots & 0 & 0 & -1 & 0 \\ 0 & 0 & -1 & \vdots & 0 & 0 & 0 & 1 \\ 0 & 1 & 0 & & 0 & -1 & 1 & -1 \end{bmatrix}$$

$$B_f = [I_3 : B_t] = \begin{array}{ccccccc} a_1 & a_4 & a_2 & & a_5 & a_6 & a_7 & a_3 \end{array} \begin{bmatrix} 1 & 0 & 0 & \vdots & 0 & -1 & -1 & 0 \\ 0 & 1 & 0 & \vdots & 1 & 1 & 0 & 0 \\ 0 & 0 & 1 & \vdots & 0 & 0 & 1 & 1 \end{bmatrix}$$

$$C_f = [C_c : I_4] = \begin{array}{ccccccc} a_1 & a_4 & a_2 & & a_5 & a_6 & a_7 & a_3 \end{array} \begin{bmatrix} 0 & -1 & 0 & \vdots & 1 & 0 & 0 & 0 \\ 1 & -1 & 0 & \vdots & 0 & 1 & 0 & 0 \\ 1 & 0 & -1 & \vdots & 0 & 0 & 1 & 0 \\ 0 & 0 & -1 & \vdots & 0 & 0 & 0 & 1 \end{bmatrix}$$

Note that the last row of $A$, corresponding to vertex $v_5$, has been removed to form $A_f$. We form linear combinations of the rows of $C_f$ to create 10 rows of $C$, representing all of the cut-sets of the digraph in Figure 11.24.

$$C = \begin{array}{c} \\ \\ \\ \\ \\ \\ \\ \\ \\ \\ \end{array} \begin{array}{ccccccc} a_1 & a_4 & a_2 & a_5 & a_6 & a_7 & a_3 \\ \left[ \begin{array}{ccccccc} 0 & -1 & 0 & 1 & 0 & 0 & 0 \\ 1 & -1 & 0 & 0 & 1 & 0 & 0 \\ 1 & 0 & -1 & 0 & 0 & 1 & 0 \\ 0 & 0 & -1 & 0 & 0 & 0 & 1 \\ -1 & 0 & 0 & 1 & -1 & 0 & 0 \\ 0 & -1 & 0 & 0 & 1 & -1 & 1 \\ 0 & 0 & 1 & -1 & 1 & -1 & 0 \\ 0 & 0 & 0 & 1 & 1 & 1 & -1 \\ 0 & -1 & 1 & 0 & 1 & -1 & 0 \\ 1 & 0 & 0 & 0 & 0 & 1 & 1 \end{array} \right] \end{array} \begin{array}{l} c_1 \\ c_2 \\ c_3 \\ c_4 \\ c_1 - c_2 \\ c_2 - c_3 + c_4 \\ c_2 - c_1 - c_3 \\ c_1 - c_2 + c_3 - c_4 \\ c_2 - c_3 \\ c_3 - c_4 \end{array}$$

The above facts lead to the following observations.

1. Given $A_r$, we can construct $B_f$ and $C_f$.

2. Given $B_r$, we can construct $C_f$.

3. Given $C_f$, we can construct $B_f$.

**Semipath matrix:** The semipath matrix $P(u, v) = [p_{ij}]$, of a digraph $D = (V, A)$, where $u, v \in V$, is the matrix with each row representing a distinct semipath from $u$ to $v$ and the columns representing the arcs of $D$, in which $p_{ij} = 1$, if the $i$th semipath contains the $j$th arc, $p_{ij} = -1$ if the $i$th semipath contains the converse of the $j$th arc, and $p_{ij} = 0$ otherwise.

The matrix $P(v_3, v_5)$ for the digraph of Figure 11.25 is

$$P(v_3, v_5) = \begin{array}{c} \\ \\ \end{array} \begin{array}{cccccc} a_1 & a_2 & a_3 & a_4 & a_5 & a_6 \\ \left[ \begin{array}{cccccc} 0 & 0 & 1 & 0 & 0 & 0 \\ 0 & 1 & 0 & 0 & -1 & 0 \\ 1 & 0 & 0 & 1 & 0 & 1 \end{array} \right] \end{array}$$

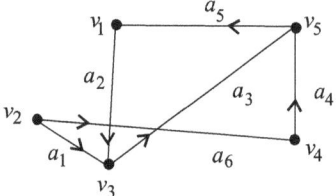

**Fig. 11.25**

We have the following observations about $P$.

1. If $P(u, v)$ contains a column of all zeros, then the vertex that it represents does not belong to any of the semipaths between $u$ and $v$.

2. If $P(u, v)$ contains a column of all unit entries, then the vertex that it represents belongs to every semipath between $u$ and $v$.

3. The number of non-zero entries in any row of $P(u, v)$ equals the number of arcs in the semipath represented by the row.

**Adjacency matrix of a digraph:**   Let $G$ be a digraph with $n$ vertices and with no parallel arcs. The adjacency matrix $X = [x_{ij}]$ of the digraph $G$ is an $n \times n$ $(0, 1)$-matrix defined by

$$x_{ij} = \begin{cases} 1, & \text{if there is an arc directed from ith vertex to jth vertex,} \\ 0, & \text{otherwise.} \end{cases}$$

**Example**   Consider the digraph $D$ of Figure 11.26.

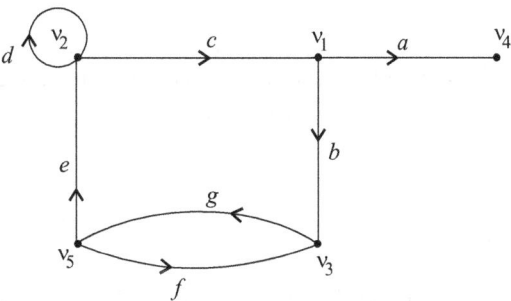

**Fig. 11.26**

The adjacency matrix of $D$ is

$$X = \begin{array}{c} \\ v_1 \\ v_2 \\ v_3 \\ v_4 \\ v_5 \end{array} \begin{array}{ccccc} v_1 & v_2 & v_3 & v_4 & v_5 \\ \begin{bmatrix} 0 & 0 & 1 & 1 & 0 \\ 1 & 1 & 0 & 0 & 0 \\ 0 & 0 & 0 & 0 & 1 \\ 0 & 0 & 0 & 0 & 0 \\ 0 & 1 & 1 & 0 & 0 \end{bmatrix} \end{array}$$

We have the following observations about the adjacency matrix X of a digraph D.

1. X is a symmetric matrix if and only if $D$ is a symmetric digraph.

2. Every non-zero element on the main diagonal element represents a loop at the corresponding vertex.

3. The parallel arcs cannot be represented by $X$ and therefore $X$ is defined only for a digraph without parallel arcs.

4. The sum of each row equals the outdegree of the corresponding vertex and the sum of each column equals the indegree of the corresponding vertex. The number of non-zero entries of X equals the number of arcs in *D*.

5. Permutation of any rows together with a permutation of the corresponding columns does not alter the digraph and thus the permutation corresponds to a reordering of the vertices. Therefore, two digraphs are isomorphic if and only if their adjacency matrices differ only by such permutations.

6. If *X* is the adjacency matrix of a digraph *D*, then the transposed matrix $X^T$ is the adjacency matrix of a digraph $D^*$ obtained by reversing the direction of every arc in *D*.

7. For any $(0-1)$-matrix *Q* of order *n*, there exists a unique digraph *D* of *n* vertices such that *Q* is the adjacency matrix of *D*.

**Connectedness and adjacency matrix:**     A digraph is disconnected if and only if its vertices can be ordered in such a way that its adjacency matrix *X* can be expressed as the direct sum of two square submatrices $X_1$ and $X_2$ as

$$X = \begin{bmatrix} X_1 & : & 0 \\ .. & : & .. \\ 0 & : & X_2 \end{bmatrix}. \tag{11.7.iii}$$

This partitioning is possible if and only if the vertices in the submatrix $X_1$ have no arc going to or coming from the vertex of $X_2$.

Similarly, a digraph is weakly connected if and only if its vertices can be ordered in such a way that its adjacency matrix can be expressed as

$$X = \begin{bmatrix} X_1 & : & 0 \\ .. & : & .. \\ X_{21} & : & X_2 \end{bmatrix} \tag{11.7.iv}$$

or $\quad X = \begin{bmatrix} X_1 & : & X_{12} \\ .. & : & .. \\ 0 & : & X_2 \end{bmatrix},$ $\tag{11.7.v}$

where $X_1$ and $X_2$ are square submatrices.

Form (11.7.iv) represents the case when there is no arc going from the subdigraph corresponding to $X_1$ to the one corresponding to $X_2$. Form (11.7.v) represents the case when there is no arc going from the subdigraph corresponding to $X_2$ to the subdigraph corresponding to $X_1$.

Since a strongly connected digraph is neither disconnected nor weakly connected, a digraph is strongly connected if and only if the vertices of *D* cannot be ordered such that its adjacency matrix *X* is expressible in the form (11.7.iii), or (11.7.iv), or (11.7.v).

**Theorem 11.22**  $[X^k]_{ij}$ is the number of different arc sequences of $k$ arcs from the $i$th vertex to the $j$th vertex.

**Proof**  Induct on $k$. The result is trivially true for $k = 1$. Assume the result holds for $[X^{k-1}]_{ij}$. Now,

$$[X^k]_{ij} = [X^{k-1}X]_{ij} = \sum_{r=1}^{n} [X^{k-1}]_{ir}[X]_{rj} = \sum_{r=1}^{n} [X^{k-1}]_{ir}x_{rj} \qquad (11.22.1)$$

$$= \sum_{r=1}^{n} (\text{number of all directed arc sequences of length } k-1 \text{ from}$$

$$\text{vertex } i \text{ to } r) \, x_{rj},$$

by induction hypothesis.

In (11.22.1), $x_{rj} = 1$ or 0, according as there is an arc from $r$ to $j$. Therefore, a term in the sum (11.22.1) is non zero if and only if there is an arc sequence of length $k$ from $i$ to $j$, whose last arc is from $r$ to $j$. If the term is non-zero, its value equals the number of such arc sequences from $i$ to $j$ through $r$. This holds for every vertex $r$, $1 \leq r \leq n$. Thus (11.22.1) is equal to the number of all possible arc sequences from $i$ to $j$. $\qquad \square$

It is to be noted that $[X^k]_{ij}$ gives the number of all arc sequences from vertex $i$ to $j$ and these arc sequences can be of the following types.

1. Directed paths from $i$ to $j$, that is, those arc sequences in which no vertex is traversed more than once.

2. Directed walks from $i$ to $j$, that is, those directed arc sequences in which a vertex may be traversed more than once, but no arc is traversed more than once.

3. Those arc sequences in which an arc may also be traversed more than once.

## 11.8 Number of Arborescences

We now give a formula for counting the number of spanning arborescence in a labeled, connected digraph (which of course is simple). First, we have the following definition.

**Kirchoff matrix:**  For a digraph (simple) $D$ of $n$ vertices, the Kirchoff matrix is an $n \times n$ matrix $K(D) = [k_{ij}]$ defined by

$$k_{ij} = \begin{cases} d^-(v_i), & i = j, \quad \text{in degree } of \text{ the ith vertex}, \\ -x_{ij}, & i \neq j \quad (i, j)th \text{ entry in the adjacency matrix}, \text{ with a negative sign}. \end{cases}$$

**Example**   Consider the digraph $D$ given in Figure 11.27.

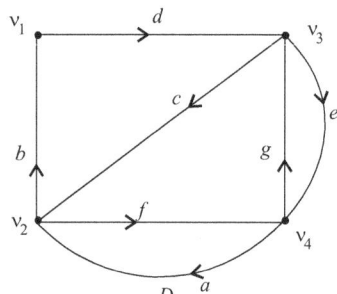

**Fig. 11.27**

The Kirchoff matrix of $D$ is

$$
K(D) = \begin{array}{c} \\ v_1 \\ v_2 \\ v_3 \\ v_4 \end{array}
\begin{array}{cccc}
v_1 & v_2 & v_3 & v_4 \\
\end{array}
\left[
\begin{array}{cccc}
1 & 0 & -1 & 0 \\
-1 & 2 & 0 & -1 \\
0 & -1 & 2 & -1 \\
0 & -1 & -1 & 2
\end{array}
\right]
$$

Clearly, the sum of the entries in each column in $K$ is equal to zero, so that the $n$ rows are linearly independent. Thus, $\det K = 0$.

**Theorem 11.23**   A digraph (simple) $D$ of $n$ vertices and $n-1$ arcs is an arborescence rooted at $v_1$ if and only if the $(1, 1)$ cofactor of $K(D)$ is equal to 1.

**Proof**

a. Let $D$ be an arborescence with $n$ vertices and rooted at vertex $v_1$. Relabel the vertices as $v_1, v_2, \ldots, v_n$ such that vertices along every path from the root $v_1$ have increasing indices. Permute the rows and columns of $K(D)$ to conform with this relabelling.

Since the indegree of $v_1$ equals zero, the first column contains only zeros. Other entries in $K(D)$ are

$$
k_{ij} = \begin{cases}
0, & i > j, \\
-x_{ij}, & i < j, \\
1, & i = j, \ i > 1.
\end{cases}
$$

Then the $K$ matrix of an arborescence rooted at $v_1$ is of the form

$$K(D) = \begin{bmatrix} 0 & -x_{12} & -x_{13} & -x_{14} & .. & -x_{1n} \\ 0 & 1 & -x_{23} & -x_{24} & .. & -x_{2n} \\ 0 & 0 & 1 & -x_{34} & .. & -x_{3n} \\ 0 & 0 & 0 & 1 & .. & \\ \vdots & & & & .. & \\ 0 & 0 & 0 & 0 & .. & 1 \end{bmatrix}.$$

Clearly, the cofactor of the $(1, 1)$ entry is 1, that is, $\det K_{11} = 1$.

b. Conversely, let $D$ be a digraph of $n$ vertices and $n-1$ arcs, and let $(1, 1)$ cofactor of its $K$ matrix be equal to 1, that is, $\det K_{11} = 1$.

Since $\det K_{11} \neq 0$, every column in $K_{11}$ has at least one non-zero entry. Therefore,

$$d^-(v_i) \geq 1, \text{ for } i = 2, 3, \ldots, n.$$

There are only $n-1$ arcs to go around, therefore

$$d^-(v_i) = 1, \text{ for } i = 2, 3, \ldots, n, \text{ and } d^-(v_1) = 0.$$

Since no vertex in $D$ has an indegree of more than one, if $D$ can have any cycle at all, it has to be a directed cycle. Suppose that such a directed cycle exists, which passes through vertices $v_{i_1}, v_{i_2}, \ldots, v_{i_r}$. Then the sum of the columns $i_1, i_2, \ldots, i_r$ in $K_{11}$ is zero. This is because each of these columns contains exactly two non-zero entries, as 1 on the main diagonal and $a-1$ for the incoming arc from the vertex preceding it in the directed cycle. Thus, the $r$ columns in $K_{11}$ are linearly dependent. So, $\det K_{11} = 0$, a contradiction. Therefore, $D$ has no cycles.

If $D$ has $n-1$ arcs and no cycles, it must be a tree. Since in this tree $d-(v_1) = 0$ and $d^-(v_i) = 1$, for $i = 2, 3, \ldots, n$,

$D$ is an arborescence rooted at vertex $v_1$.

The arguments in (a) and (b) are valid for an arborescence rooted at any vertex $v_q$. Any reordering of the vertices in $D$ corresponds to identical permutations of rows and columns in $K(D)$. Such permutations do not alter the value or sign of the determinant. ❏

**Theorem 11.24** If $K(D)$ is the Kirchoff matrix of a (simple) digraph $D$, then the value of the $(q, q)$ cofactor of $K(D)$ is equal to the number of arborescences in $D$ rooted at the vertex $v_q$.

**Proof** The proof depends on the result of Theorem 11.23 and on the fact that the determinant of a square matrix is a linear function of its columns. In particular, if $P$ is a square matrix consisting of $n$ column vectors, each of dimension $n$, that is

$$P = [p_1, p_2, \ldots, (p_i + p'_i), \ldots, p_n],$$

then $\det P = \det [p_1, p_2, \ldots p_i, \ldots, p_n] + \det [p_1, p_2, \ldots, p'_i, \ldots, p_n].$ \hfill (11.24.1)

In digraph $D$, suppose that vertex $v_j$ has indegree $d_j$. The $j$th column of $K(D)$ can be regarded as the sum of $d_j$ different columns, each corresponding to a digraph in which $v_j$ has indegree one. And then (11.24.1) can be repeatedly applied. After this, splitting of columns can be carried out for each $j$, $j \neq q$ and $\det K_{qq}(D)$ can be expressed as a sum of determinants of subdigraphs, that is

$$\det K_{qq}(D) = \sum_{D'} \det K_{qq}(D'),$$ \hfill (11.24.2)

where $D'$ is a subdigraph of $D$ with the following properties.

1. Every vertex in $D'$ has an indegree of exactly one except $v_q$.
2. $D'$ has $n-1$ vertices and hence $n-1$ arcs.

From Theorem 11.23,

$$\det K_{qq}(D') = \begin{cases} 1, & \textit{if and only if } D' \textit{ is an arborescence rooted at } q, \\ 0, & \textit{otherwise}. \end{cases}$$

Thus, the summation in (11.24.2) carried over all $D''$s equals the number of arborescences rooted at $v_q$. \hfill $\square$

## 11.9  Tournaments

A tournament is an orientation of a complete graph. Therefore, in a tournament each pair of distinct vertices $v_i$ and $v_j$ is joined by one and only one of the oriented arcs $(v_i, v_j)$ or $(v_j, v_i)$. If the arc $(v_i, v_j)$ is in $T$, then we say $v_i$ dominates $v_j$ and is denoted by $v_i \rightarrow v_j$. The relation of dominance thus defined is a complete, irreflexive, antisymmetric binary relation. Figure 11.28 displays all tournaments on three and four vertices.

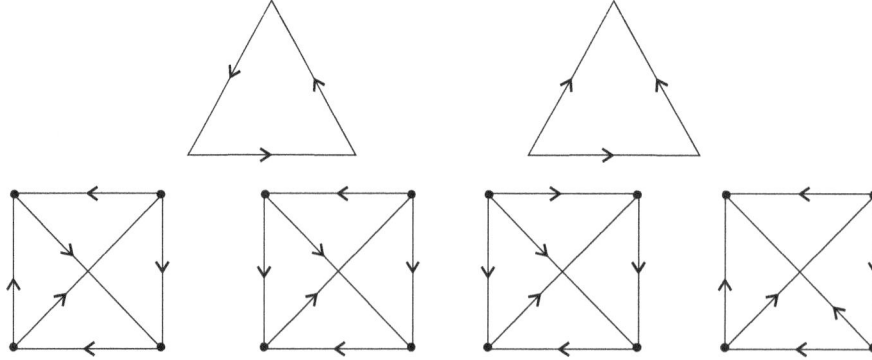

**Fig. 11.28**

**Definition:** A *triple* in a tournament $T$ is the subdigraph induced by any three vertices. A triple $(u, v, w)$ in $T$ is said to be *transitive* if whenever $(u, v) \in A(T)$ and $(v, w) \in A(T)$, then $(u, w) \in A(T)$. That is, whenever $u \to v$ and $v \to w$, then $u \to w$.

**Definition:** A *bipartite tournament* is an orientation of a complete bipartite graph. A *k-partite tournament* is an orientation of a complete $k$-partite graph. Figure 11.29 displays a bipartite and a tripartite tournament.

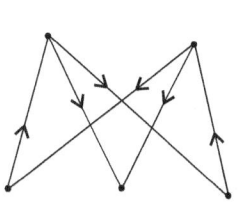 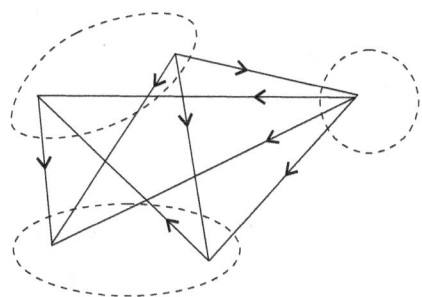

**Fig. 11.29**

**Theorem 11.25** If $v$ is a vertex having maximum outdegree in the tournament $T$, then for every vertex $w$ of $T$ there is a directed path from $v$ to $w$ of length at most 2.

**Proof** Let $T$ be a tournament with $n$ vertices and let $v$ be a vertex of maximum outdegree in $T$. Let $d^+(v) = m$ and let $v_1, v_2, \ldots, v_m$ be the vertices in $T$ such that there are arcs from $v$ to $v_i$, $1 \leq i \leq m$. Since $T$ is a tournament, there are arcs from the remaining $n - m - 1$ vertices, say $u_1, u_2, \ldots, u_{n-m-1}$ to $v$. That is, there are arcs from $u_j$ to $v$, $1 \leq j \leq n - m - 1$ (Fig. 11.30).

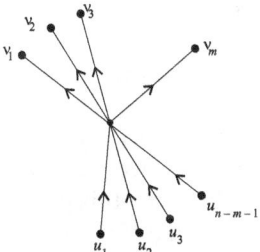

**Fig. 11.30**

Then for each $i$, $1 \leq i \leq m$, the arc from $v$ to $v_i$ gives a directed path of length 1 from $v$ to $v_i$. We now show that there is a directed path of length 2 from $v$ to $u_j$ for each $j$, $1 \leq j \leq n - m - 1$.

Given such a vertex $u_j$, if there is an arc from $v_i$ to $u_j$ for some $i$, then $vv_iu_j$ is a directed path of length 2 from $v$ to $u_j$. Now, let there be a vertex $u_k$, $1 \leq k \leq n - m - 1$, such that no vertex $v_i, 1 \leq i \leq m$, has an arc from $v_i$ to $u_k$. Since $T$ is tournament, there is an arc from $u_k$ to each of the $m$ vertices $v_i$. Also, there is an arc from $u_k$ to $v$ and therefore $d^+(u_k) \geq m + 1$.

This contradicts the fact that $v$ has maximum outdegree with $d^+(v) = m$. Thus, each $u_j$ must have an arc joining it from some $v_i$ and the proof is complete by using the directed path $vv_iu_j$. ❏

Let $T$ be a tournament with $n$ vertices and let $v$ be any vertex of $T$. Then $T - v$ is the digraph obtained from $T$ by removing $v$ and all arcs incident with $v$. Clearly, any two vertices of $T - v$ are joined by exactly one arc, since these two vertices are joined by exactly one arc in $T$. Thus, $T - v$ is again a tournament.

**Definition:** A *directed Hamiltonian path* of a digraph $D$ is the directed path in $D$ that includes every vertex of $D$ exactly once.

The following result, due to Redei [216], shows that a tournament contains a direct Hamiltonian path.

**Theorem 11.26 (Redei [216])**    Every tournament $T$ has a directed Hamiltonian path.

**Proof** Let $T$ be a tournament with $n$ vertices. We induct on $n$. When $n = 1$, 2, or 3, the result is trivially true (Fig. 11.31).

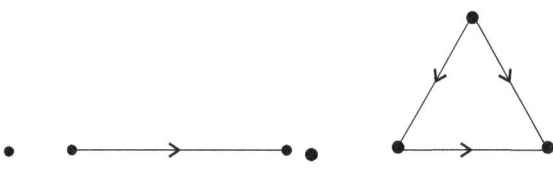

**Fig. 11.31**

Now, let $n \geq 4$. Assume that the result is true for all tournaments with fewer than $n$ vertices. Let $v$ be any vertex of $T$. Then $T - v$ is a tournament with $n - 1$ vertices and by induction hypothesis has a directed Hamiltonian path, say $P = v_1v_2 \ldots v_{n-1}$.

In case there is an arc from $v$ to $v_1$, then $P_1 = vv_1 v_2 \ldots v_{n-1}$ is a directed Hamiltonian path in $T$. Similarly if there is an arc from $v_{n-1}$ to $v$, then $P_2 = v_1 v_2 \ldots v_{n-1}v$ is a directed Hamiltonian path in $T$.

Now, assume that there is no arc from $v$ to $v_1$ and no arc from $v_{n-1}$ to $v$. Then there is at least one vertex $w$ on the path $P$ with the property that there is an arc from $w$ to $v$ and $w$ is not $v_{n-1}$. Let $v_i$ be the last vertex on $P$ having this property, so that the next vertex $v_{i+1}$ does not have this property. Then there is an arc from $v_i$ to $v$ and an arc from $v$ to $v_{i+1}$, as shown in Figure 11.32. Thus $Q = v_1v_2 \ldots v_iv \, v_{i+1} v_{i+2} \ldots v_{n-1}$ is a directed Hamiltonian path in $T$. ❏

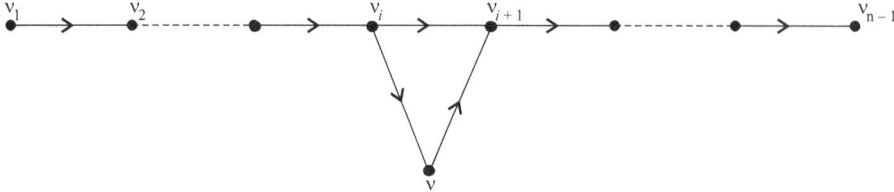

**Fig. 11.32**

**Definition:** A *directed Hamiltonian cycle* in a digraph $D$ is a directed cycle which includes every vertex of $D$. If $D$ contains such a cycle, then $D$ is called *Hamiltonian*.

The next two results are due to Camion [44].

**Theorem 11.27 (Camion [44])** A strongly connected tournament $T$ on $n$ vertices contains cycles of length $3, 4, \ldots, n$.

**Proof** First we show that $T$ contains a cycle of length three. Let $v$ be any vertex of $T$. Let $W$ denote the set of all vertices $w$ of $T$ for which there is an arc from $v$ to $w$. Let $Z$ denote the set of all vertices $z$ of $T$ for which there is an arc from $z$ to $v$. We note that $W \cap Z = \varphi$, since $T$ is a tournament.

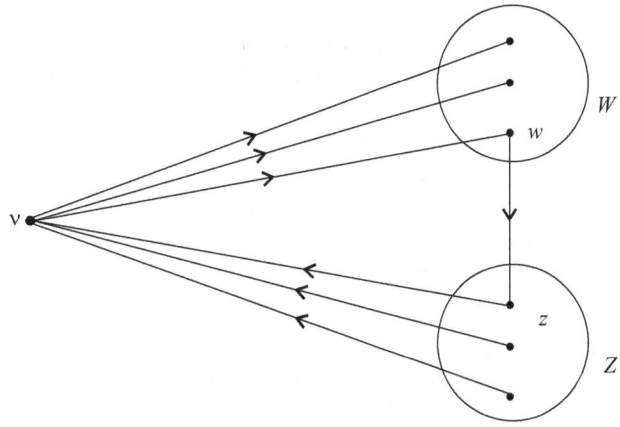

**Fig. 11.33**

Since $T$ is strongly connected, $W$ and $Z$ are both nonempty. For, if $W$ is empty, then there is no arc going out of $v$, which is impossible because $T$ is strongly connected and the same argument can be used for $Z$. Again, because $T$ is strongly connected, there is an arc in $T$ going from some $w$ in $W$ to some $z$ in $Z$. This gives the directed cycle $v\,w\,z\,v$ of length 3 (Fig. 11.33).

Now induct on $n$. Assume $T$ has a cycle $C$ of length $k$, where $k < n$ and $k \geq 3$ and let this cycle be $v_1 v_2 \ldots v_k v_1$. We show that $T$ has a cycle of length $k+1$.

Let there be a vertex $v$ not on the cycle $C$, with the property that there is an arc from $v$ to $v_i$ and an arc from $v_j$ to $v$ for some $v_i, v_j$ on $C$. Then there is a vertex $v_i$ on $C$ with an arc from $v_{i-1}$ to $v$ and an arc from $v$ to $v_i$. Therefore, $C_1 = v_1 v_2 \ldots v_{i-1} v v_i v_{i+1} \ldots v_k v_1$ is a cycle of length $k+1$ (Fig. 11.34).

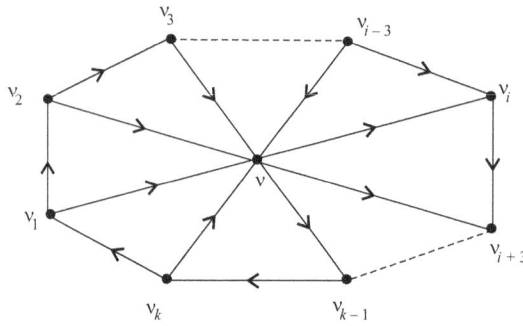

**Fig. 11.34**

If no vertex exists with the above property, then the set of vertices not contained in the cycle can be divided into two distinct sets $W$ and $Z$, where $W$ is the set of vertices $w$ such that for each $i$, $1 \le i \le k$, there is an arc from $v_i$ to $w$ and $Z$ is the set of vertices $z$ such that for each $i$, $1 \le i \le k$, there is an arc from $z$ to $v_i$. If $W$ is empty then the vertices of $C$, and the vertices of $Z$ together make up all the vertices in $T$. But, by definition of $Z$, there is no arc from a vertex on $C$ to a vertex in $Z$, a contradiction, because $T$ is strongly connected. Thus, $W$ is nonempty. A similar argument shows that $Z$ is nonempty. Again, since $T$ is strongly connected, there is an arc from some $w$ in $W$ to some $z$ in $Z$. Then $C_1 = v_1 \, w \, z \, v_3 v_4 \ldots v_k v_1$ is a cycle of length $k+1$ (Fig. 11.35). This completes the proof.                                    ❑

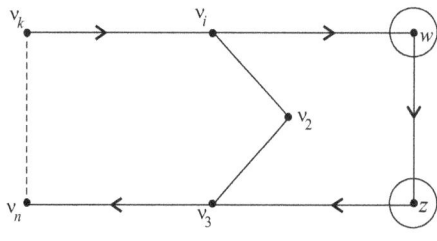

**Fig. 11.35**

**Theorem 11.28 (Camion)**    A tournament $T$ is Hamiltonian if and only if it is strongly connected.

**Proof**    Let $T$ have $n$ vertices. If $T$ is strongly connected, then by Theorem 11.27, $T$ has a cycle of length $n$. Such a cycle is a Hamiltonian cycle, since it includes every vertex of $T$. Hence $T$ is Hamiltonian.

Conversely, let $T$ be Hamiltonian with Hamiltonian cycle $C = v_1 v_2 \ldots v_n v_1$. Then given any $v_i, v_j, i \ge j$, in the vertex set of $T$, $v_j v_{j+1} \ldots v_i$ is a path $P_1$ from $v_j$ to $v_i$ while

$v_i v_{i+1} \ldots v_{n-1} v_n v_1 \ldots v_{j-1} v_j$ is a path $P_2$ from $v_i$ to $v_j$ (Fig. 11.36). Thus each vertex is reachable from any other vertex and so $T$ is strongly connected.                                    ❑

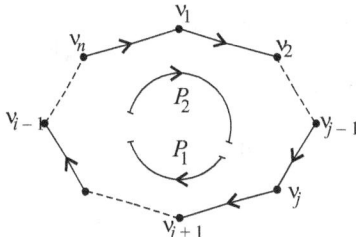

**Fig. 11.36**

## 11.10   Exercises

1. Prove that the converse of a strong digraph is also strong.

2. Show that $D^*$, the condensation of any digraph $D$, is cyclic.

3. Prove that the converse of a unilateral digraph is unilateral.

4. Show that the transmitters, receivers and isolates of a digraph D retain their properties in $D^*$.

5. Prove that the only acyclic digraph is $K_1 \cong K_1^*$.

6. Prove that an acyclic digraph without isolates has a transmitter and a receiver.

7. Show that every vertex $v$ of a non-trivial digraph $D$ has total even degree if and only if $D$ is the union of arc-disjoint cycles.

8. Prove that every arc in a digraph belongs either to a directed cycle or a directed cut-set.

9. Prove that the digraph $D = (V, A)$ with $d_v^+ > 0$ for all $v \in V$, has a cycle.

10. Prove that the digraph distance satisfies the triangle inequality.

11. Prove that every Eulerian digraph is strong. Is the converse true?

12. If $E|G|$ is the number of Euler lines in an n-vertex Euler digraph $D$, show that $2^{n-1}.E|G|$ is the number of Euler lines in $L(D)$.

13. If $D$ is a digraph with an odd number of vertices and if each vertex of $D$ has an odd outdegree, prove that there is an odd number of vertices of $D$ with odd indegree.

14. Prove by induction on n that for each $n \geq 1$, there is a simple digraph $D$ with $n$ vertices $v_1, v_2, \ldots, v_n$ such that $d_{v_i}^+ = i - 1$ and $d_{v_i}^- = n - i$ for each $i = 1, 2, \ldots, n$.

15. Prove that no strictly weak digraph contains a vertex whose removal results in a strong digraph.

16. There exists a digraph with outdegree sequence $[s_1, s_2, \ldots, s_n]$, where $n-1 \geq s_1 \geq s_2 \geq \ldots \geq s_n$ and indegree sequences $[t_1, t_2, \ldots, t_n]$ where every $t_j \leq n-1$ if and only if $\sum s_i = \sum t_i$ and for each integer $k < n$,

$$\sum_{i=1}^{k} s_i \leq \sum_{i=1}^{k} \min\{k-1, t_i\} + \sum_{i=k+1}^{n} \min\{k, t_i\}.$$

17. If $X$ is the adjacency matrix of the edge digraph of a complete symmetric digraph, then $X^2 + X$ has all entries 1.

18. Let $T$ be any tournament. Prove that the converse of $T$ and the complement of $T$ are isomorphic.

19. Prove that a tournament is transitive if and only if it has a unique Hamiltonian path.

20. Prove that if a simple digraph D has a cycle of length three, then it is not transitive.

21. Prove that a tournament is transitive if and only if it has no directed cycles.

# 12.  The Four Colour Theorem

*Map of Madhya Pradesh and adjoining states in India, circa 2000*

## Introduction

The famous four colour theorem seems to have been first proposed by Möbius in 1840, later by DeMorgan and the Guthrie brothers in 1852 and again by Cayley in 1878. The problem of proving this theorem has a distinguished history, details of which abound in the literature. The statement of the theorem may be introduced as follows. In colouring a geographical map it is customary to give different colours to any two countries that have a segment of their boundaries in common. It has been found empirically that any map, no matter how many countries it contains nor how they are situated, can be so coloured by using only four different colours. The map of India requires four colours in the states bordering Madhya Pradesh. The fact that no map was ever found whose colouring requires more than four colours suggests the mathematical theorem.

**The four colour theorem:**   For any subdivision of the plane into non-overlapping regions, it is always possible to mark each of the regions with one of the numbers 0, 1, 2, 3, in such a way that no two adjacent regions receive the same number.

**Steps of the proof**    We shall outline the strategy of the new proof given in this chapter. In section 12.1 on *Map Colouring*, we define maps on the sphere and their proper colouring. For purposes of proper colouring it is equivalent to consider maps on the plane and furthermore, only maps which have exactly three edges meeting at each vertex. Lemma 12.1 proves the six colour theorem using Euler's formula, showing that any map on the plane may be properly coloured by using at most six colours. We may then make the following basic definitions.

- Define $N$ to be the minimal number of colours required to properly colour any map from the class of all maps on the plane.

- Based on the definition of $N$, select a specific map $m(N)$ on the plane, which requires no fewer than $N$ colours to be properly coloured.

- Based on the definition of the map $m(N)$, select a proper colouring of the regions of the map $m(N)$ using the $N$ colours $0, 1, \ldots, N-1$.

The whole proof works with the fixed number $N$, the fixed map $m(N)$ and the fixed proper colouring of the regions of the map $m(N)$. In Section 12.2, we define *Steiner Systems* and prove Tits' inequality and its consequence that if a Steiner system $S(N+1, 2N, 6N)$ exists, then $N$ cannot exceed 4. Now the goal is to demonstrate the existence of such a Steiner system. In Section 12.3, we define *Eilenberg Modules*. The regions of the map $m(N)$ are partitioned into disjoint, nonempty equivalence classes $0, 1, \ldots, N-1$ according to the colour they receive. This set is given the structure of the cyclic group $\mathbf{Z}_N = \{0, 1, \ldots, N-1\}$ under addition modulo $N$. We regard $\mathbf{Z}_N$ as an Eilenberg module for the symmetric group $S_3$ on three letters and consider the split extension $\mathbf{Z}_N]S_3$ corresponding to the trivial representation of $S_3$. By Section 12.4 on *Hall Matchings*, we choose a common system of coset representatives for the left and right cosets of $S_3$ in the full symmetric group on $|\mathbf{Z}_N]S_3|$ letters. In Section 12.5, for each such common representative and for each ordered pair of elements of $S_3$, on *Riemann Surfaces*, we establish a certain action of the two-element cyclic group on twelve copies of the partitioned map $m(N)$ by using the twenty-fourth root function of the sheets of the complex plane. In Section 12.6, we give the details of the *Main Construction* using this action. The $6N$ elements of $\mathbf{Z}_N]S_3$ are regarded as the set of points and Lemma 12.23 builds the blocks of $2N$ points with every set of $N+1$ points contained in a unique block. This constructs a Steiner system $S(N+1, 2N, 6N)$ which implies by Tits' inequality that $N$ cannot exceed 4, completing the proof.

## 12.1   Map Colouring

A *map* on the sphere is a subdivision of the surface into finitely many regions. A map is regarded as *properly coloured* if each region receives a colour and no two regions having a whole segment of their boundaries in common receive the same colour. Since deformations of the regions and their boundary lines do not affect the proper colouring of a map, we shall confine ourselves to maps whose regions are bounded by simple closed polygons. For purposes of proper colouring, this is equivalent to considering maps drawn on the plane. Any map on the sphere may be represented on the plane by boring a small hole through the interior of one of the regions and deforming the resulting surface until it is flat. Conversely, by a reversal of this process, any map on the plane may be represented on the sphere. Furthermore, it suffices to consider *3-regular maps*, i.e., maps with exactly three edges meeting at each vertex, by the following argument. Replace each vertex at which more than three edges meet by a small circle and join the interior of each such circle to one of the regions meeting at the vertex. A new map is obtained which is 3-regular. If this new map can be properly coloured by using at most $n$ colours, then by shrinking the circles down to points, the desired colouring of the original map using at most $n$ colours is obtained. Example of a map that requires four colours to be properly coloured is shown in Figure 12.1.

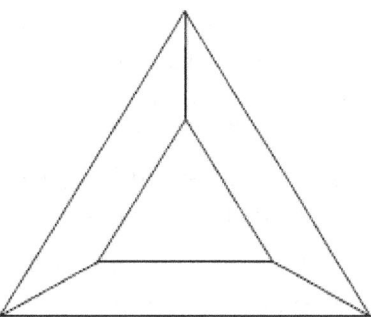

**Fig. 12.1**   *A map that requires four colours to be properly coloured*

**Lemma 12.1**   Any map on the sphere can be properly coloured by using at most six colours.

**Proof**   Assume that the given map is 3-regular. First show that there must be at least one region whose boundary is a polygon with fewer than six sides as follows. Let $E$ be the number of edges, $V$ the number of vertices, $F$ the number of regions and $F_n$ the number of regions whose boundary is a polygon with $n$ sides in the given map. Then

$$F = F_2 + F_3 + F_4 + \ldots$$

$$2E = 3V = 2F_2 + 3F_3 + 4F_4 + \ldots$$

since a region bounded by $n$ edges has $n$ vertices and each vertex belongs to three regions. By Euler's formula $V - E + F = 2$

$$\Rightarrow \qquad\qquad\qquad\qquad 6V - 6E + 6F \ = \ 12$$
$$\Rightarrow \qquad\qquad\qquad\qquad 4E - 6E + 6F \ = \ 12$$
$$\Rightarrow \qquad\qquad\qquad\qquad\qquad 6F - 2E \ = \ 12$$
$$\Rightarrow \quad 6(F_2 + F_3 + F_4 + \ldots) - (2F_2 + 3F_3 + 4F_4 + \ldots) \ = \ 12$$
$$\Rightarrow \qquad 4F_2 + 3F_3 + 2F_4 + F_5 + 0 + \text{negative terms} \ = \ 12.$$

Hence at least one of $F_2$, $F_3$, $F_4$, $F_5$ must be positive. Now, if a region $R$ with fewer than six sides is removed from the map and the resulting map coloured with six colours inductively, there is always a colour left for $R$.                                                                    ❏

By Lemma 12.1 the minimal number of colours required to properly colour any map from the class of all maps on the sphere is a well-defined natural number. We may now make the following basic definition.

**Definition**

- Define $N$ to be the minimal number of colours required to properly colour any map from the class of all maps on the sphere. That is, given any map on the sphere, no more than $N$ colours are required to properly colour it and there exists a map on the sphere which requires no fewer than $N$ colours to be properly coloured.

- Based on the definition of $N$, select a specific map $m(N)$ on the sphere, which requires no fewer than $N$ colours to be properly coloured.

- Based on the definition of the map $m(N)$, select a proper colouring of the regions of the map $m(N)$ using the $N$ colours $0, 1, \ldots, N-1$.

The natural number $N$, the map $m(N)$ and the proper colouring of the regions of $m(N)$ is fixed for all future reference. By the example shown in Figure 12.1 and Lemma 12.1, $4 \leq N \leq 6$. The goal is to show that $N \leq 4$.

## 12.2  Steiner Systems

A Steiner system $S(t, k, v)$ is a set $P$ of points together with a set $B$ of blocks such that

- There are $v$ points.

- Each block consists of $k$ points.

- Every set of $t$ points is contained in a unique block.

Note that by definition $t$, $k$, $v$ are non-negative integers with $t \leq k \leq v$. Steiner systems with $v = k$ (only one block that contains all the points) or $k = t$ (every $k$-element subset of points is a block) are called *trivial*. An example of a nontrivial Steiner system is $S(5, 8, 24)$ due to Witt, whose blocks are known as Golay codewords of weight eight. The group of automorphisms of $S(5, 8, 24)$ (permutations of points which permute blocks) is the largest of the Mathieu groups, $M_{24}$.

**Lemma 12.2 (J. Tits)**   If there exists a nontrivial Steiner system $S(t,k,v)$ then

$$v \geq (t+1)(k-t+1).$$

**Proof**   First show that there exists a set $X_0$ of $t+1$ points that is not contained in any block as follows. Suppose that for every set $X$ of $t+1$ points there is a block $B_X$ that contains it. Then this block $B_X$ must be the unique block containing $X$, since $X$ has more than $t$ points. Let $b$ denote the total number of blocks. Count in two ways the number of pairs $(X, B_X)$ where $X$ is a set of $t+1$ points and $B-X$ is the unique block containing it. One finds that

$$\binom{v}{t+1} \overset{=b}{} \binom{k}{t+1}$$

Count in two ways the number of pairs $(Y, B_Y)$ where $Y$ is a set of $t$ points and $B_Y$ is the unique block containing it, by definition of a Steiner system. One finds

$$\binom{v}{t} \overset{=b}{} \binom{k}{t}$$

Hence

$$\frac{\binom{v}{t+1}}{\binom{k}{t+1}} = \frac{\binom{k}{t+1}}{\binom{k}{t}} = b$$

and it follows that $b = 1$ and $k = v$, contradicting the hypothesis that the Steiner system is nontrivial. Now choose a fixed set $X_0$ of $t+1$ points that is not contained in any block. For each set $Z$ of $t$ points contained in $X_0$ there is a unique block $B_Z$ containing $Z$. Each such $B_Z$ has $k-t$ points not in $X_0$ and any point not in $X_0$ is contained in atmost one such $B_Z$ since two such blocks already have $t-1$ points of $X_0$ in common. The union of the blocks $B_Z$ contains $(t+1) + (t+1)(k-t)$ points and this number cannot exceed the total number of points $v$.   ❑

Recall the definition from section 12.1 that $N$ is the minimal number of colours required to properly colour any map from the class of all maps on the sphere and $m(N)$ is a specific map which requires all of the $N$ colours to properly colour it. The regions of the map $m(N)$ have been properly coloured using the $N$ colours $0, 1, \ldots, N-1$. From the map $m(N)$ and its fixed proper colouring, we shall construct a Steiner system $S(N+1, 2N, 6N)$ by defining the points and blocks in a certain way. The next lemma shows that this construction would force $N \leq 4$.

**Lemma 12.3**   Referring to the definition of $N$ in section 12.1, if there exists a Steiner system $S(N+1, 2N, 6N)$, then $N \leq 4$.

**Proof**   Since $4 \leq N \leq 6$ by definition, the Steiner system is nontrivial if it exists. By Lemma 12.2, $6N \geq (N+1+1)(2N-N-1+1) = (N+2)N$. Hence $6 \geq N+2$ and it follows that $4 \geq N$.                                                                                     ❑

Now the goal is to demonstrate the existence of the Steiner system $S(N+1, 2N, 6N)$ based upon the definition of the map $m(N)$.

## 12.3   Eilenberg Modules

Let $G$ be a group with identity element $e$ and let $Z$ denote the integers. The *integral group algebra* $(\mathbf{Z}G, +, \cdot)$ is a ring whose elements are formal sums

$$\sum_{g \in G} n_g g$$

with $g$ in $G$ and $n_g$ in $\mathbf{Z}$ such that $n_g = 0$ for all but a finite number of $g$. Addition and multiplication in $\mathbf{Z}G$ are defined by

$$\sum_{g \in G} n_g g + \sum_{g \in G} m_g g = \sum_{g \in G} (n_g + m_g) g$$

$$\sum_{g \in G} n_g g \cdot \sum_{g \in G} m_g g = \sum_{g \in G} \sum_{n \in G} h \in G (n_{gh} - 1 m_h) g$$

The element $n$ of $\mathbf{Z}$ is identified with the element $n \cdot e$ of $\mathbf{Z}G$ and the element $g$ of $G$ is identified with the element $1 \cdot g$ of $\mathbf{Z}G$, so that $\mathbf{Z}$ and $G$ are to be regarded as subsets of $\mathbf{Z}G$. The underlying additive abelian group $(\mathbf{Z}G, +)$ is the direct sum of copies of the integers $Z$ indexed by elements of $G$. If $Q$ is a subgroup of $G$ then $\mathbf{Z}Q$ is a subring of $\mathbf{Z}G$ in a natural way. For each element $g$ of $G$, the right multiplication $R(g) : G \to G$; $x \to xg$ and the left multiplication $L(g) : G \to G$; $x \to gx$ are permutations of the set $G$. Denote the group of all permutations of the set $G$ by Sym(G). Then

$$R : G \to Sym(G); \ g \to R(g)$$

$$L^{-1} : G \to Sym(G); \ g \to L^{-1}(g) = L(g^{-1})$$

are embeddings of the group $G$ in Sym(G). The images $R(G), L^{-1}(G)$ are called the *Cayley right and left regular representations of $G$*, respectively. The subgroup of Sym(G) generated by the set $R(G) \cup L^{-1}(G) = \{R(g), L(g^{-1}) | g \in G\}$ is called the *combinatorial multiplication group Mlt(G)* of $G$. There is an exact sequence of groups

$$\begin{array}{ccc} & \Delta & T \\ 1 \to C(G) & \to G \times G \to & Mlt(G) \to 1 \end{array}$$

where $T(x, y) = R(x)L(y^{-1})$ and $\Delta c = (c, c)$ for an element $c$ of the center $C(G)$ of $G$. If $Q$ is a subgroup of $G$ then the *relative combinatorial multiplication group* $Mlt_G(Q)$ of $Q$ in $G$ is the subgroup of $Mlt(G)$ generated by the set $R(Q) \cup L^{-1}(Q) = \{R(q), L^{-1}(q) | q \in Q\}$. The orbits of the action of $Mlt_G(Q)$ on $G$ are the double cosets $QgQ$ of the subgroup $Q$ in $G$. The stabilizer of the identity element $e$ is the subgroup of $Mlt_G(Q)$ generated by the set $\{T(q) = R(q)L^{-1}(q) | q \in Q\}$. A *representation* of the group $Q$ is usually defined as a module, i.e., an abelian group $(M, +)$, for which there is a homomorphism $T : Q \to Aut(M, +)$ showing how $Q$ acts as a group of automorphisms of the module. Another approach due to Eilenberg views a module $M$ for the group $Q$ as follows. The set $M \times Q$ equipped with the multiplication

$$(m_1, q_1) \cdot (m_2, q_2) = (m_1 + m_2 T(q_1), q_1 q_2)$$

becomes a group $M]Q$ known as the *split extension of M by Q*. There is an exact sequence of groups

$$\begin{array}{ccc} & \imath & \pi \\ 1 \to M & \to M]Q \to & Q \to 1 \end{array}$$

With $\imath : M \to M]Q$; $m \to (m, e)$ and $\pi : M]Q \to Q$; $(m, q) \to q$ split by $0 : Q \to M]Q$; $q \to (0, q)$. The group action $T$ is recovered from the split extension $M]Q$ by $mT(q)\imath = m\imath R((0, q))L^{-1}((0, q))$ for $m$ in $M$ and $q$ in $Q$. In this context we shall call $M$ an *Eilenberg module for the group Q*. For example, the trivial representation for the group $Q$ is obtained by defining $T : Q \to Aut(M, +)$; $q \to 1_M$, the identity automorphism of $(M, +)$ and the corresponding split extension is the group direct product $M \times Q$. The Cayley right regular representation for the group $Q$ is obtained by defining

$$T : Q \to Aut(ZQ, +); \quad q \to ( \sum_{g \in Q} n_g g \to \sum_{g \in Q} n_g g R(q)).$$

Here, $T(q) = R(q)L^{-1}(q)$ with $L^{-1}(q)$ is acting trivially on the module elements and $R(q)$ is acting as the usual right multiplication. The split extension $ZQ]Q$ has multiplication given by

$$(m_1, q_1) \cdot (m_2, q_2) = (m_1 + m_2 R(q_1), q_1 q_2)$$

for $m_1, m_2$ in $ZQ$ and $q_1, q_2$ in $Q$.

Referring to the definition in section 12.1, $N$ is the minimal number of colours required to properly colour any map from the class of all maps on the sphere and $m(N)$ is a specific map that requires all of $N$ colours to be properly coloured. Note that $m(N)$ has been properly coloured by using the $N$ colours $0, 1, \ldots, N-1$ and this proper colouring is fixed. The set of regions of $m(N)$ is then partitioned into subsets $\underline{0}, \underline{1}, \ldots, \underline{N-1}$ where the subset $\underline{m}$ consists of all the regions which receive the colour $m$. Note that the subsets $\underline{0}, \underline{1}, \dots, \underline{N-1}$ are each nonempty (since $\boldsymbol{m}(N)$ requires all of the $N$ colours to be properly coloured) and form a partition of the set of regions of $\boldsymbol{m}(N)$ (by virtue of proper colouring). Identify the set $\{\underline{0}, \underline{1}, \dots, \underline{N-1}\}$ with the underlying set of the $N$-element cyclic group $\mathbf{Z}_N$ under addition modulo $N$. Let $S_3$ denote the symmetric group on three letters, identified with the dihedral group of order six generated by $\rho$, $\sigma$ where $|\rho| = 3$ and $|\sigma| = 2$.

**Lemma 12.4**    $(\mathbf{Z}_N, +)$ is an Eilenberg module for the group $S_3$ with the trivial homomorphism

$$T_1 : S_3 \to Aut\ (\mathbf{Z}_N, +); a \to 1\ \mathbf{Z}_N$$

where $1\ \mathbf{Z}_N$ denotes the identity automorphism of $\mathbf{Z}_N$. The corresponding split extension $\mathbf{Z}_N]S_3$ has multiplication given by

$$(\underline{m}_1, \alpha_1) \cdot (\underline{m}_2, \alpha_2) = (\underline{m}_1 + \underline{m}_2, \alpha_1\alpha_2)$$

and is a group isomorphic to the direct product $\mathbf{Z}_N \times S_3$.

**Proof**    follows from definition.                                                                    ❏

Referring to section 12.2, the goal is to construct a Steiner system $S(N+1, 2N, 6N)$. We shall take the point set of the Steiner system to be the underlying set of the split extension $\mathbf{Z}_N]S_3$. The following lemma is used in section 12.5.

**Lemma 12.5**    Let $(\mathbf{Z}(\mathbf{Z}_N]S_3), +)$ and $(\mathbf{Z}S_3, +)$ denote the underlying additive groups of the integral group algebras $\mathbf{Z}(\mathbf{Z}_N]S_3)$ and $\mathbf{Z}S_3$, respectively. Then $(\mathbf{Z}(\mathbf{Z}_N]S_3), +)$ is an Eilenberg module for the group $(\mathbf{Z}S_3, +)$ with the trivial homomorphism

$$T_2 : (\mathbf{Z}S_3, +) \to Aut\ (\mathbf{Z}(\mathbf{Z}_N]S_3), +);\ \sum_{\alpha \in S_3} n_\alpha \alpha \to 1\ \mathbf{Z}(\mathbf{Z}_N]S_3)$$

where $1\ \mathbf{Z}(\mathbf{Z}_N]S_3)$ denotes the identity automorphism of $(\mathbf{Z}(\mathbf{Z}_N]S_3), +)$. The corresponding split extension $\mathbf{Z}(\mathbf{Z}_N]S_3)]\mathbf{Z}S_3$ has multiplication given by

$$\left( \sum_{(\underline{m}, \beta) \in \mathbf{Z}_N]S_3} n_{(\underline{m}, \beta)}(\underline{m}, \beta), \sum_{\alpha \in S_3} n_\alpha \alpha \right)\left( \sum_{(\underline{m}, \beta) \in \mathbf{Z}_N]S_3} n'_{(\underline{m}, \beta)}(m, \beta), \sum_{\alpha \in S_3} n'_\alpha \alpha \right)$$

$$= \left( \sum_{(\underline{m}, \beta) \in \mathbf{Z}_N]S_3} (n_{\underline{m}, \beta} + n'_{(\underline{m}\beta)})(\underline{m}, \beta), \sum_{\alpha \in S_3} (n_\alpha + n'_\alpha)\alpha \right)$$

and is a group isomorphic to the direct product $\mathbf{Z}(\mathbf{Z}_N]S_3) \times \mathbf{Z}S_3, +)$.

**Proof**    follows from definition.

## 12.4 Hall Matchings

Let $\Gamma$ be a bipartite graph with vertex set $V = X \cup Y$ and edge set $E$ (every edge has one end in $X$ and the other end in $Y$). A *matching from $X$ to $Y$ in* $\Gamma$ is a subset $M$ of $E$ such that no vertex is incident with more than one edge in $M$. A matching $M$ from $X$ to $Y$ in $\Gamma$ is called *complete* if every vertex in $X$ is incident with an edge in $M$. If $A$ is a subset of $V$ then let $adj(A)$ denote the set of all vertices adjacent to a vertex in $A$.

**Lemma 12.6 (P. Hall)**    If $|adj(A)| \geq |A|$ for every subset $A$ of $X$, then there exists a complete matching from $X$ to $Y$ in $\Gamma$.

**Proof**    A matching from $X$ to $Y$ in $G$ with $|M| = 1$ always exists by choosing a single edge in $E$. Let $M$ be a matching from $X$ to $Y$ in $G$ with $m$ edges, $m < |X|$. Let $x_0 \in X$ such that $x_0$ is not incident with any edge in $M$. Since $|adj(\{x_0\})| \geq 1$, there is a vertex $y_1$ adjacent to $x_0$ by an edge in $E/M$. If $y_1$ is not incident with an edge in $M$, then stop. Otherwise let $x_1$ be the other end of such an edge. If $x_0, x_1, \ldots, x_k$ and $y_1, \ldots, y_k$ have been chosen, since $|adj(\{x_0, x_1, \ldots, x_k\})| \geq k+1$, there is a vertex $y_{k+1}$, distinct from $y_1, \ldots, y_k$, that is adjacent to at least one vertex in $\{x_0, x_1, \ldots, x_k\}$. If $y_k + 1$ is not incident with an edge in $M$ then stop. Otherwise let $x_k + 1$ be the other end of such an edge. This process must terminate with some vertex, say $y_k + 1$. Now build a simple path from $y_k + 1$ to $x_0$ as follows. Start with $y_k + 1$ and the edge in $E/M$ joining it to, say $x_{i_1}$, with $i_1 < k+1$. Then add the edge in $M$ from $x_{i_1}$ to $y_{i_1}$. By construction $y_{i_1}$ is joined by an edge in $E/M$ to some $x_{i_2}$ with $i_2 < i_1$. Continue adding edges in this way until $x_0$ is reached. One obtains a path $y_{k+1}, x_{i_1}, y_{i_1}, x_{i_2}, y_{i_2}, \ldots, x_{i_r}, y_{i_r}, x_0$ of odd length $2r+1$ with the $r+1$ edges $\{y_{k+1}, x_{i_1}\}, \{y_{i_1}, x_{i_2}\}, \ldots, \{y_{i_r}, x_0\}$ in $E/M$ and the $r$ edges $\{x_{i_1}, y_{i_1}\}, \ldots, \{x_{i_r}, y_{i_r}\}$ in $M$. Define

$$M' = (M/\{\{x_{i_1}, y_{i_1}\}, \ldots, \{x_{i_r}, y_{i_r}\}\}) \cup \{\{y_{k+1}, x_{i_1}\}, \{y_{i_1}, x_{i_2}\}, \ldots, \{y_{i_r}, x_0\}\}.$$

Then $M'$ is a matching from $X$ to $Y$ in $\Gamma$, with $|M'| = |M| - r + r + 1 = |M| + 1$. Repeating this process a finite number of times must yield a complete matching from $X$ to $Y$ in $\Gamma$.

**Lemma 12.7**    Referring to section 12.3, let $Sym(\mathbf{Z}_N]S_3)$ denote the group of all permutations of the underlying set of the split extension $\mathbf{Z}_N]S_3$ of Lemma 12.4. Then $S_3$ embeds in $Sym((\mathbf{Z}_N]S_3)$ via the Cayley right regular representation.

**Proof**    Note that $S_3 = \{(\underline{0}, \alpha) | \alpha \in S_3\}$ is a subgroup of $\mathbf{Z}_N]S_3$. Since $S_3$ embeds in $Sym(S_3)$ via the Cayley right regular representation $\alpha \to R(\alpha)$ and $Sym(S_3)$ is a subgroup of $Sym(\mathbf{Z}_N]S_3)$, the lemma follows.

**Lemma 12.8**    By Lemma 12.7, regard $S_3$ as a subgroup of $Sym(\mathbf{Z}_N]S_3)$. There exists a common system of coset representatives $\phi_1, \ldots, \phi_k$ such that $\{\phi_1 S_3, \ldots, \phi_k S_3\}$ is the family of left cosets of $S_3$ in $Sym(\mathbf{Z}_N]S_3)$ and $\{S_3 \phi_1, \ldots, S_3 \phi_k\}$ is the family of right cosets of $S_3$ in $Sym(\mathbf{Z}_N]S_3)$.

**Proof**  By Lagrange's theorem the left cosets of $S_3$ partition $Sym(\mathbf{Z}_N]S_3)$ into $k = [Sym(\mathbf{Z}_N]S_3) : S_3]$ disjoint nonempty equivalence classes of size $|S_3| = 6$. The same is true of the right cosets. Define a bipartite graph $\Gamma$ with vertices $X \cup Y$ where $X = \{\psi_1 S_3, \ldots, \psi_k S_3\}$ is the family of left cosets of $S_3$ in $Sym(\mathbf{Z}_N]S_3)$ and $Y = \{S_3 \psi'_1, \ldots, S_3 \psi'_k\}$ is the family of right cosets of $S_3$ in $Sym(\mathbf{Z}_N]S_3)$ with an edge $\{\psi_i S_3, S_3 \psi'_j\}$ if and only if $\psi_i S_3$ and $S_3 \psi'_j$ have nonempty intersection. For any subset $A = \{\psi i_1 S_3, \ldots, \psi i_r S_3\}$ of $X$, one has $\psi i_1 \in \psi i_1 S_3, \ldots, \psi i_r \in \psi i_r S_3$ and there exist distinct $j_1, \ldots, j_r$ such that $\psi i_1 \in S_3 \psi' j_1, \ldots, \psi i_r \in S_3 \psi' j_r$. Hence in the graph $\Gamma$, $|adj(A)| \geq |A|$. Hall's hypothesis of Lemma 12.6 is satisfied and there exists a complete matching from $X$ to $Y$ in $\Gamma$. This is precisely the statement that a common system of coset representatives $\phi_1, \ldots, \phi_k$ exists.

## 12.5  Riemann Surfaces

Let $C$ denote the complex plane. Consider the function $C \rightarrow C$ ; $z \rightarrow w = z^n$, where $n \geq 2$. There is a one-to-one correspondence between each sector

$$\{z | (k-1)2\pi/n < arg\ z < k2\pi/n\}(k = 1, \ldots, n)$$

and the whole $w$-plane except for the positive real axis. The image of each sector is obtained by performing a cut along the positive real axis; this cut has an upper and a lower edge. Corresponding to the $n$ sectors in the $z$-plane, take $n$ identical copies of the $w$-plane with the cut. These will be the *sheets* of the Riemann surface and are distinguished by a label $k$, which serves to identify the corresponding sector. For $k = 1, \ldots, n-1$ attach the lower edge of the sheet labelled $k$ with the upper edge of the sheet labelled $k+1$. To complete the cycle, attach the lower edge of the sheet labelled $n$ to the upper edge of the sheet labelled 1. In a physical sense, this is not possible without self-intersection but the idealised model shall be free of this discrepancy. The result of the construction is a *Riemann surface* whose points are in one-to-one correspondence with the points of the $z$-plane. This correspondence is continuous in the following sense. When $z$ moves in its plane the corresponding point $w$ is free to move on the Riemann surface. The point $w = 0$ connects all the sheets and is called the *branch point*. A curve must wind $n$ times around the branch point before it closes. Now consider the $n$-valued relation

$$z = \sqrt[n]{w}.$$

To each $w \neq 0$, there correspond $n$ values of $z$. If the $w$-plane is replaced by the Riemann surface just constructed, then each complex number $w \neq 0$ is represented by $n$ points of the Riemann surface at superposed positions. Let the point on the uppermost sheet represent the principal value and the other $n-1$ points represent the other values. Then $z = \sqrt[n]{w}$ becomes a single-valued, continuous, one-to-one correspondence of the points of the Riemann surface with the points of the $z$-plane. Now recall the definition of the map $\boldsymbol{m}(N)$ from section 12.1. The map $\boldsymbol{m}(N)$ is on the sphere. Pick a region and deform the sphere so that both 0 and $\infty$ are two distinct points inside this region when the sphere is regarded as the extended complex plane. Using the stereographic projection one obtains the map $m(N)$ on the complex plane $C$ with the region containing 0 and $\infty$ forming a "sea" surrounding the other regions which form an "island". Put this copy of $C$ on each sheet of the Riemann surface corresponding to $w = z^n$. The branch point lies in the "sea". The inverse function $z = \sqrt[n]{w}$ results in $n$ copies of the map $m(N)$ on the $z$-plane in the sectors

$$\{z \mid (k-1)2\pi/n < arg\ z < k2\pi/n\}(k = 1, \ldots, n).$$

The origin of the $z$-plane lies in the n "seas". An example with $n = 4$ is given in Figure 12.2.

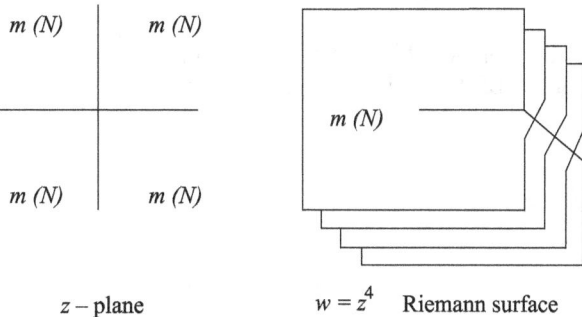

*z* – plane     $w = z^4$     Riemann surface

**Fig. 12.2** *An example with* $n = 4$

Referring to section 12.3, the full symmetric group $Sym(\mathbf{Z}_N]S_3)$ acts faithfully on the set $\mathbf{Z}_N]S_3$. The action of an element $\psi$ of $Sym(\mathbf{Z}_N]S_3)$ on an element $(\underline{m}, \alpha)$ of $\mathbf{Z}_N]S_3$ will be written as $(\underline{m}, \alpha)\psi$. This action extends to the integral group algebra $\mathbf{Z}(\mathbf{Z}_N]S_3)$ by linearity

$$\Big( \sum_{(\underline{m}\alpha) \in \mathbf{Z}_N]S_3} n_{(\underline{m}, \alpha)}(\underline{m}, \alpha)\Big) = \sum_{(\underline{m}, \alpha) \in \mathbf{Z}_N]S_3} n_{(\underline{m}, \alpha)}((\underline{m}, \alpha)\psi).$$

Referring to Lemma 12.8, fix a common coset representative $\phi_i$ of $S_3$ in $Sym(\mathbf{Z}_N]S_3)$ and fix a pair $(\beta, \gamma) \in S_3 \times S_3 = Mlt(S_3)$. There are two cases depending on whether $\beta = \gamma$ or whether $\beta \neq \gamma$.

**Case 1**  Suppose $\beta \neq \gamma$. Consider the composition of the functions

$$C \to C;\ z \to t = z^2 \text{ and } C \to C;\ t \to w = t^{12}.$$

The composite is given by the assignment

$$z \to t = z^2 \to w = t^{12} = Z^{24}.$$

There are twenty-four superposed copies of the map $m(N)$ on the $w$-Riemann surface corresponding to the sectors

$$\{z|(k-1)2\pi/24 < arg\ z < k2\pi/24\}(k = 1, \ldots, 24)$$

on the $z$-plane. These are divided into two sets. The first set consists of twelve superposed copies of the map $m(N)$ corresponding to the sectors

$$\{z(k-1)2\pi/24 < arg\ z < k2\pi/24\}(k = 1, \ldots, 12)$$

of the upper half of the $z$-plane which comprise the upper sheet of the $t$-Riemann surface. The second set consists of twelve superposed copies of the map $m(N)$ corresponding to the sectors

$$\{z|(k-1)2\pi/24 < arg\ z < k2\pi/24\}(k = 13, \ldots, 24)$$

of the lower half of the $z$-plane which comprise the lower sheet of the $t$-Riemann surface. Figure 12.3 shows sheets of the $t$-Riemann surface.

**Fig. 12.3** *Sheets of the t-Riemann surface*

Label the twelve sectors of the upper sheet of the $t$-Riemann surface by elements of $\mathbf{Z}(\mathbf{Z}_N]S_3)]\mathbf{Z}S_3$ as shown in Figure 12.4.

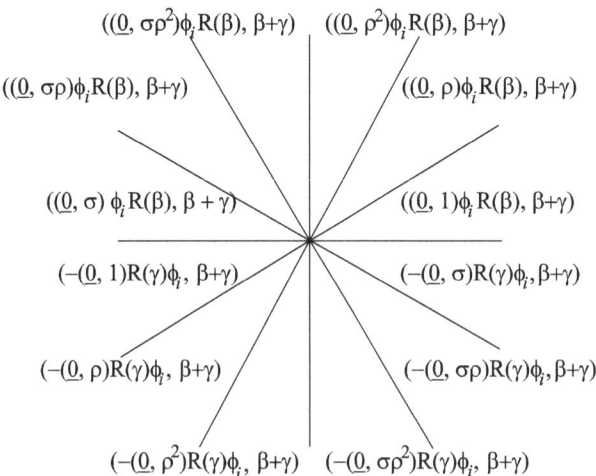

**Fig. 12.4** *Upper sheet of the t-Riemann surface*

Label the twelve sectors of the lower sheet of the $t$-Riemann surface by elements of $\mathbf{Z}(\mathbf{Z}_N]S_3)]\mathbf{Z}S_3$ as shown in Figure 12.5.

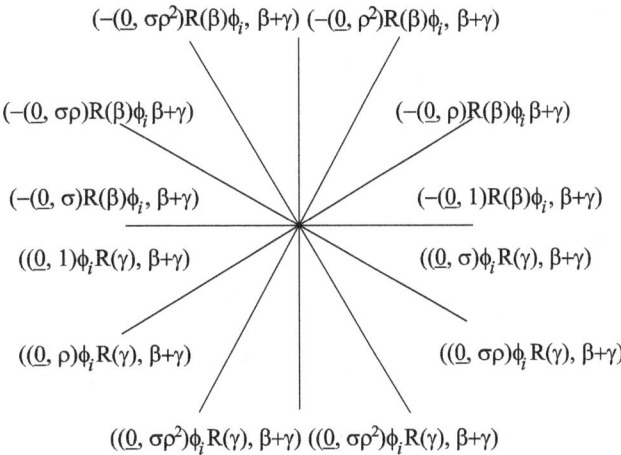

$(-(\underline{0}, \sigma\rho^2)R(\beta)\phi_i, \beta+\gamma)$ $(-(\underline{0}, \rho^2)R(\beta)\phi_i, \beta+\gamma)$

$(-(\underline{0}, \sigma\rho)R(\beta)\phi_i, \beta+\gamma)$ $(-(\underline{0}, \rho)R(\beta)\phi_i, \beta+\gamma)$

$(-(\underline{0}, \sigma)R(\beta)\phi_i, \beta+\gamma)$ $(-(\underline{0}, 1)R(\beta)\phi_i, \beta+\gamma)$

$((\underline{0}, 1)\phi_i R(\gamma), \beta+\gamma)$ $((\underline{0}, \sigma)\phi_i R(\gamma), \beta+\gamma)$

$((\underline{0}, \rho)\phi_i R(\gamma), \beta+\gamma)$ $((\underline{0}, \sigma\rho)\phi_i R(\gamma), \beta+\gamma)$

$((\underline{0}, \sigma\rho^2)\phi_i R(\gamma), \beta+\gamma)$ $((\underline{0}, \sigma\rho^2)\phi_i R(\gamma), \beta+\gamma)$

**Fig. 12.5** *Lower sheet of the t-Riemann surface*

Referring to section 12.2, the regions of the map $m(N)$ have been partitioned into disjoint, nonempty equivalence classes $0, 1, \ldots, N-1$ and this set of equivalence classes forms the underlying set of the cyclic group $\mathbf{Z}_N$. Hence there are twelve copies of $\mathbf{Z}_N$ on the upper sheet and twelve copies of $\mathbf{Z}_N$ on the lower sheet of the $t$-Riemann surface. The copies of $\mathbf{Z}_N$ are indexed by the elements of $\mathbf{Z}(\mathbf{Z}_N]S_3)]\mathbf{Z}S_3$ which label the sectors on a particular sheet. The branch point of the $t$-Riemann surface is labelled by the element $(0, \beta + \gamma)$ of $\mathbf{Z}(\mathbf{Z}_N]S_3)]\mathbf{Z}S_3$ where 0 denotes the zero element of $\mathbf{Z}(\mathbf{Z}_N]S_3)$.

**Lemma 12.9**  Referring to Lemma 12.8, fix a common representative $\phi_i$ of the left and right cosets of $S_3$ in $Sym(\mathbf{Z}_N]S_3)$. Fix a pair $(\beta, \gamma) \in S_3 \times S_3$ with $\beta \neq \gamma$. Referring to Lemma 12.5, define a subset $T_{(\beta, \gamma)}$ of $\mathbf{Z}(\mathbf{Z}_N]S_3)]\mathbf{Z}S_3$ as follows.

$$T_{(\beta, \gamma)} = \{((\underline{m}, \alpha), \beta + \gamma)|(\underline{m}, \alpha) \in \mathbf{Z}_N]S_3\}$$

$$\cup$$

$$\{(0, \beta + \gamma)\}$$

$$\cup$$

$$\{(-(\underline{m}, \alpha), \beta + \gamma)|(\underline{m}, \alpha) \in \mathbf{Z}_N]S_3\}.$$

Referring to the preceding discussion, consider the composite function

$$z \to t = z^2 \to w = t^{12} = z^{24}$$

of the complex $z$-plane to the $w$-Riemann surface. There is a copy of the set $T_{(\beta,\gamma)}$ on the upper sheet and a copy of the set $T_{(\beta,\gamma)}$ on the lower sheet of the $t$-Riemann surface according to the labels of the sectors in Figures 12.4 and 12.5 with the branch point labelled by the element $(0,\beta+\gamma)$ of both copies. The rotation of the $z$-plane by $p$ radians induces a permutation

$$p : T_{(\beta,\gamma)} \to T_{(\beta,\gamma)}$$

given by

$$(-(\underline{m},\ \alpha)R(\gamma)\phi_i,\ \beta+\gamma)p = ((\underline{m},\ \alpha)\phi_i R(\gamma),\beta+\gamma)$$

$$(0,\ \beta+\gamma)p = (0,\ \beta+\gamma)$$

$$((\underline{m},\ \alpha)\phi_i R(\beta),\ \beta+\gamma)p = (-(\underline{m},\ \alpha)R(\beta)\phi_i,\ \beta+\gamma)$$

for all $(\underline{m},\ \alpha) \in Z_N]S_3$, such that each point of the copy of $T_{(\beta,\gamma)}$ on the upper sheet moves continuously along a circular curve that winds exactly once around the branch point, to the point superposed directly below it on the copy of $T_{(\beta,\gamma)}$ on the lower sheet of the $t$-Riemann surface.

**Proof**  $T_{(\beta,\gamma)}$ is seen to be a well-defined subset of $Z(Z_N]S_3)]ZS_3$ by setting the appropriate coefficients to zero in a typical element as described in Lemma 12.5. Each of $R(\gamma)\phi_i$, $\phi_i R(\gamma)$, $\phi_i R(\beta)$ and $R(\beta)\phi_i$ are permutations of the set $Z_N]S_3$ and the rotation of the $z$-plane by $p$ radians clearly induces a permutation $p$ of the set $T_{(\beta,\gamma)}$ as described.  ❑

**Lemma 12.10**  Referring to Lemma 12.9, let $Sym(T_{(\beta,\gamma)})$ denote the full permutation group of the set $T_{(\beta,\gamma)}$. Let $<p>$ denote the cyclic subgroup of $Sym(T_{(\beta,\gamma)})$ generated by $p$. Then $<p>$ is nontrivial and acts faithfully on the set $T_{(\beta,\gamma)}$.

**Proof**  If $p=1$, then $(-(\underline{0},1)R(\gamma)\phi_i,\ \beta+\gamma) = (-(\underline{0},1)R(\gamma)\phi_i,\beta+\gamma)p = ((\underline{0},1)\phi_i R(\gamma),\ \beta+\gamma)$ which implies that $-(\underline{0},1)R(\gamma)\phi_i = (0,1)\phi_i R(\gamma)$ in $Z(Z_N]S_3)$. This is impossible since $1 \neq 1$ in $Z$. Hence $p \neq 1$. Since the full permutation group $Sym(T_{(\beta,\gamma)})$ acts faithfully on $T_{(\beta,\gamma)}$, so does its subgroup $<p>$.  ❑

**Lemma 12.11**  Referring to Lemma 12.9 and Lemma 12.10, let $1 : C \to C$; $z \to z$ denote the identity and $\pi : C \to C$; $z \to -z$ denote the rotation through an angle of $p$ radians of the $z$-plane. Then the two-element cyclic group $\{1,\pi\}$ acts faithfully on the set $<p>$ as follows: $p^n \cdot 1 = p^n$ and $p^n \cdot \pi = p^{1-n}$, for all $n$ in $Z$.

**Proof**  The set $\{1,\pi\}$ forms a two-element cyclic group $<\pi>$ under function composition. To show that $\{1,\pi\}$ acts on $<p>$ as defined, observe that $(p^n \cdot \pi) \cdot \pi = (p^{1-n}) \cdot \pi = p^{1-(1-n)} = p^{1-1+n} = p^n = p^n \cdot 1 = p^n \cdot (\pi\pi)$, for all $n$ in $Z$. To show that the action is faithful, let $\theta \in \{1,\pi\}$. If $\theta$ belongs to the kernel of the action, then $p^n \cdot \theta = p^n$ for all $n$ in $Z$, so that $p \cdot \theta = p$ which implies that $\theta = 1$, since $p \neq 1$ by Lemma 12.10.  ❑

**Lemma 12.12** Putting together Lemma 12.9, Lemma 12.10 and Lemma 12.11, there is a well-defined action of the two-element cyclic group $\{1, \pi\}$ on the set $T_{(\beta, \gamma)}$ given by

$$((\underline{m}, \alpha)\phi_i R(\gamma),\ \beta + \gamma) \cdot 1 = ((\underline{m}, \alpha)\ \phi_i R(\gamma),\ \beta + \gamma)$$

$$(0, \beta + \gamma) \cdot 1 = (0, \beta + \gamma)$$

$$(-(\underline{m}, \alpha)R(\beta)\phi_i,\ \beta + \gamma) \cdot 1 = (-(\underline{m}, \alpha)R(\beta)\phi_i,\ \beta + \gamma)$$

and

$$((\underline{m}, \alpha)\phi_i R(\gamma),\ \beta + \gamma) \cdot \pi = (-(\underline{m}, \alpha)R(\gamma)\phi_i,\ \beta + \gamma)$$

$$(0, \beta + \gamma) \cdot \pi = (0, \beta + \gamma)$$

$$(-(\underline{m}, \alpha)R(\beta)\phi_i,\ \beta + \gamma) \cdot \pi = ((\underline{m}, \alpha)\phi_i R(\beta),\ \beta + \gamma)$$

for all $(\underline{m}, \alpha)$ in $\mathbf{Z}_N]S_3$. This action is faithful.

**Proof** For each $x \in T_{(\beta, \gamma)}$, let $Orb(x) = \{xp^n | n \in \mathbf{Z}\}$ denote the orbit of $x$ under $< p >$. The collection $\{Orb(x) | x \in T_{(\beta, \gamma)}\}$ forms a partition of the set $T_{(\beta, \gamma)}$ as follows. Let $x, y \in T_{(\beta, \gamma)}$. If $z \in Orb(x) \cap Orb(y)$ then $z = xp^n = yp^m$ for some $m, n \in \mathbf{Z}$. This implies $xp^{n-m} = y \Rightarrow y \in Orb(x) \Rightarrow Orb(y) \subseteq Orb(x)$ and $yp^{m-n} = x \Rightarrow x \in Orb(y) \Rightarrow Orb(x) \subseteq Orb(y)$. Hence $Orb(x) = Orb(y)$. Also, each $x \in T_{(\beta, \gamma)}$ belongs to an orbit, namely $x \in Orb(x)$. Hence the orbits are disjoint, nonempty and their union is all of the set $T_{(\beta, \gamma)}$. For each fixed $x \in T_{(\beta, \gamma)}$, define

$$\pi : Orb(x) \to Orb(x); (xp^n)\pi = xp^{1-n}$$

Then, $\pi$ is well-defined since $xp^n = xp^m \Rightarrow xp^{n-m} = x \Rightarrow xp^{-m} = xp^{-n} \Rightarrow xp^{1-m} = xp^{1-n} \Rightarrow (xp^n)\pi = (xp^m)\pi$. Also, $\pi$ is a permutation of $Orb(x)$ with $\pi^2 = 1$ since for each $xp^n \in Orb(x)$ we have $((xp^n)\pi)\pi = (xp^{1-n})\pi = xp^{1-(1-n)} = xp^n$. Now define

$$\pi : T_{(\beta, \gamma)} \to T_{(\beta, \gamma)}; (xp^n)\pi = xp^{1-n}$$

orbit by orbit. Then, since $\{Orb(x) | x \in T_{(\beta, \gamma)}\}$ forms a partition of $T_{(\beta, \gamma)}$, $\pi$ is a well-defined permutation of $T_{(\beta, \gamma)}$ with $\pi^2 = 1$, the identity permutation of $T_{(\beta, \gamma)}$. Hence, using the definition of $p$ in Lemma 12.9, we obtain an action of the two-element cyclic group $\{1, \pi\}$ on $T_{(\beta, \gamma)}$ as follows. For all $(\underline{m}, \alpha)$ in $\mathbf{Z}_N]S_3$, define

$$((-(\underline{m}, \alpha)R(\gamma)\phi_i,\ \beta + \gamma)p) \cdot 1 = (-(\underline{m}, \alpha)R(\gamma)\phi_i,\ \beta + \gamma)p$$

$$((0, \beta + \gamma)p) \cdot 1 = (0, \beta + \gamma)p$$

$$(((\underline{m}, \alpha)\phi_i R(\beta),\ \beta + \gamma)p) \cdot 1 = ((\underline{m}, \alpha)\phi_i R(\beta),\ \beta + \gamma)p$$

and

$$((-(\underline{m},\ \alpha)R(\gamma)\phi_i,\ \beta+\gamma)p)\cdot\pi\ =(-(\underline{m},\ \alpha)R(\gamma)\phi_i,\ \beta+\gamma)$$

$$((0,\ \beta+\gamma)p)\cdot\pi\ =(0,\ \beta+\gamma)$$

$$(((\underline{m},\ \alpha)\phi_iR(\beta),\ \beta+\gamma)p)\cdot\pi\ =((\underline{m},\ \alpha)\phi_iR(\beta),\ \beta+\gamma)$$

Now, using the definition of $p$ in Lemma 12.9, the action of $\{1,\pi\}$ on $T_{(\beta,\gamma)}$ may be rewritten as

$$((\underline{m},\ \alpha)\phi_iR(\gamma),\ \beta+\gamma)\cdot 1\ =((\underline{m},\ \alpha)\phi_iR(\gamma),\ \beta+\gamma)$$

$$(0,\ \beta+\gamma)\cdot 1\ =(0,\ \beta+\gamma)$$

$$(-(\underline{m},\ \alpha)R(\beta)\phi_i,\ \beta+\gamma)\cdot 1=(-(\underline{m},\ \alpha)R(\beta)\phi_i,\ \beta+\gamma)$$

and

$$((\underline{m},\ \alpha)\phi_iR(\gamma),\ \beta+\gamma)\cdot\pi\ =(-(\underline{m},\ \alpha)R(\gamma)\phi_i,\ \beta+\gamma)$$

$$(0,\ \beta+\gamma)\cdot\pi\ =(0,\ \beta+\gamma)$$

$$(-(\underline{m},\ \alpha)R(\beta)\phi_i,\ \beta+\gamma)\cdot\pi\ =((\underline{m},\ \alpha)\phi_iR(\beta),\ \beta+\gamma)$$

for all $(\underline{m},\ \alpha)$ in $\mathbf{Z}_N]S_3$, as in the statement of this lemma. To verify that the action of $\{1,\ \pi\}$ on $T_{(\beta,\gamma)}$ is faithful, note that

$$1:T_{(\beta,\ \gamma)}\rightarrow T_{(\beta,\ \gamma)};\ x\rightarrow x$$

$$\pi:T_{(\beta,\ \gamma)}\rightarrow T_{(\beta,\ \gamma)};\ x\rightarrow x\pi$$

are permutations of the set $T_{(\beta,\gamma)}$. If $\theta\in\{1,\ \pi\}$ and $\theta$ belongs to the kernel of the action then $x\theta=x$ for all $x\in T_{(\beta,\ \gamma)}$. Then $\theta=1$, since $\pi$ moves $((\underline{0},\ 1)\phi_iR(\gamma),\ \beta+\gamma)$ to $(-(\underline{0},\ 1)R(\gamma)\phi_i,\ \beta+\gamma)$ which are distinct elements of $\mathbf{Z}(\mathbf{Z}_N]S_3)]\mathbf{Z}S_3$.

**Case 2**   Suppose $\beta=\gamma$. Note that in the labelling of the sectors of the sheets of the $t$-Riemann surface in Figures. 12.4 and 12.5, $R(\beta)=R(\gamma)$ and $\beta+\gamma=2\beta$ in the group algebra $\mathbf{Z}S_3$.                                                                                             $\square$

**Lemma 12.13**   Referring to Lemma 12.8, fix a common representative $\phi_i$ of the left and right cosets of $S_3$ in $Sym(\mathbf{Z}_N]S_3)$. Fix a pair $(\beta,\ \beta)\in S_3\times S_3$. Referring to Lemma 12.5, define a subset $T_{(\beta,\ \beta)}$ of $\mathbf{Z}(\mathbf{Z}_N]S_3)\mathbf{Z}S_3$ as follows.

$$T_{(\beta,\beta)} = \{((\underline{m}, \alpha), 2\beta)|(\underline{m}, \alpha) \in \mathbf{Z}_N]S_3\}$$

$$\cup$$

$$\{(0, 2\beta)\}$$

$$\cup$$

$$\{(-(\underline{m}, \alpha), 2\beta)|(\underline{m}, \alpha) \in \mathbf{Z}_N]S_3\}.$$

Referring to the preceding discussion, consider the composite function

$$z \to t = z^2 \to w = t^{12} = z^{24}$$

of the complex $z$-plane to the $w$-Riemann surface. There is a copy of the set $T_{(\beta,\beta)}$ on the upper sheet and a copy of the set $T_{(\beta,\beta)}$ on the lower sheet of the $t$-Riemann surface according to the labels of the sectors in Figures 12.4 and 12.5. Note that in this case $R(\beta) = R(\gamma)$ and $\beta + \gamma = 2\beta$ with the branch point labelled by the element $(0, 2\beta)$ of both copies. The rotation of the $z$-plane by $p$ radians induces a permutation

$$p : T_{(\beta,\beta)} \to T_{(\beta,\beta)}$$

given by

$$(-(\underline{m}, \alpha)R(\beta)\phi_i, 2\beta)p = ((\underline{m}, \alpha)\phi_i R(\beta), 2\beta)$$

$$(0, 2\beta)p = (0, 2\beta)$$

$$((\underline{m}, \alpha)\phi_i R(\beta), 2\beta)p = (-(\underline{m}, \alpha)R(\beta)\phi_i, 2\beta)$$

for all $(\underline{m}, \alpha) \in \mathbf{Z}_N]S_3$, such that each point of the copy of $T_{(\beta,\beta)}$ on the upper sheet moves continuously along a circular curve that winds exactly once around the branch point, to the point superposed directly below it on the copy of $T_{(\beta,\beta)}$ on the lower sheet of the $t$-Riemann surface. Then, $p = p^{-1}$ so that $< p > = \{1, p\}$ is a two-element cyclic subgroup of the full permutation group $Sym(T_{(\beta,\beta)})$ and $< p >$ acts faithfully on the set $T_{(\beta,\beta)}$.

**Proof** As in the proof of Lemma 12.9, $T_{(\beta,\beta)}$ is seen to be a well-defined subset of $Z(\mathbf{Z}_N]S_3)]ZS_3$ by setting the appropriate coefficients to zero in a typical element as described in Lemma 12.5. Both $\phi_i R(\beta)$, $R(\beta)\phi_i$ are permutations of the set $\mathbf{Z}_N]S_3$ and the rotation of the $z$-plane by $p$ radians clearly induces a permutation $p$ of the set $T_{(\beta,\beta)}$ as described. Furthermore, it is clear from the definition that $p = p - 1$ by chasing elements of $T_{(\beta,\beta)}$. Then, $< p > = \{1, p\}$ is a subgroup of $Sym(T_{(\beta,\beta)})$ and $< p >$ acts faithfully on the set $T_{(\beta,\beta)}$. $\qquad\square$

**Lemma 12.14**   Referring to Lemma 12.13, let $1 : C \to C; z \to z$ denote the identity and $\pi : C \to C; z \to -z$ denote the rotation through an angle of $p$ radians of the $z$-plane. Then there is a well-defined action of the two-element cyclic group $\{1, \pi\}$ on the set $T_{(\beta, \beta)}$ given by

$$((\underline{m}, \alpha)\phi_i R(\beta), 2\beta) \cdot 1 = ((\underline{m}, \alpha)\phi_i R(\beta), 2\beta)$$

$$(0, 2\beta) \cdot 1 = (0, 2\beta)$$

$$(-(\underline{m}, \alpha)R(\beta)\phi_i, 2\beta) \cdot 1 = (-(\underline{m}, \alpha)R(\beta)\phi_i, 2\beta)$$

and

$$((\underline{m}, \alpha)\phi_i R(\beta), 2\beta) \cdot \pi = (-(\underline{m}, \alpha)R(\beta)\phi_i, 2\beta)$$

$$(0, 2\beta) \cdot \pi = (0, 2\beta)$$

$$(-(\underline{m}, \alpha)R(\beta)\phi_i, 2\beta) \cdot \pi = ((\underline{m}, \alpha)\phi_i R(\beta), 2\beta)$$

for all $(\underline{m}, \alpha)$ in $\mathbf{Z}_N]S_3$. This action is faithful.

**Proof**   The isomorphism $1 \to 1$, $p \to \pi$ of the two-element cyclic groups $\{1, p\}$ and $\{1, \pi\}$ establishes the lemma.                                                                                                    ❏

## 12.6  Main Construction

Let us review the final goal with a brief résumé. In section 12.1, we have defined $N$ to be the minimal number of colours required to properly colour any map from the class of all maps on the sphere. We know that $4 \leq N \leq 6$. We have chosen a specific map $\boldsymbol{m}(N)$ on the sphere that requires all of the $N$ colours 0, 1, ..., $N - 1$ to properly colour it. The map $\boldsymbol{m}(N)$ has been properly coloured and the regions of $\boldsymbol{m}(N)$ partitioned into disjoint, nonempty equivalence classes 0, 1, ..., $N - 1$ according to the colour they receive. The set $\{0, 1, ..., N - 1\}$ is endowed with the structure of the cyclic group $\mathbf{Z}_N$ under addition modulo $N$. In section 12.3, we have built the split extension $\mathbf{Z}_N]S_3$. The underlying set $\mathbf{Z}_N]S_3$ of cardinality $6N$ is taken to be the point set of a Steiner system $S(N + 1, 2N, 6N)$ which will be constructed in this section. We are required to define the blocks of size $2N$ and show that every set of $N + 1$ points is contained in a unique block. Once this goal is achieved, Lemma 12.3 shows that $N \leq 4$.

**Lemma 12.15**   Let $\mathbf{Z}_N]S_3$ denote the split extension defined in Lemma 12.4 and $\mathrm{Sym}(\mathbf{Z}_N]S_3)$ denote the full permutation group on the set $\mathbf{Z}_N]S_3$.

Define

$$\mu : Sym(\mathbf{Z}_N]S_3) \to Sym(\mathbf{Z}_N]S_3)$$

by

$$\Psi = R(\gamma)\phi_i \to \phi_i(\gamma) = \Psi^\mu.$$

Then, $\mu$ is a bijection of the set $Sym(\mathbf{Z}_N]S_3)$ with itself.

**Proof**  Referring to Lemma 12.8, $\mu$ is well-defined since each $\Psi \in Sym(\mathbf{Z}_N]S_3)$ may be written uniquely as $\Psi = R(\gamma)\phi_i$ for some $\gamma \in S_3$ and some $\phi_i$. Then, $\mu$ is a surjection because for any $\Psi \in Sym(\mathbf{Z}_N]S_3)$ one may also write $\Psi = \phi_i R(\gamma)$ uniquely for some $\gamma \in S_3$ and some $\phi_i$, whence $R(\gamma)\phi_i \to \phi_i R(\gamma) = \Psi$. Since $Sym(\mathbf{Z}_N]S_3)$ is a finite set, $\mu$ must be a bijection by counting.  ❑

**Lemma 12.16**  Define the set $G$ as follows.

$$G = \left\{ \begin{pmatrix} \Psi \\ \Psi^\mu \end{pmatrix} \;\middle|\; \Psi \in \mathrm{Sym}(Z_N]S_3) \right\} = \left\{ \begin{pmatrix} R(\gamma)\phi_i \\ \phi_i R(\gamma) \end{pmatrix} \;\middle|\; \begin{matrix} \gamma \in S_3 \\ i=1,\,...,\,k \end{matrix} \right\}$$

Define multiplication in $G$ as follows:

$$\begin{bmatrix} \Psi_1 \\ (\Psi_1^\mu) \end{bmatrix} \begin{bmatrix} \Psi_2 \\ (\Psi_2^\mu) \end{bmatrix} = \begin{bmatrix} \Psi_1\Psi_2 \\ (\Psi_1\Psi_2)^\mu \end{bmatrix}$$

i.e.,

$$\begin{bmatrix} R(\gamma_1)\phi i_1 \\ \phi i_1 R(\gamma_1) \end{bmatrix} \begin{bmatrix} R(\gamma_2)\phi i_2 \\ \phi i_2 R(\gamma_2) \end{bmatrix} = \begin{bmatrix} R(\gamma_3)\phi i_3 \\ \phi i_3 R(\gamma_3) \end{bmatrix},$$

where $R(\gamma_3)\phi_3$ is the unique expression for $R(\gamma_1)\phi i_1 R(\gamma_2)\phi i_2$ according to the right coset decomposition of $S_3$ in $Sym(\mathbf{Z}_N S_3)$. Then, $G$ is a group.

**Proof**  Referring to Lemma 12.8 and Lemma 12.15, the set $G$ is well-defined by the decomposition of $Sym(Z_N]S_3)$ into the left and right cosets of $S_3$ by the $\phi_1$. Define

$$\mu' : Sym(\mathbf{Z}_N]S_3) \to G; \; \Psi \to \begin{bmatrix} \Psi \\ \Psi^\mu \end{bmatrix}.$$

Then, $\mu'$ is a well-defined bijection of the set $Sym(\mathbf{Z}_N]S_3)$ with $G$ since m is a bijection by Lemma 12.15. The definition of multiplication in $G$ mirrors the multiplication in $Sym(\mathbf{Z}_N]S_3)$ via $\mu'$ and is designed to make $G$ a group and $\mu'$ an isomorphism.  ❑

**Lemma 12.17**    Consider the set $\mathbf{Z}_N]S_3$ and let

$$\begin{bmatrix} \Psi \\ \Psi^\mu \end{bmatrix} = \begin{bmatrix} R(\gamma)\phi_i \\ \phi_i R(\gamma) \end{bmatrix} \in G.$$

Define

$$\uparrow \begin{bmatrix} \Psi \\ \Psi^\mu \end{bmatrix} : \mathbf{Z}_N]s_3 \to \mathbf{Z}_N]S_3 \text{ by}$$

$$(\underline{m}, \alpha) \to (\underline{m}, \alpha) \uparrow \begin{bmatrix} \Psi \\ \Psi^\mu \end{bmatrix} = (\underline{m}, \alpha) \uparrow \begin{bmatrix} R(\gamma)\phi_i) \\ \phi_i R(\gamma) \end{bmatrix} = (\underline{m}, \alpha)R(\gamma)\phi_i.$$

Define

$$\downarrow \begin{bmatrix} \Psi \\ \Psi^\mu \end{bmatrix} : \mathbf{Z}_N]S_3 \text{ by}$$

$$(\underline{m}, \alpha) \to (\underline{m}, \alpha) \downarrow \begin{bmatrix} \Psi \\ \Psi^\mu \end{bmatrix} = (\underline{m}, \alpha) \downarrow \begin{bmatrix} R(\gamma)\phi_i \\ \phi_i R(\gamma) \end{bmatrix} = (\underline{m}, \alpha)\phi_i R(\gamma).$$

Then

$$(\underline{m}, \alpha) \uparrow \begin{bmatrix} \Psi \\ \Psi^\mu \end{bmatrix} = (\underline{m}, \alpha) \downarrow \begin{bmatrix} \Psi \\ \Psi^\mu \end{bmatrix} \text{ for all } (\underline{m}, \alpha) \in \mathbf{Z}_N]S_3.$$

Both $\uparrow$ and $\downarrow$ are well-defined, faithful and $|\mathbf{Z}_N]S_3|$-transitive right actions of the group $G$ on the set $\mathbf{Z}_N]S_3$.

**Proof**    Referring to Lemma 12.12 and Lemma 12.12, put $\beta = 1$. Working in the set $T_(1, \gamma)$, for each $(\underline{m}, \alpha) \in \mathbf{Z}_N]S_3$, we have

$$\left((\underline{M}, \alpha) \downarrow \begin{bmatrix} \Psi \\ \Psi^\mu \end{bmatrix}, 1+\gamma\right)$$

$$= \left((\underline{m}, \alpha) \downarrow \begin{bmatrix} R(\gamma)\phi_i \\ \phi_i R(\gamma) \end{bmatrix}, 1+\gamma\right)$$

$$= \left((\underline{m}, \alpha)\phi_i R(\gamma), 1+\gamma\right)$$

$$= \left(-(\underline{m}, \alpha)R(\gamma)\phi_i, 1+\gamma\right)\pi$$

$$= \left(-(\underline{m}, \alpha\gamma)\phi_i, 1+\gamma\right)\pi$$

$$= \left(-(\underline{m}, \alpha\gamma)R(1)\phi_i, 1+\gamma\right)\pi$$

$$= ( (\underline{m}, \ \alpha\gamma)\phi_i R(1), \ 1+\gamma)$$

$$= ( (\underline{m}, \ \alpha\gamma)\phi_i, \ 1+\gamma)$$

$$= ( (\underline{m}, \ \alpha)R(\gamma)\phi_i, \ 1+\gamma)$$

$$= ( (\underline{m}, \ \alpha) \uparrow \begin{bmatrix} R(\gamma)\phi_i, \\ \phi_i R(\gamma) \end{bmatrix}, \ 1+\gamma)$$

$$= ( (\underline{m}, \ \alpha) \uparrow \begin{bmatrix} \Psi \\ \Psi^\mu \end{bmatrix}, \ 1+\gamma)$$

using the action of the two-element cyclic group $\{1, \ \pi\}$ on the set $T_{(1, \ \gamma)^n}$ according to Lemma 12.12 and Lemma 12.12. Hence

$$(\underline{m}, \ \alpha) \uparrow \begin{bmatrix} \Psi \\ \Psi^\mu \end{bmatrix} = (\underline{m}, \ alpha) \downarrow \begin{bmatrix} \Psi \\ \Psi^\mu \end{bmatrix} \text{ for all } (\underline{m}, \ \alpha) \in Z_N]S_3.$$

Since the action $\uparrow$ is the usual action of $Sym(Z_N]S_3)$ on the set $Z_N]S_3$, it is faithful and $|Z_N]S_3|$ -transitive. By the last equality, so is the $\downarrow$ action. $\qquad \square$

**Lemma 12.18** Let $(\underline{m}_1, \ \alpha_1), \ \cdots, \ (\underline{m}_r, \ \alpha_r)$ be any $r$ distinct elements of $Z_N]S_3$ and let $(\underline{n}_1, \ \beta_1), \ \ldots, \ (\underline{n}_s, \ \beta_s)$ be any $s$ distinct elements of $Z_N]S_3$. Let

$$H_{r, s} = \left\{ \begin{pmatrix} \Psi \\ \Psi^\mu \end{pmatrix} \in G \ \left| \ \begin{array}{l} (\underline{m}_i, \alpha_i) \uparrow \begin{pmatrix} \Psi \\ \Psi^\mu \end{pmatrix} = (\underline{m}_i, \ \alpha_i) \text{ for } i = 1, \ \ldots, \ r \\ \text{and} \\ (\underline{n}_j, \beta_i) \downarrow \begin{pmatrix} \Psi \\ \Psi^\mu \end{pmatrix} = (\underline{n}_j, \ \beta_i) \text{ for } j = 1, \ \ldots, s \end{array} \right. \right\},$$

then $H_{r, s}$ is a subgroup of $G$.

**Proof**   Note that if $\Psi = R(\gamma)\phi_i = 1$ then $\phi_i = R(\gamma)^{-1}$, so that

$$\Psi^\mu = \phi_i R(\gamma) = R(\gamma)^{-1} R(\gamma) = 1.$$

Then

$$\begin{bmatrix} 1 \\ 1^\mu \end{bmatrix} \in H_{r, s}$$

since

$$(\underline{m}_i, \alpha_i) \uparrow \begin{bmatrix} 1 \\ 1^\mu \end{bmatrix} = (\underline{m}_i, \ \alpha_i)1 = (\underline{m}_i, \alpha_i)$$

for $i = 1, \ldots, r$ and

$$(\underline{n}_j, \beta_j) \downarrow \begin{bmatrix} 1 \\ 1^\mu \end{bmatrix} = (\underline{n}_j, \beta_j) 1^\mu = (\underline{n}_j, \beta_j) 1 = (\underline{n}_j, \beta_j)$$

for $j = 1, \ldots, s$. If

$$\begin{bmatrix} \Psi_1 \\ \Psi_1^\mu \end{bmatrix} \text{ and } \begin{bmatrix} \Psi_2 \\ \Psi_2^\mu \end{bmatrix} \in H_{r,s},$$

then

$$(\underline{m}_i, \alpha_i) \uparrow \left[ \begin{bmatrix} \Psi_1 \\ \Psi_1^\mu \end{bmatrix} \begin{bmatrix} \Psi_2 \\ \Psi_2^\mu \end{bmatrix} \right] = (\underline{m}_i, \alpha_i) \uparrow \begin{bmatrix} \Psi_1 \\ \Psi_1^\mu \end{bmatrix} \begin{bmatrix} \Psi_2 \\ \Psi_2^\mu \end{bmatrix}$$

$$= (\underline{m}_i, \alpha_i) \uparrow \begin{bmatrix} \Psi_2 \\ \Psi_2^\mu \end{bmatrix} = (\underline{m}_i, \alpha_i) \text{ for } i = 1, \ldots, r$$

and

$$(\underline{n}_j, \beta_j) \downarrow \left[ \begin{bmatrix} \Psi_1 \\ \Psi_1^\mu \end{bmatrix} \begin{bmatrix} \Psi_2 \\ \Psi_2^\mu \end{bmatrix} \right] = (\underline{n}_j, \beta_j) \downarrow \begin{bmatrix} \Psi_1 \\ \Psi_1^\mu \end{bmatrix} \downarrow \begin{bmatrix} \Psi_2 \\ \Psi_2^\mu \end{bmatrix}$$

$$= (\underline{n}_j, \beta_j) \downarrow \begin{bmatrix} \Psi_2 \\ \Psi_2^\mu \end{bmatrix} = (\underline{n}_j, \beta_j) \text{ for } j = 1, \ldots, s.$$

Hence

$$\begin{bmatrix} \Psi_1 \\ \Psi_1^\mu \end{bmatrix} \begin{bmatrix} \Psi_2 \\ \Psi_2^\mu \end{bmatrix} \in H_{r,s}$$

Since $G$ is finite, $H_{r,s}$ is a subgroup of $G$. ❑

Note that $\mathbf{Z}_N$ is embedded as the subgroup $\{(\underline{m}, 1) | \underline{m} \in |Z_N\}$ in $\mathbf{Z}_N] S_3$ and $S_3$ is embedded as the subgroup $\{(\underline{0}, \alpha) | \alpha \in S_3\}$ in $\mathbf{Z}_N] S_3$. Since $\mathbf{Z}_N] S_3 = Z_N \times S_3$ is the direct product of groups by Lemma 4, both $\mathbf{Z}_N$ and $S_3$ are normal subgroups. Recall the notation

$$S_3 = <\sigma, \rho> = \{1, \rho, \rho^2, \sigma, \sigma\rho, \sigma\rho_2\}.$$

**Lemma 12.19**   Define

$$H = \left\{ \begin{pmatrix} \Psi \\ \Psi^\mu \end{pmatrix} \in G \; \middle| \; \begin{array}{l} (\underline{m}, 1) \uparrow \begin{pmatrix} \Psi \\ \Psi^\mu \end{pmatrix} = (\underline{m}, 1) \; \text{ for all } \underline{m} \in Z_N \\ \text{and} \\ (\underline{0}, \sigma) \downarrow \begin{pmatrix} \Psi \\ \Psi^\mu \end{pmatrix} = (\underline{0}, \sigma) \end{array} \right\}$$

Then given

$$\begin{bmatrix} \Psi \\ \Psi\mu \end{bmatrix} \in H,$$

either

$$(\underline{m}, \alpha) \downarrow \begin{bmatrix} \Psi \\ \Psi\mu \end{bmatrix} = (\underline{m}, \alpha) \text{ for all } (\underline{m}, \alpha) \in Z_N]S_3$$

or

$$(\underline{m}, \alpha) \downarrow \begin{bmatrix} \Psi \\ \Psi\mu \end{bmatrix} = (\underline{m}, \alpha^\sigma) \text{ for all } (\underline{m}, \alpha) \in Z_N]S_3.$$

**Proof** $H$ is a well-defined subgroup of $G$ according to Lemma 12.18. Let

$$\begin{bmatrix} \Psi \\ \Psi\mu \end{bmatrix} = \begin{bmatrix} R(\gamma)\phi_i \\ \phi_i R(\gamma) \end{bmatrix} \in H$$

and $(\underline{m}, \alpha) \in \mathbf{Z}_N]S_3$ be given. Referring to Lemmas 12.12 and 12.14, put $\beta = \gamma^{-1}\alpha\gamma$. Working in the set $T_(\beta, \gamma)$, we have

$$\left( (\underline{m}, \alpha) \downarrow \begin{bmatrix} \Psi \\ \Psi\mu \end{bmatrix}, \beta + \gamma \right)$$

$$= \left( (\underline{m}, \alpha) \downarrow \begin{bmatrix} R(\gamma)\phi_i \\ \phi_i R(\gamma) \end{bmatrix}, \beta + \gamma \right)$$

$$= ((\underline{m}, \alpha)\phi_i R(\gamma), \beta + \gamma)$$

$$= (-(\underline{m}, \alpha)R(\gamma), \phi_i\beta + \gamma)\pi$$

$$= (-(\underline{m}, \alpha\gamma)\phi_i, \beta + \gamma)\pi$$

$$= (-(\underline{m}, \gamma\beta)\phi_i, \beta + \gamma)\pi$$

$$= (-(\underline{m}, \gamma)R(\beta)\phi_i, \beta + \gamma)\pi$$

$$= ((\underline{m}, \gamma)\phi_i R(\beta), \beta + \gamma)$$

$$= ((\underline{m}, 1)R(\gamma)\phi_i R(\beta), \beta + \gamma)$$

$$= ((\underline{m}, 1)R(\beta), \beta + \gamma)$$

$$= ((\underline{m}, \beta),\, \beta + \gamma)$$

using the definition of $H$ and the action of the two-element cyclic group $\{1,\, \pi\}$ on the set $T_{(\beta,\gamma)}$. Hence

$$(\underline{m},\, \alpha) \downarrow \begin{bmatrix} \Psi \\ \Psi^\mu \end{bmatrix} = (\underline{m},\, \beta) = (\underline{m},\, \gamma^{-1}\alpha\gamma) = (\underline{m},\, \alpha^\gamma).$$

Now since

$$(\underline{0},\, \sigma) = (\underline{0},\, \sigma) \downarrow \begin{bmatrix} \Psi \\ \Psi^\mu \end{bmatrix} = (\underline{0},\, \sigma^\gamma)$$

by hypothesis, we have $\sigma = \sigma^\gamma$. Hence $\gamma\sigma = \sigma\gamma$ so that either $\gamma = 1$ or $\gamma = \sigma$.

**Lemma 12.20**　Let $H$ be the subgroup of $G$ defined in Lemma 12.19. Then, $H$ is a nontrivial group of involutions of the set $\mathbf{Z}_N ]S_3$. In particular, every nontrivial element of $H$ is of order 2.

**Proof**　Define

$$\Psi : \mathbf{Z}_N ]S_3 \to \mathbf{Z}_N ]S_3;\ (\underline{m}, \alpha) \to (\underline{m},\, \alpha^\sigma).$$

Then

$$(\underline{m},\, 1) \uparrow \begin{bmatrix} \Psi \\ \Psi^\mu \end{bmatrix} = (\underline{m},\, 1)\Psi = (\underline{m},\, 1^\sigma) = (\underline{m},\, 1) \text{ for all } \underline{m} \in \mathbf{Z}_N$$

and

$$(\underline{0},\, \sigma) \downarrow \begin{bmatrix} \Psi \\ \Psi^\mu \end{bmatrix} = (\underline{0},\, \sigma) \uparrow \begin{bmatrix} \Psi \\ \Psi^\mu \end{bmatrix} = (\underline{0},\, \sigma)\Psi = (\underline{0},\, \sigma^\sigma) = (\underline{0},\, \sigma).$$

Now $\Psi^1 \neq 1$, so

$$\begin{bmatrix} 1 \\ 1^\mu \end{bmatrix} \neq \begin{bmatrix} \Psi \\ \Psi^\mu \end{bmatrix} \in H,$$

hence $H$ is nontrivial. To show that each nontrivial element of $H$ is of order 2, let

$$\begin{bmatrix} \Psi \\ \Psi^\mu \end{bmatrix} = \begin{bmatrix} R(\gamma)\phi_i \\ \phi_i R(\gamma) \end{bmatrix} \in H$$

Then, by the proof of Lemma 12.19, $\gamma = 1$ or $\gamma = \sigma$. In particular, $\gamma^2 = 1$. Hence, for any $(\underline{m}, \alpha) \in \mathbf{Z}_N ]S_3$

$$(\underline{m},\ \alpha) \downarrow \left[\begin{matrix} R(\gamma)\phi^i \\ \phi_i R(\gamma) \end{matrix}\right]^2$$

$$= (\underline{m},\ \alpha) \downarrow \left[\begin{matrix} R(\gamma)\phi^i \\ \phi_i R(\gamma) \end{matrix}\right] \downarrow \left[\begin{matrix} R(\gamma)\phi^i \\ \phi_i R(\gamma) \end{matrix}\right]$$

$$= (\underline{m},\ \alpha^\gamma) \downarrow \left[\begin{matrix} R(\gamma)\phi^i \\ \phi_i R(\gamma) \end{matrix}\right]$$

$$= (\underline{m},\ (\alpha^\gamma)^\gamma) = (\underline{m},\ \alpha).$$

Since the $\downarrow$ action of $G$ on the set $\mathbf{Z}_N]S_3$ is faithful,

$$\left[\begin{matrix} R(\gamma)\phi_i \\ \phi_i R(\gamma) \end{matrix}\right]^2 = \left[\begin{matrix} 1 \\ 1^\mu \end{matrix}\right],$$

the identity element of $G$.

**Lemma 12.21** Denote the right cosets of $\mathbf{Z}_N$ in $\mathbf{Z}_N]S_3$ by

$$\mathbf{Z}_N,\ \mathbf{Z}_{N\rho},\ \mathbf{Z}_{N\rho}^2,;\ \mathbf{Z}_{N\sigma},\ \mathbf{Z}_{N\sigma\rho},\ \mathbf{Z}_{N\sigma\rho}^2.$$

Define Fix$\downarrow(H) =$

$$\left\{ (\underline{m},\ \alpha) \in \mathbf{Z}_N]S_3 \left| (\underline{m}\alpha) \downarrow \left[\begin{matrix} \Psi \\ \Psi^\mu \end{matrix}\right] = (\underline{m}\alpha) \text{ for all } \left[\begin{matrix} \Psi \\ \Psi^\mu \end{matrix}\right] \in H \right. \right\}.$$

Then Fix $\downarrow(H) = \{(\underline{m},\ \alpha) \in \mathbf{Z}_N]S_3 | a = 1 \text{ or } \alpha = \sigma$. The $\downarrow$ action of a nontrivial element of $H$ transposes the coset $\mathbf{Z}_N\rho$ with the coset $\mathbf{Z}_N\rho^2$ and transposes the coset $\mathbf{Z}_N\sigma\rho$ with the coset $\mathbf{Z}_N\sigma\rho^2$.

**Proof** By Lemmas 12.19 and 12.20, the elements

$$\left[\begin{matrix} \Psi \\ \Psi^\mu \end{matrix}\right] \in H$$

are of two kinds:

(i)  $(\underline{m},\ \alpha) \downarrow \left[\begin{matrix} \Psi \\ \Psi^\mu \end{matrix}\right] = (\underline{m},\ \alpha) \text{ for all } (\underline{m},\ \alpha) \in \mathbf{Z}_N]S_3$

in which case

$$\left[\begin{matrix} \Psi \\ \Psi^\mu \end{matrix}\right] = \left[\begin{matrix} 1 \\ 1^\mu \end{matrix}\right]$$

the identity element of H, and

(ii)   $(\underline{m}, \alpha) \downarrow \begin{bmatrix} \Psi \\ \Psi\mu \end{bmatrix} = (\underline{m}, \alpha^\sigma)$ for all $(\underline{m}, \alpha) \in \mathbf{Z}_N]S_3$

in which case

$$\begin{bmatrix} 1 \\ 1\mu \end{bmatrix} \neq \begin{bmatrix} \Psi \\ \Psi\mu \end{bmatrix}$$

is an element of order 2 in $H$. In the second case, compute according to the cosets of $\mathbf{Z}_N$ in $\mathbf{Z}_N]S_3$:

$$(\underline{m}, 1) \downarrow \begin{bmatrix} \Psi \\ \Psi\mu \end{bmatrix} = (\underline{m}, 1^\sigma) = (\underline{m}, 1) \text{ for all } \underline{m} \in \mathbf{Z}_N$$

$$(\underline{m}, \sigma) \downarrow \begin{bmatrix} \Psi \\ \Psi\mu \end{bmatrix} = (\underline{m}, \sigma^\sigma) = (\underline{m}, \sigma) \text{ for all } \underline{m} \in \mathbf{Z}_N$$

$$(\underline{m}, \rho) \downarrow \begin{bmatrix} \Psi \\ \Psi\mu \end{bmatrix} = (\underline{m}, \rho^\sigma) = (\underline{m}, \rho_2) \text{ for all } \underline{m} \in \mathbf{Z}_N$$

$$(\underline{m}, \rho^2) \downarrow \begin{bmatrix} \Psi \\ \Psi\mu \end{bmatrix} = (\underline{m}, (\rho^2)^\sigma) = (\underline{m}, \rho) \text{ for all } \underline{m} \in \mathbf{Z}_N$$

$$(\underline{m}, \sigma\rho) \downarrow \begin{bmatrix} \Psi \\ \Psi\mu \end{bmatrix} = (\underline{m}, (\sigma\rho)^\sigma) = (\underline{m}, \sigma\rho^2) \text{ for all } \underline{m} \in \mathbf{Z}_N$$

$$(\underline{m}, \sigma\rho^2) \downarrow \begin{bmatrix} \Psi \\ \Psi\mu \end{bmatrix} = (\underline{m}, (\sigma\rho^2)^\sigma) = (\underline{m}, \sigma\rho) \text{ for all } \underline{m} \in \mathbf{Z}_N$$

and the lemma follows.

**Lemma 12.22**   Let $Norm_G(H)$ denote the normaliser of $H$ in $G$. The action$\downarrow$ of $G$ on $\mathbf{Z}_N]S_3$ restricts to an action $\downarrow$ of $Norm_G(H)$ on Fix$\downarrow$ $(H)$ which is $(|Z_N|+1)$-transitive.

**Proof**   Let

$$\begin{bmatrix} \Psi \\ \Psi\mu \end{bmatrix} \in G.$$

First show that

$$\text{Fix} \downarrow \left[ \begin{bmatrix} \Psi \\ \Psi\mu \end{bmatrix}^{-1} H \begin{bmatrix} \Psi \\ \Psi\mu \end{bmatrix} \right] = \text{Fix} \downarrow (H) \downarrow \begin{bmatrix} \Psi \\ \Psi\mu \end{bmatrix}$$

as follows.

$$(\underline{m},\,\alpha) \in Fix \downarrow \left[\begin{bmatrix} \Psi \\ \Psi\mu \end{bmatrix}^{-1} H \begin{bmatrix} \Psi \\ \Psi\mu \end{bmatrix}\right]$$

$$\Leftrightarrow (\underline{m},\,\alpha) \downarrow \left[\begin{bmatrix} \Psi \\ \Psi\mu \end{bmatrix}^{-1} \begin{bmatrix} \Psi_* \\ \Psi_*\mu \end{bmatrix} \begin{bmatrix} \Psi \\ \Psi\mu \end{bmatrix}\right] = (\underline{m},\,\alpha) \text{ for all } \begin{bmatrix} \Psi_* \\ \Psi_*\mu \end{bmatrix} \in H$$

$$\Leftrightarrow (\underline{m},\alpha) \downarrow \left[\begin{bmatrix} \Psi \\ \Psi\mu \end{bmatrix}^{-1} \begin{bmatrix} \Psi_* \\ \Psi_*\mu \end{bmatrix}\right] = (\underline{m},\alpha) \downarrow \begin{bmatrix} \Psi \\ \Psi \end{bmatrix}^{-1} \text{ for all } \begin{bmatrix} \Psi_* \\ \Psi_*\mu \end{bmatrix} \in H$$

$$\Leftrightarrow (\underline{m},\,\alpha) \downarrow \begin{bmatrix} \Psi \\ \Psi\mu \end{bmatrix}^{-1} \downarrow \begin{bmatrix} \Psi_* \\ \Psi_*\mu \end{bmatrix} = (\underline{m},\,\alpha) \downarrow \begin{bmatrix} \Psi \\ \Psi\mu \end{bmatrix}^{-1} \text{ for all } \begin{bmatrix} \Psi_* \\ \Psi_*\mu \end{bmatrix} \in H$$

$$\Leftrightarrow (\underline{m},\,\alpha) \downarrow \begin{bmatrix} \Psi \\ \Psi\mu \end{bmatrix} \in \text{Fix} \downarrow (H) \Leftrightarrow (\underline{m},\,\alpha) \in \text{Fix} \downarrow (H) \downarrow \begin{bmatrix} \Psi \\ \Psi\mu \end{bmatrix}.$$

Now let

$$\begin{bmatrix} \Psi \\ \Psi\mu \end{bmatrix} \in Norm_G(H) = \left\{ \begin{bmatrix} \Psi \\ \Psi\mu \end{bmatrix} \in G \left| \begin{bmatrix} \Psi \\ \Psi\mu \end{bmatrix}^{-1} H \begin{bmatrix} \Psi \\ \Psi\mu \end{bmatrix} = H \right. \right\}.$$

Then

$$\text{Fix}\downarrow (H) = Fix \downarrow \left[\begin{bmatrix} \Psi \\ \Psi\mu \end{bmatrix}^{-1} H \begin{bmatrix} \Psi \\ \Psi\mu \end{bmatrix}\right] = \text{Fix}\downarrow (H) \downarrow \begin{bmatrix} \Psi \\ \Psi\mu \end{bmatrix}$$

showing that the action restricts to an action of $Norm_G(H)$ on Fix $\downarrow (H)$. Now to show that the action $\downarrow$ of $Norm_G(H)$ on Fix$\downarrow (H)$ $=\mathbf{Z}_N \cup \mathbf{Z}_N\sigma$ is $(|\mathbf{Z}_N|+1)$-transitive, label the elements of $\mathbf{Z}_N$ as $(\underline{m}_1,\alpha_1),\ldots,(\underline{m}_N,\alpha_N)$ and label $(\underline{0},\sigma) = (\underline{m}_{N+1},\,\alpha_{N+1})$. Let $(\underline{m}_1^*,\,\alpha_1^*),\ldots,(\underline{m}_{N+1}^*,\,\alpha_{N+1}^*)$ be any $|\mathbf{Z}_N|+1$ distinct points of Fix$\downarrow (H)$. It is enough to show that there exists

$$\begin{bmatrix} \Psi \\ \Psi\mu \end{bmatrix} \in Norm_G(H)$$

such that

$$(\underline{m}_i^*,\,a_i^*) \downarrow \begin{bmatrix} \Psi \\ \Psi\mu \end{bmatrix} = (\underline{m}_i,\,\alpha_i) \text{ for } i = 1,\ldots,N+1.$$

Now there exists

$$\begin{bmatrix} \Psi \\ \Psi\mu \end{bmatrix} \in G$$

such that

$$(\underline{m}_i^*, \alpha_i^*) \downarrow \begin{bmatrix} \Psi \\ \Psi\mu \end{bmatrix} = (\underline{m}_i, \alpha_i) \text{ for } i = 1, \ldots, N+1.$$

Hence

$$(\underline{m}_i^*, \alpha_i^*) = (\underline{m}_i, \alpha_i) \downarrow \begin{bmatrix} \Psi \\ \Psi\mu \end{bmatrix}^{-1} \text{ for } I = 1, \ldots, N+1.$$

Note that for every

$$\begin{bmatrix} \Psi \\ \Psi\mu \end{bmatrix} \in H$$

and for $i = 1, \ldots, N+1$:

$$(\underline{m}_i, \alpha_i) \downarrow \left( \begin{bmatrix} \Psi \\ \Psi\mu \end{bmatrix}^{-1} \begin{bmatrix} \Psi_* \\ \Psi_*\mu \end{bmatrix} \begin{bmatrix} \Psi \\ \Psi\mu \end{bmatrix} \right)$$

$$= (\underline{m}_i, \alpha_i) \downarrow \begin{bmatrix} \Psi \\ \Psi\mu \end{bmatrix}^{-1} \downarrow \begin{bmatrix} \Psi \\ \Psi\mu \end{bmatrix} \downarrow \begin{bmatrix} \Psi \\ \Psi\mu \end{bmatrix}$$

$$= (\underline{m}_i^*, \alpha_i^*) \downarrow \begin{bmatrix} \Psi_* \\ \Psi_*\mu \end{bmatrix} \downarrow \begin{bmatrix} \Psi \\ \Psi\mu \end{bmatrix}$$

$$= (\underline{m}_i^*, \alpha_i^*) \downarrow \begin{bmatrix} \Psi \\ \Psi\mu \end{bmatrix} = (\underline{m}_i, \alpha_i).$$

Hence

$$\begin{bmatrix} \Psi \\ \Psi\mu \end{bmatrix}^{-1} \begin{bmatrix} \Psi_* \\ \Psi_*\mu \end{bmatrix} \begin{bmatrix} \Psi \\ \Psi\mu \end{bmatrix} \in H \text{ for all } \begin{bmatrix} \Psi_* \\ \Psi_*\mu \end{bmatrix} \in H \Rightarrow \begin{bmatrix} \Psi \\ \Psi\mu \end{bmatrix} \in \mathit{Norm}_G(H). \qquad \square$$

**Lemma 12.23**  There exists a Steiner system $S(N+1, 2N, 6N)$, where the points are the elements of the set $\mathbf{Z}_N]S_3$ and the set of blocks is

$$\left\{ \mathrm{Fix} \downarrow (H) \downarrow \begin{bmatrix} \Psi \\ \Psi\mu \end{bmatrix} \, \middle| \, \begin{bmatrix} \Psi \\ \Psi\mu \end{bmatrix} \in G \right\}.$$

**Proof** There are $6N = |\mathbf{Z}_N]S_3|$ points. Each block, for a fixed

$$\begin{bmatrix} \Psi \\ \Psi\mu \end{bmatrix} \in G$$

contains $2N = |\mathbf{Z}_N \cup \mathbf{Z}_N\sigma| = |\text{Fix} \downarrow (H)| =$

$$\left| Fix \downarrow (H) \downarrow \begin{bmatrix} \Psi \\ \Psi\mu \end{bmatrix} \right|$$

points. Label the elements of $\mathbf{Z}_N$ as $(\underline{m}_1, \alpha_1), \ldots, (\underline{m}_N, \alpha_N)$ and label $(\underline{0}, \sigma) = (\underline{m}_{N+1}, \alpha_{N+1})$. Let $(\underline{m}_1^*, \alpha_1^*), \ldots, (\underline{m}_{N+1}^*, \alpha_{N+1}^*)$ be any $N+1$ distinct points of $\mathbf{Z}_N]S_3$. Then there exists

$$\begin{bmatrix} \Psi \\ \Psi\mu \end{bmatrix} \in G$$

such that

$$(\underline{m}_i, \alpha_i) \downarrow \begin{bmatrix} \Psi \\ \Psi\mu \end{bmatrix} = (\underline{m}_i^*, \alpha_i^*) \text{ for } i = 1, \ldots, N+1.$$

Hence there is at least one block, namely $\text{Fix} \downarrow (H)$, that contains the points $(\underline{m}_1^*, \alpha_1^*), \ldots, (\underline{m}_{N+1}^*, \alpha_{N+1}^*)$. It remains to show that this is the unique block that contains the points $(\underline{m}_1^*, \alpha_1^*), \ldots, (\underline{m}_{N+1}^*, \alpha_{N+1}^*)$. Suppose $(\underline{m}_1^*, \alpha_1^*), \ldots, (\underline{m}_{N+1}^*, \alpha_{N+1}^*)$ are contained in

$$\text{Fix} \downarrow (H) \downarrow \begin{bmatrix} \Psi_* \\ (\Psi_*\mu) \end{bmatrix} \text{ for some } \begin{bmatrix} \Psi_* \\ (\Psi_*\mu) \end{bmatrix} \in G.$$

Then there exist points $(\underline{m}_1^{**}, \alpha_1^{**}), \ldots, (\underline{m}_{N+1}^{**}, \alpha_{N+1}^{**})$ in $\text{Fix} \downarrow (H)$ such that

$$(\underline{m}_i^*, \alpha_i^*) = (\underline{m}_i^{**}, \alpha_i^{**}) \downarrow \begin{bmatrix} \Psi \\ \Psi\mu \end{bmatrix} \text{ for } i = 1, \ldots, N+1.$$

By Lemma 12.22, there exists

$$\begin{bmatrix} \Psi_{**} \\ \Psi_{**}\mu \end{bmatrix} \in \text{Norm}_G(H)$$

such that

$$(\underline{m}_i^{**}, \alpha_i^{**}) = (\underline{m}_i, \alpha_i) \downarrow \begin{bmatrix} \Psi_{**} \\ \Psi_{**}\mu \end{bmatrix} \text{ for } i = 1, \ldots, N+1.$$

Hence for $i = 1, \ldots, N+1$

$$(\underline{m}_i,\, \alpha_i) \downarrow \begin{bmatrix} \Psi \\ \Psi^\mu \end{bmatrix}$$

$$= (\underline{m}_i^*,\, \alpha_i^*)$$

$$= (\underline{m}_i^{**},\, \alpha_i^{**}) \downarrow \begin{bmatrix} \Psi_* \\ \Psi_{*}{}^\mu \end{bmatrix}$$

$$= (\underline{m}_i,\, \alpha_i) \downarrow \begin{bmatrix} \Psi_{**} \\ \Psi_{**}{}^\mu \end{bmatrix} \downarrow \begin{bmatrix} \Psi_* \\ \Psi_{*}{}^\mu \end{bmatrix}$$

$$\Leftrightarrow (\underline{m}_i,\, \alpha_i) = (\underline{m}_i,\, \alpha_i) \downarrow \left[ \begin{bmatrix} \Psi_{**} \\ \Psi_{**}{}^\mu \end{bmatrix} \begin{bmatrix} \Psi_* \\ \Psi_{*}{}^\mu \end{bmatrix} \begin{bmatrix} \Psi \\ \Psi^\mu \end{bmatrix} \right]^{-1} \text{ for } i = 1, \ldots, N+1.$$

Then, by Lemma 12.17

$$(\underline{m}_i,\, \alpha_i) = (\underline{m}_i,\, \alpha_i) \uparrow \left[ \begin{bmatrix} \Psi_{**} \\ \Psi_{**}{}^\mu \end{bmatrix} \begin{bmatrix} \Psi_* \\ \Psi_{*}{}^\mu \end{bmatrix} \begin{bmatrix} \Psi \\ \Psi^\mu \end{bmatrix} \right]^{-1} \text{ for } i = 1, \ldots, N,$$

and

$$(\underline{m}_{N+1},\, \alpha_{N+1}) = (\underline{m}_{N+1},\, \alpha_{N+1}) \downarrow \left[ \begin{bmatrix} \Psi_{**} \\ \Psi_{**}{}^\mu \end{bmatrix} \begin{bmatrix} \Psi_* \\ \Psi_{*}{}^\mu \end{bmatrix} \begin{bmatrix} \Psi \\ \Psi^\mu \end{bmatrix} \right]^{-1}.$$

Hence

$$\begin{bmatrix} \Psi_{**} \\ \Psi_{**}{}^\mu \end{bmatrix} \begin{bmatrix} \Psi_* \\ \Psi_{*}{}^\mu \end{bmatrix} \begin{bmatrix} \Psi \\ \Psi^\mu \end{bmatrix}^{-1} \in H.$$

Now, $H$ is a subgroup of $Norm_G(H)$

$$\Leftrightarrow \begin{bmatrix} \Psi_{**} \\ \Psi_{**}{}^\mu \end{bmatrix} \begin{bmatrix} \Psi_* \\ \Psi_{*}{}^\mu \end{bmatrix} \begin{bmatrix} \Psi \\ \Psi^\mu \end{bmatrix}^{-1} \in Norm_G(H)$$

$$\Leftrightarrow \begin{bmatrix} \Psi_* \\ \Psi_{*}{}^\mu \end{bmatrix} \begin{bmatrix} \Psi \\ \Psi^\mu \end{bmatrix}^{-1} \in \begin{bmatrix} \Psi_{**} \\ \Psi_{**}{}^\mu \end{bmatrix}^{-1} Norm_G(H) = Norm_G(H)$$

$$\left[\left[\begin{array}{c}\Psi_*\\\Psi_{*\mu}\end{array}\right]^{-1}\left[\begin{array}{c}\Psi\\\Psi\mu\end{array}\right]\right]H\left[\left[\begin{array}{c}\Psi_*\\\Psi_{*\mu}\end{array}\right]^{-1}\left[\begin{array}{c}\Psi\\\Psi\mu\end{array}\right]^{-1}\right]=H$$

$$\Leftrightarrow\left[\begin{array}{c}\Psi^*\\\Psi_{*\mu}\end{array}\right]\left[\begin{array}{c}\Psi\\\Psi\mu\end{array}\right]^{-1}H\left[\begin{array}{c}\Psi\\\Psi\mu\end{array}\right]\left[\begin{array}{c}\Psi_*\\\Psi_{*\mu}\end{array}\right]^{-1}=H$$

$$\Leftrightarrow\left[\begin{array}{c}\Psi\\\Psi\mu\end{array}\right]^{-1}H\left[\begin{array}{c}\Psi\\\Psi\mu\end{array}\right]=\left[\begin{array}{c}\Psi_*\\\Psi_{*\mu}\end{array}\right]^{-1}H\left[\begin{array}{c}\Psi_*\\\Psi_{*\mu}\end{array}\right].$$

Now, using the first fact in the proof of Lemma 12.22

$$\text{Fix}\downarrow(H)\downarrow\left[\begin{array}{c}\Psi_*\\\Psi_{*\mu}\end{array}\right]$$

$$=\text{Fix}\downarrow\left[\left[\begin{array}{c}\Psi_*\\\Psi_{*\mu}\end{array}\right]^{-1}H\left[\begin{array}{c}\Psi_*\\\Psi_{*\mu}\end{array}\right]\right]$$

$$=\text{Fix}\downarrow\left[\left[\begin{array}{c}\Psi\\\Psi\mu\end{array}\right]^{-1}H\left[\begin{array}{c}\Psi\\\Psi\mu\end{array}\right]\right]$$

$$=\text{Fix}\downarrow(H)\downarrow\left[\begin{array}{c}\Psi\\\Psi\mu\end{array}\right]$$

This establishes the uniqueness of the block.

**Theorem 12.1.** Any map on the sphere can be properly coloured by using at most four colours.

**Proof** Referring to section 12.1, we have defined $N$ to be the minimal number of colours required to properly colour any map from the class of all maps on the sphere. Based on the definition of $N$, we have selected a specific map $m(N)$ on the sphere that requires no fewer than $N$ colours to be properly coloured. Based on the definition of the map $m(N)$ we have selected a proper colouring of its regions using the $N$ colours $0, 1, \ldots, N-1$. Working with the fixed number $N$, the fixed map $m(N)$, and the fixed proper colouring of the regions of the map $m(N)$, Lemma 12.23 has explicitly constructed a Steiner system $S(N+1, 2N, 6N)$. Now Lemma 12.3 implies that $N$ cannot exceed four.

# 13. Graph Algorithms

An *algorithm* is a problem-solving method suitable for implementation as a computer program. While designing algorithms we are typically faced with a number of different approaches. For small problems, it hardly matters which approach we use, as long as it is one that solves the problem correctly. However, there are many problems, including some problems in graph theory, for which the known algorithms take so long to compute the solution that they are practically useless. A *polynomial-time algorithm* is one whose number of computational steps is always bounded by a polynomial function of the size of the input. The class of all such problems that have polynomial-time algorithms is denoted by **P**. For some problems, there are no known polynomial-time algorithms but these problems do have *nondeterministic polynomial-time algorithms*: try all candidates for solutions simultaneously and for each given candidate, verify whether it is a correct solution in polynomial-time. The class of all such problems is denoted by **NP**. Clearly $\mathbf{P} \subseteq \mathbf{NP}$. On the other hand, there are problems that are known to be in *NP* and are such that any polynomial-time algorithm for them can be transformed into a polynomial-time algorithm for every problem in **NP**. Such problems are called **NP**-*complete*. Thus, if anybody ever finds a polynomial-time algorithm for an **NP**-complete problem, he or she would have proved that $\mathbf{P} = \mathbf{NP}$. One of the greatest unresolved problems in mathematics and computer science today is whether $\mathbf{P} = \mathbf{NP}$ or $\mathbf{P} \neq \mathbf{NP}$ [62].

In this chapter we present algorithms for five well-known graph theory problems. First, we discuss Dijkstra's algorithm for computing shortest paths in graphs. This algorithm is polynomial-time and hence the problem is in **P**. Dijkstra's algorithm is widely used in practice, for example, as part of the TCP/IP protocol suite for routing internet traffic over shortest paths. The second is Prim's algorithm for computing minimal spanning trees in graphs. This algorithm is obtained by a trivial modification of Dijkstra's algorithm, is also polynomial-time and hence the problem again is in **P**. Prim's algorithm is also widely used in practice, for example, in distribution problems and broadcasting data over computer networks. Third, we present Fleury's algorithm for finding Eulerian circuits (cycles) in graphs. The critical point here is to decide whether a certain pair of subgraphs partition the graph into two nontrivial components and the trick is to compute a tree using Prim's algorithm and check whether the tree spans the graph. Fleury's algorithm is also polynomial-time and hence the problem of finding Eulerian circuits is also in **P**. Fourth, we show how to construct the De Bruijn graphs and sequences using Eulerian circuits in polynomial-time. These graphs have recently found important applications in multihop and fault tolerant computer networks. The fifth and final algorithm is for finding Hamiltonian circuits (cycles) in graphs. The algorithm is an example of a nondeterministic polynomial-time algorithm. The problem of finding a Hamiltonian circuit in a graph is an example of a **NP**-complete problem.

All algorithms are implemented in C++ and tested using Microsoft Visual C++ [158].

# 13.1 Dijkstra's Algorithm for Shortest Paths

In 1956, Edsger W. Dijkstra [65, 153], was the main programmer at the Burroughs Corporation in Amsterdam, where the construction of one of the earliest computers (the AR-MAC) was on the verge of completion. In order to celebrate its inauguration they needed a nice demonstration. It should solve a problem that could be easily stated to a predominantly lay audience. For the purpose of the demonstration, Dijkstra drew a slightly simplified map of the Dutch railroad system; someone in the audience could ask for the shortest connection between, say, Harlingen and Maastricht, and the ARMAC would print out the shortest route town by town. The demonstration was a great success; Dijkstra reminisces that he could show that the inversion of source and destination could influence the computation time required. The speed of the ARMAC and the size of the map were such that one-minute computations always sufficed.

The general problem is to find shortest paths from one specified vertex to all other vertices in a weighted graph, where the edge weights correspond to distances between towns. Together, these paths will form a (not necessarily minimal) spanning tree. The main idea of Dijkstra's algorithm is the following; if $P$ is a shortest path from $u$ to $z$ and $P$ contains $v$, then the portion of the path $P$ from $u$ to $v$ must be a shortest path from $u$ to $v$. This suggests that we should determine optimal routes from $u$ to every other vertex $z$ in increasing order of the distance $d(u,z)$ from $u$ to $z$. We maintain a current tentative distance and use this to update the remaining tentative distances in tabular form. The algorithm works equally well for directed graphs. A formal description of Dijkstra's algorithm is the following.

### 13.1.1 Dijkstra's algorithm to compute distances from $u$

**Input**  A weighted graph (or digraph) and a starting vertex $u$. The weight of edge $xy$ is $w(xy)$; let $w(xy) = \infty$ if $xy$ is not an edge.

**Idea**  Maintain the set $S$ of vertices to which the shortest route from $u$ is known, enlarging $S$ to include all the vertices. To do this maintain also a tentative distance $t(z)$ from $u$ to each $z$ not in $S$; this is the length of the shortest path found yet from $u$ to $z$.

**Initialisation**  Let $S = \{u\}$; $d(u,u) = 0$; $t(z) = w(uz)$ for all $z \neq u$.

**Iteration**  Select a vertex $v$ outside $S$ such that $t(v)$ is minimum in the set $\{t(z)|z \notin S\}$. Add $v$ to $S$. For each edge $vz$ with $z \notin S$, update $t(z)$ to $\min\{t(z), d(u,v) + w(vz)\}$.

**Termination**    Continue the iteration until $S = V(G)$ or until $t(z) = \infty$ for every $z \notin S$. In the first case, all shortest paths from $u$ have been found. In the latter case, the remaining vertices are unreachable from $u$.

We give an implementation of Dijkstra's algorithm in C++ below.

**dijkstra.cpp**

```cpp
#include <iostream>
#include <fstream>
#include <string>
#include <vector>
#include <set>

using namespace std;

ifstream infile(''graph.txt'');
ofstream outfile(''shortest_paths.txt'');

int main()
cout<<''Dijkstra's Algorithm.''<<endl;
 int m,n,i,j,k;
 //Read adjacency matrix of weights from graph.txt
 infile>>n;
 vector< vector< float> > weight;
 float val;
 for(i=0; i<n; i++)
 {
  vector< float > row;
  for(j=0; j<n; j++)
  {
   infile>>val;
   row.push_back(val);
  }
  weight.push_back(row);
 }

 //Initialize Table
 const float infinity=1000000;
```

```cpp
vector<bool> known;
for(i=0; i<n; i++) known.push_back(false);
vector<float> d;
d.push_back(0);
for(i=1; i<n; i++) d.push_back(infinity);
vector<int> p;
for(i=0; i<n; i++) p.push_back(-1);
//Print Initial Table
 outfile<<endl<<''INITIAL TABLE:''<<endl;
 outfile<<endl<<''Vertex  : t'';
 for(i=0; i<n; i++) outfile<<i<<'\t';
 outfile<<endl<<''Known   :\t'';
 for(i=0; i<n; i++) outfile<<known[i]<<'\t';
 outfile<<endl<<''Distance:\t'';
 for(i=0; i<n; i++) outfile<<d[i]<<'\t';
 outfile<<endl<<''Path    :\t'';
 for(i=0; i<n; i++) outfile<<p[i]<<'\t';
 outfile<<endl;
//Iteration
for(k=0; k<n; k++)
{
 //Find min of d for unknown vertices
 int min=0;
 while(known[min]==true)min++;
 for(i=0; i<n; i++)
 if(known[i]==false && d[i]<d[min])min=i;
 //Update Table
 known[min]=true;
 for (int j=0; j<n; j++)
 {
  if(weight[min][j]!=0 &&
   d[j]>d[min]+weight[min][j] &&
   known[j]==false)
  {
   d[j]=d[min]+weight[min][j];
   p[j]=min;
  }
```

```
    }
    //Print Table
    outfile<<endl<<''TABLE No.''<<k<<":''<<endl;
    outfile<<endl<<''Vertex  :\t'';
    for(i=0; i<n; i++) outfile<<i<<'\t';
    outfile<<endl<<''Known    :\t'';
    for(i=0; i<n; i++) outfile<<known[i]<<'\t';
    outfile<<endl<<''Distance:\t'';
    for(i=0; i<n; i++) outfile<<d[i]<<'\t';
    outfile<<endl<<''Path     :\t'';
    for(i=0; i<n; i++) outfile<<p[i]<<'\t';
    outfile<<endl;
  }
  //Find shortest paths and spanning tree
  outfile<<endl<<endl<<''SHORTEST PATHS:''<<endl;
  set< vector<int> > span;
  for(i=0; i<n; i++)
{
 outfile<<''Path='';
 vector<int> temp;
 m=i;
 while(m!=-1)
 {
  temp.push_back(m);
  m=p[m];
 }
 outfile<<temp[temp.size()-1]<<" ";
 for(j=temp.size()-2; j>=0; j--)
 {
 outfile<<temp[j]<<" ";
 vector<int> edge;
 edge.push_back(temp[j+1]);
 edge.push_back(temp[j]);
 span.insert(edge);
 }
outfile<<''Distance=''<<d[i]<<endl;
 }
```

```
//Print spanning tree
outfile<<endl<<''SPANNING TREE:''<<endl;
set< vector<int>>::iterator it;
for(it=span.begin(); it!=span.end(); it++)
outfile<<''('' <<(*it)[0]<<'',''<<(*it)[1]<<")"<<" ";
cout<<''See shortest_paths.txt for results.''  <<endl;system(''PAUSE'');
return 0;}
```

**Example 13.1.2** Consider the Petersen graph with vertices labelled as shown in Figure 13.1.

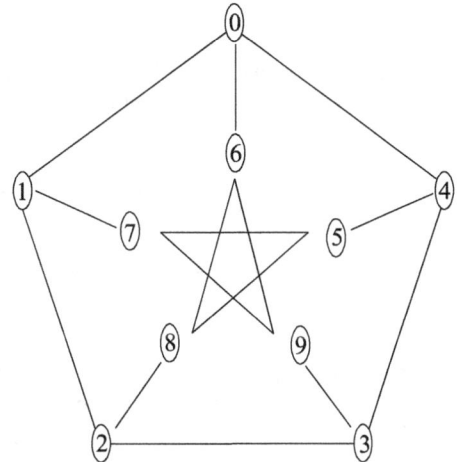

**Fig. 13.1** *The Petersen graph with labelled vertices*

Make a text file "graph.txt" as shown below and save it in the same directory as the program "dijkstra.cpp". The first line of the file "graph.txt" is the number of vertices and the rest is the adjacency matrix of the labelled Petersen graph shown in Figure 13.1.

**graph.txt**

```
10
0   1   0   0   1   0   1   0   0   0
1   0   1   0   0   0   0   1   0   0
0   1   0   1   0   0   0   0   1   0
0   0   1   0   1   0   0   0   0   1
1   0   0   1   0   1   0   0   0   0
0   0   0   0   1   0   0   1   1   0
1   0   0   0   0   0   0   0   1   1
0   1   0   0   0   1   0   0   0   1
0   0   1   0   0   1   1   0   0   0
0   0   0   1   0   0   1   1   0   0
```

Now compile and run the program "dijkstra.cpp". The following output file "shortest_paths.txt" is produced in the program's directory. The output shows shortest paths from the vertex labelled 0 to all other vertices:

**shortest_paths.txt**

INITIAL TABLE

| Vertex | : | 0 | 1 | 2 | 3 | 4 | 5 | 6 | 7 | 8 | 9 |
|---|---|---|---|---|---|---|---|---|---|---|---|
| known | : | 0 | 0 | 0 | 0 | 0 | 0 | 0 | 0 | 0 | 0 |
| Distance | : | 0 | $\infty$ | $\infty$ | $\infty$ | $\infty$ | $\infty$ | $\infty$ | $\infty$ | $\infty$ | $\infty$ |
| Path | : | $-1$ | $-1$ | $-1$ | $-1$ | $-1$ | $-1$ | $-1$ | $-1$ | $-1$ | $-1$ |

TABLE No.0:

| Vertex | : | 0 | 1 | 2 | 3 | 4 | 5 | 6 | 7 | 8 | 9 |
|---|---|---|---|---|---|---|---|---|---|---|---|
| Known | : | 1 | 0 | 0 | 0 | 0 | 0 | 0 | 0 | 0 | 0 |
| Distance | : | 0 | 1 | $\infty$ | $\infty$ | 1 | $\infty$ | 1 | $\infty$ | $\infty$ | $\infty$ |
| Path | : | $-1$ | 0 | $-1$ | $-1$ | 0 | $-1$ | 0 | $-1$ | $-1$ | $-1$ |

TABLE No.1:

| Vertex | : | 0 | 1 | 2 | 3 | 4 | 5 | 6 | 7 | 8 | 9 |
|---|---|---|---|---|---|---|---|---|---|---|---|
| Known | : | 1 | 1 | 0 | 0 | 0 | 0 | 0 | 0 | 0 | 0 |
| Distance | : | 0 | 1 | 2 | $\infty$ | 1 | $\infty$ | 1 | 2 | $\infty$ | $\infty$ |
| Path | : | $-1$ | 0 | 1 | $-1$ | 0 | $-1$ | 0 | 1 | $-1$ | $-1$ |

TABLE No.2:

| Vertex | : | 0 | 1 | 2 | 3 | 4 | 5 | 6 | 7 | 8 | 9 |
|---|---|---|---|---|---|---|---|---|---|---|---|
| Known | : | 1 | 1 | 0 | 0 | 1 | 0 | 0 | 0 | 0 | 0 |
| Distance | : | 0 | 1 | 2 | 2 | 1 | 2 | 1 | 2 | $\infty$ | $\infty$ |
| Path | : | $-1$ | 0 | 1 | 4 | 0 | 4 | 0 | 1 | $-1$ | $-1$ |

TABLE No.3:

| Vertex | : | 0 | 1 | 2 | 3 | 4 | 5 | 6 | 7 | 8 | 9 |
|---|---|---|---|---|---|---|---|---|---|---|---|
| Known | : | 1 | 1 | 0 | 0 | 1 | 0 | 1 | 0 | 0 | 0 |
| Distance | : | 0 | 1 | 2 | 2 | 1 | 2 | 1 | 2 | 2 | 2 |
| Path | : | $-1$ | 0 | 1 | 4 | 0 | 4 | 0 | 1 | 6 | 6 |

TABLE No.4:

| Vertex | : | 0 | 1 | 2 | 3 | 4 | 5 | 6 | 7 | 8 | 9 |
|---|---|---|---|---|---|---|---|---|---|---|---|
| Known | : | 1 | 1 | 1 | 0 | 1 | 0 | 1 | 0 | 0 | 0 |
| Distance | : | 0 | 1 | 2 | 2 | 1 | 2 | 1 | 2 | 2 | 2 |
| Path | : | $-1$ | 0 | 1 | 4 | 0 | 4 | 0 | 1 | 6 | 6 |

TABLE No.5:

| Vertex | : | 0 | 1 | 2 | 3 | 4 | 5 | 6 | 7 | 8 | 9 |
|---|---|---|---|---|---|---|---|---|---|---|---|
| Known | : | 1 | 1 | 1 | 1 | 1 | 0 | 1 | 0 | 0 | 0 |
| Distance | : | 0 | 1 | 2 | 2 | 1 | 2 | 1 | 2 | 2 | 2 |
| Path | : | $-1$ | 0 | 1 | 4 | 0 | 4 | 0 | 1 | 6 | 6 |

TABLE No.6:

| Vertex | : | 0 | 1 | 2 | 3 | 4 | 5 | 6 | 7 | 8 | 9 |
|---|---|---|---|---|---|---|---|---|---|---|---|
| Known | : | 1 | 1 | 1 | 1 | 1 | 1 | 1 | 0 | 0 | 0 |
| Distance | : | 0 | 1 | 2 | 2 | 1 | 2 | 1 | 2 | 2 | 2 |
| Path | : | $-1$ | 0 | 1 | 4 | 0 | 4 | 0 | 1 | 6 | 6 |

TABLE No.7:

| Vertex | : | 0 | 1 | 2 | 3 | 4 | 5 | 6 | 7 | 8 | 9 |
|---|---|---|---|---|---|---|---|---|---|---|---|
| Known | : | 1 | 1 | 1 | 1 | 1 | 1 | 1 | 1 | 0 | 0 |
| Distance | : | 0 | 1 | 2 | 2 | 1 | 2 | 1 | 2 | 2 | 2 |
| Path | : | −1 | 0 | 1 | 4 | 0 | 4 | 0 | 1 | 6 | 6 |

TABLE No.8:

| Vertex | : | 0 | 1 | 2 | 3 | 4 | 5 | 6 | 7 | 8 | 9 |
|---|---|---|---|---|---|---|---|---|---|---|---|
| Known | : | 1 | 1 | 1 | 1 | 1 | 1 | 1 | 1 | 1 | 0 |
| Distance | : | 0 | 1 | 2 | 2 | 1 | 2 | 1 | 2 | 2 | 2 |
| Path | : | −1 | 0 | 1 | 4 | 0 | 4 | 0 | 1 | 6 | 6 |

TABLE No.9:

| Vertex | : | 0 | 1 | 2 | 3 | 4 | 5 | 6 | 7 | 8 | 9 |
|---|---|---|---|---|---|---|---|---|---|---|---|
| Known | : | 1 | 1 | 1 | 1 | 1 | 1 | 1 | 1 | 1 | 1 |
| Distance | : | 0 | 1 | 2 | 2 | 1 | 2 | 1 | 2 | 2 | 2 |
| Path | : | −1 | 0 | 1 | 4 | 0 | 4 | 0 | 1 | 6 | 6 |

**Shortest paths**

Path=0 Distance=0

Path=0 1 Distance=1

Path=0 1 2 Distance=2

Path=0 4 3 Distance=2

Path=0 4 Distance=1

Path=0 4 5 Distance=2

Path=0 6 Distance=1

Path=0 1 7 Distance=2

Path=0 6 8 Distance=2

Path=0 6 9 Distance=2

**Spanning tree**

(0, 1) (0, 4) (0, 6) (1, 2) (1, 7) (4, 3) (4, 5) (6, 8) (6, 9)

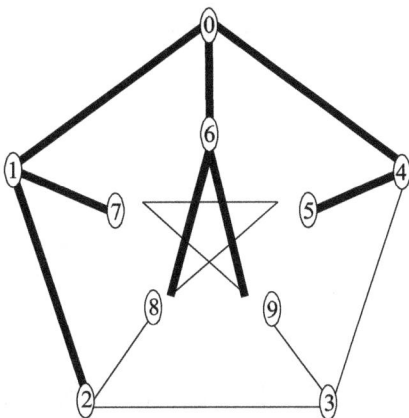

**Fig 13.2** *Shortest paths from vertex 0 and a spanning tree*

The output of Dijkstra's algorithm for the Petersen graph with starting vertex 0 is shown in Figure 13.2. In this case, the spanning tree is minimal as the reader may verify. This is not always the case as the next example shows.

**Example 13.1.3** Consider the weighted and directed graph shown in Figure 13.3.

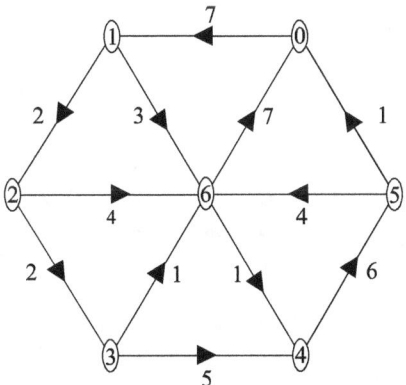

**Fig. 13.3** *A weighted and directed graph*

Make a text file "graph.txt" as shown below and save it in the same directory as the program "dijkstra.cpp". The first line of the file "graph.txt" is the number of vertices and the rest is the matrix of weights of the labelled directed graph shown in Figure 13.3 above.

**graph.txt**

```
7
0  7  0  0  0  0  0
0  0  2  0  0  0  3
0  0  0  2  0  0  4
0  0  0  0  5  0  1
0  0  0  0  0  6  0
1  0  0  0  0  0  4
7  0  0  0  1  0  0
```

Now run the program "dijkstra.cpp". The following output file "shortest_paths.txt" is produced in the program's directory. The output shows shortest paths from the vertex labelled 0 to all other vertices:

**shortest_paths.txt**

INITIAL TABLE:

| Vertex | : | 0 | 1 | 2 | 3 | 4 | 5 | 6 |
|---|---|---|---|---|---|---|---|---|
| Known | : | 0 | 0 | 0 | 0 | 0 | 0 | 0 |
| Distance | : | 0 | $\infty$ | $\infty$ | $\infty$ | $\infty$ | $\infty$ | $\infty$ |
| Path | : | $-1$ | $-1$ | $-1$ | $-1$ | $-1$ | $-1$ | $-1$ |

TABLE No.0:

| Vertex | : | 0 | 1 | 2 | 3 | 4 | 5 | 6 |
|---|---|---|---|---|---|---|---|---|
| Known | : | 1 | 0 | 0 | 0 | 0 | 0 | 0 |
| Distance | : | 0 | 7 | $\infty$ | $\infty$ | $\infty$ | $\infty$ | $\infty$ |
| Path | : | $-1$ | 0 | $-1$ | $-1$ | $-1$ | $-1$ | $-1$ |

TABLE No.1:

| Vertex | : | 0 | 1 | 2 | 3 | 4 | 5 | 6 |
|---|---|---|---|---|---|---|---|---|
| Known | : | 1 | 1 | 0 | 0 | 0 | 0 | 0 |
| Distance | : | 0 | 7 | 9 | $\infty$ | $\infty$ | $\infty$ | 10 |
| Path | : | $-1$ | 0 | 1 | $-1$ | $-1$ | $-1$ | 1 |

TABLE No.2:

| Vertex | : | 0 | 1 | 2 | 3 | 4 | 5 | 6 |
|---|---|---|---|---|---|---|---|---|
| Known | : | 1 | 1 | 1 | 0 | 0 | 0 | 0 |
| Distance | : | 0 | 7 | 9 | 11 | ∞ | ∞ | 10 |
| Path | : | −1 | 0 | 1 | 2 | −1 | −1 | 1 |

TABLE No.3:

| Vertex | : | 0 | 1 | 2 | 3 | 4 | 5 | 6 |
|---|---|---|---|---|---|---|---|---|
| Known | : | 1 | 1 | 1 | 0 | 0 | 0 | 1 |
| Distance | : | 0 | 7 | 9 | 11 | 11 | ∞ | 10 |
| Path | : | −1 | 0 | 1 | 2 | 6 | −1 | 1 |

TABLE No.4:

| Vertex | : | 0 | 1 | 2 | 3 | 4 | 5 | 6 |
|---|---|---|---|---|---|---|---|---|
| Known | : | 1 | 1 | 1 | 1 | 0 | 0 | 1 |
| Distance | : | 0 | 7 | 9 | 11 | 11 | ∞ | 10 |
| Path | : | −1 | 0 | 1 | 2 | 6 | −1 | 1 |

TABLE No.5:

| Vertex | : | 0 | 1 | 2 | 3 | 4 | 5 | 6 |
|---|---|---|---|---|---|---|---|---|
| Known | : | 1 | 1 | 1 | 1 | 1 | 0 | 1 |
| Distance | : | 0 | 7 | 9 | 11 | 11 | 17 | 10 |
| Path | : | −1 | 0 | 1 | 2 | 6 | 4 | 1 |

TABLE No.6:

| Vertex | : | 0 | 1 | 2 | 3 | 4 | 5 | 6 |
|---|---|---|---|---|---|---|---|---|
| Known | : | 1 | 1 | 1 | 1 | 1 | 1 | 1 |
| Distance | : | 0 | 7 | 9 | 11 | 11 | 17 | 10 |
| Path | : | −1 | 0 | 1 | 2 | 6 | 4 | 1 |

**Shortest paths**

Path=0 Distance=0

Path=0 1 Distance=7

Path=0 1 2 Distance=9

Path=0 1 2 3 Distance=11

Path=0 1 6 4 Distance=11

Path=0 1 6 4 5 Distance=17

Path=0 1 6 Distance=10

**Spanning tree**

(0, 1) (1, 2) (1, 6) (2, 3) (4, 5) (6, 4)

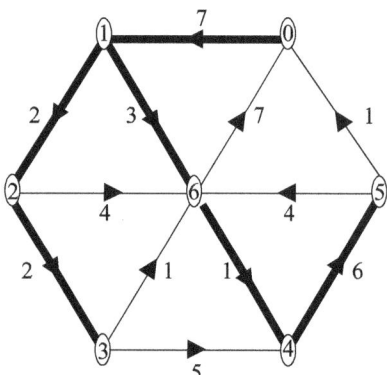

**Fig. 13.4** *Shortest paths from vertex 0 and a spanning tree*

The output of Dijkstra's algorithm for the graph of Figure 13.3 with starting vertex 0 is shown in Figure 13.4. In this case, the spanning tree has total weight 21 and is *not* minimal. Indeed, it is possible to find a spanning tree of total weight 19 in this graph, as the next section will show.

## 13.2  Prim's Algorithm for Minimal Spanning Tree

Almost simultaneously in 1957, while Dijkstra's algorithm was just becoming widely known, R.C. Prim [211, 249] of Bell Labs discovered an algorithm for finding a minimal spanning tree in a weighted graph. Surprisingly, a minor modification of Dijkstra's algorithm allows us to find a minimal spanning tree. A formal description of Prim's algorithm is the following.

### 13.2.1 Prim's algorithm to compute a minimal spanning tree from u

**Input**    A weighted graph and a starting vertex $u$. The weight of edge $xy$ is $w(xy)$; let $w(xy) = \infty$ if $xy$ is not an edge.

**Idea**    Maintain the set $S$ of vertices to which the shortest path from $u$ is known, enlarging $S$ to include all the vertices. To do this maintain also a tentative distance $t(z)$ from $u$ to each $z$ not in $S$; this is the length of the shortest path found yet from $u$ to $z$.

**Initialisation**    Let $S = \{u\}$; $d(u, u) = 0$; $t(z) = w(uz)$ for all $z \neq u$.

**Iteration**    Select a vertex $v$ outside $S$ such that $t(v)$ is minimum in the set $\{t(z)|z \notin S\}$. Add $v$ to $S$. For each edge $vz$ with $z \notin S$, update $t(z)$ to $\min\{t(z), w(vz)\}$. ( Note this step is different from the corresponding step in Dijkstra's algorithm).

**Termination**    Continue the iteration until $S = V(G)$ or until $t(z) = \infty$ for every $z \notin S$. In the first case, all shortest paths from $u$ have been found. Together, they yield a minimal spanning tree. In the latter case, the remaining vertices are unreachable from $u$ and the shortest paths together will not span the graph.

We give an implementation of Prim's algorithm in C++ below.

**prim.cpp**

```cpp
#include <iostream>
#include <fstream>
#include <string>
#include <vector>
#include <set>

using namespace std;

ifstream infile(''graph.txt'');
ofstream outfile(''minimal_spanning_tree.txt'');

int main()
{
 cout<<''Prim's Algorithm.''<<endl;
 int m,n,i,j;
 //Read adjacency matrix of weights from graph.txt
 infile>>n;
```

```cpp
vector< vector<float> > weight; {float }val;
for(i=0; i<n; i++)
{
 vector<{float}> row;
  for(j=0; j<n; j++)
  {
   infile>>val;
   row.push_back(val);
  }
  weight.push_back(row);
}
//Initialize Table
const float infinity=1000000;
vector<bool> known;
for(i=0; i<n; i++) known.push_back(false);
vector<float> d;
d.push_back(0);
for(i=1; i<n; i++) d.push_back(infinity);
vector<int> p;
for(i=0; i<n; i++) p.push_back(-1);
//Print Table
 outfile<<endl<<''INITIAL TABLE:''<<endl;
 outfile<<endl<<''Vertex  :\t'';
 for(i=0; i<n; i++) outfile<<i<<'\t';
 outfile<<endl<<''Known   :\t'';
 for(i=0; i<n; i++) outfile<<known[i]<<'\t';
 outfile<<endl<<''Distance:\t'';
 for(i=0; i<n; i++) outfile<<d[i]<<'\t';
 outfile<<endl<<''Path    :\t'';
 for(i=0; i<n; i++) outfile<<p[i]<<'\t';
 outfile<<endl;
//Iteration
for(m=0; m<n; m++)
{
 //Find min of d for unknown vertices
 int min=0;
 while(known[min]==true)min++;
```

```
for(i=0; i<n; i++)
if(known[i]==false && d[i]<d[min])min=i;
//Update Table
known[min]=true;
for(j=0; j<n; j++)
{
  if(weight[min][j]!=0 &&
   d[j]>weight[min][j] &&
   known[j]==false)
  {
   d[j]=weight[min][j];
   p[j]=min;
  }
}
//Print Table
outfile<<endl<<endl<<''TABLE No.''<<m<<'':''<<endl;
outfile<<endl<<''Vertex  :\t'';
for(i=0; i<n; i++) outfile<<i<<'\t';
outfile<<endl<<''Known   :\t'';
for(i=0; i<n; i++) outfile<<known[i]<<'\t';
outfile<<endl<<''Distance:\t'';
for(i=0; i<n; i++) outfile<<d[i]<<'\t';
outfile<<endl<<''Path    :\t'';
for(i=0; i<n; i++) outfile<<p[i]<<'\t';
outfile<<endl;
}
//Print minimal spanning tree
outfile<<endl<<''MINIMAL SPANNING TREE:''<<endl;
for(i=1; i<n; i++)
outfile<<''(''<<i<<'',''<<p[i]<<'') '';
cout<<''See minimal spanning tree.txt.''<<endl;
system(''PAUSE'');
return 0;
}
```

**Example 13.2.1**    Consider the weighted graph shown in Figure 13.5.

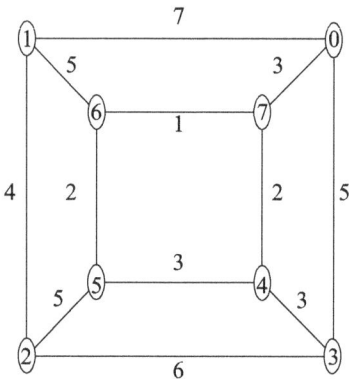

**Fig. 13.5** *A weighted graph*

Make a text file "graph.txt" as shown below and save it in the same directory as the program "prim.cpp". The first line of the file "graph.txt" is the number of vertices and the rest is the matrix of weights of the labelled graph shown in Figure 13.5.

**graph.txt**

```
8
0   7   0   5   0   0   0   3
7   0   4   0   0   0   5   0
0   4   0   6   0   5   0   0
5   0   6   0   3   0   0   0
0   0   0   3   0   3   0   2
0   0   5   0   3   0   2   0
0   5   0   0   0   2   0   1
3   0   0   0   2   0   1   0
```

Now compile and run the program "prim.cpp". The following output file "minimal_spanning_tree.txt" is produced in the program's directory. The output shows a minimal spanning tree for the graph.

**minimal_spanning_tree.txt**

INITIAL TABLE:

| Vertex | : | 0 | 1 | 2 | 3 | 4 | 5 | 6 | 7 |
|---|---|---|---|---|---|---|---|---|---|
| Known | : | 0 | 0 | 0 | 0 | 0 | 0 | 0 | 0 |
| Distance | : | ∞ | ∞ | ∞ | ∞ | ∞ | ∞ | ∞ | ∞ |
| Path | : | −1 | −1 | −1 | −1 | −1 | −1 | −1 | −1 |

TABLE No.0:

| Vertex | : | 0 | 1 | 2 | 3 | 4 | 5 | 6 | 7 |
|---|---|---|---|---|---|---|---|---|---|
| Known | : | 1 | 0 | 0 | 0 | 0 | 0 | 0 | 0 |
| Distance | : | 0 | 7 | ∞ | 5 | ∞ | ∞ | ∞ | 3 |
| Path | : | −1 | 0 | −1 | 0 | −1 | −1 | −1 | 0 |

TABLE No.1:

| Vertex | : | 0 | 1 | 2 | 3 | 4 | 5 | 6 | 7 |
|---|---|---|---|---|---|---|---|---|---|
| Known | : | 1 | 0 | 0 | 0 | 0 | 0 | 0 | 1 |
| Distance | : | 0 | 7 | ∞ | 5 | 2 | ∞ | 1 | 3 |
| Path | : | −1 | 0 | −1 | 0 | 7 | −1 | 7 | 0 |

TABLE No.2:

| Vertex | : | 0 | 1 | 2 | 3 | 4 | 5 | 6 | 7 |
|---|---|---|---|---|---|---|---|---|---|
| Known | : | 1 | 0 | 0 | 0 | 0 | 0 | 1 | 1 |
| Distance | : | 0 | 5 | ∞ | 5 | 2 | 2 | 1 | 3 |
| Path | : | −1 | 6 | −1 | 0 | 7 | 6 | 7 | 0 |

TABLE No.3:

| Vertex | : | 0 | 1 | 2 | 3 | 4 | 5 | 6 | 7 |
|---|---|---|---|---|---|---|---|---|---|
| Known | : | 1 | 0 | 0 | 0 | 1 | 0 | 1 | 1 |
| Distance | : | 0 | 5 | ∞ | 3 | 2 | 2 | 1 | 3 |
| Path | : | −1 | 6 | −1 | 4 | 7 | 6 | 7 | 0 |

TABLE No.4:

| Vertex | : | 0 | 1 | 2 | 3 | 4 | 5 | 6 | 7 |
|---|---|---|---|---|---|---|---|---|---|
| Known | : | 1 | 0 | 0 | 0 | 1 | 1 | 1 | 1 |
| Distance | : | 0 | 5 | 5 | 3 | 2 | 2 | 1 | 3 |
| Path | : | −1 | 6 | 5 | 4 | 7 | 6 | 7 | 0 |

TABLE No.5:

| Vertex   | : | 0  | 1 | 2 | 3 | 4 | 5 | 6 | 7 |
|----------|---|----|---|---|---|---|---|---|---|
| Known    | : | 1  | 0 | 0 | 1 | 1 | 1 | 1 | 1 |
| Distance | : | 0  | 5 | 5 | 3 | 2 | 2 | 1 | 3 |
| Path     | : | −1 | 6 | 5 | 4 | 7 | 6 | 7 | 0 |

TABLE No.6:

| Vertex   | : | 0  | 1 | 2 | 3 | 4 | 5 | 6 | 7 |
|----------|---|----|---|---|---|---|---|---|---|
| Known    | : | 1  | 1 | 0 | 1 | 1 | 1 | 1 | 1 |
| Distance | : | 0  | 5 | 4 | 3 | 2 | 2 | 1 | 3 |
| Path     | : | −1 | 6 | 1 | 4 | 7 | 6 | 7 | 0 |

TABLE No.7:

| Vertex   | : | 0  | 1 | 2 | 3 | 4 | 5 | 6 | 7 |
|----------|---|----|---|---|---|---|---|---|---|
| Known    | : | 1  | 1 | 1 | 1 | 1 | 1 | 1 | 1 |
| Distance | : | 0  | 5 | 4 | 3 | 2 | 2 | 1 | 3 |
| Path     | : | −1 | 6 | 1 | 4 | 7 | 6 | 7 | 0 |

**Minimal spanning tree**

[(1, 6) (2, 1) (3, 4) (4, 7) (5, 6) (6, 7) (7, 0)

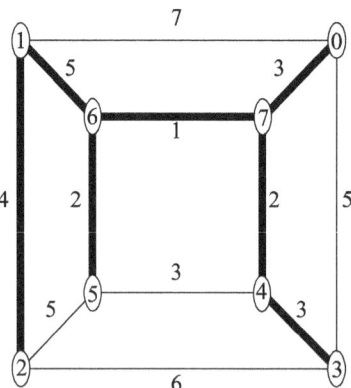

**Fig. 13.6** *A minimal spanning tree for the weighted graph*

The output of Prim's algorithm for the graph of Figure 13.5 with starting vertex 0 is shown in Figure 13.6. The spanning tree has total weight 20 and is of minimal total weight compared to all other spanning trees.

**Example 13.2.1**  It makes sense to try to compute a minimal spanning tree for a directed graph. Consider the weighted directed graph of Figure 13.3. We claimed that the spanning tree found by Dijkstra's algorithm of total weight 21 is *not* minimal. Let us run the program for Prim's algorithm on the file graph.txt of Example 13.1.3. We obtain the following output.

**minimal_spanning_tree.txt**

INITIAL TABLE:

| Vertex   | : | 0  | 1  | 2  | 3  | 4  | 5  | 6  |
|----------|---|----|----|----|----|----|----|----|
| Known    | : | 0  | 0  | 0  | 0  | 0  | 0  | 0  |
| Distance | : | 0  | ∞  | ∞  | ∞  | ∞  | ∞  | ∞  |
| Path     | : | −1 | −1 | −1 | −1 | −1 | −1 | −1 |

TABLE No.0:

| Vertex   | : | 0  | 1  | 2  | 3  | 4  | 5  | 6  |
|----------|---|----|----|----|----|----|----|----|
| Known    | : | 1  | 0  | 0  | 0  | 0  | 0  | 0  |
| Distance | : | 0  | 7  | ∞  | ∞  | ∞  | ∞  | ∞  |
| Path     | : | −1 | 0  | −1 | −1 | −1 | −1 | −1 |

TABLE No.1:

| Vertex   | : | 0  | 1  | 2  | 3  | 4  | 5  | 6  |
|----------|---|----|----|----|----|----|----|----|
| Known    | : | 1  | 1  | 0  | 0  | 0  | 0  | 0  |
| Distance | : | 0  | 7  | 2  | ∞  | ∞  | ∞  | 3  |
| Path     | : | −1 | 0  | 1  | −1 | −1 | −1 | 1  |

TABLE No.2:

| Vertex   | : | 0  | 1  | 2  | 3  | 4  | 5  | 6  |
|----------|---|----|----|----|----|----|----|----|
| Known    | : | 1  | 1  | 1  | 0  | 0  | 0  | 0  |
| Distance | : | 0  | 7  | 2  | 2  | ∞  | ∞  | 3  |
| Path     | : | −1 | 0  | 1  | 2  | −1 | −1 | 1  |

TABLE No.3:

| Vertex   | : | 0  | 1  | 2  | 3  | 4  | 5  | 6  |
|----------|---|----|----|----|----|----|----|----|
| Known    | : | 1  | 1  | 1  | 1  | 0  | 0  | 0  |
| Distance | : | 0  | 7  | 2  | 2  | 5  | ∞  | 1  |
| Path     | : | −1 | 0  | 1  | 2  | 3  | −1 | 3  |

TABLE No.4:

| Vertex | : | 0 | 1 | 2 | 3 | 4 | 5 | 6 |
|---|---|---|---|---|---|---|---|---|
| Known | : | 1 | 1 | 1 | 1 | 0 | 0 | 1 |
| Distance | : | 0 | 7 | 2 | 2 | 1 | ∞ | 1 |
| Path | : | −1 | 0 | 1 | 2 | 6 | −1 | 3 |

TABLE No.5:

| Vertex | : | 0 | 1 | 2 | 3 | 4 | 5 | 6 |
|---|---|---|---|---|---|---|---|---|
| Known | : | 1 | 1 | 1 | 1 | 1 | 0 | 1 |
| Distance | : | 0 | 7 | 2 | 2 | 1 | 6 | 1 |
| Path | : | −1 | 0 | 1 | 2 | 6 | 4 | 3 |

TABLE No.6:

| Vertex | : | 0 | 1 | 2 | 3 | 4 | 5 | 6 |
|---|---|---|---|---|---|---|---|---|
| Known | : | 1 | 1 | 1 | 1 | 1 | 1 | 1 |
| Distance | : | 0 | 7 | 2 | 2 | 1 | 6 | 1 |
| Path | : | −1 | 0 | 1 | 2 | 6 | 4 | 3 |

**Minimal spanning tree**

(1, 0) (2, 1) (3, 2) (4, 6) (5, 4) (6, 3)

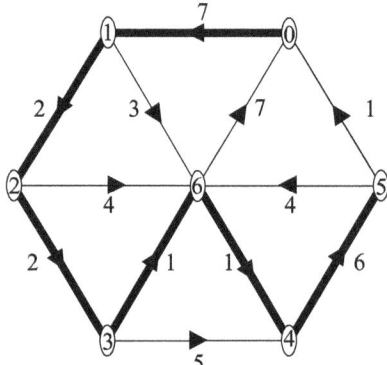

**Fig. 13.7** *A minimal spanning tree for the weighted directed graph*

The output of Prim's algorithm for the directed weighted graph of Figure 13.3 with starting vertex 0 is shown in Figure 13.7. The spanning tree has total weight 19 and is of minimal total weight compared to all other spanning trees, in particular, the spanning tree of total weight 21 found by Dijkstra's algorithm.

## 13.3   Fleury's Algorithm for Eulerian Circuit

The problem of finding an Eulerian circuit in a graph (possibly with multiple edges) has been studied since Leonhard Euler's solution [74] to the problem of the seven bridges of Königsberg in 1736 (see Chapter 3). Lewis Carroll [263], we are told in a biography by his nephew, was fond of asking little children to draw, in one stroke, without lifting the pen off the paper, Figure 13.8.

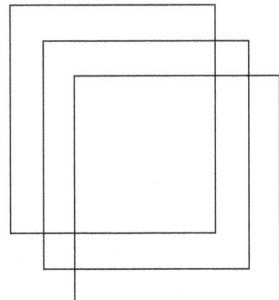

**Fig. 13.8** *Lewis Carroll's three-square graph*

If we define all intersections of line segments as vertices then we obtain a planar graph with 18 vertices, known as Lewis Carroll's three-square graph. Drawing the figure in one stroke is equivalent to finding an Eulerian circuit in the graph. Since all vertices have even degree 2 or 4, we know that there *must* be such an Eulerian circuit. In Example 13.3.1, we show how to find this Eulerian circuit in a systematic way. Lucas [153] describes an algorithm for finding an Eulerian trail (or circuit, if one exists) due to Fleury.

### 13.3.1   Fleury's algorithm

**Input**   A graph $G$ with one nontrivial component and at most two odd vertices.

**Initialisation**   If G has an odd vertex, start at an odd vertex. Else start at any vertex.

**Iteration**   From the current vertex, traverse any remaining edge whose deletion from the remaining graph does not leave a graph with two nontrivial components. To check this, run Prim's algorithm and make sure that the resulting tree spans the graph.

**Termination**   Stop when there are no more edges left to traverse.

We give an implementation of Fleury's algorithm in C++ below. Note that this implementation is for simple graphs, but we can easily modify it to handle graphs with multiple edges.

## fleury.cpp

```cpp
#include <iostream>
#include <fstream>
#include <string>
#include <vector>

using namespace std;

ifstream infile(''graph.txt'');
ofstream outfile(''eulerian_circuit.txt'');
bool euler(vector<vector<int> > edge);
bool fleury(vector<vector<int> > edge, vector<int> del);
vector<vector<int> >erase(vector<vector<int> > edge, vector<int> del );
bool empty(vector<vector<int> > edge);

int main()
 {
 cout<<''Fleury's Algorithm.''<<endl;
 int n, i, j;
 //Read adjacency matrix of edges from graph.txt
 infile>>n;
 vector<vector<int> > edge; int val;
 for(i=0; i<n; i++)
 {
  vector<int> row;
  for(j=0; j<n; j++)
  {
   infile>>val; row.push_back(val);
  }
  edge.push_back(row);
 }
 cout<<''Read graph from file graph.txt...''<<endl;
 if(euler(edge))
 {
 cout<<''Finding Eulerian circuit...''<<endl;
 vector<int> circuit; int current=0;
```

```
circuit.push_back(current); cout<<current<<'' ";
while(!empty(edge))
{
 for(i=0; i<n; i++)
 {
  int previous=current;
  if(edge[current][i]==1)
  {
   vector<int> del;
   del.push_back(current);
   del.push_back(i);
   if(fleury(edge, del))
   {
    edge=erase(edge,del); current=i;
    circuit.push_back(current);
    cout<<current<<" ";
    break;
   }
  }
 }
}
for(i=0; i<circuit.size(); i++)
outfile<<circuit[i]<<" ";
 cout<<endl<<''See circuit.txt for results.''<<endl;
}
 else
 cout<<''No Eulerian circuit.''<<endl;
 system(''PAUSE''); return 0;
}

bool euler(vector<vector<int> > edge)
{
 for(int i=0; i<edge.size(); i++)
 {
  int deg=0;
  for(int j=0; j<edge[0].size(); j++)
  deg+=edge[i][j];
```

```
  if(deg%2!=0) return false;
 }
 return true;
}
bool fleury(vector<vector<int> > edge, vector<int> del )
{
 int n, i, j, k;
 if(del[0]==del[1]) return false;
 vector<vector<int> > edged=edge;
 edged[del[0]][del[1]]=0; edged[del[1]][del[0]]=0;
 n= edged[0].size();
 //Initialize Table
 const int infinity=1000000;
 vector<bool> known;
 for(i=0; i<n; i++) known.push_back(false);
 vector<int> d; d.push_back(0);
 for(i=1; i<n; i++) d.push_back(infinity);
 vector<int> p;
 for(i=0; i<n; i++) p.push_back(-1);
 //Iteration
 for(k=0; k<n; k++)
 {
  //Find min of d for unknown vertices
  int min=0;
  while(known[min]==true)min++;
  for(i=0; i<n; i++)
  if(known[i]==false && d[i]<d[min])min=i;
  //Update Table
  known[min]=true;
  for(j=0; j<n; j++)
  {
   if(edged[min][j]!=0 && d[j]>edged[min][j] &&
     known[j]==false)

   {
    d[j]=edged[min][j]; p[j]=min;
   }
```

```
 }
}
bool ok=true;
//Find if resulting graph has two nontrivial //components
it for(i=1; i<n; i++)
{
 if(p[i]==-1)
 for (int j=0; j<n; j++)
 if(edged[i][j]!=0)
 {ok=false; break;}
}
 return ok;
}
vector<vector<int> > erase(vector< vector<int> > edge, vector<int> del )
{
 vector< vector<int> > edged=edge;
 edged[del[0]][del[1]]=0; edged[del[1]][del[0]]=0;
 return edged;
}
bool empty(vector<vector<int> > edge)
{
for(int i=0; i<edge.size(); i++)
for(int j=0; j<edge[0].size(); j++)
if(edge[i][j]==1)
return false; return true;
}
```

**Example 13.3.1**  Consider the following labelled graph given in Figure 13.9 corresponding to Lewis Carroll's three-square puzzle.

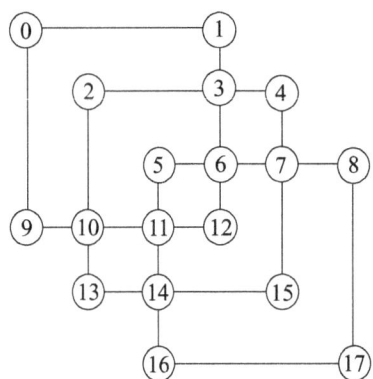

**Fig. 13.9** *Labelled graph corresponding to Lewis Carroll's puzzle*

Make a text file "graph.txt" corresponding to Figure 13.9.

**graph.txt**

```
18
0 1 0 0 0 0 0 0 0 1 0 0 0 0 0 0 0 0
1 0 0 1 0 0 0 0 0 0 0 0 0 0 0 0 0 0
0 0 0 1 0 0 0 0 0 0 1 0 0 0 0 0 0 0
0 1 1 0 1 0 1 0 0 0 0 0 0 0 0 0 0 0
0 0 0 1 0 0 0 1 0 0 0 0 0 0 0 0 0 0
0 0 0 0 0 0 1 0 0 0 0 1 0 0 0 0 0 0
0 0 0 1 0 1 0 1 0 0 0 0 1 0 0 0 0 0
0 0 0 0 1 0 1 0 1 0 0 0 0 0 0 1 0 0
0 0 0 0 0 0 0 1 0 0 0 0 0 0 0 0 0 1
1 0 0 0 0 0 0 0 0 0 1 0 0 0 0 0 0 0
0 0 1 0 0 0 0 0 0 1 0 1 0 1 0 0 0 0
0 0 0 0 0 1 0 0 0 0 1 0 1 0 1 0 0 0
0 0 0 0 0 0 1 0 0 0 0 1 0 0 0 0 0 0
0 0 0 0 0 0 0 0 0 0 1 0 0 0 1 0 0 0
0 0 0 0 0 0 0 0 0 0 0 1 0 1 0 1 1 0
0 0 0 0 0 0 0 1 0 0 0 0 0 0 1 0 0 0
0 0 0 0 0 0 0 0 0 0 0 0 0 0 1 0 0 1
0 0 0 0 0 0 0 0 1 0 0 0 0 0 0 0 1 0
```

Now compile and run the program "fleury.cpp". The following output file "eulerian_circuit.txt" is produced in the program's directory.

0 1 3 2 10 11 5 6 3 4 7 6 12 11 14 15 7 8 17 16 14 13 10 9 0 The output shows the final Eulerian circuit in the given graph. If we trace the actual computation of this Eulerian circuit step by step, we obtain a solution to Carroll's puzzle, as shown in Figure 13.10.

**Fig. 13.10** *Solution to Lewis Carroll's three-square puzzle*

## 13.4 De Bruijn Graphs

To see a good application of Eulerian circuits, we briefly consider the *rotating drum problem*, described as follows. Suppose the head of a rotating drum is divided into $2^4 = 16$ sectors, with each sector labelled with either a zero or a one, as shown in Figure 13.11.

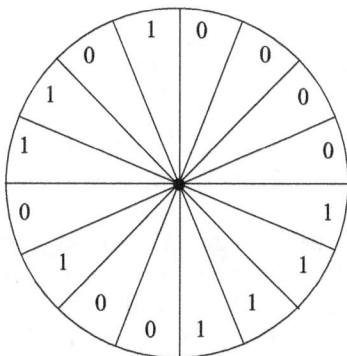

**Fig. 13.11** *Labelled rotating drum*

Can the sectors be labelled in such a way that the labels of any four consecutive sectors uniquely determine the position of the drum? This means that the 16 possible quadruples of consecutive binary labels on the drum should be the binary representations of the integers 0 to 15. This question was studied by N.G. de Bruijn [41] in 1946 and thus the resulting binary circular sequences and their corresponding graphs given below are called *De Bruijn sequences* and *De Bruijn graphs*, respectively.

We define a directed graph called $G_4$ (the De Bruijn graph of order 4) as follows. The vertices are all 3-bit binary strings $x_1x_2x_3$ (each $x_i$ is either zero or one). Thus there are 8 vertices. Each vertex $x_1x_2x_3$ has directed edges to the vertices $x_2x_30$ and $x_2x_31$. The directed edge $(x_1x_2x_3, x_2x_3x_4)$ is labelled $e_j$, where $j = x_1 2^3 + x_2 2^2 + x_3 2^1 + x_4 2^0$ is the unique binary representation of the integer $j$. Thus, there are 16 directed edges. The graph is shown in the following Figure 13.12.

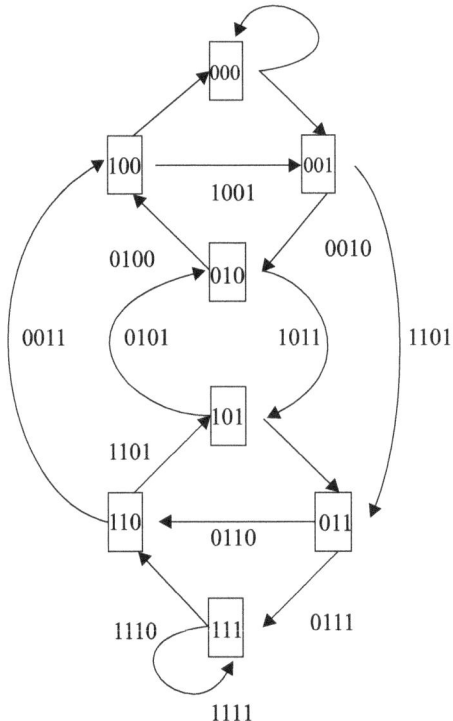

**Fig. 13.12** $G_4$, *the De Bruijn graph of order 4*

Now, $G_4$ is Eulerian since the in-degree and out-degree of every vertex is 2. Using Fleury's algorithm we find the following Eulerian circuit in $G_4$: 000, 000, 001, 011, 111, 111, 110, 100, 001, 010, 101, 011, 110, 101, 010, 100, 000. This Eulerian circuit corresponds to the De Bruijn sequence 0000111100101101, to be read circularly, with the required property. This is exactly the labeling on the rotating drum shown in Figure 13.11.

In general, we may build the De Bruijn graph of order $n$, denoted by $G_n$, and find circular De Bruijn sequences of length $n$ corresponding to Eulerian circuits in $G_n$. Among other things, this technique has recently been used to design large scale multihop and fault tolerant computer networks [239].

## 13.5  Hamiltonian Circuits

A concept that is similar to Eulerian circuits but in reality quite different is that of Hamiltonian circuit. A hamiltonian circuit in a graph is a simple closed path that passes through each vertex exactly once. In the mid 19th century, Sir William Rowan Hamilton [102] tried to popularise the exercise of finding such a circuit in the graph of a dodecahedron. While we have seen a polynomial-time algorithm for Eulerian circuits, no such algorithm is known for Hamiltonian circuits. We give an implementation of a nondeterministic algorithm in C++ below.

**hamilton.cpp**

```
#include <iostream>
#include <fstream>
#include <string>
#include <vector>
#include <algorithm>

using namespace std;

ifstream infile(``graph.txt'');
ofstream outfile(``hamiltonian_circuits.txt'');
int main()
 {
  cout<<``Algorithm for Hamiltonian circuits.''<<endl;
  //Read adjacency matrix of from graph.txt
  int i, j, k, l, m, n;
  infile>>n;
  vector< vector<bool> > graph;
  bool edge;
  for(i=0; i<n; i++)
  {
   vector<bool> row;
```

```
  for(j=0; j<n; j++)
  {
   infile>>edge;
   row.push_back(edge);
  }
  graph.push_back(row);
 }
 int count=0;
 vector<int> vertex;
 for(k=0; k<n; k++)
 vertex.push_back(k);
 cout<<''\nStarting search...\n'';
 bool found=false, circuit=true;
while(next_permutation(vertex.begin(),vertex.end())
)
 {
  if(vertex[0]!=0) break;
  for(l=0; l<n; l++)
   circuit=
     circuit*graph[vertex[l%n]][vertex[(l+1)%n]];
  switch(circuit)
  {
   case true:
    found=true;
    count++;
    cout<<endl
      <<count
      <<'' Hamiltonian Circuits found:''<<endl;
    outfile<<endl
          <<count
          <<'' Hamiltonian Circuits found:''
          <<endl;
   for(m=0; m<n; m++)
   {
    cout<<vertex[m]<<'' '';
    outfile<<vertex[m]<<'' '';
   }
```

```
   cout<<endl; outfile<<endl;
   break;
   default:
   break;
  }
 circuit=true;
}
if(!found)
{
 cout<<''\nNo Hamiltonian Circuits found.\n'';
 outfile<<''\nNo Hamiltonian Circuits found.\n'';
}
cout<<''\nSee hamilton_circuits.txt.''
    <<endl;
 system(''PAUSE'');
 return 0;
}
```

**Example 13.5.1**   We consider the graphs of the five platonic solids, all of which are known to be Hamiltonian. First, consider the simplest of the platonic graphs, the tetrahedron:

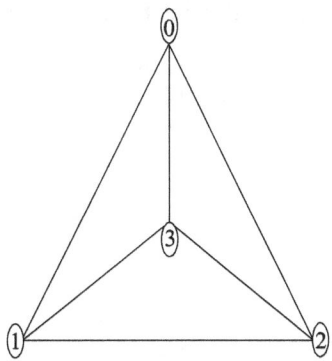

**Fig 13.13** *Labelled graph of the tetrahedron*

Make a text file "graph.txt" as shown below and save it in the same directory as the program "hamilton.cpp". The first line of the file "graph.txt" is the number of vertices and the rest is the adjacency matrix of the labelled tetrahedron graph shown in Figure 13.13.

**graph.txt**

```
4
0   1   1   1
1   0   1   1
1   1   0   1
1   1   1   0
```

Now compile and run the program. We find five Hamiltonian circuits: 0 1 3 2, 0 2 1 3, 0 2 3 1, 0 3 1 2, 0 3 2 1. Next, consider the graph of the octahedron:

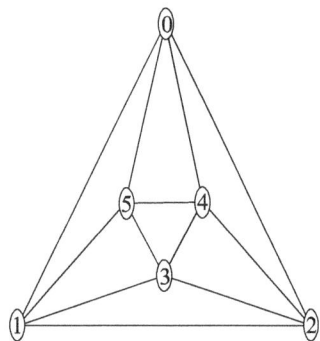

**Fig. 13.14** *Labelled graph of the octahedron*

The file "graph.txt" for the octahedron graph is shown below:

**graph.txt**

```
6
0   1   1   0   1   1
1   0   1   1   0   1
1   1   0   1   1   0
0   1   1   0   1   1
1   0   1   1   0   1
1   1   0   1   1   0
```

Now run the program. We find 31 Hamiltonian circuits: 0 1 2 3 5 4, 0 1 2 4 3 5, 0 1 3 2 4 5, 0 1 3 5 4 2, 0 1 5 3 2 4, 0 1 5 3 4 2, 0 1 5 4 3 2, 0 2 1 3 4 5, 0 2 1 3 5 4, 0 2 1 5 3 4, 0 2 3 1 5 4, 0 2 3 4 5 1, 0 2 4 3 1 5, 0 2 4 3 5 1, 0 2 4 5 3 1, 0 4 2 1 3 5, 0 4 2 3 1 5, 0 4 2 3 5 1, 0 4 3 2 1 5, 0 4 3 5 1 2, 0 4 5 1 3 2, 0 4 5 3 1 2, 0 4 5 3 2 1, 0 5 1 2 3 4, 0 5 1 3 2 4, 0 5 1 3 4 2, 0 5 3 1 2 4, 0 5 3 4 2 1, 0 5 4 2 3 1, 0 5 4 3 1 2, 0 5 4 3 2 1.

Next, consider the graph of the cube.

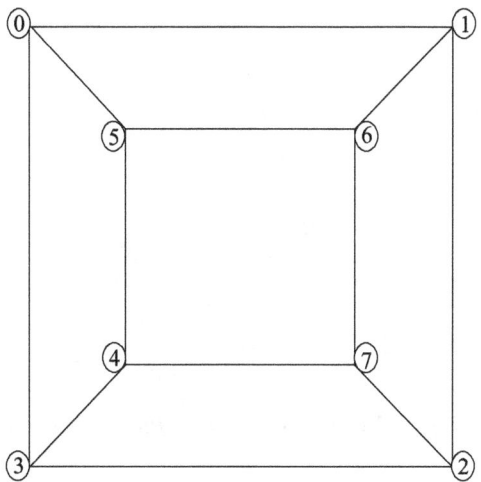

**Fig. 13.15** *Labelled graph of the cube*

**graph.txt**

```
8
0   1   0   1   0   1   0   0
1   0   1   0   0   0   1   0
0   1   0   1   0   0   0   1
1   0   1   0   1   0   0   0
0   0   0   1   0   1   0   1
1   0   0   0   1   0   1   0
0   1   0   0   0   1   0   1
0   0   1   0   1   0   1   0
```

Now run the program. We find 12 Hamiltonian circuits: 0 1 2 3 4 7 6 5, 0 1 2 7 6 5 4 3, 0 1 6 5 4 7 2 3, 0 1 6 7 2 3 4 5, 0 3 2 1 6 7 4 5, 0 3 2 7 4 5 6 1, 0 3 4 5 6 7 2 1, 0 3 4 7 2 1 6 5, 0 5 4 3 2 7 6 1, 0 5 4 7 6 1 2 3, 0 5 6 1 2 7 4 3, 0 5 6 7 4 3 2 1. Next, consider the graph of the icosahedrons, given in Figure 13.16.

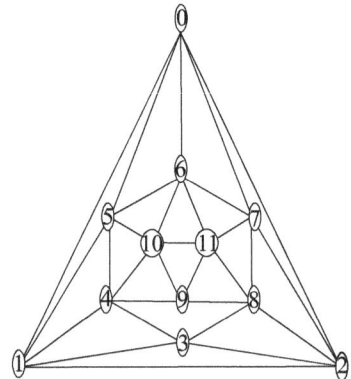

**Fig. 13.16** *Labelled graph of the icosahedron*

**graph.txt**

---

```
12
0  1  1  0  0  1  1  1  0  0  0  0
1  0  1  1  1  1  0  0  0  0  0  0
1  1  0  1  0  0  0  1  1  0  0  0
0  1  1  0  1  0  0  0  1  1  0  0
0  1  0  1  0  1  0  0  0  1  1  0
1  1  0  0  1  0  1  0  0  0  1  0
1  0  0  0  0  1  0  1  0  0  1  1
1  0  1  0  0  0  1  0  1  0  0  1
0  0  1  1  0  0  0  1  0  1  0  1
0  0  0  1  1  0  0  0  1  0  1  1
0  0  0  0  1  1  1  0  0  1  0  1
0  0  0  0  0  0  1  1  1  1  1  0
```

---

Now, it gets interesting. The program runs for quite a few minutes! We find 2560 Hamiltonian circuits: 0 1 2 3 4 5 6 10 9 8 11 7, ..., 0 7 11 10 9 8 3 2 1 4 5 6. Finally, consider the graph of the dodecahedron, the original inspiration for Hamilton [212], given in Figure 13.17.

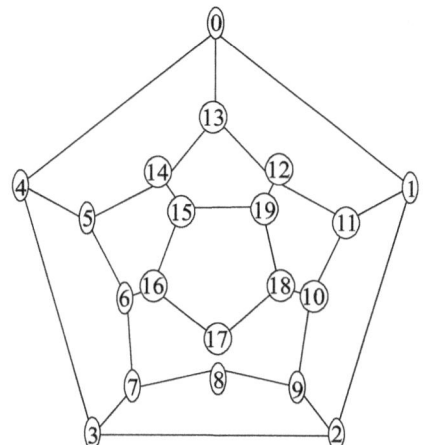

**Fig. 13.17** *Labelled graph of the dodecahedron*

The file "graph.txt" for the dodecahedron graph is shown below:

**graph.txt**

```
20
0  1  0  0  1  1  0  0  0  0  0  0  0  0  0  0  0  0  0  0
1  0  1  0  0  0  1  0  0  0  0  0  0  0  0  0  0  0  0  0
0  1  0  1  0  0  0  1  0  0  0  0  0  0  0  0  0  0  0  0
0  0  1  0  1  0  0  0  1  0  0  0  0  0  0  0  0  0  0  0
1  0  0  1  0  0  0  0  0  1  0  0  0  0  0  0  0  0  0  0
1  0  0  0  0  0  0  0  0  0  1  1  0  0  0  0  0  0  0  0
0  1  0  0  0  0  0  0  0  0  0  1  1  0  0  0  0  0  0  0
0  0  1  0  0  0  0  0  0  0  0  0  1  1  0  0  0  0  0  0
0  0  0  1  0  0  0  0  0  0  0  0  0  1  1  0  0  0  0  0
0  0  0  0  1  0  0  0  0  0  1  0  0  0  1  0  0  0  0  0
0  0  0  0  0  1  0  0  0  1  0  0  0  0  0  0  0  0  0  1
0  0  0  0  0  1  1  0  0  0  0  0  0  0  0  1  0  0  0  0
0  0  0  0  0  0  1  1  0  0  0  0  0  0  0  0  1  0  0  0
0  0  0  0  0  0  0  1  1  0  0  0  0  0  0  0  0  1  0  0
0  0  0  0  0  0  0  0  1  1  0  0  0  0  0  0  0  0  1  0
0  0  0  0  0  0  0  0  0  0  1  1  0  0  0  0  1  0  0  1
0  0  0  0  0  0  0  0  0  0  0  1  1  0  0  1  0  1  0  0
0  0  0  0  0  0  0  0  0  0  0  0  1  1  0  0  1  0  1  0
0  0  0  0  0  0  0  0  0  0  0  0  0  1  1  0  0  1  0  1
0  0  0  0  0  0  0  0  0  0  1  0  0  0  0  1  0  0  1  0
```

Depending on the speed of your computer, the program will now run for quite a few hours, and perhaps even days. One of the Hamiltonian circuits we found was: 0 1 2 3 4 5 6 7 8 9 10 11 12 19 18 17 16 15 14 13.

# 14. Score Structure in Digraphs

Landau [145] associated with each tournament an ordered sequence of non-negative integers, its score structure, formed by listing the vertex outdegrees in non-decreasing order. Since then the concept of score structure has been extended to various other classes of digraphs, namely oriented graphs and semicomplete graphs. The score structure property has been of great help in the study of some structural properties of digraphs.

## 14.1 Score Sequences in Tournaments

**Definition:** In a tournament $T$, the score $s(v_i)$, or simply $s_i$ of a vertex $v_i$ is the number of arcs directed away from $v_i$ and the score sequence $S(T)$ is formed by listing the vertex scores in non-decreasing order. Clearly, $0 \leq s_i \leq \frac{n(n-1)}{2}$. Further, no two scores can be zero and no two scores can be $n-1$. Tournament score sequences have also been called score structures [145], score vectors [165] and score lists [29].

One interpretation of a tournament is as a competition where $n$ participants play each other once in a match that cannot end in a tie and score one point for each win. Player $v$ is represented in the tournament by vertex $v$ and an arc from $u$ to $v$ means that $u$ defeats $v$. Then player $v$ obtains a total of $d_v^+$ points in the competition and the vertex scores can be ordered to obtain the score sequence of the tournament. We use $u \rightarrow v$ to denote the both, an arc from $u$ to $v$ and the fact that $u$ dominates $v$. A result of Ryser [227] states that an n-tournament can be obtained from any other having the same score sequence by a sequence of arc reversals of 3-cycles.

Now, we give the characterisation of score sequences of tournaments which is due to Landau [145]. This result has attracted quite a bit of attention as nearly a dozen different proofs appear in the literature. Early proofs tested the readers patience with special choices of subscripts, but eventually such gymnastics were replaced by more elegant arguments. Many of the existing proofs are discussed in a survey by Reid [221] and the proof we give here is due to Thomassen [242]. Further, two new proofs can be found in [89].

**Theorem 14.1 (Landau [145])**   A sequence of non-negative integers $S = [s_i]_1^n$ in non-decreasing order is a score sequence of a tournament if and only if for each subset $I \subseteq [n] = \{1, 2, \ldots, n\}$,

$$\sum_{i \in I} s_i \geq \binom{|I|}{2}, \tag{14.1.1}$$

with equality when $|I| = n$.

Because of the monotonicity assumption $s_1 \leq s_2 \leq \ldots \leq s_n$, the inequalities (14.1.1), known as the Landau inequalities, are equivalent to

$$\sum_{i=1}^k s_i \geq \binom{k}{2},$$

for $1 \leq k \leq n$, with equality for $k = n$.

## Proof

*Necessity*   If a sequence of non-negative integers $[s_i]_1^n$ in the non-decreasing order is the score sequence of an $n$-tournament $T$, then the sum of the first $k$ scores in the sequence counts exactly one each arc in the subtournamnent $W$ induced by $\{v_1, v_2, \ldots, v_k\}$ plus each arc from $W$ to $T - W$ . Therefore the sum is at least $\frac{k(k-1)}{2}$, the number of arcs in W. Also, since the sum of the scores of the vertices counts each arc of the tournament exactly once, the sum of the scores is the total number of arcs, that is, $\frac{n(n-1)}{2}$.

*Sufficiency (Thomassen)*   Let $n$ be the smallest integer for which there is a non-decreasing sequence $S$ of non-negative integers satisfying Landau's conditions (14.1.2), but for which there is no $n$-tournament with score sequence $S$. Among all such $S$, pick one for which $s_1$ is as small as possible.

First consider the case where for some $k < n$,

$$\sum_{i=1}^k s_i = \binom{k}{2}.$$

By the minimality of $n$, the sequence $S_1 = [s_1, s_2, \ldots, s_k]$ is the score sequence of some tournament $T_1$. Further,

$$\sum_{i=1}^m (s_{k+i} - k) = \sum_{i=1}^{m+k} s_i - \binom{k}{2} - mk \geq \binom{m+k}{2} - \binom{k}{2} - mk = \binom{m}{2},$$

for each $m$, $1 \leq m \leq n-k$, with the equality when $m = n-k$. Therefore, by the minimality of $n$, the sequence $S_2 = [s_{k+1} - k, s_{k+2} - k, \ldots, s_n - k]$ is the score sequence of some tournament

$T_2$. By forming the disjoint union of $T_1$ and $T_2$, and adding all arcs from $T_2$ to $T_1$, we obtain a tournament with score sequence $S$.

Now, consider the case where each inequality in 14.1.2 is strict when $k < n$ (in particular $s_1 > 0$). Then the sequence $S_3 = [s_1 - 1, s_2, \ldots, s_{n-1}, s_n + 1]$ satisfies (14.1.2) and by the minimality of $s_1$, $S_3$ is the score sequence of some tournament $T_3$. Let $u$ and $v$ be the vertices with scores $s_n + 1$ and $s_1 - 1$ respectively. Since the score of $u$ is larger than that of $v$, $T_3$ has a path $P$ from $u$ to $v$ of length $\leq 2$. By reversing the arcs of $P$, we obtain a tournament with score sequence $S$, a contradiction. $\qquad \square$

Landau's theorem is the tournament analog of the Erdos-Gallai theorem for graphical sequences. A tournament analog of the Havel-Hakimi theorem for graphical sequences is the following result, the proof of which can be found in Reid and Beineke [218].

**Theorem 14.2** A non-decreasing sequence $[s_i]_1^n$ of non-negative integers, $n \geq 2$, is the score sequence of an n-tournament if and only if the new sequence

$$[s_1, s_2, \ldots, s_m, s_{m+1} - 1, \ldots, s_{n-1} - 1]$$

where $m = s_n$, when arranged in non-decreasing order, is the score sequence of some $(n-1)$-tournament.

**Definition:** A tournament is *strongly connected* or strong if for every two vertices $u$ and $v$ there is a path from $u$ to $v$ and a path from $v$ to $u$. A *strong component* of a tournament is a maximal strong subtournament.

The following extension of Theorem 14.1, characterises strong components. The proof is straightforward and consequently omitted.

**Theorem 14.3** A non-decreasing sequence $[s_i]_1^n$ of non-negative integers is the score sequence of a strong n-tournament if and only if

$$\sum_{i=1}^{k} s_i > \binom{k}{2}, \ 1 \leq k \leq n, \text{ and } \sum_{i=1}^{n} s_i = \binom{n}{2}.$$

We have the following observation from Theorem 14.3. Let $S = [s_i]_1^n$ be a score sequence of an $n$-tournament $T$ with vertex set $V = \{1, 2, \ldots, n\}$. Let $\sum_{i=1}^{p} s_i = \binom{p}{2}$, $\sum_{i=1}^{q} s_i = \binom{q}{2}$ and $\sum_{i=1}^{k} s_i > \binom{k}{2}$, for $p + 1 \leq k \leq q - 1$, where $0 \leq p < q \leq n$.

Then the subtournament induced by $\{p+1, \ldots, q\}$ is a strong component of $T$ with score sequence $[s_{p+1} - p, s_{p+2} - p, \ldots, s_q - p]$.

We say that $S$ is strong if $T$ is strong and the strong components of $S$ are the score sequences of the strong components of $T$. Theorem 14.3 shows that the strong components of $S$ are determined by the successive values of $k$ for which

$$\sum_{i=1}^{k} s_i = \binom{k}{2},$$

that is, the successive values of $k$ for which equality holds in condition (14.1.2).

For example, consider the score sequence

$$S = [1, 1, 1, 4, 4, 5, 5, 7, 9, 9, 10, 11, 11],$$

$$\sum_{i=1}^{k} s_i = \binom{k}{2} \text{ for } k = 3, 7, 8, \text{ and } 13.$$

Therefore the strong components of $S$ are, in ascending order,

$$[1, 1, 1], [1, 1, 2, 2], [0], \text{ and } [1, 1, 2, 3, 3].$$

The next result due to Brualdi and Shen [40] shows that the score sequence of an $n$-tournament satisfies inequalities (14.4.1) below, which are individually stronger than the inequalities (14.1.1), although collectively the two sets of inequalities are equivalent.

**Theorem 14.4 (Brualdi and Shen [40])**    A sequence $S = [s_i]_1^n$ of non-negative integers in non-decreasing order is a score sequence of a tournament if and only if for every subset $I \subseteq [n]$,

$$\sum_{i \in I} s_i \geq \frac{1}{2} \sum_{i \in I} (i - 1) + \frac{1}{2} \binom{|I|}{2}, \tag{14.4.1}$$

with equality when $|I| = n$.

**Proof**    The sufficiency (14.4.1) imply the inequalities (14.1.1).

We prove that the score sequence $S$ of a tournament satisfies (14.4.1). For any subset $I \subseteq [n]$, define

$$f(I) = \sum_{i \in I} s_i - \frac{1}{2} \sum_{i \in I} (i - 1) - \frac{1}{2} \binom{|I|}{2}.$$

Choose $I$ firstly to have $f(I)$ minimum and secondly to have $|I|$ minimum.

Claim that $I = \{i : 1 \leq i \leq |I|\}$.

Otherwise, there exists $i \notin I$ and $j \in I$ such that $j = i+1$. Then $s_i \leq s_j$. Since

$$s_j - \frac{1}{2}(j + |I| - 2) = f(I) - f(I - \{j\}) < 0$$

and $s_i - \frac{1}{2}(i + |I| - 1) = f(I \cup \{i\}) - f(I) \geq 0$,

$$\frac{1}{2}(i + |I| - 1) \leq s_i \leq s_j < \frac{1}{2}(j + |I| - 2) = \frac{1}{2}(i + |I| - 1),$$

which is a contradiction. This proves the claim.

Thus, $f(I) = \sum_{i=1}^{|I|} s_i - \frac{1}{2}\sum_{i=1}^{|I|}(i-1) - \frac{1}{2}\binom{|I|}{2} = \sum_{i=1}^{|I|} s_i - \binom{|I|}{2} \geq 0,$

where the inequality follows from 14.1. By the choice of the subset I, Theorem 14.4 follows. ❑

**Remark**   Clearly, the equality can occur often in (14.4.1). For example, equality holds for regular tournaments of odd order $n$ (with score sequence $\left[\frac{n-1}{2}, \ldots, \frac{n-1}{2}\right]$ ), whenever $|I| = k$, and $I = \{n-k+1, \ldots, n\}$.

Further, Theorem 14.4 is best possible in the sense that, for any real $\epsilon > 0$, the inequality

$$\sum_{i \in I} s_i \geq \left(\frac{1}{2} + \epsilon\right)\sum_{i \in I}(i-1) + \left(\frac{1}{2} - \epsilon\right)\binom{|I|}{2}$$

fails for some $I$ and some tournaments (for example, regular tournaments).

The following set of upper bounds for $\sum_{i \in I} s_i$ is equivalent to the set of lower bounds for $\sum_{i \in I} s_i$ in Theorem 14.4.

**Corollary 14.1** A sequence $S = [s_i]_1^n$ of non-negative integers in non-decreasing order is a score sequence of a tournament if and only if for any subset $I \subseteq [n]$,

$$\sum_{i \in I} s_i \leq \frac{1}{2}\sum_{i \in I}(i-1) + \frac{1}{4}|I|(2n - |I| - 1),$$

with equality when $|I| = n$.

**Proof**   Let $J = [n] - I$. Then,

$$\sum_{i \in I} s_i = \frac{1}{2} \sum_{i \in I} (i-1) + \frac{1}{4} |I| (2n - |I| - 1),$$

with equality when $|I| = n$, if and only if

$$\sum_{i \in J} s_i = \binom{n}{2} - \sum_{i \in I} s_i \geq \frac{1}{2} \sum_{i \in J} (i-1) + \frac{1}{2} \binom{|J|}{2},$$

with equality when $|J| = n$. Therefore Corollary 14.1 follows from Theorem 14.4.    ❑

**Corollary 14.2**   If $S = [s_i]_1^n$ is a score sequence of a tournament, then for each $i$, $\frac{i-1}{2} \leq s_i \leq \frac{n+i-2}{2}$.

**Proof**   Choose $I = \{i\}$. Then the result follows immediately from Theorem 14.4 and Corollary 14.1.    ❑

The next result by Brualdi and Shen [40] shows that when equality occurs in the inequalities (14.4.1), there are implications concerning the strong connectedness and regularity of every tournament with score sequence $S$. For any integers $r$ and $s$ with $r \leq s$, $[r, s]$ denotes the set of all integers between $r$ and $s$.

**Theorem 14.5 (Brualdi and Shen [40])** If $S = [s_i]_1^n$ is a tournament score sequence and if

$$\sum_{i \in I} s_i = \frac{1}{2} \sum_{i \in I} (i-1) + \frac{1}{2} \binom{|I|}{2}$$

for some $I \subseteq [n]$, then one of the following holds.

i. $I = [1, |I|]$ and $\displaystyle\sum_{i=1}^{|I|} s_i = \binom{|I|}{2}$.

ii. $I = [t, t + |I| - 1]$ for some $t$, $2 \leq t \leq n - |I| + 1$,

$$\sum_{i=1}^{t+|I|-1} s_i = \binom{t+|I|-1}{2} \text{ and } s_i = (t + |I| - 2)/2, \text{ for all } i \leq t + |I| - 1.$$

iii. $I = [1, r] \cup [r+t, t + |I| - 1]$, for some $r$ and $t$ such that $1 \leq r \leq |I| - 1$ and $2 \leq t \leq n - |I| + 1$,

$$\sum_{i=1}^{t+|I|-1} s_i = \binom{t+|I|-1}{2}$$

and $s_i = (r + t + |I| - 2)/2$ for all $i$, $r + 1 \leq i \leq t + |I| - 1$.

**Remark** Conditions (i), (ii) and (iii) of Theorem 14.5 are equivalent to the assertion that every tournament with the score sequence $S$ has one of the three structures shown in Figure 14.1. The notation $T_r$ is used to denote a subtournament on $r$ vertices and the double arrows mean that all the arcs between the two parts go in that direction.

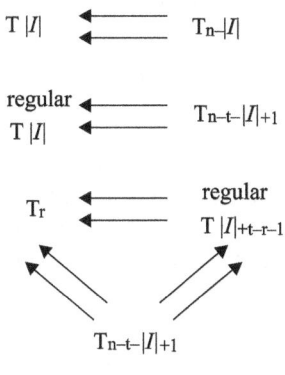

**Fig. 14.1**

The next result due to Bjelica [26] gives a criterion for score segments and subsequences with arbitrary positions of scores.

**Theorem 14.6 (Bjelica [26])** If $[t_i]_1^m$ is a sequence of non-negative integers in non-decreasing order and $[s_i]_1^n$ is a score sequence of an $n$-tournament $T$ with $m < n$, then the following properties are equivalent.

i. $\displaystyle\sum_{i=1}^{j} t_i \geq \binom{j}{2}$,    $1 \leq j \leq m$,

ii. $t_j = s_j$,           $1 \leq j \leq m$, for some $T$,

iii. $t_j = s_{k+j}$,        $1 \leq j \leq m$, for some $T$ and $k$,

iv. $t_j = s_{k_j}$,         $1 \leq j \leq m$, for some $T$ and $k_1 < k_2 < \ldots < k_m$.

The following result due to Bjelica and Lakic [27] gives the conditions for a set of integers to be the subset of scores with prescribed positions in some score sequence.

Denote $b(x) = \binom{x}{2}$, $X(k) = \displaystyle\sum_{i=1}^{k} x_i$.

**Theorem 14.7 (Bjelica and Lakic [27])** Let $0 \leq t_1 \leq t_2 \leq \ldots \leq t_m$ and $0 < k_1 < k_2 < \ldots < k_m$ be two sequences of integers. Then there exists an $n$-tournament $T$ with score sequence $s_1 \leq s_2 \leq \ldots \leq s_n$ such that $t_j = s_{k_j}$, $1 \leq j \leq m$, if and only if

$$\sum_{i=1}^{j}(k_i - k_{i-1})t_i \geq \binom{k_j}{2}, \quad 1 \leq j \leq m,\ k_0 = 0. \tag{14.7.1}$$

The size of the tournament can be $k_m$ if and only if in (14.7.1) the equality holds for $j = m$.

**Proof**

*Necessity*   If, for some tournament, we have $t_j = s_{k_j}$, where $1 \leq j \leq m$, then monotonicity of the score sequence and the Landau theorem give

$$\sum_{i=1}^{j}(k_i - k_{i-1})t_i \sum_{i=1}^{j}(k_i - k_{i-1})S_{k_i} \geq \sum_{i=1}^{k_j} S_i \geq \binom{k_j}{2}, \quad 1 \leq j \leq m$$

*Sufficiency*   Let some sequences $t$ and $k$ satisfy (14.7.1). Define the sequence $u_1 \leq u_2 \leq \ldots \leq u_{k_m}$ which includes the sequence $t$ as subsequence $u_k = t_j$, $k_{j-1} < k < k_j$, $1 \leq j \leq m$, and we prove that it satisfies property (*i*) from Theorem 14.6. In the following minorisation, we apply piecewise linearity of $U$, inequalities (14.7.1) and convexity of binomial function

$$bU(k) = U(k_{j-1}) + (k - k_{j-1})t_{k_j} = U(k_{j-1}) + (k - k_{j-1})\frac{U(k_j) - U(k_{j-1})}{k_j - k_{j-1}}$$

$$= \frac{k_j - k}{k_j - k_{j-1}}U(k_{j-1}) + \frac{k - k_{j-1}}{k_j - k_{j-1}}U(k_j)$$

$$\geq \frac{k_j - k}{k_j - k_{j-1}}b(k_{j-1}) + \frac{k - k_{j-1}}{k_j - k_{j-1}}b(k_j)$$

$$\geq b\left(\frac{k_j - k}{k_j - k_{j-1}}k_{j-1} + \frac{k - k_{j-1}}{k_j - k_{j-1}}k_j\right) = b(k).$$

By property (*ii*) from Theorem 14.6, there exists an $n$-tournament $T$ with beginning score segment $u$. Hence, scores from the sequence $t$ appear on the prescribed positions $k$.   ❑

## 14.2   Frequency Sets in Tournaments

**Definition:**   The number of times that a particular score occurs in a score sequence of a tournament is called the *frequency* of that score. A set of distinct positive integers $F = \{f_1, f_2, \ldots, f_k\}$ is a *frequency set* if there exists a tournament $T$ such that the set of frequencies of the scores in $T$ is exactly $F$. Note that in such a case $T$ has order at least $f_1 + f_2 + \ldots + f_k$. For example, the reversal of the orientation of three vertex disjoint arcs in a regular 7-tournament results in a 7-tournament with score sequence [2, 2 , 2, 3, 4, 4, 4] and frequency set $F = \{1, 3\}$.

Define $N(F)$ to be the smallest $m$ such that there exists a tournament on $m$ vertices with frequency set $F$. Clearly, $N(f_1, f_2, \ldots, f_n) \geq \sum_{i=1}^{n} f_i$. An almost regular tournament is an even order tournament in which the scores of the vertices are all as nearly equal as possible.

We have the following observations due to Alspach and Reid [3].

**Lemma 14.1 (Alspach and Reid [3])**   If $f_1 < f_2 < \ldots < f_n$, $n \geq 2$, are positive integers, $f_k$ is odd, and $N(f_1, f_2, \ldots, f_{k-1}, f_{k+1}, \ldots, f_n) = \left( \sum_{i=1}^{n} f_i \right) - f_k$, then

$$ N(f_1, f_2, \ldots, f_n) = \left( \sum_{i=1}^{n} f_i \right). $$

**Proof**   Let $R$ be a tournament on $\left( \sum_{i=1}^{n} f_i \right) - f_k$ vertices with frequency set $\{f_1, f_2, \ldots, f_{k-1}, f_{k+1}, \ldots, f_n\}$. Let $Q$ be a regular tournament on $f_k$ vertices. Let $T$ be the tournament obtained from disjoint copies of $Q$ and $R$ and in which every vertex of $Q$ dominates every vertex of $R$. It is clear that $T$ has $\left( \sum_{i=1}^{n} f_i \right)$ vertices and frequency set $\{f_1, f_2, \ldots, fn\}$.   ❑

**Lemma 14.2**   Let $R$ be a regular tournament of order $r$ and let $Q_1, Q_2, \ldots, Q_k$ be almost regular tournaments of orders $q_1 < q_2 < \ldots < q_k$. Then there exists a tournament $T$ of order $r + \sum_{i=1}^{k} q_i$ containing disjoint copies of $R, Q_1, Q_2, \ldots, Q_k$ as subtournaments such that $<R \cup Q_i>$ is regular of order $r + q_i$ and each vertex of $Q_i$ dominates each vertex of $Q_j$ when $i > j$.

The following constructive criterion due to Alspach and Reid [3] shows that every set $F$ of positive integers is the frequency set of some tournament and determines the least order $N(F)$ of such a tournament. The proof is omitted as it is lengthy and can be found in [3].

**Theorem 14.8 (Alspach and Reid [3])**   Let $f_1 < f_2 < \ldots < f_n$ be positive integers, at least one of which is odd. Then

$$ N(f_1, f_2, \ldots, f_n) = \sum_{i=1}^{n} f_i \tag{14.8.1} $$

unless either

    i. $n = 2$, $f_1 \not\equiv f_2 \ (\mathrm{mod}\ 2)$, $\gcd \{f_1, f_2\} = 1$, and $f_2 \neq 2$, in which case $N(f_1, f_2) = 2f_1 + f_2$

$$\tag{14.8.2}$$

    or

    ii. $n = 2$, $f_1 = 1$, $f_2 = 2$, in which case

    $N(1, 2) = 5$. $\tag{14.8.3}$

**Definition:**   Given the set of even integers $F = \{f_1, f_2, \ldots, f_n\}$, $e(F)$ denotes the largest power of 2 that divides every $f_i$, i.e., $e(F) = 2^m$ if $2^m / f_i$, $i = 1, 2, \ldots, n$, but $2^{m+1} \times f_j$ for some $j$.

**Lemma 14.3:**   If $T$ is a tournament with $r$ vertices and frequency set $\{f_1, f_2, \ldots, f_n\}$, then for each $k \geq 1$ there is a tournament with $kr$ vertices and frequency set $\{kf_1, kf_2, \ldots, kf_n\}$.

**Proof**   For each $i$, $1 \leq i \leq k$, let $T_i$ be a copy of $T$ with vertices $u_{i1}, u_{i2}, \ldots, u_{ir}$. Orient the arcs between $T_i$ and $T_j, i < j$, so that $u_{ik}$ exactly dominates the $r/2$ vertices $u_{j,k+1}, u_{j,k+2}, \ldots, u_{j,(k+r)/2}$ where the second subscripts are interpreted modulo $r$. Every vertex has had its scores increased by $r(k-1)/2$ and since we started with $k$ copies of $T$, the resulting tournament has $kr$ vertices and frequency set $\{kf_1, kf_2, \ldots, kf_n\}$.                                                                    ❑

The following result is also due to Alspach and Reid [3].

**Theorem 14.9 (Alspach and Reid [3])**   Let $f_1 < f_2 < \ldots < f_n$ be even positive integers, let $e(f_1, f_2, \ldots, f_n) = 2^m$ and let $f_i = 2^m k_i$, $1 \leq i \leq n$. Then

i.   if an even number of the $k_i's$ is odd, then

$$N(f_1, f_2, \ldots, f_n) = \sum_{i=1}^{n} f_i \text{ and}$$

ii.   if an odd number of the $k_i's$ is odd and $j$ is the smallest index for which $k_i$ is odd, then

$$N(f_1, f_2, \ldots, f_n) = f_j + \sum_{i=1}^{n} f_i.$$

**Proof**   Since $2^m$ is the highest power of 2 which is a factor of all the $f_i's$, at least one of the $k_i's$ is odd. Hence, in case (i) there are at least two odd $k_i's$ and $n \geq 2$. By Theorem 14.8, there is a tournament $T$ on $\sum_{i=1}^{n} k_i$ vertices with frequency set $\{k_1, k_2, \ldots, k_n\}$. By Lemma 14.3 with $k = 2^m$, there is a tournament on $\sum_{i=1}^{n} f_i$ vertices and frequency set $\{f_1, f_2, \ldots, f_n\}$. This proves (i).

Now, consider Case (ii) with $(k_1, k_2) \neq (1, 2)$. First, we show that $N(f_1, f_2, \ldots, f_n) = \sum_{i=1}^{n} f_i$ is impossible. Let $\alpha_i$ be the score occurring with frequency $f_i$. Then

$$\sum_{i=1}^{n} \alpha_i f_i = \binom{f_1 + f_2 + \ldots + f_n}{2} = \frac{(f_1 + f_2 + \ldots + f_n)(f_1 + f_2 + \ldots + f_n - 1)}{2}.$$

Clearly, the left hand side is divisible by $2^m$, while the right hand side is not. Thus, $N(f_1, f_2, \ldots, f_n) > \sum_{i=1}^{n} f_i$. Also, $N(f_1, f_2, \ldots, f_n) > f_k + \sum_{i=1}^{n} f_i$, for $k < j$, because of the same problem regarding divisibility by $2^m$. Thus,

$$N(f_1, f_2, \ldots, f_n) \geq f_j + \sum_{i=1}^{n} f_i.$$

If $n \geq 3$, by Theorem 14.8, there is a tournament on $\sum\limits_{i=1}^{n} k_i$ vertices with frequency set $K = \{k_1, k_2, \ldots, k_n\}$. As $k_j$ is odd, by Lemma 14.3 there is a tournament on $k_j + \sum\limits_{i=1}^{n} k_i$ vertices with frequency set $K$. If $n = 2$ and $k_1$ is odd, there is a tournament on $k_1 + k_1 + k_2$ vertices with frequency set $K$ (use Lemma 14.1 if $gcd\{k_1, k_2\} > 1$). If $k_1$ is even and and $k_2$ is odd, again there is a tournament with $k_1 + k_2 + k_2$ vertices and frequency set $K$ by application of Theorem 14.8 and Lemma 14.1 if $gcd\{k_1, k_2\} > 1$, or by Lemma 14.2 if $gcd\{k_1, k_2\} = 1$. In case $n = 1$ and $k_1$ is odd, then an almost regular tournament of order $2k_1$ has frequency set $F$. In all of the above cases employ Lemma 14.3 with $k = 2^m$ to obtain the result in (ii).

Finally, let $F = \{2^m, 2^{m+1}\}, m \geq 1$. As above, $N(2^m, 2^{m+1}) \geq 2^{m+2}$. Any tournament with $2^m$ scores of $2^{m+1} - 2$, $2^m$ scores of $2^{m+1} + 2$ and $2^{m+1}$ scores of $2^{m+1}$ shows that $N(2^m, 2^{m+1}) = 2^{m+2}$ as required in (ii). $\qquad\square$

## 14.3 Score Sets in Tournaments

**Definition:** The set of distinct scores in a tournament $T$ is called the score set of $T$.

It is easy to see that every singleton set $\{k\}$, $k \geq 0$, is a score set. Reid [217] showed that a nonempty set $S$ of non-negative integers is a score set whenever $|S| = 1, 2, 3$ or whenever $S$ is either an arithmetic or geometric progression. Reid conjectured that any nonempty set $S$ of non-negative integers is a score set. Hager [96] proved the conjecture for $|S| = 4, 5$. This problem proved to be more resistant than issue of frequency set discussed above.

If the set $\{x_1, x_2, \ldots, x_k\}$ is the score set of some $n$-tournament $T$, then there are multiplicities $m_1, m_2, \ldots, m_k$ (positive integers) such that $x_i$ occurs as a score exactly $m_i$ times in $T$. These $m_i$ are not necessarily distinct, so they are not the frequencies discussed above. Therefore, by Landau's Theorem, $\{x_1, x_2, \ldots, x_k\}$ is the score set of some $n$-tournament $T$ if and only if there exist positive integers $m_1, m_2, \ldots, m_k$ such that

$$\sum_{i=1}^{j} m_i x_i \geq \binom{\sum\limits_{i=1}^{j} m_i}{2}, \text{ for } 1 \leq j \leq k, \text{ with equality for } j = k.$$

Consequently, the connection to tournaments is removed, and Reid's conjecture becomes strictly an arithmetical supposition. Yan [271] proved the conjecture by pure arithmetical analysis.

S. Pirzada and T. A. Naikoo [200] proved by construction that if $s_1, s_2, \ldots, s_p$ are $p$ non-negative integers with $s_1 < s_2 < \ldots < s_p$, then there exists a tournament with score set $S = \left\{ s_1, \sum\limits_{i=1}^{2} s_i, \ldots, \sum\limits_{i=1}^{p} s_i \right\}$. More results on score sets in tournaments can be found in [197, 198, 200]. Also, the reconstruction of complete tournament can be seen in [119, 120]. The concept of scores in hyper tournament can be found in Zhou et.at [272]. The literature on kings in tournaments can be found in [130, 131, 132, 133, 155, 182, 184, 219, 220].

## 14.4  Lexicographic Enumeration and Tournament Construction

**Definition:** Let $[s_i]_1^n$ be any sequence of integers. The (transitive) deviation sequence of $[s_i]_1^n$ is defined to be the sequence $[d(i)] = [s_i - i + 1]$ and $d(i)$ is called the *deviation* of $s_i$. It is easy to see that $[s_i]_1^n$ is non-decreasing if and only if $d(i) - d(i+1) \leq 1$ for each $i < n$. Also, for each $k = 1, 2, \ldots, n$, $\sum_{i=1}^{k} d(i) = \sum_{i=1}^{k} s_i - \binom{k}{2}$. From this, it follows that a sequence $[s_i]_1^n$ of non-negative integers in non-decreasing order is a score sequence if and only if its deviation sequence $[d(i)]_1^n$ satisfies $\sum_{i=1}^{k} d(i) \geq 0$ for $k = 1, 2, \ldots, n$ with equality for $k = n$.

Let $[s_i]_1^n$ and $[s_i']_1^n$ be score sequences of length $n$. Then, say that $[s_i]_1^n$ precedes $[s_i']_1^n$ if there exists a positive integer $k \leq n$ such that $s_i = s_i'$ for each $1 \leq i \leq k$ and $s_k < s_k'$. They are equal if equality holds for all $i$. In symbols, $[s_i] \leq [s_i']$ means $[s_i]$ proceeds $[s_i']$. Further, $[s_i']$ is the successor of $[s_i]$ if they are distinct, $[s_i] \leq [s_i']$, and $[s_i'] \leq [s_i'']$ whenever $[s_i] \leq [s_i'']$. An enumeration of all score sequences of a given length with the property that the successor of any score sequence follows it immediately in the list is called a *lexicographic enumeration*. Clearly, $[0, 1, 2, \ldots, n-1]$ is not the successor of any score sequence of length $n$ and thus it is always the first in the lexicographic enumeration. Also, $[s_i]$ has no successor if and only if $s_n - s_1 \leq 1$.

If we know the first sequence in a lexicographic enumeration, then we can complete the work provided we know how to get the successor of any given sequence. The following algorithm due to Gervacio [85] gives the successor $[s_i']$ of $[s_i]$, if it exists.

## Algorithm

1. Determine the maximum $k$ such that $s_n - s_k \geq 2$.

2. Let $s_i' = s_i$ for all $i < k$.

3. Let $s_k' = s_k + 1$.

4. Let $s_j' = s_k + 1$ until $\sum_{i=1}^{j} s_i' < \binom{j}{2}$.

5. Let $t$ be the minimum $j$ such that $\sum_{i=1}^{j} s_i' < \binom{j}{2}$, set $s_t' = \binom{t}{2} - \sum_{i=1}^{t-1} s_i'$.

6. Let $s_i' = i - 1$ for all $i$, $t < i \leq n$.

**Constructing a tournament:** One method of constructing a tournament with a given score sequence can be found in [16]. The method we give here is due to Gervacio [85]. First, we have the following observations.

**Lemma 14.4** Let $[s_i]_1^n$ be a score sequence with deviation sequence $[d(i)]$.

a. If max $\{d(i)\} = M > 0$, then for each $1 \leq k \leq M$, there exists a vertex $u$ such that $d(u) = k$.

b. If min $\{d(i)\} = m < 0$, then for each $-1 \geq k \geq m$, there exists a vertex $u$ such that $d(u) = k$.

## Proof

a. Clearly, the result is true for $k = M$. Now, we show that it is true for all $1 \leq k \leq M$ by induction on $k$. Assume the result to be true for $k+1$, i. e., there exists a vertex $u$ such that $d(u) = k+1$. Since $d(n) \leq 0$, there exists $t > u$ such that $d(u) = d(t) > d(t+1)$. Since $d(t) \leq d(t+1)+1, d(t+1) = d(u)-1 = k$. Hence (a) holds.

b. This can be proved by using the argument as in (a).

**Lemma 14.5** Let $[s_i]_1^n$ be a score sequence with deviation sequence $[d(i)]$. If $c$ is the number of negative terms, then $c \geq max\{d(i)\}$.

**Proof** Let $p = $ max $\{d(i)\}$. If $p = 0$, then $c = 0$ and the result holds. For $p > 0$, we have the following cases.

**Case 1** There exists a vertex $k$ such that $d(k) < 0$ and $|d(k)| \geq p$. By Lemma 14.4, $c \geq |d(k)| \geq p$.

**Case 2** For each non-negative deviation $d(k)$, $|d(k)| < p$.

Let $q = \max\{|d(i)| : d(i) < 0\}$ and let $c < p$. Then using Lemma 14.4, $\displaystyle\sum_{d(i)<0} |d(i)| \leq 1+2+\ldots+q$ $+(p-q)q$. But $\displaystyle\sum_{d(i)<0} |d(i)| = \sum_{d(i)>0} |d(i)|$, and by Lemma 14.4, $\displaystyle\sum_{d(i)>0} |d(i)| \geq 1+2+\ldots+p$. Hence, $\dfrac{q(q+1)}{2+(p-q)q} > \dfrac{p(p+1)}{2}$.

This gives the quadratic inequality $p^2 - (2q-1) + q(q-1) < 0$ and this implies that $q-1 < p < q$, which is absurd, since $p$ and $q$ are integers. Hence, $c \geq p$.

Now, we describe and validate the above algorithm.

## Construction algorithm

Let $[s_i]_1^n$ be a score sequence with deviation sequence $[d(i)]_1^n$. First take $n$ vertices arranged horizontally and labelled $1, 2, \ldots, n$ from left to right.

**Step 1** Subdivide $[d(i)]$ into maximal non-increasing segments and denote by $p$ the number of segments in the subdivision. Let $n_i$ be the number of negative deviations in the $i$th segment, counting from left to right.

**Step 2** Let $j$ be the last integer such that $d(j) > 0$. If no such $j$ exists, go to step 6. Else, determine the least integer $q$ such that $\displaystyle\sum_{i \leq q} n_i \geq d(j)$. For each $i$ in the segments to the left of the $q$th segment such that $d(i) < 0$, let $d'(i) = d(i)+1$ and draw the arc $ji$.

**Step 3**    Let $\sigma = d(j) - \sum_{i<q} n_i$ and choose a smallest (negatively largest) deviations $d(i)$ in the $q$th segment. For each such $d(i)$, let $d'(i) = d(i) + 1$ and draw the arc $ji$. Let $d'(j) = 0$.

**Step 4**    For all other deviations $d(i)$ not changed in the preceding steps, let $d'(i) = d(i)$.

**Step 5**    If $[d'(i)] \neq [0]$, go to step 1 using $[d'(i)]$ in place of $[d(i)]$.

**Step 6**    Whenever $u < v$ and there is no arc between $u$ and $v$, draw the arc $vu$.

**Step 7**  The resulting digraph is a tournament with score sequence $[s_i]_1^n$.

Now, we analyse the algorithm to verify its validity. Clearly, step 1 can always be carried out. Step 2 can be done in view of Lemma 14.5. Step 3 can be implemented because of step 2. Obviously, step 4 can be done, and after this step, $[d'(i)]$ satisfies $d'(i) - d'(i+1) \leq 1$ for all $1 \leq i \leq n$. Let $D$ be the digraph formed when $[d'(i)] = [0]$. Then for each vertex $i$ in $D$,

$$d_i^+(D) = \begin{cases} d(i), & \text{if } d(i) \geq 0 \\ 0, & \text{if } d(i) < 0 \end{cases}$$

and     $$d_i^-(D) = \begin{cases} d(i), & \text{if } d(i) \geq 0 \\ -d(i), & \text{if } d(i) < 0 \end{cases}$$

Let $T$ be the tournament formed after step 6 and let $i$ be any vertex of $T$. If $d(i) \geq 0$, then $s_i = d_i^+ = d_i^+(D) + i - 1 = s_i$. If $d(i) < 0$, then $d_i^- = d_i^-(D) + n - i = -d(i) + n - i$, and thus $s_i = d_i^+ = (n-1) - d_i^- = d(i) + i - 1 = s_i$.

**Example**    Let $[s_i] = [1, 1, 2, 2]$. Then, $[d(i)] = [1, 0, 0, -1]$.

The resulting digraph after using above algorithm upto $[d'(i)] = [0]$ is shown in Figure 14.2. To get the tournament, add all arcs $ij$ $(i > j)$.

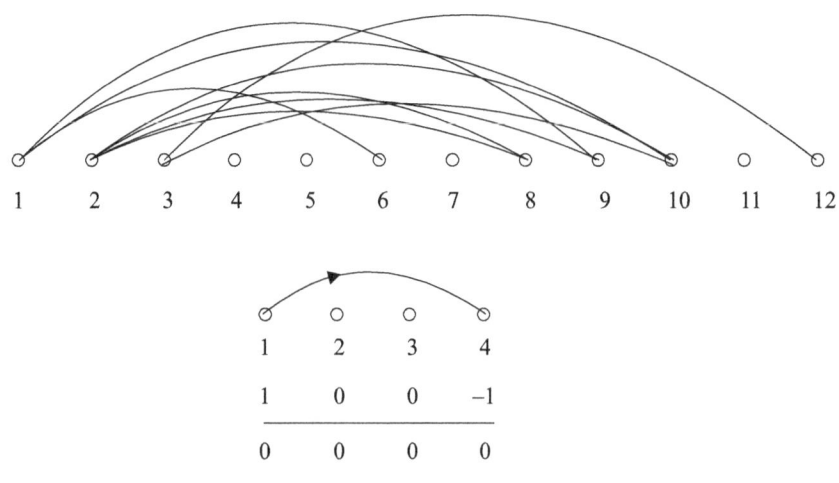

**Fig. 14.2**

## 14.5 Simple Score Sequences in Tournaments

**Definition:** A score sequence is simple (uniquely realisable) if it belongs to exactly one tournament. Every score sequence of tournaments with fewer than five vertices is simple, but the score sequence [1, 2, 3, 3, 3] is not simple, since the tournaments in Figure 14.3 are not isomorphic.

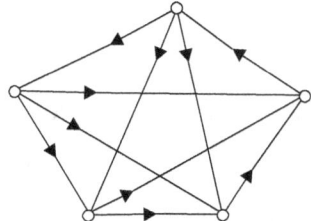 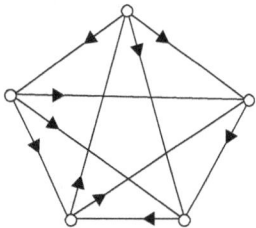

**Fig. 14.3**

We have the following observations.

**Lemma 14.6** A score sequence $S$ is simple if and only if every strong component of $S$ is simple.

The following result due to Avery [8] gives a condition for determining simple score sequence in tournaments.

**Theorem 14.10 (Avery [8])** A strong score sequence is simple if it is one of [0], [1, 1, 1], [1, 1, 2, 2] or [2, 2, 2, 2, 2].

**Proof** Let $S$ be a strong score sequence. Consider the following three cases.

**Case 1** $S$ has at least three different scores, say $s_u$, $s_v$ and $s_w$. Then there is a tournament having score sequence $S$ with $(u, v, w)$ as a cyclic triple. For suppose the tournament $T$ has score sequence $S$ and that the triple $(u, v, w)$ is not cyclic in $T$. Without loss of generality, let $T$ have arcs $u \to v \to w \leftarrow u$. As $T$ is strong, there is a path from $w$ to $u$ (which may include the vertex $v$) that together with $u \to w$ forms a cycle. Reverse the orientation of every arc of this cycle to form a tournament $T_1$ with the same score sequence $S$. Then $T_1$ has arcs $u \to v \to w \to u$, so that $(u, v, w)$ is a cyclic triple.

Now, reverse this cyclic triple to form the tournament $T_2$ also with score sequence $S$. Then, $T_2$ is not isomorphic to $T_1$. For consider the number of upsets, which are the arcs for which the score of the dominant vertex is less than the score of the dominated vertex. Without loss of generality, let $s_u$ be maximal in $T_1$. Then reversing $u \to v \to w \to u$, to form $T_2$ increases the number of upsets if $s_u > s_v > s_w$ and decreases the number if $s_u < s_v < s_w$. Therefore $T_1$ and $T_2$ are not isomorphic and thus $S$ is not simple.

**Case 2** $S$ has exactly two different scores, say

$s_1 = \ldots = s_k < s_{k+1} = \ldots = s_n$, for some $k$.

Put $U = \{1, \ldots, k\}$ and $V = \{k+1, \ldots, n\}$. Now,

$$ks_1 + (n-k)s_n = \sum_{i=1}^{n} s_i = \frac{n(n-1)}{2} \text{ and } s_1 \geq 1, \text{ since } S \text{ is strong. If } k = 1,$$

then $(n-1)s_n < s_1 + (n-1)s_n = \frac{n(n-1)}{2} < ns_n$.

Thus, $\frac{(n-1)}{2} < s_n < \frac{n}{2}$, which is impossible. Therefore, $k \geq 2$ and so $|U| \geq 2$. Similarly, $|V| \geq 2$.

Let $T$ be a tournament with score sequence $S$. Suppose that the sub-tournament $T(U)$ induced by $U$ has score sequence

$$S(T(U)) = [r_1, r_2, \ldots, r_k], \text{ where } r_1 \leq r_2 \leq \ldots \leq r_k.$$

As $r_1 \leq r_2 \leq \ldots \leq r_k$ and $s_1 = s_2 = \ldots = s_k$, then vertex 1 dominates as many vertices of $V$ as any of $\{2, \ldots, k-1\}$ and vertex $k$ is dominated by as many vertices as any of $\{1, \ldots, k-1\}$. Since $T$ is strong, there are arcs from $U$ to $V$ and from $V$ to $U$. Thus 1 dominates some vertex $v_1$ of $V$ and $k$ is dominated by some vertex $v_2$ of $V$, where $v_2$ is not necessarily different from $v_1$.

Since $r_k \geq r_1$, there is a path $k \rightarrow j \rightarrow 1$ in $T(U)$ from $k$ to 1 of length at most two. Similarly, as $s_{v_1} = s_{v_2}$, there is a path $v_1 \rightarrow w \rightarrow v_2$ in $T$ from $v_1$ to $v_2$ of length at most two. It follows that $T$ has a closed walk $1 \rightarrow v_1 \rightarrow w \rightarrow v_2 \rightarrow k \rightarrow j \rightarrow 1$ of length at most 6.

The orientation of the arcs of this closed walk can be reversed to form another tournament $T_1$ with score sequence $S$. Let $S(T_1(U))$ denote the score sequence of the subtournament $T_1$ induced by $U$. Now in forming $T_1(U)$ from $T(U)$, the score of vertex 1 is increased by 1, the score of $k$ is decreased by 1, the score of $j$ is unchanged if $j \neq 1$ or $k$, and the score of $w$ is unchanged if $w \in U$.

Therefore, $S(T_1(U)) \neq S(T(U))$ unless $r_k = r_1 + 1$. But if $S(T_1(U)) \neq S(T(U))$, then $T_1(U)$ is not isomorphic to $T(U)$, in which case $T_1$ is not isomorphic to $T$ and $S$ is not simple. So $S$ can be simple only if $r_k = r_1 + 1$, that is if $k$ is even, $k = 2p$ say, with $r_1 = \ldots = r_p = p-1$ and $r_{p+1} = \ldots = r_{2p} = \ldots = r_{2p}(r_k) = p$.

Let $S(T(U))$ be of this form. Then for some value of $q$, each vertex of $\{1, \ldots, p\}$ dominates $q$ vertices of $V$ and each vertex of $\{p+1, \ldots, 2p\}$ dominates $q-1$ vertices of $V$, so that $s_1 = \ldots = s_{2p} = p+q-1$. Now, $q \geq 1$, since $S$ is strong. If $q = |V|$, then of the $2pq$ arcs between $U$ and $V$, $p(2q-1)$ are from $U$ to $V$. So there are only $p$ arcs from $V$ to $U$. Thus, the average score in $T$ of the vertices of $V$ is $\frac{(q-1)}{2} + \frac{p}{q} < (q-1) + p = s_1$, which contradicts the definition of $V$, so $q < |V|$. Therefore, each vertex of $U$ is dominated by atleast one vertex of $V$.

Now, let $p \geq 2$, so that $|U| \geq 4$ and let the vertex 2 be dominated by vertex $v_2$ of $V$. Then arguing the same way as in the general case, there is a closed walk $1 \rightarrow v_1 \rightarrow w \rightarrow v_2 \rightarrow 2 \rightarrow$

$j \to 1$ in $T$ of length at most six, where $v_1, v_2 \in V$ and $j \in U$. These arcs can be reversed to form a tournament $T_1$ with score sequence $S$.

Now, $S(T_1(U)) \neq S(T(U))$, since the least score in $T(U)$ is $p - 1$, whereas vertex 2 has a score of $p - 2$ in $T_1(U)$. Thus $T_1(U)$ is not isomorphic to $T(U)$, so $T_1$ is not isomorphic to $T$ and $S$ is not simple. Hence $S$ is not simple if $p \geq 2$.

Altogether, if $|U| \geq 3$, then $S$ is not simple. Similarly, if $|V| \geq 3$, then $S$ cannot be simple. Since $|U| \geq 2$ and $|V| \geq 2$, the only remaining possibility is $|U| = |V| = 2$. The only strong score sequence of type 2 with $|U| = |V| = 2$ is $[1, 1, 2, 2]$, which is simple. Therefore, in case 2, the only simple strong score sequence is $[1, 1, 2, 2]$. The tournament with this score sequence is shown in Figure 14.4(a).

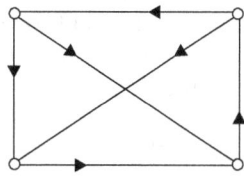

**Fig. 14.4(a)**

**Case 3**  All scores of S are equal, so that the number of vertices $n$, is odd. Let $n = 2p + 1$. Define $T_1$ to be the tournament with vertices $\{0, 1, \ldots, 2p\}$ in which for all $v$ in $\{0, 1, \ldots, 2p\}$, the vertex $v$ dominates vertices $v+1, v+2, \ldots, v+p$ (reduced modulo $2p + 1$). Then $T_1$ has all scores equal.

Let $p \geq 3$, so that $n \geq 7$. Reverse the arcs of the cycle $1 \to p \to 2p \to 1$ in $T_1$ to form a tournament $T_2$ with the same score sequence and containing the cycle $1 \to 2 \to p \to 1$. Clearly, $T_1$ and $T_2$ are not isomorphic, since for each vertex $v$ in $T_1$ the vertices adjacent from $v$ (its outset) induce a transitive subtournament, whereas the outset of vertex 0 in $T_2$ induces a subtournament with the cyclic triple $(1, 2, p)$. Thus $S$ is not simple unless $p \leq 2$, in which case it is one of $[0]$, $[1, 1, 1]$ or $[2, 2, 2, 2, 2]$. These strong score sequences are simple. Thus the only simple strong score sequences of type 3 are $[0]$, $[1, 1, 1]$ and $[2, 2, 2, 2, 2]$. The respective associated regular tournaments are displayed in Figure 14.4(b). Let $p \geq 3$, so that $n \geq 7$. Reverse the arcs of the cycle $1 \to p \to 2p \to 1$ in $T_1$ to form a tournament $T_2$ with the same score sequence and containing the cycle $1 \to 2 \to p \to 1$. Clearly, $T_1$ and $T_2$ are not isomorphic, since for each vertex $v$ in $T_1$ the vertices adjacent from $v$ (its outset) induce a transitive subtournament, whereas the outset of vertex 0 in $T_2$ induces a subtournament with the cyclic triple $(1, 2, p)$. Thus $S$ is not simple unless $p \leq 2$, in which case it is one of $[0]$, $[1, 1, 1]$ or $[2, 2, 2, 2, 2]$. These strong score sequences are simple. Thus the only simple strong score sequences of type 3 are $[0]$, $[1, 1, 1]$ and $[2, 2, 2, 2, 2]$. The respective associated regular tournaments are displayed in Figure 14.4(b).

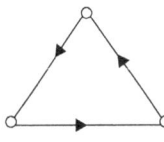 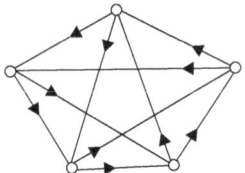

**Fig. 14.4(b)**

**Corollary 14.3**   The score sequence $S$ is simple if and only if every strong component of $S$ is one of $[0], [1, 1, 1] , [1, 1, 2, 2]$ or $[2, 2, 2, 2, 2]$.

Hence it is possible to decide whether a given score sequence $S$ is simple by using Theorem 14.10 to determine the strong components of $S$ and then applying Corollary 14.3.

Let $s(n)$ denote the number of simple score sequences of order $n$. It is easy to show that $s(n)$ satisfies the following recurrence relation, which can be used to evaluate $s(n)$.

**Theorem 14.11**   $s(n) = s(n-1) + s(n-3) + s(n-4) + s(n-5)$, where $s(k) = 0$ if $k < 0$, and $s(0) = 1$.


# 14.6   Score Sequences of Self-Converse Tournaments

We know, the converse of an $n$-tournament $T_n$ is the tournament $T_n'$ obtained by reversing the orientation of all the arcs in $T_n$. A tournament is called *self-converse* if $T_n \cong T_n'$. The transitive tournaments are examples of self-converse tournaments.

The following characterisation of score sequences of self-converse tournaments is due to Eplett [71].

**Theorem 14.12 (Eplett [71])**   A score sequence $[s_i]_1^n$ is the score sequence of a self-converse tournament if and only if

$$s_i + s_{n+1-i} = n - 1, \tag{14.12.1}$$

for $1 \le i \le n$ .

## Proof

*Necessity*   An automorphism of a tournament $T$ is a permutation of the vertices that preserves the orientation of all arcs of $T$ and an anti-automorphism is a permutation that reverses all arcs. Let the group of all automorphisms of $T$ be denoted by $G = G(T)$ and the group of all automorphisms and anti-automorphisms of $T$ be denoted by $H = H(T)$. Clearly, $T$ is self-converse if and only if $G$ is a proper subgroup of $H$. In particular, if $T$ is self-converse, then $H$ is a group of order $2m$, $m$ an odd integer, so there must be a self-inverse anti-automorphism (with one fixed vertex when $n$ is odd). Hence the vertices of a self-converse tournament $T$ can be labelled so that the permutation $r : i \to n + 1 - i$, for $1 \le i \le n$ is an anti-automorphism.

*Sufficiency* Let the score sequence $[s_i]_1^n$ satisfy (14.12.1). Consider first the case when $n = 2p + 1$, where $p \geq 2$. Clearly,

$$\sum_{i=1}^{k} s_i \geq \binom{k}{2},$$ (14.12.2)

for $1 \leq k \leq n$, with equality when $k = n$.

Define the functions $Q(k)$ and $R(k)$ as follows.

$$R(k) = \sum_{i=1}^{k} s_i - \binom{k}{2} \text{ and } Q(k) = \sum_{i=1}^{k} (s_i - 1) - \binom{k}{2}.$$

Then, $R(k) \geq 0$ for $i \leq k \leq m - 1$ and assume that

$$\min \{Q(k) : 1 \leq k \leq m - 1\} = -r$$

for some positive integer $r$. The possibility that $Q(k) \geq 0$ for $1 \leq k \leq m - 1$ will be considered later. Let

$$\min \{k : Q(k) = -r\} = j_r.$$

Since, $Q(k) = R(k) - k$, (14.12.3)

it follows that $1 \leq r \leq j_r \leq p - 1$. Define $j_1, \ldots, j_{r-1}$ as follows: $j_i$ is the smallest integer $t$ such that $1 \leq t < j_{i+1}$ and $Q(t) = -i$, if such an integer exists, otherwise $j_i = j_{i+1} - 1$. Following (14.12.3), it is easy to see that $1 \leq j_1 < \ldots < j_r \leq p - 1$. Let $J = \{j_i : 1 \leq i \leq r\}$.

Consider the $2p - 1$ modified scores $[s_1', \ldots, s_{p-1}', s_{p+1}', s_{p+3}', \ldots, s_{2p+1}']$, where $s_i' = s_i$ and $s_{2p+2-i}' = s_{2p+2-i} - 2$ if $i \in J$ and $s_i' = s_i - 1$ for the remaining relevant values of $i$. If $Q(k) \geq 0$, for $1 \leq k \leq p - 1$, then only the second part of the definition applies. These scores satisfy conditions (14.12.1) and (14.12.2). To show that

$$s_1' \leq \ldots \leq s_{p-1}' \leq s_{p+1}' \leq \ldots \leq s_{2p-1}',$$

it suffices to consider the cases where $s_k' = s_k$ and $s_{k+1}' = s_{k+1}' - 1$. In such cases, $k = j_i$ for some $i$ and $k + 1 < j_{i+1}$. Consequently, $s_k - k = Q(k) - Q(k-1) < 0$ and $s_{k+1} - (k+1) = Q(k+1) - Q(k) \geq 0$, from the definition of $j_i$. Thus, $s_k < s_{k+1}$, and the required inequality certainly holds. Therefore, assume as an induction hypothesis that there exists a self-converse tournament $T_{2p-1}$ with $2p - 1$ vertices labelled $1, \ldots, p-1, p+1, p+3, \ldots, 2p+1$ with scores $s_1', \ldots, s_p' - 1, s_{p+1}', s_{p+3}', \ldots, s_{2p+1}'$. Further, assume that the mapping $r : i \to 2m + 2 - i$, for $1 \leq i \leq p - 1$ and $i = p + 1$, defines an anti-automorphism.

Now, we propose to join two vertices labelled $p$ and $p + 2$ to $T_{2p-1}$ and orient arcs joining these vertices to each other and to the vertices of $T_{2p-1}$ in such a way that the resulting tournament $T_{2p+1}$ is a self-converse tournament with scores $s_1, \ldots, s_{2p+1}$. Orient the arcs as follows.

$(p, j_i), (p+2, j_i), (2p+2-j_i, p+2), (2p+2-j_i, p)$ for $1 \le i \le r$. This takes care of all arcs incident with the vertices $j_i$ and $2_p + 2 - j_i$ and these vertices clearly have the required score now. If $Q(k) \ge 0$ for $1 \le k \le m-1$, then this step does not apply. Further more, vertex $p+1$ is dominated by $p$, and dominates $p+2$.

Now, assume that $s_p = r + 2\upsilon + 1 + \delta$, where $\delta = 0$ or 1. Notice that $2\upsilon + \delta \ge 0$. Replacing $j_r$ by $j$ (without loss of generality), then

$$0 \le R(j+1) = R(j) + s_{i+1} - (j+1) = Q(j) + s_{j+1} - 1 \le -r + s_p - 1 = 2\upsilon + \delta.$$

Also, the fact that $s_p \le s_{p+1} = p$ implies that $\upsilon \le 2\upsilon \le p + 1 - r - \delta$.

To continue the definition of $T_n$, orient the arcs as follows. $(p,i), (p, 2p+2-i), (i, p+2), (2p+2-i, p+2)$ for $i$ running through the first $\upsilon$ positive integers not in $J$, $(i, p), (2p+2-i, p), (p+2, i), (p+2, 2p+2-i)$ for the next $p-1-r-\upsilon$ positive integers not in $J$. Finally, $p$ dominates or is dominated by $p+2$ according as $\delta = 1$ or 0. It is easy to verify that the tournament $T_n$ defined in this way is indeed a self-converse tournament with scores $s_1, \ldots, s_n$ as required. This suffices to prove the theorem for odd values of $n$ by induction.

The proof for even $n$ is similar.

## 14.7 Score Sequences of Bipartite Tournaments

A *bipartite tournament* $T$ is an orientation of a complete bipartite graph. Clearly, the vertex set of $T$ is the union of two disjoint nonempty sets $X$ and $Y$, and arc set of $T$ comprises exactly one of the pairs $(x, y)$ or $(y, x)$ for each $x \in X$ and each $y \in Y$. If the orders of $X$ and $Y$ are $m$ and $n$ respectively, $T$ is said to be an $m \times n$ bipartite tournament.

A bipartite tournament may be used to represent competition between two teams and each player competes against everyone on the opposing team. The *score* $s_v$ of vertex $v$ is the number of vertices it dominates and for a bipartite tournament there is a pair of score sequences, one sequence for each set. For example, the bipartite tournament in Figure 14.5 has sequence [4, 3, 2, 0] and [2, 2, 2, 1].

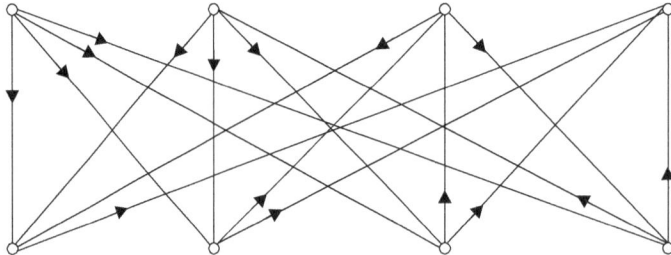

**Fig. 14.5**

**Definition** A bipartite tournament is *reducible* if there is a nonempty proper subset of its vertex set to which there are no arcs from the other vertices, otherwise irreducible. A *component* is a maximal irreducible sub-bipartite tournament. A *non-trival component* contains at least two vertices one from each partite set.

A bipartite tournament is consistent if it contains no directed cycles. It can be easily seen that a bipartite tournament is consistent if and only if, for $v$ and $w$ in the same partite set, $v$ dominates every vertex which $w$ dominates if its score is at least that of $w$.

The *converse* of a bipartite tournament is obtained by reversing the direction of all its arcs and a bipartite tournament is *self-converse* if it is isomorphic to its converse.

Now, assume that the partite sets of bipartite tournaments have a fixed ordering with $X$ first and $Y$ second. Then a given bipartite tournament $T$ has associated with it two bipartite graphs (on the same sets of vertices) in a natural way, one graph containing those edges corresponding to the arcs directed from $X$ to $Y$, the other from $Y$ to $X$. For example, two graphs of the bipartite tournament $T$ are shown in Figure 14.6. The two graphs are relative complements as bipartite graphs.

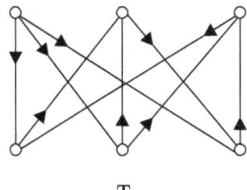 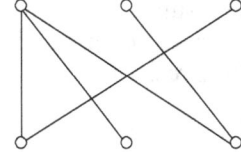 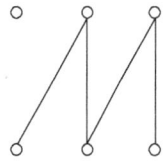

T

**Fig. 14.6**

Clearly, the pairs of score sequences of bipartite tournaments and pairs of degree sequences of bipartite graphs are equivalent.

**Lemma 14.7** Let $A = [a_1, \ldots, a_m]$ and $B = [b_1, \ldots, b_n]$ be sequences of integers and let $\overline{A} = [n - a_1, \ldots, n - a_m]$ and $\overline{B} = [m - b_1, \ldots, m - b_n]$. Then the following are equivalent.

1. $A$ and $B$ are the score sequences of a bipartite tournament.

2. $\overline{A}$ and $\overline{B}$ are the score sequences of a bipartite tournament.

3. $A$ and $\overline{B}$ are the degree sequences of a bipartite graph.

4. $\overline{A}$ and $B$ are the degree sequences of a bipartite graph.

The following observation can be easily established.

**Lemma 14.8** If $v$ and $v'$ are vertices in the same partite set of a bipartite tournament $T$, if $s_v \leq s'_v$, and if there is a vertex $w$ which is dominated by $v$ and which dominates $v'$, then there is another vertex $w'$ which is dominated by $v'$ and which dominates $v$, that is, $v \to w \to v' \to w' \to v$ is a 4-cycle.

The following result is due to Gale [84].

**Theorem 14.13 (Gale [84])** If $A = [a_1, \ldots, a_m]$ and $B = [b_1, \ldots, b_n]$ are sequences of non-negative integers in non–decreasing order, then $A$ and $B$ are the score sequences of some bipartite tournament if and only if the sequences $A' = [a_1, \ldots, a_{m-1}]$ and $B' = [b_1, \ldots, b_{a_m}, b_{a_m+1} - 1, \ldots, b_n - 1]$ are.

**Proof** First assume that $A'$ and $B'$ are the score sequences of a bipartite tournament $T'$. To the first partite set of $T'$, add a new vertex $v$ with arcs directed from it to vertices (in the second set) with scores $b_1, \ldots, b_{a_m}$, and to it from the others. The result is a bipartite tournament with score sequences $A$ and $B$.

For the converse, it is sufficient to show that if $A$ and $B$ are the score sequences of a bipartite tournament, then in one realisation, a vertex (in the first set) of score am dominates vertices of scores $b_1, \ldots, b_{a_m}$. Among the bipartite tournament realisations of $A$ and $B$, let $T$ be the one in which a vertex $x$ of score $a_m$ is such that the sum $S$ of the scores of the vertices it dominates is as small as possible. Let $S > \sum\limits_{j=1}^{a_m} b_j$. Then there exists vertices $y$ and $y'$ such that $x \to y'$, $y \to x$ and $s_y < s'_y$. By Lemma 14.8, T has a 4-cycle $x \to y' \to x' \to y \to x$, and if its arcs are reversed, the result is a bipartite tournament with the same sequences, but in which score sum of the vertices dominated by $x$ is less than before. Since the sum was assumed to be minimised, the result follows.                                                                            ❏

Theorem 14.3 gives a natural construction for a canonical tournament $T^*(A, B)$ from a given pair of score sequences $A$ and $B$. The only point which needs clarification is getting $B'$ into non-decreasing order, i.e., we must specify dominance when a vertex $v_i$ must dominate some but not all vertices $y$ with a particular score. This is done by forming $B'$ as follows. Let $h$ and $k$ denote the smallest and largest integers $j$ for which $b_j = b_{a_m}$. Let $A' = [a_1, \ldots, a_{m-1}]$, and $B' = [b'_1, \ldots, b'_n]$ with

$$b'_j = \begin{cases} b_j, & \text{for } 1 \leq j < h \text{ and } h + k - a_m \leq j \leq k, \\ b_j - 1, & \text{otherwise} \end{cases}$$

This reduction and the resulting construction is illustrated by the following example, starting with sequences $A = [1, 1, 3, 5, 5]$ and $B = [1, 1, 2, 3, 4, 4]$ .

$A = [1, 1, 3, 5, 5]$ $\qquad\qquad\qquad$ $B = [1, 1, 2, 3, 4, 4]$

$A_1 = [1, 1, 3, 5]$ $\qquad\qquad\qquad$ $B_1 = [1, 1, 2, 3, 3, 4]$

$\qquad\qquad\qquad\qquad\qquad\qquad$ ($x_5$ dominates $y_1, y_2, y_3, y_4, y_6$)

$A_2 = [1, 1, 3]$ $\qquad\qquad\qquad$ $B_2 = [1, 1, 2, 3, 3]$

$\qquad\qquad\qquad\qquad\qquad\qquad$ ($x_4$ dominates $y_1, y_2, y_3, y_4, y_5$)

$A_3 = [1, 1]$ $\qquad\qquad\qquad\qquad$ $B_3 = [1, 1, 2, 2, 2, 2]$

$\qquad\qquad\qquad\qquad\qquad\qquad$ ($x_3$ dominates $y_1, y_2, y_3$)

$$A_4 = [1] \qquad\qquad\qquad B_4 = [0, 1, 1, 1, 1, 1]$$

$$(x_2 \text{ dominates } y_2)$$

$$A_5 = \varphi \qquad\qquad\qquad B_5 = [0, 0, 0, 0, 0, 0]$$

$$(x_1 \text{ dominates } y_1)$$

Figure 14.7 shows the $X$ to $Y$ arcs resulting from this construction.

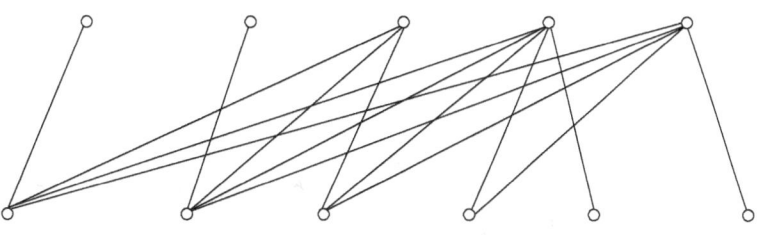

**Fig. 14.7**

A canonical tournament $T^*(A, B)$ has the following special property. In the subtournament $T_{r,n}^*$ induced by $\{x_1, \ldots, x_r\}$ and $Y$, if $x_r^* \to y_i^*$ and $y_j^* \to x_r^*$, then $s_{y_i^*} \le s_{y_j^*}$, that is, $b_i \le b_j$.

The next result is due to Beineke and Moon [20].

**Theorem 14.14 (Beineke and Moon [20])**    If two bipartite tournaments have the same score sequences, then each can be transformed into the other by successively reversing the arcs of 4-cycles.

It can be noted that Theorem 14.14 does not imply that all bipartite tournaments with given score sequences have the same number of 4-cycles (they need not), although the corresponding statement does hold for 3-cycles in tournaments.

The next result first established by Moon [162] and then in the present form by Beineke and Moon [20] gives a simple criterion for determining whether a pair of sequences are realisable as scores.

**Theorem 14.15 (Moon [162])**    A pair of sequences $A$ and $B$ of non-negative integers in non-decreasing order are the score sequences of some bipartite tournament if and only if

$$\sum_{i=1}^{k} a_i + \sum_{j=1}^{l} b_j \ge kl$$

for $1 \le k \le m$ and $1 \le l \le n$, with equality when $k = m$ and $l = n$.

Further more, the bipartite tournament is irreducible if and only if the inequality is strict except when $k = m$, and $l = n$.

**Proof**    In any bipartite tournament $T$, the combined scores of any collection of $k$ vertices from the first set and $l$ from the second must be at least $kl$, so that the inequalities certainly hold. Further, if $T$ irreducible, the inequality is strict unless $k = m$ and $l = n$.

*Sufficiency*   If $A$ and $B$ satisfy the inequalities, we show that $A'$ and $B'$ satisfy the inequalities reordered as in construction of 14.13. It is easily seen that $A'$ and $B'$ are then in non-decreasing order, and further, their combined sum is

$$\sum_1^{m-1} a'_i + \sum_1^h b'_j = mn - a_m - (n - a_m) = (m-1)n.$$

For a fixed value of $k(1 \leq k \leq m-1)$, assume there is a value of $l$ for which the inequality does not hold. And let $h$ denote the least such that

$$\sum_1^k a'_i + \sum_1^h b'_j < kh ..$$

It follows from the minimality of $h$ that $b'_h < k$, whence $b_h \leq k$. Now, let $p$ and $q$ be the least and greatest values of $j$ for which $b_j = b_{a_m}$ and set $r = \max (h, q)$. Since the first $p-1$ values of $b_j$ were unchanged, we have $h \geq p$ and thus $b_h = \ldots = b_r$. Finally, let s denote the number of $j \leq h$ such that $b'_j = b_{j-1}$. If $h \leq q$, then $s \leq q - a_m$, and if $h > q$, then $s = (h-q) + (q-a_m) = h - a_m$. In either case, $a_m + s \leq r$. Therefore,

$$\sum_1^{k+1} a_i + \sum_1^r b_j = \sum_1^k a'_i + \sum_1^h b'_j + \sum_{h+1}^r b_j + a_{k+1} + s < kh + (r-h)b_h + a_m + s$$
$$\leq kh + (r-h)k + r < (k+1)r,$$

which is a contradiction. Therefore $A'$ and $B'$ satisfy the inequalities, as required.

It is easily seen that if the strict inequalities hold for $A$ and $B$, no realisation can be reducible, completing the proof.                                                                               ❑

The next criterion derived by Ryser [227] in the context of (0, 1)-matrices with prescribed row and column sums is equivalent to prescribed degrees in bipartite graphs.

**Theorem 14.16 (Ryser [227])**   If $A$ and $B$ are sequences of non-negative integers with $A$ in non-increasing order, then $A$ and $B$ are the score sequences of some bipartite tournament if and only if

$$\sum_1^k a_i \leq \sum_1^n \min(k, m - b_j)$$

for $1 \leq k \leq m$, with equality when $k = m$. Further, the bipartite tournament is irreducible if and only if the inequality, is strict for all $k < m$ and $0 < b_j < m$ for all $j \leq n$.

**Remarks**   In general, one need not check the inequalities for all values of $h$ and $k$, but only for those for which the next value in the sequence is different. Thus in order to show that [5, 5, 5, 3, 3, 2, 2] and [6, 5, 4, 1, 1, 0] belong to a bipartite tournament, we need to check the inequalities in Theorem 14.16 for $k = 3, 5$, and 7 only.

**Theorem 14.17** If $A = [a_i]_1^m$ and $B = [b_j]_1^n$ are non-decreasing integer sequences with $0 \le a_i \le n$ and $0 \le b_j \le m$, and are such that

   a. $A_m + B_n = mn$ and

   b. $A_r + B_s \ge rs$, whenever $a_r < a_{r+1}$ and $b_s < b_{s+1}$.

Then $(A, B)$ is realisable.

(Note that for a sequence $L = [x_i]_1^p, L_q = \sum_{i=1}^{q} x_i$ for $1 \le q \le p$ and $L_0 = 0$).

**Proof** We show that the inequality

$$A_k + B_x \ge kx \tag{4.17.1}$$

holds for all $1 \le k \le m$ and $1 \le x \le n$. If this is not the case for some $k$ and $x$, let $q$ and $s$ be the smallest, and $r$ and $t$ be the largest indices such that $a_{q+1} = a_k = a_r$ and $b_{s+1} = b_x = b_t (q, s = 0)$. Now, $A_r + B_x < kx$. Claim that at least one of $A_k + B_s < ks$ and $A_k + B_t < kt$ holds. For otherwise, $(x - s)b_x < k(x - s)$ and $(t - x)b_x > k(t - x)$ which is impossible. Thus, assume (i) $A_k + B_s < ks$.

Now, by hypothesis, (ii) $A_q + B_s \ge qs$ and (iii) $A_r + B_s \ge rs$ (observe that if $r = m$, then $A_m + B_n = mn$ and $0 \le b_j \le m$ together imply (iii)). Then (i) and (ii) give $(k - q)a_k < (k - q)s$, while (i) and (iii) give $(r - k)a_k > (r - k)s$. These again lead to a contradiction. The case $A_k + B_t < kt$ can similarly be treated. ◻

The following result can be obtained from Theorem 14.16.

**Corollary 14.4** If $C = [c_1, \ldots, c_m]$ and $D = [d_1, \ldots, d_n]$ be two non-increasing sequences having equal sum, then the following are equivalent.

   i. $\sum_1^k c_i \le \sum_1^n \min(k, d_j)$, for $k = 1, \ldots, m$.

   ii. $\sum_1^x d_j \le \sum_1^m \min(x, c_i)$, for $x = 1, \ldots, n$.

Now, different bipartite tournaments can have the same score sequences and they can differ only within irreducible components. That is, they must have the same numbers of components the same numbers of vertices within components, the same scores within components, and the same dominance between components. The general procedure here is to find a dominating component (a component is called dominating if it has no incoming arcs), delete its vertices and repeat. While an ordinary tournament has precisely one dominating component, the situation in the bipartite case is slightly different. It is described in the following result [19] and is a direct consequence of Theorem 14.15.

**Theorem 14.18** Let $A = [a_1, \ldots, a_m]$ and $B = [b_1, \ldots, b_n]$ be score sequences (in non-decreasing order) of a reducible $m \times n$ bipartite tournament.

i. If $a_m = n$ or $b_n = m$, then there is a corresponding trivial dominating component, consisting of one vertex which dominates all the vertices in the other partite set.

ii. Otherwise, if $k$ and $x$ are the largest indices with $k < m$ and $x < n$ such that $\sum\limits_{i=1}^{k} a_i + \sum\limits_{j=1}^{x} b_j < kx$, then the non-trivial dominating component consists of all the vertices in the two partite sets with scores exceeding $a_k$ and $b_k$ respectively.

The following results can be found in [11].

**Theorem 14.19**  If $A = [a_i]_1^m$ and $B = [b_j]_1^n$ are realisable pair of score sequences with $0 < a_i < n$ and $0 < b_j < m$ and if $|a_i - a_k| \leq 1$ for any $i, k = 1, 2, \ldots, m$, then any bipartite tournament with score sequences $A$ and $B$ is irreducible.

**Proof**  Assume that $T$ is a reducible bipartite tournament on partite sets $X$ and $Y$ with score sequences $A$ and $B$ respectively. Since $0 < a_i < n$ and $0 < b_j < m$, $T$ has at least two non trivial components, say $C$ and $C'$, with $C$ being the dominating one. If $x_i \in C \cap X$ and $x_k \in C' \cap X$, then $x_i$ dominates all the vertices in $Y$ dominated by $x_k$. Also, there exists $y_j \in C \cap Y$ and $y_l \in C' \cap Y$ such that $x_i \to y_j \to x_k$ and $x_i \to y_l \to x_k$. Thus, $a_i =$ score $(x_i) \geq$ score $(x_k) + 2 = a_k + 2$. This contradicts the hypothesis, and the result follows.  ❑

**Theorem 14.20**  Let $A = [a_i]_1^m$ (in non-decreasing order) and $B = [b^n]$ be sequences such that $(A, B)$ is realisable. Let $a_k, a_x$ be two entries in $A$ with $a_k > 0$ and $a_x < n$. Define a new sequence $A' = [a_i']_1^m$ as follows.

$$a_k' = a_k - 1, \ a_x' = a_x + 1$$

and $a_i' = a_i$, for $i \neq k, x$. Then $(A', B)$ is realisable.

**Proof:**  It follows immediately from Theorem 14.17.  ❑

**Theorem 14.21**  If $(A, B)$ is irreducible, i. e., $(A, B)$ is realisable and all its realisation are irreducible, and if $A'$ is obtained from $A$ by adding 1 to some entry and $B'$ is obtained from $B$ by subtracting 1 from some entry, then $(A', B')$ is realisable.

## 14.8  Uniquely Realisable (Simple) Pairs of Score Sequences

If $(A, B)$ is realisable, let $T$ denote a realisation on partite sets $X$ and $Y$. If a pair $(A, B)$ is realisable and all its realisations are isomorphic, then $(A, B)$ is called uniquely realisable.

The following observations can be found in [11].

**Lemma 14.9** Let $(A, B)$ be uniquely realisable. For any entry $a_i$ in $A$ and $b_j$ in $B$, let $X_i$ and $Y_j$ be the subsets of $X$ and $Y$ consisting of vertices of scores $a_i$ and $b_j$ respectively. Then any cycle in $T$ contains the same number of arcs from $X_i$ to $Y_j$ as from $Y_j$ to $X_i$.

**Proof** If this were not the case for some cycle $Z$, then reversal of the arcs of $Z$ would produce an $(A, B)$ realisation non-isomorphic to $T$.

**Lemma 14.10** If $(A, B)$ is irreducible and uniquely realisable, then $A$ or $B$ is constant.

**Proof** If neither $A$ nor $B$ is constant, let $X_1$ be the set of vertices of minimum score in $X$ and let $X_2 = X - X_1$, and similarly define $Y_1$ and $Y_2$. Since $T$ is irreducible, every arc is contained in a cycle. Thus by Lemma 14.9, none of the four subtournaments $T(X_i, Y_i)$, $i, j = 1, 2$ is unanimous, i. e., has all its arcs directed from one partite set to the other.

Choose a vertex $x_1$ in $X_1$ of minimum score in $T(X_1, Y_1)$. Then $x_1$ dominates some $y_2 \in Y_2$. Consider two cases depending on whether or not $y_2$ dominates some vertex in $X_2$.

**Case (i)** Every vertex in $X_2$ dominates $y_2$. Let $uv$ be an arc from $Y_1$ to $X_2$. Since score $(y_2) > \text{score}(u)$ in $T$, there exists an $x \in X_1$ such that $y_2 \rightarrow x \rightarrow u$ (Fig. 14.8).

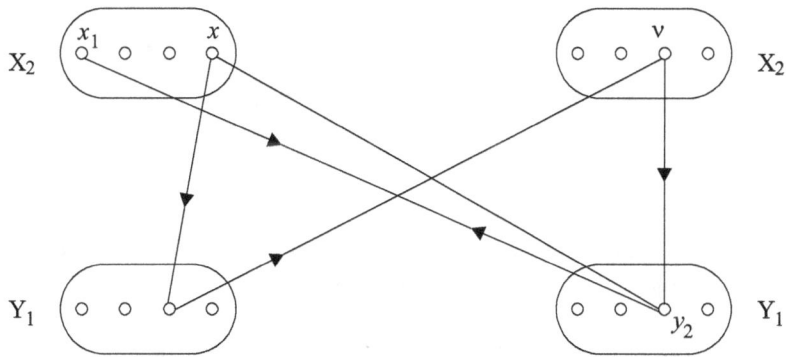

**Fig. 14.8**

But then $x \rightarrow u \rightarrow v \rightarrow y_2 \rightarrow x$ is a cycle which violates Lemma 14.9.

**Case (ii)** Some vertex $x_2 \in X_2$ is dominated by $y_2$. By the choice of $x_1$, there exists $y_1 \in Y_1$ such that $y_1 \rightarrow x_1$. If $x_2 \rightarrow y_1$, we again get a 4-cycle which has one arc from $Y_1$ to $X_1$, but no arcs in the other direction. So assume $y_1 \rightarrow x_2$ (Fig. 14.9). Since score $(y_2) > \text{score}$ $(y_1)$ in $T$, there exists an $x \in X_1$ with $y_2 \rightarrow x \rightarrow y_1$. If $x \in X_2$, we again get a forbidden cycle $x_1 \rightarrow y_2 \rightarrow x \rightarrow y_1 \rightarrow x_1$. Thus, $x \in X_1$. Likewise, there exists $y \in Y_1$ with $x_2 \rightarrow y \rightarrow x$. But then $x_1 \rightarrow y_2 \rightarrow x_2 \rightarrow y \rightarrow x \rightarrow y_1 \rightarrow x_1$ is a 6-cycle which violates Lemma 14.9.

Since all the possibilities have been exhausted, the result follows. $\quad\square$

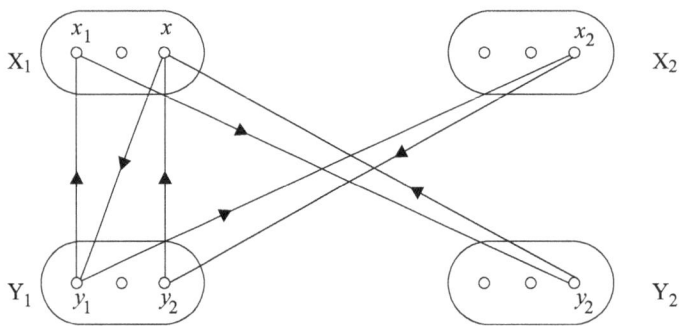

**Fig. 14.9**

**Remarks**   We note that if in a realisable pair $(A, B)$, one of the sequences has all entries as 1's (the sequence is constantly 1), then $(A, B)$ is uniquely realisable. This is illustrated in Figure 14.10, where $A = [1, 1, \ldots, 1]$ and $B = [b_1, b_2, \ldots, b_n]$ and only $X$ to $Y$ arcs are shown.

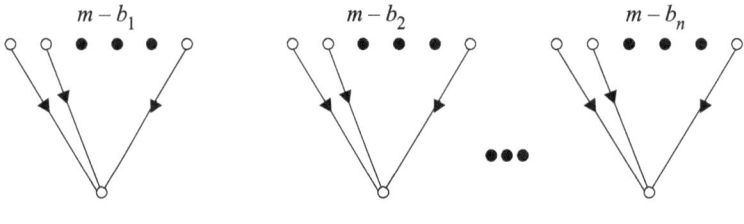

**Fig. 14.10**

Now onwards assume that none of the sequences $A$, $B$, $\overline{A}$ and $\overline{B}$ is constantly 1. Hence, the following observations [232] can be easily proved.

**Lemma 14.11**   If $(A, B)$ is irreducible and uniquely realisable, then $A$ or $B$ is non-constant.

**Lemma 14.12**   With $A$ and $B$ as above, the sequence $B$ has precisely two distinct values.

**Lemma 14.13**   If $(A, B)$ is irreducible, uniquely realisable and if $A = [a^m]$, $2 = a = n - 2$, and $B = [b^r, c^s]$, $1 = b < c = m - 1$, and $r, s > 0, r + s = n$, then $r = 1$ or $s = 1$.

**Remarks**   It has been shown that if $(A, B)$ is irreducible, uniquely realisable, then one of the sequences (or its dual) consists entirely of 1's, or one of the sequences is constant and the other has exactly two distinct values, one of which appears precisely once.

Now, the following important result due to Bagga and Beineke [11] gives necessary and sufficient conditions for unique realisability in the irreducible case.

**Theorem 14.22 (Bagga and Beineke [11])**   An irreducible pair $(A, B)$ of score sequences is uniquely realisable if and only if one of the following holds.

I. (without loss of generality) $A = [1^m]$ and $B$ is arbitrary,

I'. the dual of ($I$), that is, $A = [(n-1)^m]$ and $B$ is arbitrary,

II. (without loss of generality) $A = [1^{m-1}, a]$ and $B = [b^n]$,

II'. the dual of II,

III. (without loss of generality) $A = [1, a^{m-1}]$ and $B = [2^n]$,

III'. the dual of III.

**Proof** The sufficiency of ($I$) has already been noted in the remarks before Lemma 14.11 where Figure 14.10 shows the unique realisation. Now, let $T$ be a realisation of $A = [1^{m-1}, a]$ and $B = [b^n]$ on partite sets $X = \{x_1, x_2, \ldots, x_m\}$ and $Y = \{y_1, y_2, \ldots, y_n\}$ respectively. If (say) $x_m$ has score $a$ and it dominates (say) $y_1, y_2, \ldots, y_n$, then $T - x_m$ has score sequences $A_1 = [1^{m-1}]$ and $B_1 = [(b-1)^{n-a}, b^a]$. Thus by ($I$), $(A_1, B_1)$ is uniquely realisable. The unique realisability of $(A, B)$ follows. This proves the sufficiency of (II). The proof of (III) is similar and the dual cases follow by the remarks before Lemma 14.11.

For proving necessity, induct on $m+n$. Since $(A,B)$ is irreducible, so $m, n \geq 2$. If (say) $m = 2$, then $B = [1^n]$, and the result follows. Now, assume that the result holds for all irreducible and uniquely realisable pairs of score sequences with combined length less than $m+n$, and consider such a pair $(A, B)$ with $|A| = m$ and $|B| = n (m, n \geq 3)$.

Assume that $A$ and $B$ are not of the type ($I$) or ($I'$). Then by the remarks after Lemma 14.13, we have, without loss of generality, $A = [a^m]$ and $B = [b^{n-1}, c]$, with $1 < a < n-1, 1 \leq b, c \leq m-1$ and $b \neq c$.

If $y$ is the vertex of score $c$ in a realisation $T$ of $(A, B)$, then $T - y$ has score sequences $A_1 = [(a-1)^{m-c}, a^c]$ and $B_1 = [b^{n-1}]$. Now, the unique realisability of $(A, B)$ implies that of $(A_1, B_1)$. Also, by Theorem 14.19, $(A_1, B_1)$ is irreducible. Thus, by the induction hypothesis, $A_1$ and $B_1$ belong to one of the six given types. Consider these cases one by one.

i. If $B_1 = [1^{n-1}]$, then $B = [1^{n-1}, c]$, so that $A$ and $B$ are of type (II).

ii. If $B_1 = [(m-1)^n - 1]$, then $A$ and $B$ belong to ($II'$).

iii. If $B_1 = [b^{n-1}]$ and $A_1 = [1^{m-1}, a]$, then $c = 1$ and $a = 2$, so that $A = [2^m]$ and $B = [1, b^{n-1}]$. This is of type (III).

iv. If $B_1 = [b^{n-1}]$ and $A_1 = [d, (n-2)^{m-1}]$, we get $A$ and $B$ of type ($III'$).

v. If $B_1 = [2^{n-1}]$ and $A_1 = [1, a^{m-1}]$, then $b = 2$, $a = 2$ and $c = m-1$. Thus, $A = [2^m]$ and $B = [2^{n-1}, m-1]$. Using Moon's theorem, we get $2m + 2(n-1) + (m-1) = mn$, so that $m = \frac{2n-3}{n-3}$. It follows that $n = 6$ and $m = 3$. But then $A$ and $B$ are both constant, a contradiction. Therefore this case is not possible.

vi. The possibility of $B_1 = [(m-2)^{n-1}]$ and $A_1 = [a^{m-1}, n-2]$ follows by duality.

This exhausts all the possibilities and hence by induction, the result is completely proved.

❑

Now, assume that $(A, B)$ is a realisable pair and $Q_1, Q_2, \ldots, Q_p$ are the irreducible components of a realisation $T$ of $(A, B)$. Also, let $Q_k$ has score sequences $A_k$ and $B_k$, $1 \le k \le p$. Then $(A, B)$ is uniquely realisable if and only if $(A_k, B_k)$ is uniquely realisable for all $k$.

## 14.9 Score Sequences of Oriented Graphs

An oriented graph is a digraph with no symmetric pairs of directed arcs and with no loops. Avery [233] extended the concept of score structure to all oriented graphs.

**Definition:** Let $D$ be an oriented graph with vertex set $V = \{1, 2, \ldots, n\}$, and let $d^+(v)$ and $d^-(v)$ be the outdegree and indegree respectively of vertex $v$. Then, the *score* of vertex $v$ denoted by $a_v$ is defined as $a_v = n - 1 + d^+(v) - d^-(v)$ with $0 \le a_v \le 2n - 2$. The sequence of scores is called the *score list*, and $A = [a_1, a_2, \ldots, a_n]$ arranged in non-decreasing order is called the *score sequence* of $D$.

Any oriented graph can be interpreted as the result of a round robin competition in which ties (draws) are allowed, that is, the participants play each other once, with an arc from $u$ to $v$ if and only if $u$ defeats $v$. A player receives two points for each win and one point for each tie, as is frequently the case in sports such as soccer, ice hockey and cricket. With this scoring system , player $v$ obtains a total of $av$ points. An arc from $u$ to $v$ denoted by $u \to v$ is written as $u(1-0)v$, and $u(0-0)v$ means that neither $u \to v$ nor $v \to u$.

**Definition:** A *triple* in an oriented graph is an induced subdigraph with three vertices. A *cyclic triple* is an intransitive triple of the form $u \to v \to w \to u$. Any triple can be of the form $u(x_1 - x_2)v(y_1 - y_2)w(z_1 - z_2)u$, where $0 \le x_i, y_i, z_i \le 1$ with $0 \le \sum x_i, \sum y_i, \sum z_i \le 1$.

The following result [9] extends a result of Ryser [234] which showed that if two tournaments have the same score structure, then each can be transformed to the other by successively reversing the arcs of appropriate cyclic triples.

**Theorem 14.23** Let $D$ and $D'$ be two oriented graphs with the same score sequence. Then $D$ can be transformed to $D'$ by successively transforming appropriate triples in one of the following ways.

Either (a) by changing a cyclic triple $u(1-0)v(1-0)w(1-0)u$ to a transitive triple $u(0-0)v(0-0)w(0-0)u$, which has the same score sequence, or vice versa.

or (b) by changing an intransitive triple $u(1-0)v(1-0)w(0-0)u$ to a transitive triple $u(0-0)v(0-0)w(0-1)u$, which has the same score sequence, or vice versa.

The following result due to Avery [9] gives a constructive condition for a non-negative sequence in non-decreasing order to be a score sequence of some oriented graph. A short proof of this result is due to S.Pirzada et. al. [199].

**Theorem 14.24 (Avery [9])** A sequence of non-negative integers in non-decreasing order is the score sequence of an oriented graph if and only if

$$\sum_{i=1}^{k} a_i \geq k(k-1)$$

for $1 \leq k \leq n$, with equality for $k = n$.

The following result can be found in [9].

**Theorem 14.25**  Among all oriented graphs with a given score sequence, those with the fewest arcs are transitive.

**Proof**  Let $A$ be a score sequence and let $D$ be a realisation of $A$ that is not transitive. Then $D$ has an intransitive triple. There are two types of intransitive triples, a cyclic triple, which can be transformed by operation (a) of Theorem 14.23 to a triple with the same score sequence and three arcs fewer, and a triple $u(1-0)v(1-0)w(0-0)u$, which can be transformed by operation (b) of Theorem 14.23 to a triple with the same score sequence and one arc fewer. So in either case, we obtain a realisation of $A$ with fewer arcs.  ❑

The next result [9] provides a useful recursive test of whether a given sequence of non-negative integers is the score list of an oriented graph. We note that a transmitter is a vertex with indegree zero.

**Theorem 14.26**  Let $A$ be a sequence of $n$ integers between 0 and $2n-2$ inclusive and let $A'$ be obtained from $A$ by deleting the greatest entry $2n-2-r$ say, and reducing each of the greatest $r$ remaining entries in $A$ by one. Then $A$ is a score list if and only if $A'$ is a score list.

**Proof**  Clearly, in a transitive oriented graph, any vertex of greatest score is a transmitter.

Let $A'$ be a score list of some oriented graph $D'$. Then an oriented graph $D$ with score list $A$ can be obtained by adding a transmitter that is adjacent to just those vertices of whose scores are not reduced in going from $A$ to $A'$.

For the converse, we show that there is an oriented graph with score list $A$ in which a transmitter $v$ with score $2n-2-r$ is adjacent to the (other) $n-1-r$ vertices with least scores. By Theorem 14.25, there is a transitive oriented graph $D$ with score list $A$, in which a vertex $v$ with greatest score $2n-2-r$ is a transmitter. Let $U$ be the set of $r$ vertices, apart from $v$, with the greatest scores in $A$, and let $W$ be the set $V - \{v \cup U\}$.

Let $v$ be adjacent in $D$ to vertices $u_1, u_2, \ldots, u_k$ of $U$. Then there are exactly $k$ vertices, say $w_1, w_2, \ldots, w_k$ of $W$ not adjacent from $v$. Now, $u_i$ cannot be adjacent to $w_i$, since $D$ is transitive. Neither can $w_i$ be adjacent to $u_i$, since taken together with the transitivity of $D$ this implies that the score of $w_i$ is greater than the score of $u_i$, which is contrary to the assumption. Thus $w_i(0-0)u_i$ for all $i$.

Now, transforming all triples $v(1-0)u_i(0-0)w_i(0-0)v$ to triples $v(0-0)u_i(0-1)w_i(0-1)v$, the vertex scores remain unchanged. This forms an (not necessarily transitive) oriented graph $D_1$ with score list $A$ in which the transmitter $v$ is adjacent to all vertices of $W$ and none of $U$, as required.  ❑

Theorem 14.26 provides an algorithm for determining whether a given non-decreasing sequence A of non-negative integers is a score sequence, and for constructing a corresponding oriented graph. At each stage, we form $A'$ according to Theorem 14.26, such that scores of $A'$ are also non-decreasing. If $a_n = 2n - 2 - r$, this means deleting $a_n$ and reducing the $r$ greatest remaining entries by one each to form $A' = |a'_1, a'_2, \ldots, a'_{n-1}|$ while ensuring that this is also non-decreasing. Arcs of an oriented graph are defined by $n \to v$ if and only if $a'_v = a_v$ If this procedure is applied recursively, then first it tests whether $A$ is a score sequence and if $A$ is a score sequence, an oriented graph $\Delta(A)$ with score sequence $A$ is constructed.

**Example**   Let $n = 5$, $A = [2, 4, 4, 4, 6]$.

| Stage | A | B | Arcs of $\Delta(A)$ |
|---|---|---|---|
| 1 | [2, 4, 4, 4, 6] | [2, 3, 3, 4] | $5 \to 4, 5 \to 1$ |
| 2 | [2, 3, 3, 4] | [2, 2, 2] | $4 \to 1$ |
| 3 | [2, 2, 2] | [1, 1] | |
| 4 | [1, 1] | [0] | |

Thus, $A$ is a score sequence.

We have the following observations [9] about $\Delta(A)$.

**Theorem 14.27**   The oriented graph $\Delta(A)$ is transitive for any score sequence $A$.

**Theorem 14.28**   There is no oriented graph with score sequence $A$ which has fewer arcs than $\Delta(A)$.

One more method of constructing oriented graph with a given score sequence can be found in *S*. Pirzada [188].

The following is an equivalent statement of Theorem 14.24. A sequence of non-negative integers $A = [a_i]_1^n$ in non-decreasing order is a score sequence of an oriented graph if and only if for each subset $I \subseteq [n] = \{1, 2, \ldots, n\}$

$$\sum_{i \in I} a_i \geq 2 \binom{|I|}{2}$$

with equality for $|I| = n$.

The following inequalities for scores in oriented graphs can be found in [235].

**Theorem 14.29**   A sequence $A = [a_i]_1^n$ of non-negative integers in non-decreasing order is a score sequence of an oriented graph if and only if for every subset, $I = [n]$,

$$\sum_{i \in I} a_i \geq \sum_{i \in I}(i - 1) + \binom{|I|}{2}$$

with equality when $I = [n]$

**Theorem 14.30** A sequence $A = [a_i]_1^n$ of non-negative integers in non-decreasing order is a score sequence if and only if for any subset $I \subseteq [n]$,

$$\sum_{i \in I} a_i \le \sum_{i \in I} (i-1) + \frac{1}{2}|I|(2n - |I| - 1),$$

with equality for $I = [n]$.

**Theorem 14.31** If $A = [a_i]_1^n$ is a score sequence of an oriented graph, then for each $i$, $i - 1 \le a_i \le n + i - 2$.

A necessary condition for a score sequence in oriented graphs to be self-converse, can be found in [192].

## 14.10 Score Sets in Oriented Graphs

**Definition:** The set $A$ of distinct scores of vertices in an oriented graph $D$ is called the *score set* of $D$.

**Definition:** A digraph $D$ is *diregular* if $d_v^+ = d_v^- = k$ holds for each vertex $v$ in $D$. In case of an oriented graph $D$ with $n$ vertices, $a_v = n - 1 + d_v^+ - d_v^-$, for each vertex $v$ in $D$, and when $d_v^+ = d_v^- = k$ (say), then $a_v = n - 1 + k - k = n - 1$ for each $v$ in oriented graph $D$. Thus an oriented graph $D$ with $n$ vertices is diregular if $a_v = n - 1$, for all $v$ in $D$.

Now, we have the following result, the proof of which is obvious.

**Lemma 14.14** The number of vertices in an oriented graph with at least two distinct scores does not exceed its largest score.

The following result is given by *S. Pirzada* and *T. A. Naikoo* [196].

**Theorem 14.32 (a) (Pirzada and Naikoo [196])** Let $A = \{a, ad, ad^2, \ldots, ad^n\}$, where $a$ and $d$ are positive integers with $a > 0$ and $d > 1$. Then there exists an oriented graph $D$ with score set $A$, except for $a = 1$, $d = 2$, $n > 0$ and for $a = 1$, $d = 3$, $n > 0$.

**Proof** We induct on $n$. For $n = 0$, there is a positive integer $a > 0$, so that $a + 1 > 0$. Let $D$ be a diregular oriented graph having $a + 1$ vertices. Then, $a_v = a + 1 - 1 = a$, for all $v \in D$. Therefore score set of $D$ is $A = \{a\}$. This proves the result for $n = 0$. If $n = 1$, then there are positive integers $a$ and $d$ with $a > 0$ and $d > 1$, and for $a = 1$, $d \ne 2, 3$.

Now, three cases arise: (I) $a > 1$, $d > 2$, (II) $a > 1$, $d = 2$ and (III) $a = 1$, $d > 3$.

(I) Let $a > 1$, $d > 2$. Therefore, $a + 1 > 0$. Let $D_1$ be a diregular oriented graph having $a + 1$ vertices. Then, $a_v = a + 1 - 1 = a$, for all $v \in D_1$.

Now, $ad - 2|V(D_1)| + 1 = ad2(a+1) + 1 = ad - 2a - 1 \ge 3a - 2a - 1 = a - 1 > 0$, as $d \ge 3$ and $a > 1$. That is, $ad - 2|V(D_1)| + 1 > 0$. Let $D_2$ be a diregular oriented graph having

$ad - 2|V(D_1)| + 1$ vertices. Then, $a_u = ad - 2|V(D_1)| + 1 - 1 = ad - 2|V(D_1)|$, for all $u \in D_2$.

Let there be an arc from every vertex of $D_2$ to each vertex of $D_1$, so that we get an oriented graph $D$ (which includes $D_1$ to $D_2$ together with all the new arcs from $D_2$ and $D_1$) having $|V(D_1)| + |V(D_2)| = a + 1 + ad - 2|V(D_1)| + 1 = a + 1 + ad - 2(a + 1) + 1 = ad - a$ vertices with $a_v = a$, for all $v \in D_1$, and $a_u = ad - 2|V(D_1)| + 2|V(D_1)| = ad$, for all $u \in D_2$. Therefore score set of $D$ is $A = \{a, ad\}$.

(II) Assume $a > 1$, $d = 2$. First take $a = 2$, $d = 2$. Then, $ad = 4 > 0$. Let $D$ be an oriented graph having $ad = 4$ vertices, say, $v_1, v_2, v_3$, and $v_4$ in which $v_1 \to v_3$ and $v_2 \to v_4$, so that $a_{v_1} = a_{v_2} = 2 + 4 - 2 = 4 = ad$, and $a_{v_3} = a_{4_4} - 2 = 2 = a$ Therefore $D$ is an oriented graph having $ad$ vertices with score set $A = \{a, ad\}$.

Now, take $a > 2$, $d = 2$. Let $D_1$ be a diregular oriented graph having 2 vertices, say, $v_1$ and $v_2$. Then, $a_{v_i} = 2 - 1 = 1$ for all $v_i \in D_1$, where $1 \leq i \leq 2$.

Again, $a > 2$ or $a - 2 > 0$. Let $D_2$ be a diregular oriented graph having $a - 2$ vertices, say, $v_3$, $v_4$, ..., $v_a$. Then, $a_{v_j} = a - 2 - 1 = a - 3$, for all $v_j \in D_2$, where $3 \leq j \leq a$.

Let there be an arc from every vertex of $D_2$ to each vertex of $D_1$, so that we get an oriented graph $D_3$ (which includes $D_1$ and $D_2$ together with all the new arcs from $D_2$ to $D_1$) having $2 + a - 2 = a$ vertices with $a_{v_i} = 1$, for all $v_i \in D_1$, where $1 \leq i \leq 2$, and $a_{v_j} = a - 3 + 2(2) = a + 1$, for all $v_j \in D_2$, where $3 \leq j \leq a$.

Again, $a > 2 > 0$. Let $D_4$ be a diregular oriented graph having $a$ vertices, say, $w_1$, $w_2$, ..., $w_a$. Then, $a_{w_k} = a - 1$, for all $w_k \in D_4$, where $1 \leq k \leq a$.

Let there be $a$ arcs from $a$ distinct vertices of $D_4$ to $a$ distinct vertices of $D_3$ ($w_q \to v_q$, for all $q = 1, 2, ..., a$), so that we get an oriented graph $D$ (which includes $D_3$ and $D_4$ together with all the new arcs from $D_4$ to $D_3$) having $a + a = 2a = ad$ vertices with $a_{v_i} = 1 + a - 1 = a$, for all $v_i \in D_3$, where $1 \leq i \leq 2$, $= a + 1 + a - 1 = 2a$, for all $v_j \in D_3$, where $3 \leq j \leq a$, and $a_{w_k} = a - 1 + 2(1) + a - 1 = 2a$, for all $w_k \in D_4$, where $1 \leq k \leq a$. Therefore score set of $D$ is $A = \{a, 2a\} = \{a, ad\}$.

(III) Finally, let $a = 1$, $d > 3$. Therefore, $a + 1 > 0$. Let $D_1$ be a diregular oriented graph having $a + 1$ vertices. Then, $a_v = a + 1 - 1 = a$, for all $v \in D_1$.

Now, $ad - 2|V(D_1)| + 1 = ad - 2(a + 1) + 1 = ad - 2a - 1 \geq 4a - 2a - 1 = 2a - 1 > 0$, as $d \geq 4$ and $a = 1$; i.e., $ad - 2|V(D_1)| + 1 > 0$. Then as in (I), we have an oriented graph $D$ having $ad - a$ vertices with score set $A = \{a, ad\}$.

Hence in all these cases, we get an oriented graph $D$ with score set $A = \{a, ad\}$. This shows that the result is also true for $n = 1$.

Assume, the result to be true for all $p \geq 1$. We show that the result is true for $p + 1$.

Let $a$ and $d$ be positive integers with $a > 0$ and $d > 1$, and for $a = 1$, $d \neq 2, 3$. Therefore by induction hypothesis, there exists an oriented graph $D_1$ having $|V(D_1)|$ vertices with score set $\{a, ad, ad^2, ..., ad^p\}$.

Once again, we have either (I) $a > 1$, $d > 2$, or (II) $a > 1$, $d = 2$, or (III) $a = 1$, $d > 3$. Obviously, for $d > 1$, in all the above possibilities, $ad^{p+1} \geq 2ad^p$, and the score set of

$D_1$, namely, $\{a, ad, ad^2, \ldots, ad^p\}$ has at least two distinct scores for $p \leq 1$. Therefore by Lemma 14.14, $|V(D_1)| \leq ad^p$. Hence, $ad^{p+1} \geq 2|V(D_1)|$, or $ad^{p+1} - 2|V(D_1)| + 1 > 0$.

Let $D_2$ be a diregular oriented graph having $ad^{p+1} - 2|V(D_1)| + 1$ vertices. Then, $a_v = ad^{p+1} - 2|V(D_1)| + 1 - 1 = ad^{p+1} - 2|V(D_1)|$, for all $v \in D_2$.

Let there be an arc from every vertex of $D_2$ to each vertex of $D_1$, so that we get an oriented graph $D$ (which includes $D_1$ and $D_2$ together with all the new arcs from $D_2$ to $D_1$) having $|V(D_1)| + |V(D_2)|$ vertices with $a, ad, ad^2, \ldots, ad^p$ as the scores of the vertices of $D_1$, and $a_v = ad^{p+1} - 2|V(D_1)| + 2|V(D_1)| = ad^{p+1}$, for all $v \in D_2$. Therefore score set of $D$ is $A = \{a, ad, ad^2, \ldots, ad^p, ad^{p+1}\}$, proving the result for $p + 1$. Hence the result follows. ❑

That no oriented graph exists when either $a = 1$, $d = 2$, $n > 0$ or $a = 1$, $d = 3$, $n > 0$, is proved in the following theorem.

**Theorem 14.32 (b)**    There exists no oriented graph with score set $A = \{a, ad, ad^2, \ldots, ad^n\}$, $n > 0$, when either (i) $a = 1$, $d = 2$, or (ii) $a = 1$, $d = 3$.

We now have the following result [201].

**Theorem 14.32 (c)**    If $a_1, a_2, \ldots, a_n$ are $n$ non-negative integers with $a_1 < a_2 < \ldots < a_n$. Then, there exists an oriented graph $D$ with score set $A = \{a_1', a_2', \ldots, a_n'\}$, where

$$a_i' = \begin{cases} a_{i-1} + a_i + 1, & \text{for } i > 1, \\ a_i, & \text{for } i = 1. \end{cases}$$

**Remarks**

1. From Theorem 14.32 ( c ), it follows that every singleton set of non-negative integers is a score set of some oriented graph.

2. As we have shown in Theorem 14.32(b), i. e., the sets $\{1, 2, 2^2, \ldots, 2^n\}$ and $\{1, 3, 3^2, \ldots, 3^n\}$ cannot be the score sets of any oriented graph for $n > 0$. It follows, therefore, that the above results cannot be generalised to conclude that any set of non-negative integers forms the score set of some oriented graph. However, there can be other special classes of non-negative integers which can form the score set of an oriented graph, and the problem needs further investigations.

Pirzada and Naikoo [195] have obtained some results on degree frequencies in oriented graphs. More results on scores, score sets and kings in oriented graphs can be seen in [199, 201, 203, 207].

## 14.11  Uniquely Realisable (Simple) Score Sequences in Oriented Graphs

An oriented graph $D$ is *reducible* if it is possible to partition its vertices into two nonempty sets $V_1$ and $V_2$ in such a way that there is an arc from every vertex of $V_2$ to each vertex of

$V_1$. Let $D_1$ and $D_2$ be induced digraphs having vertex sets $V_1$ and $V_2$ respectively. Then $D$ consists of $D_1$ and $D_2$ and arcs from every vertex of $D_2$ to each vertex of $D_1$. We write $D = [D_1, D_2]$. If this is not possible, then the oriented graph $D$ is irreducible. Let $D_1, D_2, \ldots, D_k$ be irreducible oriented graphs with disjoint vertex sets. Then $D = [D_1, D_2, \ldots, D_k]$ denotes the oriented graph having all arcs of $D_i$, $1 \le i \le k$ and there are arcs from every vertex of $D_j$ to each vertex of $D_i$, $1 \le i < j \le k$. Here $D_1, D_2, \ldots, D_k$ are called *irreducible components* of $D$. Such a decomposition is called as *irreducible component decomposition* of $D$, which is unique.

**Definition:** A score sequence $A$ is said to be *irreducible* if all the oriented graphs $D$ with score sequence $A$ are irreducible.

In case of ordinary tournaments, the score sequence used to decide whether a tournament $T$ having score sequence $S$, is strong or not. This is not true in case of oriented graphs. For example, the oriented graphs $D_1$ and $D_2$ in Figure 14.11, both have score sequence $[2, 2, 2]$ but $D_1$ is strong and $D_2$ is not.

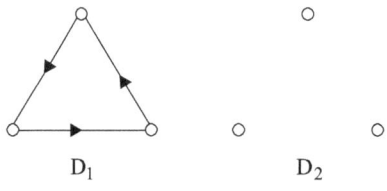

$$D_1 \qquad\qquad D_2$$

**Fig. 14.11**

The following result due to S. Pirzada [186] characterises irreducible oriented graphs.

**Theorem 14.33 (S. Pirzada [186])**    Let $D$ be an oriented graph having score sequence. Then $D$ is irreducible if and only if, for $k = 1, 2, \ldots, n-1$

$$\sum_{i=1}^{k} a_i \ge k(k-1) \tag{14.33.1}$$

and

$$\sum_{i=1}^{k} a_i = n(n-1) \tag{14.33.2}$$

**Proof**    Suppose $D$ is an irreducible oriented graph having score sequence $[a_i]_1^n$. Condition (14.33.2) holds since Theorem 14.24 has already established it for any oriented graph. To verify inequalities (14.33.1) we observe that for any integer $k < n$, the subdigraph induced by any set of $k$ vertices has a sum of scores $k(k-1)$. Since $D$ is irreducible, there must be an arc from at least one of these vertices to one of the other $n-k$ vertices, or there is no arc from these $k$ vertices to other $n-k$ vertices. Thus, for $1 \le k \le n-1$

$$\sum_{i=1}^{k} a_i > k(k-1).$$

For the converse, suppose conditions (14.33.1) and (14.33.2) hold, we know by Theorem 14.24 that there exists an oriented graph $D$ with these scores. Assume that such an oriented graph $D$ is irreducible. Let $D = [D_1, D_2, \ldots, D_k]$ be the irreducible component decomposition of $D$. If $m$ is the number of vertices in $D_1$, then $m < n$, and the following equation holds,

$$\sum_{i=1}^{k} a_i = m(m-1),$$

which is a contradiction. This proves the converse part. $\qquad\qquad\qquad\qquad$ ❏

The following result result given by Pirzada [186] can be proved easily.

**Theorem 14.34** Let $D$ be an oriented graph with score sequence $A = [a_i]_1^n$. Suppose that , $\sum_{i=1}^{k} a_i = p(p-1)$, $\sum_{i=1}^{q} a_i = q(q-1)$ and for $p+1 \leq k \leq q-1$, where $0 \leq p < q \leq n$. Then the subdigraph induced by the vertices $v_{p+1}, v_{p+2}, \ldots, v_q$ is an irreducible component of $D$ with score sequence $[a_{p+1} - 2p, \ldots, a_q - 2p]$.

Now, $A$ is irreducible if $D$ is irreducible and the irreducible components of $A$ are the score sequences of the irreducible components of $D$. Theorem 14.34 shows that the irreducible components of $A$ are determined by the successive values of $k$ for which

$$\sum_{i=1}^{k} a_i = k(k-1), \ 1 \leq k \leq n. \tag{14.33.3}$$

We illustrate it with the following example.

Let $A = [1, 2, 3, 8, 8, 8, 13, 13]$. Equation (14.33.3) is satisfied for $k = 3, 6, 8$. Thus irreducible components of $S$ are $[1, 2, 3]$, $[2, 2, 2]$ and $[1, 1]$ in ascending order.

**Definition:** A score sequence is *simple* if it belongs to exactly one oriented graph. We characterise simple score sequences of oriented graphs. First we have the following observation.

**Lemma 14.15** The score sequence $A$ of an oriented graph is simple if and only if every irreducible component of $A$ is simple.

The following result due to S. Pirzada [186] determines which irreducible score sequences are simple.

**Theorem 14.35 (S. Pirzada [186])** Let $A$ be an irreducible score sequence. Then $A$ is simple if and only if it is one of $[0]$, or $[1, 1]$.

**Proof** Suppose $A$ is an irreducible score sequence and let $D$ be an oriented graph having score sequence $A$. We have three cases to consider. (1) $D$ has $n \geq 3$ vertices, (2) $D$ has two vertices, (3) $D$ has one vertex.

**Case (1)**    $D$ has $n \geq 3$ vertices. Since $A$ is irreducible, there exist vertices $u, v$ and $w$ such that $D$ has a cyclic triple $u(1-0)v(1-0)w(1-0)u$; or an intransitive triple $u(1-0)v(1-0)w(0-0)u$; or a transitive triple $u(0-0)v(0-0)w(0-1)u$; or a transitive triple $u(0-0)v(0-0)w(0-0)$.

Now, if $D$ contains the cyclic triple $u(1-0)v(1-0)w(1-0)u$, it can be changed to the transitive triple $u(0-0)v(0-0)w(0-0)u$ to form an oriented graph with the same score sequence, or vice versa. So the number of arcs in $D$ and $D'$ is different. If $D$ contains the intransitive triple $u(1-0)v(1-0)w(0-0)u$, we can transform it to the transitive triple $u(0-0)v(0-0)w(0-0)u$, to form an oriented graph having the same score sequence, or vice versa. Here, also the number of arcs in $D$ and $D'$ is different. Since in every case the number of arcs in $D$ and is not same, $D$ is not isomorphic to $D'$ . Thus $A$ is not simple.

**Case (2)**    $D$ has two vertices. Then, $A = [1, 1]$ is the only irreducible score sequence and it belongs to exactly one oriented graph, namely $u(0-0)v$.

**Case (3)**    $D$ has just one vertex. Then, $A = [0]$ which is obviously simple.

Hence $[0]$ and $[1, 1]$ are the only irreducible score sequences that are simple.    ❏

**Corollary 14.5**    The score sequence $A$ is simple if and only if every irreducible component of $A$ is one of $[0]$, or $[1, 1]$.

## 14.12   Score Sequences in Oriented Bipartite Graphs

An *oriented bipartite graph* is the result of assigning a direction to each edge of a simple bipartite graph. Let $X = \{x_1, x_2, \ldots, x_m\}$ and $Y = \{y_1, y_2, \ldots, y_n\}$ be the partite sets of an oriented bipartite graph $D$. For any vertex $u$ in $D$, let $d^+(u)$ and $d^-(u)$ be the outdegree and indegree respectively. Define $a_x = n + d^+(x) - d^-(x)$ and $b_y = m + d^+(y) - d^-(y)$ as the scores of $x$ in $X$ and $y$ in $Y$ respectively. Clearly, $0 \leq a_x \leq 2n$ and $0 \leq b_y \leq 2m$. Then the sequences $A = [a_i]_1^m$ and $B = [b_j]_1^n$ in non-decreasing order are called a *pair of score sequences* of $D$. An arc from $x$ to $y$, that is, $x \to y$ is denoted by $x(1-0)$, and $x(0-0)$ means neither $x \to y$ nor $y \to x$.

**Definition:**    A *tetra* in an oriented bipartite graph is an induced subdigraph with two vertices from each partite set. Define tetras of the form $x(1-0)y(1-0)x'(1-0)y(1-0)x$ and $x(1-0)y(1-0)x'(1-0)y'(0-0)x$ to be of $\alpha$ *-type*, and all other tetras to be of $\beta$ *-type*. An oriented bipartite graph is said to be of $\alpha$ -type or $\beta$ -type according as all of its tetras are of $\alpha$ -type or $\beta$ -type respectively (Fig. 14.12).

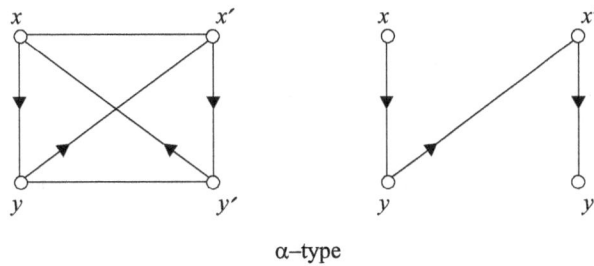

α–type

**Fig. 14.12**

We have the following simple observation.

**Theorem 14.36**   Among all oriented bipartite graphs with a given pair of score sequences, those with the fewest arcs are of $\beta$ -type.

**Proof**   Let $(A, B)$ be a given pair of score lists and let $D$ be a realisation of $(A, B)$ that is not $\beta$ -type. Then $D$ has a tetra of $\alpha$-type: $x(1-0)y(1-0)x'(1-0)y'(1-0)x$ or $x(1-0)y(1-0)x'(1-0)y'(0-0)x$. Since $x(1-0)y(1-0)x'(1-0)y'(1-0)x$ can be changed to $x(0-0)y(0-0)x'(0-0)y'(0-0)x$ with the same score sequences and four arcs fewer , and $x(1-0)y(1-0)x'(1-0)y'(0-0)x$ can be changed to $x(0-0)y(0-0)x'(0-0)y'(0-1)x$ with the same score sequences and two arcs fewer, so in either case we can obtain a realisation of $(A, B)$ with fewer arcs.   □

A *transmitter* is a vertex with indegree zero. In a $\beta$ -type oriented bipartite graph with score sequences $A = [a_1, a_2, \ldots, a_m]$ and $B = [b_1, b_2, \ldots, b_n]$ either the vertex with score $a_m$, or the vertex with score $b_n$, or both may act as transmitter.

The next result due to Pirzada, Merajudin and Yin Jianhua [194] provides a useful recursive test to find whether a pair of lists is realisable.

**Theorem 14.37 (Pirzada, Merajudin and Yin Jianhua [194])**   Suppose $A = [a_1, a_2, \ldots, a_m]$ and $B = [b_1, b_2, \ldots, b_n]$ be two sequences of non-negative integers in nondecreasing order. Let $A'$ be obtained from $A$ by deleting one entry $a_m$ and $B'$ be obtained from $B$ by reducing $2n - a_m$ greatest entries of $B$ by 1 each provided $a_m \geq n$ and $b_n \leq 2m - 1$. Then $A$ and $B$ are the score sequences of some oriented bipartite graph if and only if $A'$ and $B'$ are also score sequences of some oriented bipartite graph.

Theorem 14.37 provides an algorithm for determining whether a given pair of sequences $(A, B)$ of non-negative integers in nondecreasing order is a pair of score sequences and for constructing a corresponding oriented bipartite graph. Suppose $A = [a_1, a_2, \ldots, a_m]$ and $B = [b_1, b_2, \ldots, b_n]$ be a pair of score sequences of an oriented bipartite graph with parts $X = \{x_1, x_2, \ldots, x_m\}$ and $Y = \{y_1, y_2, \ldots, y_n\}$, where $a_m \geq n$, $b_n \leq 2m - 1$. Deleting $a_m$ and reducing $2n - a_m$ greatest entries of $B$ by 1 each to form $B' = [b'_1, b'_2, \ldots, b'_n]$.. Then arcs are defined by $x_m \to y_j$ for which $b_j = b'_j$. Now, if at least one of the conditions $a_m \geq n$ or $b_n \leq 2m - 1$ does not hold, then we delete $b_n$ (obviously $b_n \geq m$, $a_m \leq 2n - 1$) and reduce $2m - b_n$ greatest entries of $A$ by 1 each to form $A = [a'_1, a'_2, \ldots, a'_m]$. In this case arcs are defined by $y_n \to x_i$ for which $a_i = a'_i$. If this method is applied successively, then it tests whether $(A, B)$ is a pair of score sequences and an oriented bipartite graph $\Delta(A, B)$ with score sequences $(A, B)$ is constructed.

We can interpret this algorithm in the following way. Let $A = [a_1, a_2, ..., a_m]$ and $B = [b_1, b_2, ..., b_n]$ be a pair of score sequences of an oriented bipartite graph, where $a_m \geq n$, $b_n \leq 2m - 1$. Let $p$ and $q$ denote the smallest and largest integers $j$ for which $b_j = b_{a_m-n}$. Let $A'[a_1, a_2, ..., a_{m-1}]$ as before and let then $B' = [b'_1, b'_2, ..., b'_n]$, then

$$b'_j = \begin{cases} b_j, & for\ 1 \leq j \leq p-1\ and\ p+q-(a_m-n) \leq j \leq q, \\ b_j - 1, & otherwise. \end{cases}$$

We illustrate this reduction and the resulting construction with the following example, beginning with lists $A_1$ and $B_1$. The oriented bipartite graph constructed is shown in Figure 14.13.

$$
\begin{array}{lll}
A_1 = [4, 4, 5] & B_1 = [1, 1, 4, 5] & x_3 \rightarrow y_2 \\
A_2 = [4, 4] & B_2 = [0, 1, 3, 4] & y_4 \rightarrow x_1, x_2 \\
A_3 = [4, 4] & B_3 = [0, 1, 3] & x_2 \rightarrow y_1 \\
A_4 = [4] & B_4 = [0, 0, 2] & y_3 \rightarrow x_1 \\
A_5 = [4] & B_5 = [0, 0] & x_1 \rightarrow y_1, y_2
\end{array}
$$

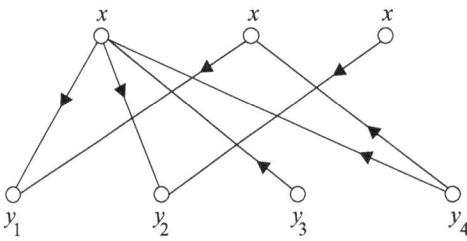

**Fig. 14.13**

Let $D_i$ be the oriented bipartite graphs with disjoint parts $X_i$ and $Y_i$ for $1 \leq i \leq t$. Let $X = U_{i=1}^t X_i$ and $Y = U_{i=1}^t Y_i$. $D = [D_1, D_2, ..., D_t]$ denotes the oriented bipartite graph with parts $X$ and $Y$, obtained from $D_i$ for $1 \leq i \leq t$ such that the arcs of $D$ are the arcs of $D_i$ and each vertex of $Y_j$ is adjacent to every vertex of $X_i$ for $j < i$ and each vertex of $X_i$ is adjacent to every vertex of $Y_j$ for $i < j$.

The next result [194] gives a criterion for determining whether a pair of sequences are realisable as scores.

**Theorem 14.38**   Let $A = [a_1, a_2, ..., a_m]$ and $B = [b_1, b_2, ..., b_n]$ be a pair of non-negative integers in non-decreasing order. Then $A$ and $B$ are scores sequences of some oriented bipartite graph if and only if

$$\sum_{i=1}^{k} a_i + \sum_{j=1}^{t} b_j \geq 2kl \tag{14.3.1}$$

for $1 \leq k \leq m$ and $1 \leq l \leq n$, with equality when $k = m$ and $l = n$.

**Proof**  The necessity of the condition follows from the fact that the oriented subbipartite graph induced by $k$ vertices from the first part and $l$ vertices from the second part has a sum of scores $2kl$.

For sufficiency, assume that $A = [a_1, a_2, \ldots, a_m]$ and $B = [b_1, b_2, \ldots, b_n]$ are the sequences of non-negative integers in non-decreasing order satisfying the conditions (14.38.1) but are not score sequences of any oriented bipartite graph. Let these sequences be chosen in such a way that $m$ and $n$ are the smallest possible and $a_1$ is the least with that choice of $m$ and $n$. We have two cases to consider.

**Case 1**  Suppose equality in (14.38.1) holds for some $k \leq m$ and $l < n$, so that

$$\sum_{i=1}^{k} a_i + \sum_{j=1}^{l} b_j = 2kl.$$

By the minimality of $m$ and $n$, $A_1 = [a_1, a_2, \ldots, a_k]$ and $B_1 = [b_1, b_2, \ldots, b_l]$ are the scores of some oriented bipartite graph $D_1$. Let

$$A_2 = [a_{k+1} - 2l, a_{k+2} - 2l, \ldots, a_m - 2l] \text{ and } B_2 = [b_{l+1} - 2k, b_{l+2} - 2k, \ldots, b_n - 2k].$$

Now,

$$\sum_{i=1}^{p} (a_{k+i} - 2l) + \sum_{j=1}^{q} (b_{l+j} - 2k) = \sum_{i=1}^{k+p} a_i + \sum_{j=1}^{l+q} b_j - \sum_{i=1}^{k} a_i + \sum_{j=1}^{t} b_j - 2pl - 2qk$$

$$\geq 2(k+p)(l+q) - 2kl - 2pl - 2qk = 2pq$$

for each $p$, $1 \leq p \leq m - k$ and each $q$, $1 \leq q \leq n - 1$ , with equality for $p = m - k$ and $q = n - l$. So by the minimality of $m$ and $n$, the sequences $A_2$ and $B_2$ form the scores of some oriented bipartite graph $D_2$. Let $X_i$ and $Y_i$ be the parts of $D_i$ for $1 \leq i \leq 2$, and let $X_1 = \{x_1, x_2, \ldots, x_k\}$, $X_2 = \{x_{k+1}, x_{k+2}, \ldots, x_m\}$, $Y_1 = \{y_1, y_2, \ldots, y_l\}$ and $Y_2 = \{y_{l+1}, y_{l+2}, \ldots, y_n\}$. Define an oriented bipartite graph $D$ with parts $X = X_1 \cup X_2$ and $Y = Y_1 \cup Y_2$ such that arcs of $D$ are the arcs of $D_i$ and each vertex of $X_2$ is adjacent to every vertex of $Y_1$ and each vertex of $Y_2$ is adjacent to every vertex of $X_1$. Thus we get an oriented bipartite graph $D$ with score sequences $A$ and $B$, which is a contradiction.

**Case 2**  Let us suppose that the strict inequality holds in (14.38.1) for $k \neq m$ and $l \neq n$. Assume that $a_1 > 0$. Let $A_1 = [a_1 - 1, a_2, \ldots, a_{m-1}, a_m + 1]$ and $B_1 = [b_1, b_2, \ldots, b_n]$ so that $A_1$ and $B_1$ satisfy the condition 14.38.1. Thus by the minimality of $a_1$, $A_1$ and $B_1$ are the score sequences of some oriented bipartite graph $D_1$ with parts $X_1$ and $Y_1$. Let $a_{x_1} = a_1 - 1$ and $a_{x_m} = a_m + 1$. Since $a_{x_m} \geq a_{x_1}$, there exists a vertex $y \in Y_1$ such that either $x_m(1-0)y(1-0)x_1$, or $x_m(0-0)y(1-0)x_1$, or $x_m(1-0)y(0-0)x_1$, or $x_m(0-0)y(0-0)x_1$ in $D_1$ and if these are changed to $x_m(0-0)y(0-0)x_1$, or $x_m(0-1)y(0-0)x_1$, or $x_m(0-0)y(0-1)x_1$, or $x_m(0-1)y(0-1)x_1$ respectively, the result is an oriented bipartite graph with score sequences $A$ and $B$ which is again a contradiction. This completes the proof of the theorem.

The characterisation of scores of oriented tripartite graphs can be found in [190] and scores (marks) in other types of digraphs can be found in [188, 196, 191].

## 14.13 Score Sequences of Semi Complete Digraphs

**Definition:** A *semi complete digraph* is a digraph with no directed loops and at least one arc between every pair of distinct vertices. Clearly, a tournament is a semi complete digraph in which there is exactly one arc between every pair of distinct vertices. Therefore every semi complete digraph contains at least one tournament on the same vertex set and is contained in the complete symmetric digraph on the same vertex set. The *score* of a vertex $v$ in a semi complete digraph $D$ is the outdegree of $v$.

The following result is due to Reid and Zhang [222].

**Theorem 14.39 (Reid and Zhang [222])** A sequence of non-negative integers $S = [s_i]_1^n$ in non-decreasing order is a score sequence of some semi complete digraph of order $n$ if and only if

$$\sum_{i=1}^k \geq \binom{k}{2} \text{ and } s_k \leq n - 1, \tag{14.39.1}$$

for all $k, 1 \leq k \leq n$.

### Proof

*Necessity* If $S$ is a score sequence of some semi complete digraph $D$ of order $n$, then any $k$ vertices of $D$ induce a semi complete digraph of order $k$ which, in turn, contains a tournament $W$ of order $k$. Therefore the sum of the scores in $D$ of these $k$ vertices is at least the sum of their scores in $W$ which is the total number of arcs in $W$, $\binom{k}{2}$. Also, a vertex of $D$ can dominate at most all of the other vertices, so no score in $S$ can exceed $n - 1$. Thus the conditions (14.39.1) are necessary.

We require the following result for proving sufficiency.

**Lemma 14.16** If $S = [s_i]_1^n, n \geq 1$, is a sequence of integers in non-decreasing order satisfying (14.39.1), then there exists a tournament $T$ with score sequence $s' = [s_i']_1^n$, such that $s_i' \leq s_i$ for $1 \leq i \leq n$.

**Proof** Define an order $\preceq$ on all non-decreasing sequences of integers satisfying (14.39.1)(thus including all sequences satisfying conditions (14.1.2)) as follows. If $B = [b_1, b_2, \ldots, b_n]$ and $m$ is the smallest index for which $b_m = b_n$ ( $= \max \{b_i : 1 \leq i \leq n\}$), then $B$ (strictly) covers the sequence $A = [a_1, a_2, \ldots, a_n]$, where $A$ and $B$ are identical such that $a_m = b_m - 1$. Note that if $m > 1$, then $b_{m-1} < b_m = b_{m+1} = \ldots = b_n$ and if $m = 1$, then $b_1 = b_2 = \ldots = b_n \geq (n-1)/2$. Also, if $B$ covers $A$, then $\sum_{i=1}^n a_i = \left( \sum_{i=1}^n b_i \right) - 1$.

This implies, by Landau's theorem, that if $S$ satisfies (14.39.1), then $S$ is the score sequence for some tournament if and only if $S$ covers no sequence satisfying (14.39.1). And, if $B$ is not the score sequence for any tournament, then $B$ covers exactly one sequence satisfying (14.39.1). For two non-decreasing sequences of integers $X$ and $Y$ satisfying (14.39.1), define $X \preceq Y$ if either $X = Y$, or there is a sequence $X_0 = X, X_1, X_2, \ldots, X_{j-1}, X_j = Y$ of non-decreasing sequences of integers each satisfying conditions (14.39.1) such that $X_i$ covers $X_{i-1}, 1 \le i \le j$.

Now, let $S = [s_i]_1^n$ be a sequence of integers in non-decreasing order satisfying conditions (14.39.1). Induct on the integer $e(S) \equiv \left( \sum_{i=1}^n s_i \right) - \binom{n}{2}$. If $e(S) = 0$, then by Landau's theorem, $S$ itself is a score sequence for some tournament $T$. If $e(S) > 0$, then by the remarks above, $S$ covers exactly one sequence $Z = [z_1, z_2, \ldots, z_n]$ satisfying (14.39.1), such that $z_i \le s_i$, for $1 \le i \le n$, and $e(Z) = \left( \sum_{i=1}^n z_i \right) - \binom{n}{2} = \left( \sum_{i=1}^n s_i \right) - \binom{n}{2} - 1 = e(s) - 1$. By the induction hypothesis applied to $Z$, there is a score sequence $S' = [s_i']_1^n$ for some tournament $T$ such that $s_i' \le z_i$, for $1 \le i \le n$. By the transitivity of $\preceq$, we have $s_i' \le s_i$, for $1 \le i \le n$, so and $T$ suffice for $S$, as required. ❏

*Sufficiency of Theorem 14.39*   Let $S = [s_i]_1^n$, $n \ge 1$, be a sequence of integers in non-decreasing order, satisfying conditions (14.39.1). By Lemma 14.16, there is a tournament $T$ of order $n$ with score sequence $S'$, where $S' \preceq S$ In $T$ denote the vertex with score $s_i'$ by $v_i$, $1 \le i \le n$. Since $v_i$ has indegree $n - 1 - s_i' \ge n - 1 - s_i$ , arcs can be added from $v_i$ to any $n - 1 - s_i$ vertices in the inset of $v_i$ in $T$ so as to produce a semi complete digraph $D$ with score sequence $S$.

## 14.15   Exercises

1. Prove that any $n$-tournament can be obtained from any other having the same scores by a sequence of arc reversals of 3-cycles.

2. If an $n$-tournament has every score $s_i$ satisfying, $\frac{1}{4}(n-1) \le s_i \le \frac{3}{4}(n-1)$ then show that it is irreducible.

3. Construct a proof for Theorem 14.2, and 14.6.

4. If $S = \{a, a+d, a+d+e\}$, where $a, d, e$ are non-negative integers and $de > 0$, and if $(d, e) = g$, $d = a$ and $e \le a + d - d/2g + (1/2)$, then prove $S$ is a score set of some tournament.

5. Prove that every set of three non-negative integers is a score set of some tournament.

6. If $a, b, c, d$ are four non-negative integers with $bcd > 0$, prove that there exists a tournament $T$ with score set $S = \{a, a+b, a+b+c, a+b+c+d\}$.

7. Construct a proof of Theorem 14.11.

8. Prove Lemma 14.7 and Lemma 14.8.

9. Construct a proof of Theorem 14.14.

10. Construct a proof of Theorem 14.16.

11. Construct proofs of Theorem 14.18 and 14.21.

12. If $T$ is a bipartite tournament with score sequences and satisfying $A = [a_i]_1^m$ and $B = [b_j]_1^n$ satisfying $n/4 < a_i < 3n/4$ for $1 \leq i \leq m$, and $m/4 < b_j < 3m/4$ for $1 \leq j \leq n$, then prove that $T$ is irreducible.

13. Construct proofs of Theorem 14.37 and Theorem 14.38.

# Bibliography

1. Akiyama, J. and Kano, M., Factors and factorization of graphs, J. Graph Theory 9 (1985) 1–42.

2. Alavi, A. and Behzad, M., Complementary graphs and edge-chromatic numbers, SIAM J. Apl. Math. 20 (1971) 161–163.

3. Alspach, B. and Reid, K. B., Degree frequencies in digraphs and tournaments, J. Graph Theory 2 (1978) 241–249.

4. Anderson, I., Perfect matching of a graph, J. Combin. Theory Ser. B 10 (1971) 183–186.

5. Appel, K. and Haken, W., Every planar map is four colorable, Bull. Amer. Math. Soc. 82 (1976) 711–712.

6. Appel, K. and Haken, W., Every planar map is four colorable, Part I, Discharging, Illinois J. Math. 21 (1977) 429–490.

7. Appel, K. and Haken, W., Every planar map is four colorable, Part II, Reducibility, Illinois J. Math. 21 (1977) 491–567.

8. Avery, P., Condition for a tournament score sequence to be simple, J. Graph Theory 4 (1980) 157–164.

9. Avery, P., Score sequences in oriented graphs, J. Graph Theory 15, 3 (1991) 251–257.

10. Babler, F., Uber die Zerlegung regularer streckenkomplexe ungerader Ordnung, Comment. Math. Helve. 10 (1938) 275–287.

11. Bagga, K. S. and Beineke, L.W., Uniquely realizable score lists in bipartite tournaments, Czech. Math. Journal 37, 112 (1987) 323–333.

12. Balakrishnan, R. and Paulraja, P., Chordal graphs and some of their derived graphs, Congressus Numerantium 53 (1986) 33–35.

13. Balakrishnan, R. and Ranganathan, K., A Textbook of Graph Theory, Springer-Verlag, New York (2000).

Graph Theory 443

14. Barnette, D. W. and Grunbaum, B., On Steinitz's theorem concerning convex 3-polytopes and on some properties of planar graphs, Lecture Notes in Math. 110 (1969) 27–40.

15. Behzad, M. and Chartrand, G., Introduction to the Theory of Graphs, Allyn and Bacon, Inc., Boston (1971).

16. Behzad, M. and Simpson, J. E., Eccentric sequences and eccentric sets in graphs, Discrete Math. 16 (1976) 187–193.

17. Behzad, M., Chartrand, M. G. and Lesniak, L., Foster, Graphs and Digraphs, Pindle, Weber and Schmidt Int. Series, Boston Mass. (1979).

18. Beineke, L. W., Derived graphs and digraphs, in H. Sachs, H. Voss and H. Walther (eds), Beitrage zur Graphentheorie Teubner, Leipzig (1968) 17–33.

19. Beineke, L. W., A tour through tournaments, or bipartite and ordinary tournaments, A comparative survey, Combinatorics, London Math. Soc. Lecture Note Series 52, Cambridge University Press (1981) 41–55.

20. Beineke, L.W. andMoon, J.W., On bipartite tournaments and scores, in G. Chartrand (ed.), The Theory of Applications of Graphs (1981) 55–71.

21. Berge, C., Two theorems in graph theory, Proc. Nat. Acad. Sci. U. S. A. 43 (1957) 842–844.

22. Berge, C., The Theory of Graphs and its Applications, Methuen, London (1962).

23. Berge, C., Graphs and Hypergraphs, North-Holland, Amsterdam (1973).

24. Bermond, J. C. and Thomassen, C., Cycles in digraphs—a survey, J. Graph Theory 5, 1 (1981) 1–43.

25. Biggs, N. L., Algebraic Graph Theory, Cambridge University Press (1974).

26. Bjelica, M. and Lakic, S., Criteria for scores sequences, Novi Sad J. Math. 30, 2 (2000) 11–14.

27. Bjelica, M. and Lakic, S., Criteria for sets of scores with prescribed positions, IMC Filomat (2001) 26–30.

28. Boesch, F. T., Suffel, C. and Tindell, R., The spanning subgraphs of Eulerian graphs, J. Graph Theory 1 (1977) 79–84.

29. Bollabas, B., External Graph Theory, Acad. Press, New York (1978).

30. Bollobas, B., Uniquely colorable graphs, J. Combin. Theory Ser. B 24 (1978).

31. Bollobas, B. and Catlin, P. A., Topological cliques of random graphs, J. Comb. Theory Ser. B 30 (1981) 224–227.

32. Bondy, J. A., Bounds on the chromatic number of a graph, J. Combin. Theory 7 (1969) 96–98.

33. Bondy, J. A., Properties of graphs with constraints on degrees, Studia Sci. Math. Hungar. 4 (1969) 473–475.

34. Bondy, J. A., Pancyclic graphs, J. Combin. Theory Ser. B11 (1971) 80–84.

35. Bondy, J. A., Short proofs of classical theorems, J. Graph Theory, 44 (2003) 159–165.

36. Bondy, J. A. and Chvatal, V., A method in graph theory, Discrete Math. 15 (1976) 111–135.

37. Bondy, J. A. and Halberstam, F. Y., Parity theorems for paths and cycles in graphs, J. Graph Theory 10 (1986) 107–115.

38. Bondy, J. A. and Thomassen, C., A short proof ofMeyniel's theorem, Discrete Math. 19 (1977) 195–197.

39. Brooks, R. L., On coloring the nodes of a network, Proc. Cambridge Phil. Soc. 37 (1941) 194–197.

40. Brualdi, R. A. and Shen, J., Landaus inequalities for tournament scores and a short proof of a theoremon transitive sub-tournaments, J. Graph Theory 38 (2001) 244–254.

41. de Bruijn, N.G., A combinatorial problem, Proc. Kon. Ned. Akad. V. Wetensch. 49 (1946) 758–764.

42. Buckley, F. and Harary, F., Distance in Graphs, Addison-Wesley (1990).

43. Buckley, F., Miller, Z. and Slater, P. J., On graphs containing a given graph as center, J. Graph Theory 5 (1981) 427–434.

44. Camion, P., Chemins et circuits Hamiltoniens des graphs complets, C. R. Acad. Sci. Paris 249 (1959) 2151–2152.

45. Cartwright, D. and Harary, F., On colorings of signed graphs, Elem. Math. 23 (1968) 85–89.

46. Cayley, A., A theoremon trees, Quart. J.Math. 23 (1989) 376–378; collected papers, Cambridge 13 (1877) 26–28.

47. Chang, L. C., The uniqueness and non uniqueness of the triangular association scheme, Sci. Record 3 (1959) 604–613.

48. Chartrand, G., On Hamiltonian line graphs, Trans. Amer. Math. Soc. 134 (1968) 559–566.

49. Chartrand, G. and Geller, D., Uniquely colorable graphs, J. Combin. Theory 6 (1969) 271–278.

Graph Theory 445

50. Chartrand, G., Gavlas, H., Harary F. and Schultz, M., On signed degrees in signed graphs, Czech. Math. Journal 44 (1994) 111–117.

51. Chartrand, G. and Kronck, H. V., Randomly traceable graphs, SIAM J. Appl. Math. 16 (1968) 696–700.

52. Chartrand, G. and Stewart, M. J., Isometric graphs, in Capobianco et. al., Recent Trends in Graph Theory, Springer-Verlag, Berlin, Lecture Notes in Math. 186 (1971) 63–67.

53. Chartrand, G. and Stewart, M. J., The connectivity of line graphs, Math. Ann. 182 (1969) 170–174.

54. Chartrand, G. and Schuster, S., Which graphs have unique distance trees, Amer. Math. Monthly 81 (1974) 53–56.

55. Chartrand, G. and Wall, C. E., On the Hamiltonian index of a graph, Studia Sci. Math. Hungar. 8 (1973) 43–48.

56. Chartrand, G. andWhite, A. T., Randomly traversable graphs, Elem. Math. 25 (1970) 101–107.

57. Chen, Y. L., A short proof of Kundus f-factor theorem, Discrete Math. 71 (1988) 177–179.

58. Choudam, S. A., A simple proof of the Erdos-Gallai theorem on graphic sequences, Bull. Austral. Math. Soc. 36 (1986) 67–70.

59. Chvatal, V., On Hamilton's ideals, J. Combin. Theory Ser. B 12 (1972) 163–168.

60. Clark, J. and Holton, D. A., A First Look at Graph Theory, World Scientific (1991).

61. Courant, R. and Robbins, H. E., What is Mathematics? Oxford Univ. Press, London (1941).

62. Cunningham, W. H., Matchings, matroids and extensions, Math. Programming B91 (2002) 515–542.

63. Deo, N., Graph Theory with Applications to Engineering and Computer Science, PHI, New Delhi (1974).

64. Dharwadker, A., A New Proof of the Four Colour Theorem, http: //www.dharwadker.org/ (2000).

65. Dijkstra, E. W., A note on two problems in connexion with graphs, Numer. Math. 1 (1959) 269–71.

66. Dirac, G. A., A property of 4-chromatic graphs with some remarks on critical graphs, J. London Math. Soc 27 (1952) 85–92.

67. Dirac, G. A., Short proof of Menger's graph theorem, Mathematika 13 (1966) 42–44.

68. Dirac, G.A., Some theorems on abstract graphs, Proc. London Math. Soc. 2 (1976) 111–135.

69. Dirac, G. A. and Schuster, S., A theorem of Kuratowski, Nederl. Akad. Wetensch. Proc. Ser. A 57 (1954) 343–348.

70. Edmonds, J. and Johnson, E. L., Matching, Euler tours and the Chinese postman, Math. Programming 5 (1973) 88–124.

71. Eplett,W. J. R., Self-converse tournaments, Canad. Math. Bull. 22 (1979) 23–27.

72. Erdos, P. and Fajtlowicz, S., On the conjecture of Hajos, Combinatorica 1 (1981) 141–143.

73. Erdos, P. and Gallai, T. G., Graphs with prescribed degrees of vertices, Mat. Lapok 11 (1960) 264–274.

74. Euler, L., Solutio problematics ad geometriamsitus pertinents, Comment. Academiae Sci. I Petropolitanae 8 (1736) 128–140.

75. Euler, L., Demonstratio Nonnullarum Insignium Proprietatum QuibusSolida Hedris Planis Inclusa Sunt Praedita, Novi Comm. Acad. Sci. Imp. Petropol 4 (1758) 140–160.

76. Euler, L., The Konigsberg bridges, Sci. Amer. 189 (1853) 66–70.

77. Fary, I., On straight line representation of planar graphs, Acta Sci. Math. 11 (1948) 229–233.

78. Finck, H. J. and Sachs, H., Uber eine von H. S. Wilf angegebene Schranke fur die chromatische Zahl endlicher Graphe, Math. Nachr. 39 (1969) 373–386.

79. Fleischner, H., Elementary proofs of (relatively) recent characterizations of Eulerian graphs, Discrete Applied Math. 24 (1989) 115–119.

80. Fleischner, H., Eulerian graphs and related topics, Annals of Discrete Math. 45, North-Holland, New York (1990).

81. Ford, L. R. and Fulkerson, D. R., Flows in Networks, Princeton University Press, Princeton (1962).

82. Foulds, L. R., Graph Theory Applications, Springer-Verlag, New York (1992).

83. Fulkerson, D. R., Hofman, A. J. and Mc Andrew, M. H., Some properties of graphs with multiple edges, Can. J. Math. 17 (1965) 166–177.

84. Gale, D., A theorem on flows in networks, Pacific J. Math. 7 (1957) 1073–1082.

85. Gervacio, S. V., Score sequences: lexicographic enumeration and tournament construction, Discrete Math. 72 (1988) 151–155.

Graph Theory 447

86. Grinberg, E. Ja., Plane homogenous graphs of degree three without Hamiltonian circuits (Russian), Latvian Math. Yearbook 4 (1968) 51–58.

87. Ghouila-Houri, A., Une condition sufficiante d, existence d un circuit Hamiltonian, C. R. Acad. Sci. Paris 25 (1960) 495–497.

88. Graver, J. E. and Jurkat, W. R., f -factors and related decompositions of graphs, J. Combin. Theory Ser. B 28 (1980) 66–84.

89. Griggs, J. R. and Reid, K.B., Landau's Theorem revisited, Australasian J. Combinatorics 20 (1999) 19–24.

90. Grotzsch, H., Ein Oreifarbensatz fur dreikreisfreie Netze auf der Kugel, Wiss. Z. Martin-Luther Univ. Halle-Wittenberg Math. Naturwiss. Reihe 8 (1958) 109–120.

91. Grunbaum, B., Grotzschs Theorem on 3-colorings, Michigan Math. J. 10 (1963) 303–310.

92. Grunbaum, B., Graphs and Complex, Report of the University of Washington, Seattle, Math., 572B, 1969.

93. Grunbaum, B., Problem 2 in combinatorial structures and their applications, Proc. Calgory International Conference, (1969), Gordon and Breach, New York.

94. Gupta, R. P., Bounds on the chromatic and achromatic numbers of complementary graphs, in W. T. Tutte (ed.), Recent Progress in Combinatorics, Acad. Press, New York 1969 (229–235).

95. Hadwiger, H., Uber eine Klassification der Streckenkomplexe, Vierteljschr Naturforsch Ges Zurich 88 (1943) 133–142.

96. Hager, M., On score sets for tournaments, Discrete Math. 58 (1986) 25–34.

97. Hajos, G., Uber eine Konstruktion nich n-farbarrer Graphen, Wiss. Z. Martin-Cuther Univ. Halle-wittenberg Math. Nat. Reihe 10 (1961) 116–117.

98. Hakimi, S. L., On realizability of a set of integers as degree of the vertices of a linear graph II. Uniqueness, J. Soc. Indust. Apple. Math.11 (1963) 135–147.

99. Hakimi, S., On the realizability of a set of integers as the degrees of the vertices of a graph, SIAM J.Appl.Math.10 (1962) 496–506.

100. Hall, M., Distinct representatives of subsets, Bull. Amer. Math. Soc. 3 (1952) 584–587.

101. Hall, P., On representation of subsets, J. London Math. Soc. 10 (1935) 20–30.

102. Hamilton,W.R., Memorandum respecting a new system of roots of unity, Philosophical Magazine, 12, 4 (1856), p.446. http://www.maths.tcd.ie/pub/Hist Math/People/ Hamilton/Icosian/

103. Harary, F., A characterization of block graphs, Canad. Math. Bull.6 (1963) 1–6.

104. Harary, F., Graph theory, Addison-Wesley, Reading, MA. (1969).

105. Harary, F. and Frisch, I. T., Communication, Transmission and Transportation Networks, Addison-Wesley, Reading MA. (1971).

106. Harary, F. and Hedetniemi, S., The achromatic number of a graph, J. Combin. Theory B (1970) 154–161.

107. Harary, F., Hedetniemi, S. and Robinson, R. W., Uniquely colorable graphs, J. Combin. Theory 6 (1969) 264–270.

108. Harary, F. and Nash-Williams, C. St. J. A., On Eulerian and Hamiltonian graphs and line graphs, Canad. Math. Bull. 8 (1965) 701–710.

109. Harary, F. and Norman, R. Z., The dissimilarity characteristic of Husimi trees, Ann. of Math. 58 (1953) 131–141.

110. Harary, F., Norman, R. Z. and Cartwright, D., Structural Models: An Introduction to the Theory of Directed Graphs,Wiley, New York (1965).

111. Hasselbarth, W., Die verzweightheit von graphen, Match 16 (1984) 3–17.

112. Havel, M., A remark on the existence of finite graphs, Casopis Pest. Mat. 80 (1955) 477–480.

113. Heawood, P. J., Map color theorem, Qurat. J. Pure Appl. Math. 24 (1890) 332–338.

114. Hedetniemi, S., On hereditary properties of graphs, Studia Sci. Math. Hungarica.

115. Heirholzer, C., Ueber die Moglickkeit, Einen Linienzug ohne Wiederholung and ohne Unterrechung zu umfahren, Math. Ann. 6 (1893) 30–32.

116. Hoffman, A. J., On the exceptional case in a characterization of the arcs of complete graphs, IBM J. Res. Develop. 4 (1960) 487–496.

117. Hoffman, A. J., On the uniqueness of the triangular association scheme, Ann. Math. Statist. 31 (1960) 492–497.

118. Hoffman, A. J., On the line graph of the complete bipartite graph, Ann. Math. Statist. 35 (1964) 883–885.

119. Ivanyi, A., Reconstruction of complete interval tournaments, Acta Univ. Sapientiae, Informatica, 1, 1 (2009) 71–88.

120. Ivanyi, A., Reconstruction of complete interval tournaments II, Acta Univ. Sapientiae, Mathematica, 2, 1 (2010) 47–71.

121. Jaeger, F., A note on sub-Eulerian graphs, J. Graph Theory 3 (1979) 91–93.

122. Jordan, C., Sur les assemblages de lignes, J. Reine Angew Math. 70 (1969) 185–190. Graph Theory 449

123. Jung, H. A., Zu einem isomorphiesatz von Whitney fur Graphen, Math. Ann. 164 (1966) 270–271.

124. Kano, M. and Saito, A., [a, b]-factors of graphs, Discrete math. 47 (1983) 113–116.

125. Kano, M. and Saito, A., [a, b]-factorization of a graph, J. Graph Theory 9 (1985)

129–146.

126. Kapoor, S. F., Polimeni, A. O. and Wall, C. E., Degree sets for graphs, Fund. Math. 95 (1977) 189–194.

127. Kasteleyn, P. W., Graph theory and crystal physics, in F. Harary (ed.) Chapter 2 in Graph Theory and Theoretical Physics, Acad. Press (1967) 44–110.

128. Kempe, A. B., On the geographical problem of the four colors, Amer. J. Math. 2 (1879) 193–200.

129. Kleitman, D. J. and Wang, D. L., Algorithms for constructing graphs and digraphs with given valencies and factors, Discrete Math. 6 (1973) 78–88.

130. Koh, K. M. and Tan, B. P., The set and number of kings in a multipartite tournament, Bull. Inst. Combin. Appl. 13 (1995) 15–22.

131. Koh, K. M. and Tan, B. P., Kings in multipartite tournaments, Discrete Mathematics 147 (1995) 171–183.

132. Koh, K. M. and Tan, B. P., Number of 4-kings in bipartite tournaments with no 3-kings, Discrete Math. 154 (1996) 281–287.

133. Koh, K. M. and Tan, B.P., The number of kings in multipartite tournaments, Discrete Math., 167 (1997) 411–418.

134. Kotzig, A., Cycles in a complete graph oriented in equilibrium, Mat. Fyz. Casopis SAV 16 (1966) 175–182.

135. Kotzig, A., The decomposition of a directed graph into quadratic factors consisting of cycles, Acta F. R. N. Univ. Comen. Math. 22 (1969) 27–29.

136. Konig, D., Graphen and Matrizen, Math. Fiz. Lapok 38 (1931) 116–119.

137. Konig, D., Theorie der endlichen and unendlichen Graphen, Leipzig.

138. Konig, D., Uber Graphen und ihre Anwendung auf determinantentheorie und Mengenlehre, Math. Ann. 77 (1916) 453–465.

139. Kouider, M. and Zbigniew Long, Stability number and [a, b]-factors in Graphs, J. Graph Theory 46 (2004) 254–264.

140. Krausz, J., Demonstration nouvelle d un theoreme de Whitney sur les reseaux, Mat. Fiz. Lapok 50 (1943) 75–89.

141. Kundu, S., Generalizations of the k-factor theorem, DiscreteMath. 9 (1974) 173–179.

142. Kundu, S., The Chartrand-Schuster Conjecture: Graphs with unique distance trees are regular, J. Combin, Theory Ser. B 22 (1977) 233–245.

143. Kundu, S., The k-factor conjecture is true, Discrete Math. 6 (1973) 367–376.

144. Kuratowski, C., Sur le probleme des courbes gauches en topologie, Fund. Math. 15 (1930) 271–283.

145. Landau, H. G., On dominance relations and the structure of animal socities: III. The condition for a score structure, Bull. Math. Biophys. 15 (1953) 143–148.

146. Lesniak, L., Eccentric sequences in graphs, PeriodMath. Hungar. 6 (1975) 287–293.

147. Lesniak-Foster and Williams, J. E., On spanning and dominating circuits in graphs, Canad. Math. Bull.20 (1977).

148. Lovasz, L., Subgraphswith prescribed valencies, J. Combin. Theory 8 (1970) 391–416.

149. Lovasz, L., Valencies of graphs with 1-factors, Period. Math. Hungar. 5 (1974) 149–151.

150. Lovasz, L., A remark on Mengers theorem, Acta Math. Acad. Sci. Hungar. 6 (1975) 287–293.

151. Lovasz, L., Three short proofs in graph theroy, J. Combin Theory Ser. B 19 (1975) 111–113.

152. Lovasz, L., Combinatorial Problems and Exercises, North-Holland, Amsterdam(1979).

153. Lucas, E., Récréations Mathématiques, Gauthier-Villares Paris (1891).

154. Mathematisch Centrum, This Weeks Citation Classic, CC/Number 7, February 14, 1983, http://www.garfield.library.upenn.edu/.

155. Maurer, S. B., The king chicken theorems, Math. Mag. 53 (1980) 67–80.

156. McCuaig,W., A simple proof ofMengers theorem, J. Graph Theory 8 (1984) 427–429.

157. Mckee, T. A., Recharacterizing Eulerian: Intimations of new duality, Discrete Math. 51 (1984) 237–242.

158. Menger, K., Zur allgemeinen Kurventheoril, Fund. Math. 10 (1927) 96–115.

159. Menon, V., On repeated interchange graphs, Amer. Math. Monthly 13 (1966) 986–989.

160. Meyniel, M., Une condition sufficiante d'existance d'un circuit Hamiltonian dans un graphe oriente, J. Combin. Theory Ser. B 14 (1973) 137–147.

161. Mirsky, L., Transversal Theroy, Acad. Press, New York (1971).

Graph Theory 451

162. Moon, J. W., On some combinatorial and probabilistic aspects of bipartite graphs, Ph.D Thesis, University of Alberta, Edmonton (1962).

163. Moon, J. W., On the line graph of the complete bigraph, Ann. Math. Statist. 34 (1963) 664–667.

164. Moon, J. W., Various proofs of Cayley's formula for counting trees, in F. Harary and L. Beineke (eds), A seminar on Graph Theory, Holt, Rinehart and Winston (1967) 70–78.

165. Moon, J.W., Topics on Tournaments, Holt, Rinehart and Winsten, New York (1968).

166. Muhammad A., Khan, K. K. Kayibi and S. Pirzada, Some results on tournaments, Annales Computatorica, To appear.

167. Nash-Williams, C. St. J. A., Edge-disjoint spanning trees of finite graphs, J. London Math. Soc. 36 (1961) 445–450.

168. Nash-Williams, C. St. J. A., Edge-disjoint Hamiltonian circuits in graphs with vertices of large valency, in Studies in Pure Mathematics, Acad. Press, London (1971).

169. Nash-Williams, C. St. J. A. and Tutte, W. T., More proofs of Mengers theorem, J. Graph Theory I (1977) 13–17.

170. Nebesky, L., On the line graph of the square and the square of the line graph of a connected graph, Casopis. Pset. Mat. 98 (1973) 285–287.

171. Newman, D. J., A Problem in Graph Theory, Amer. Math. Monthly, 65 (1958) 611.

172. Nordhaus, E. A. andGaddam, J.W., On complementary graphs, Amer. Math. Monthly 63 (1956) 175–177.

173. O Neil, P. V., A new proof of Menger's theorem, J. Graph Theory 2 (1978) 257–259.

174. Ore, O., A problem regarding the tracing of graphs, Elem. Math. 6 (1951) 49–53.

175. Ore, O., Graphs and matching theorems, Duke Math. J. 22 (1955) 625–639.

176. Ore, O., Note on Hamilton circuits, Amer. Math. Monthly 67 (1960) 55.

177. Ore, O., The Four Color Problem, Acad. Press, New York (1967).

178. Ore, O., Theory of Graphs, Amer. Math. Soc. Colloq. Publ. 38, Providence (1962).

179. Ore, O. and Stemple, G. J., Numerical methods in the four color problem, in W. T. Tutte (ed.) Recent Progress in Combinatorics, Acad. Press, New York (1969).

180. Parthasarathy, K. R., Basic Graph Theory, Tata McGraw-Hill, New Delhi (1994).

181. Peterson, J., Die Theories der regularen graphen, Acta Math. 15 (1891) 193–220.

182. Petrovic, V. and Thomassen, C., Kings in k-partite tournaments, Discrete Mathematics 98 (1991) 237–238.

183. Petrovic, V., Decomposition of some planar graphs into trees, Discrete Mathematics 150 (1996) 449–451.

184. Petrovic, V., Kings in bipartite tournaments, DiscreteMathematics 173 (1997) 187–196.

185. Pirzada, S., Self converse score lists in oriented graphs, J. China Univ. of Science and Tech. 26, 3 (1996) 392–394.

186. Pirzada, S., Simple score sequences in oriented graphs, Novi Sad J. Math. 33, 1 (2003) 25–29.

187. Pirzada, S., Construction of oriented graphs, in Balakrishnan, Wilson and Sethuraman (eds) Proc. National Conf. Graph Theory and Applications, Narosa, (2004) 146–149.

188. Pirzada, S., Mark sequences in multidigraphs, Discrete Mathematics and Applications 17, 1 (2007) 71–76.

189. Pirzada, S., Imbalances in digraphs, Kragujevic J. Mathematics 31 (2008) 143–14

190. Pirzada, S. and Merajuddin, Score lists in oriented tripartite graphs, Novi Sad J. Math 26, 2 (1996) 1–9.

191. Pirzada, S. and Samee, U., Seminare Loth. de Comb. 55 Art B.(2006).

192. Pirzada, S., Merajuddin and Samee, U., On oriented graph scores, Matematicki Vesnik 60 (2008) 187–191.

193. Pirzada, S., Merajuddin and Samee, U., Inequalities in oriented graph scores, Bull. Allahabad Mathematical Society 23, 2 (2008) 389–395.

194. Pirzada, S., Merajuddin and Yin Jianhua, On the scores of oriented bipartite graphs, J. Math. Study 33, 4 (2000) 245–359.

195. Pirzada, S. and Naikoo, T. A., Degree frequencies in oriented graphs, Thai J. Maths. 2 (2004) 85–96.

196. Pirzada, S. and Naikoo, T. A., Inequalities on marks in digraphs, J. Mathematical Inequalities and Applications 9, 2 (2006) 189–198.

197. Pirzada, S. and Naikoo, T. A., Score sets in tournaments, Vietnam J. Mathematics 34, 2 (2006) 157–161.

198. Pirzada, S. and Naikoo, T. A., Score sets in k-partite tournaments, J. Applied Mathematics and Computing 22, 1-2 (2006) 237–245.

199. Pirzada, S. and Naikoo, T. A., Score sets in oriented k-partite graphs, AKCE International J. Graphs and Combinatorics 3, 2 (2006).
Graph Theory 453

200. Pirzada, S. and Naikoo, T. A., Score sets for oriented graphs, Applicable Analysis and Discrete Mathematics 2, No. 1 (2008) 107–113.

201. Pirzada, S., Naikoo, T. A. and Chishti, T. A., Score sets in oriented bipartite graphs, Novi Sad J. Mathematics 36, 1 (2006) 35–45.

202. Pirzada, S., Naikoo, T. A. and Dar, F. A., Signed degree sets in signed graphs, Czech. Mathematical Journal 57, 3 (2007) 843–848.

203. Pirzada, S., Naikoo, T. A. and Shah, N. A., Score sequences in oriented graphs, J. Applied Mathematics and Computing 22, 1–2 (2007) 257–268.

204. Pirzada, S., Naikoo, T. A. and Dar, F. A., Degree sets in bipartite and 3-partite graphs, Oriental Journal of Mathematical Sciences 1, 1 (2007) 47–53.

205. Pirzada, S., Naikoo, T. A. and Dar, F. A., Signed degree sequences in signed bipartite graphs, AKCE International J. Graphs and Combinatorics 4, 2 (2007) 1–12.

206. Pirzada, S. Naikoo, T. A. and Dar, F. A., A note on signed degree sets in signed bipartite graphs, Applicable Analysis and DiscreteMathematics 2, 1 (2008) 114–117.

207. Pirzada, S. and Shah, N. A., Kings and serfs in oriented graphs, Mathematica Slovaca 58, 3 (2008) 277–288.

208. Pirzada, S. and Yin Jian Hua, Degree sequences in graphs, J. Math. Study 39, No. 1 (2006) 25–31.

209. Plesnik, J., Connectivity of regular graphs and the existence of 1-factors, Math. Casopis 22 (1972) 310–318.

210. Posa, L., A theorem concerning Hamiltonian lines, Magyar Tud. Akad. Math. Kutato Int. Kozl 7 (1962) 225–226.

211. Prim, R. C., Shortest connection networks and some generalizations, Bell System Techn. J. 36 (1957) 1389–1401.

212. Prufer, H., Beweis eines Satzes uber Permutationen, Arch-Math Physics 27 (1918) 742–744.

213. Pym, J. S., A proof of Menger's theorem, Monatsh. Math. 73 (1969) 81–83.

214. Randic, M., Characterization of atoms, molecules and classes of molecules based on path enumerations, Proc. Bremen Konferenz zur Chemie, Univ. Bremen (1978) Part II, Match No. 7 (1979) 5–64.

215. Rao, A. R. and Rao, S. B., On factorable degree sequences, J. Combin, Ser. B 13 (1972) 185–191.

216. Redei, L., Ein Kombinatorischer Satz, Acta. Litt. Szeged, 7 (1934) 39–43.

217. Reid, K. B., Score sets for tournaments, Congressus Num. 21 (1978) 607–618.

218. Reid, K. B. and Beineke, L.W., Tournaments, in L. W. Beineke and R. J. Wilson (eds), Chapter 7 in Selected Topics in Graph Theory, Acad. Press, London (1979) 169–204.

219. Reid, K. B., Tournaments with prescribed number of kings and serfs, Congressus Numerantium 29 (1980) 809–826.

220. Reid, K. B., Every vertex a king, Discrete Mathematics 38 (1982) 93–98.

221. Reid, K. B., Tournaments: Scores, kings, generalizations and special topics-a survey, Congressus Num. 115 (1996) 171–211.

222. Reid, K. B. and Zhang, C. Q., Score sequences of semi complete digraphs, Bulletin of the ICA 24 (1998) 27–32.

223. Ringel, G. and Youngs, J. W. T., Solution of the Heawood map coloring problem, Proc. Nat. Acad. Sci. USA 60 (1968) 438–445.

224. Rizzi, R., A short proof of Konig's matching theorem, J. Graph Theory 33 (2000) 138–139.

225. Robertson, N., Sanders, D. P., Seymour, P. D. and Thomas, R., The four color theorem, J. Combin. Theory Ser. B. 70 (1997) 2–44.

226. Rooij, V. A. and Wilf, H., The interchange graphs of a finite graph, Acta. Math. Acad. Sci. Hungar. 16 (1965) 263–269.

227. Ryser, H. J., Combinatorial properties of matrices of zeros and ones, Canad. J. Math.

9 (1957) 371–377.

228. Ryser, H. J., Matrices of zeros and ones in combinatorial mathematics, in Recent Advances in Matrix Theory, Univ. Wisconsin Press (1964) 103–124.

229. Saaty, T. L., Thirteen colorful variations on Guthrie's four color conjecture, Amer. Math. Monthly 79 (1972) 2–43.

230. Saaty, T. L. and Kainen, P. C., The Four Color Problem, McGraw Hill, New York (1977).

231. Sabidussi, G., Graph derivatives, Math. 2. 76 (1961) 385–401.

232. Schnyder, W., Planar graphs and poset dimension, Order 5 (1989) a. 323–343.

233. Seshu, S. and Reed, M., Linear Graphs and Electrical Networks, F. Harary(ed.), Addison-Wesley, Reading (1961) Acad Press, London (1967) 44–110.

234. Slater, P. J., A counter-example to Randic's conjecture on distance degree-sequences for trees, J. Graph Theory 6 (1982) 89–92.

235. Sierksma, G. and Hoogeveen, H., Seven criteria for integer sequences being graphic, J. Graph Theory 15 (1991) 223–231.

Graph Theory 455

236. Sumner, D. P., 1-factors and antifactor sets, J. London Math. Soc. 2 (13) (1976) 351–359.

237. Szekeres, G. and Wilf, H. S., An inequality for the chromatic number of a graph, J. Combin. Theory 4 (1968) 1–3.

238. Tait, P. G., Remarks on the coloring of maps, Proc. Roy. Soc. Edinburgh Sect. A 10 (1880) 501–503, 729.

239. Tanenbaum, A. S., Computer Networks, Prentice Hall Inc., Upper Saddle River, NJ, 1988.

240. The Mathematics Center at Bell Labs, History of Mathematics at Bell Labs, http://cm.bell-labs.com/cm/ms/center/history.html

241. Thomassen, C., Kuratowskis theorem, J. Graph Theory 5 (1981) 225–241.

242. Thomassen, C., Landaus characterization of tournament score sequences, in The Theory and Applications of Graphs, John Wiley (1981) 589–591.

243. Titov, V. K., A constructive description of some classes of graphs, Ph.D. Thesis, Moscow (1975).

244. Toida, S., Properties of an Euler graph, J. Franklin Inst. 295 (1973) 343–345.

245. Tripathi, A. and Vijay, S., A note on a theorem of Erdos and Galai, Discrete Mathematics 265 (1–3) (2003) 417–420.

246. Tripathi, A., Venugopalan, S. and West, D. B., A short proof of the Erdos-Gallai characterization of graphic lists, Discrete Math. 310 (2010) 843–844.

247. Turan, P., Eine Extremalaufgabe aus der Grapheneorie, Mat. Fiz. Lapok 48 (1941) 436–452 (1936), Reprinted Chelsea, New York (1950).

248. Tutte, W. T., A family of cubical graphs, Proc. Cambridge Phil. Soc. 21 (1946) 98–101.

249. Tutte, W. T., The factorization of linear graphs, J. London Math. Soc. 22 (1947) 107–111.

250. Tutte, W. T., The factors of graphs, Canad. J. Math. 4 (1952) 314–328.

251. Tutte, W. T., A short proof of the factor theorem for finite graphs, Canad. J. Math. 6 (1954) 347–352.

252. Tutte, W. T., On the problem of decomposing a graph into n connected factors, J. London Math. Soc. 36 (1961) 221–230.

253. Tutte, W. T., Graph factors, Combinatorica 1 (1981) 79–97.

254. Veblen, O., An application of modular equations in analysis situs, Ann. of Math. 14 (1912–13) 86–94.

255. Vergnas, M. L., An extension of Tuttes 1-factor theorem, Discrete Math. 23 (1978) 241–255.

256. Vergnas, M. L., M. Thesis, University of Paris, v1 (1972).

257. van Aardenne-Ehrenfest, T. and de Bruijn, N. G., Circuits and trees in oriented graphs, Simon Stevin 28 (1951) 203–217.

258. Vizing, V. G., On an estimate of the chromatic class of a p-graph (Russian) Diskret. Analiz. 3 (1964) 25–30.

259. Vizing, V. G., The chromatic class of a multigraph, Cybernetics 3 (1965) 32–41.

260. Wagner, K., Bemerkungen zumVierfarbenproblem, Jahresber Deutsch. Math. Verein 46 (1936) 21–22.

261. Wang, D. L. and Kleitman, D. J., On the existence of n-connected graphs with prescribed degrees (n > 2), Networks 3 (1973) 225–239.

262. Welsh, D. J. A. and Powell, M. B., An upper bound for the chromatic number of a graph and its application to time-tabling problems, Comput. J. 10 (1967) 85–86.

263. West, D. B., Introduction to Graph Theory, Prentice Hall Inc. (1996).

264. Whitney, H., A theorem on graphs, Annals Math. 32 (1931) 378–390.

265. Whitney, H., Congruent graphs and the connectivity of graphs, Amer. J. Math. 54 (1932) 150–168.

266. Whitney, H., 2-isomorphic graphs, Amer. J. Math. 55, (1933) 245–254.

267. Whitney, H., Planar graphs, Fund. Math. 21 (1933) 73–84.

268. Wiener, N., A simplification of logic of relations, Proc. Cambridge Phil. Soc. 17 (1912–14) 387–390.

269. Wilson, R. J., Introduction to Graph Theory, Oliver and Boyd, Edindurgh (1972).

270. Woodall, D. R., Sufficient conditions for circuits in graphs, Proc. London Math.Soc. 24 (1972) 739–755.

271. Yan, J., Lih, K., Kuo, D. and Chang, G. J., Signed degree sequences in signed graphs, J. Graph Theory 26 (1997) 111–117.

272. Zhou, G., Yao, T. and Zhang, K, On score sequences of k-hypertournaments, European J. Combin. 21 (8) (2000) 993–1000.

www.ingramcontent.com/pod-product-compliance
Lightning Source LLC
Chambersburg PA
CBHW081102170526
45165CB00008B/2303